T0249906

IBN AL-HAYTHAM, NEW SPHERICAL GEOMETRY AND ASTRONOMY

This volume provides a unique primary source on the history and philosophy of mathematics and science from the mediaeval Arab world. The fourth volume of *A History of Arabic Sciences and Mathematics* is complemented by three preceding volumes which focused on infinitesimal determinations and other chapters of classical mathematics.

This book includes five main works of the polymath Ibn al-Haytham (Alhazen) on astronomy, spherical geometry and trigonometry, plane trigonometry and studies of astronomical instruments on hour lines, horizontal sundials and compasses for great circles.

In particular, volume four examines:

- the increasing tendency to mathematize the inherited astronomy from Greek sources, namely Ptolemy's *Almagest*;
- the development of celestial kinematics;
- new research in spherical geometry and trigonometry required by the new kinematical theory;
- the study on astronomical instruments and its impact on mathematical research. These new historical materials and their mathematical and historical commentaries contribute to rewriting the history of mathematical astronomy and mathematics from the 11th century on.

Including extensive commentary from one of the world's foremost authorities on the subject, this fundamental text is essential reading for historians and mathematicians at the most advanced levels of research.

Roshdi Rashed is one of the most eminent authorities on Arabic mathematics and the exact sciences. A historian and philosopher of mathematics and science and a highly celebrated epistemologist, he is currently Emeritus Research Director (distinguished class) at the Centre National de la Recherche Scientifique (CNRS) in Paris, and is the former Director of the Centre for History of Medieval Science and Philosophy at the University of Paris (Denis Diderot, Paris VII). He also holds an Honorary Professorship at the University of Tokyo and an Emeritus Professorship at the University of Mansourah in Egypt.

J. V. Field is a historian of science, and is a Visiting Research Fellow in the Department of History of Art and Screen Media, Birkbeck, University of London, UK.

CULTURE AND CIVILIZATION
IN THE MIDDLE EAST

General Editor: Ian Richard Netton

Professor of Islamic Studies, University of Exeter

This series studies the Middle East through the twin foci of its diverse cultures and civilisations. Comprising original monographs as well as scholarly surveys, it covers topics in the fields of Middle Eastern literature, archaeology, law, history, philosophy, science, folklore, art, architecture and language. While there is a plurality of views, the series presents serious scholarship in a lucid and stimulating fashion.

PREVIOUSLY PUBLISHED BY CURZON

THE ORIGINS OF ISLAMIC LAW
The Qur'an, the Muwatta' and Madinan Amal
Yasin Dutton

A JEWISH ARCHIVE FROM OLD CAIRO
The History of Cambridge University's Genizah Collection
Stefan Reif

THE FORMATIVE PERIOD OF TWELVER SHI'ISM
Hadith as Discourse Between Qum and Baghdad
Andrew J. Newman

QUR'AN TRANSLATION
Discourse, Texture and Exegesis
Hussein Abdul-Raof

CHRISTIANS IN AL-ANDALUS 711–1000
Ann Rosemary Christys

FOLKLORE AND FOLKLIFE IN THE UNITED ARAB EMIRATES
Sayyid Hamid Hurriez

IBN AL-HAYTHAM, NEW SPHERICAL GEOMETRY AND ASTRONOMY

A history of Arabic sciences and mathematics

Volume 4

Roshdi Rashed

Translated by J.V. Field

Routledge
Taylor & Francis Group

LONDON AND NEW YORK

مركز دراسات الوحدة العربية
CENTRE FOR ARAB UNITY STUDIES

First published 2014
by Routledge
2 Park Square, Milton Park, Abingdon, Oxon OX14 4RN

and by Routledge
711 Third Avenue, New York, NY 10017

First issued in paperback 2017

Routledge is an imprint of the Taylor & Francis Group, an informa business

British Library Cataloguing in Publication Data
A catalogue record for this book is available from the British Library

Library of Congress Cataloging in Publication Data
A catalog record for this book has been requested

ISBN 13: 978-0-8153-4882-5 (pbk)
ISBN 13:978-0-415-58216-2 (hbk)

Typeset in Times New Roman
by Cenveo Publisher Services

This book was prepared from press-ready files supplied by the editor

CONTENTS

PART II

FOREWORD

This book is a translation of *Les Mathématiques infinitésimales du IXe au XIe siècle*, vol. V: *Ibn al-Haytham. Astronomie, géométrie sphérique et trigonométrie*. The French version, published in London in 2006, also included critical editions of all the Arabic mathematical texts that were the subjects of analysis and commentary in the volume.

The whole book has been translated, with great scholarly care, by Dr J. V. Field. The translation of the primary texts was not simply made from the French; I checked a draft English version against the Arabic. This procedure converged to give an agreed translation. The convergence was greatly helped by Dr Field's experience in the history of the mathematical sciences and in translating from primary sources. I should like to take this opportunity of expressing my deep gratitude to Dr Field for this work.

Very special thanks are due to Aline Auger (Centre National de la Recherche Scientifique), who helped me check the English translations against the original Arabic texts, prepared the camera ready copy and compiled the indexes.

Roshdi Rashed
Bourg-la-Reine, November 2013

PREFACE

This volume of *Les Mathématiques infinitésimales du IXe au XIe siècle* includes the *editio princeps* and the first translation of five works by Ibn al-Haytham, together with mathematical and historical commentary on them. The works are on astronomy and ancillary disciplines such as spherical geometry, trigonometry and studies of instruments. The reader might reasonably inquire whether the title of the volume gives a fair description of its contents: why should a book on the history of infinitesimal geometry contain works by Ibn al-Haytham on astronomy and subjects related to it? One might also ask why we present these five treatises rather than all Ibn al-Haytham's works on astronomy, and even why, after all, we present the works of Ibn al-Haytham rather than those of some other scholar.

We need to provide a brief explanation of what we are trying to achieve and the means we propose to adopt.

Like its predecessors, this volume is intended to establish the corpus of Arabic texts on infinitesimal geometry and to give an account of its history. We need to carry out this programme if we are to understand the emergence and development of ideas about analysis, not only in geometry but also in algebra, in the form in which we find it in the work of Sharaf al-Dīn al-Ṭūsī in the twelfth century. In short, we proceeded in this way in order to shed light on the origins of analytico-algebraic mathematics from the ninth century onwards. But this procedure ran up against several obstacles, of various origins. The most serious obstacles arose from the lack of historical writing on Arabic mathematics, a lack which is the more noticeable because between the ninth and twelfth century, that is, in the period that concerns us here, mathematical activity is prodigiously abundant.

It is well known that very few Arabic mathematical texts have been edited; and even fewer texts have been established in a properly scholarly manner. It is also the case that in this area historical studies worthy of the name can be counted on the fingers of one hand; and this dearth of historical research extends equally to the history of texts as well as that of concepts. It requires little experience in the history of Arabic mathematics before one comes to recognize that, even today, this area has hardly been explored at all.

The lack of historical writings (on the one hand) and (on the other) the great abundance of mathematical activity to write about, make for a situation that is inevitably awkward for any historian concerned to rely upon something other than empiricism. Plucking one flower in every field, juxtaposing names of mathematicians and the titles of their writings, and so on, have never got anyone anywhere. So a proper strategy needed to be worked out for exploring this continent of Arabic mathematics, or (at least) one of its provinces. In this book, as in the other books we have written on algebra, on number theory, on Diophantine analysis, and so on, the strategic principle is to ensure that there is the closest possible connection between investigation of the history of texts and the investigation of conceptual structures. But the great quantity of the material to which this strategy was to be applied made it necessary for us to start by determining what was the high point of mathematical activity in each of the chosen areas. For infinitesimal geometry, it is Ibn al-Haytham whose work represents that high point. He is the one who went furthest in investigating curved surfaces and curved solids; he is also the one who wrote the most important text on lunes, the one who provided the first coherent treatment of the solid angle, and so on. Having determined the highest point, we worked backwards, trying to reconstruct the tradition that led up to this point. We needed to go back to the Banū Mūsā, in the ninth century, to enable us to reconstruct the tradition, which we could then follow as far as Ibn al-Haytham, in the eleventh century. The first two volumes of *Les Mathématiques infinitésimales* were entirely concerned with that reconstruction.

Elucidating the structures of the proofs and the texts of the works that make up that tradition, and teasing out relationships between the texts, has revealed their main features. We may note, among other examples, that there are close connections between two ancient traditions, that of Archimedes and that of Apollonius; there is a far more frequent use of geometrical transformations than we find in Hellenistic mathematics; that the use of numerical calculation is carried much further than it is in the Archimedean tradition, and so on. A mere reconstruction of this tradition of infinitesimal geometry obviously does not provide a deep understanding of the formation and development of this branch of mathematics. We needed to look further afield, not only to find out what conditions made this branch of mathematics possible, but also to identify its applications. It is, in fact, often under the pressure provided by an application that new concepts are introduced and old ones modified. To understand such changes required us to set research in infinitesimal geometry within the framework of the research methods employed by the representatives of the tradition whose outlines we had established, that is by the Banū Mūsā, Thābit ibn Qurra, al-Māhānī, Ibrāhīm ibn Sinān, al-Khāzin, al-Qūhī, Ibn Sahl and al-Sijzī, up to Ibn al-Haytham. This is the task to which we devoted the third and fourth volumes of *Les Mathématiques infinitésimales*, as well as other works.[1] In the present volume we have tried to set infinitesimal geometry in historical perspective, examining it not only in the context of Ibn al-Haytham's other work in geometry, but also in the context of contributions made by other mathematicians in the same tradition, such as Ibn Sinān, al-Qūhī, and al-Sijzī, notably in regard to the geometry of conic sections. Our investigations led us to consider the conditions that made it possible for this tradition to emerge.

We also needed to examine the theoretical and technical advances that resulted from employing this type of mathematics. Adopting the same strategy as before, here we again began by examining the works of Ibn al-Haytham, before going back to look at the works of his predecessors. For Ibn al-Haytham, who was also eminent in the study of physics, was the mathematician best equipped to confront problems of applying mathematics. And he did, indeed, make magisterial contributions to the principal disciplines in applied mathematics of the time: optics, statics, astronomy and work on scientific instruments. The only exception, it seems, was acoustics.

We have already examined Ibn al-Haytham's writings on optics in several other works,[2] and we shall not return to them here. As for his statics, we are less lucky, since his writings on the subject are lost and all we have is a single reference to them, by al-Khāzinī.[3] That leaves astronomy and instruments.

It is well known that in ancient and Hellenistic culture astronomy was strongly privileged as an area for applying mathematical learning. This use of mathematics, required for constructing models of celestial motions, for making sundials and for other purposes, led to much new mathematics. We may, for example, point to spherical geometry, or methods of interpolation. So our investigation of the history of infinitesimal geometry in this period, in

---

[1] See R. Rashed and H. Bellosta, *Ibrāhīm ibn Sinān. Logique et géométrie au Xᵉ siècle*, Leiden, 2000; R. Rashed, *Œuvre mathématique d'al-Sijzī*. Vol. I: *Géométrie des coniques et théorie des nombres au Xᵉ siècle*, Les Cahiers du Mideo, 3, Louvain/Paris, 2004 and *id.*, *Geometry and Dioptrics in Classical Islam*, London, 2005.

[2] See R. Rashed, *Optique et mathématiques: Recherches sur l'histoire de la pensée scientifique en arabe*, Variorum Reprints, Aldershot, 1992 and *Geometry and Dioptrics*.

[3] Al-Khāzinī, *Kitāb Mīzān al-ḥikma*, Hyderabad, 1941; see also F. Bancel, 'Les centres de gravité d'Abū Sahl al-Qūhī', *Arabic Sciences and Philosophy*, 11.1, 2001, pp. 45–78.

particular the work of Ibn al-Haytham, necessarily led us to study his writings on mathematical astronomy. But there considerable surprises again awaited us, for several different reasons.

It is not unusual for historians of astronomy to mention Ibn al-Haytham, to stress the importance of his work and the crucial part he played in the criticism of Ptolemy's astronomy. However, on looking into the matter in more detail, one cannot but be disconcerted by the general ignorance that surrounds his works. Of the roughly twenty-five treatises that Ibn al-Haytham wrote on astronomy, only one short essay on the direction of Mecca has been the subject of a critical edition and a serious commentary;[4] his important treatise, *Doubts concerning Ptolemy*, has been the subject of an edition that can at best be described as provisional,[5] and as yet there has been no worthwhile commentary on it. Finally, the treatise on *Resolution of Doubts concerning the Winding Motion* has been published without any analysis or commentary.[6]

Thus Ibn al-Haytham's work in astronomy remains effectively unknown. And this ignorance has provided a basis for serious misapprehensions. In writing about Ibn al-Haytham's astronomy, most historians in fact rely upon a treatise called *The Configuration of the Universe*, or on another one, the *Commentary on the Almagest*. But, as we shall see, *The Configuration of the Universe* is an apocryphal work, while the *Commentary on the Almagest* is not by al-Ḥasan ibn al-Haytham but by a philosopher with a similar name, Muḥammad ibn al-Haytham.[7] Thus, even today, studies of Ibn al-Haytham's astronomy are being based on an apocryphal text, or a book that is not by him, or again on the less than solid foundation provided by the edition of the *Doubts concerning Ptolemy* to which we have already referred. The image of Ibn al-Haytham's astronomy that emerges thus contains contradictions: on the one hand it is a strictly Ptolemaic astronomy, and on the other hand, in the *Doubts*, it makes criticisms of Ptolemy. Curiously, this contradiction has not been noticed by many of those who have written about Ibn al-Haytham's

---

[4] A. Dallal, 'Ibn al-Haytham's universal solution for finding the direction of the *Qibla* by calculation', *Arabic Sciences and Philosophy*, 5.2, 1995, pp. 145–93.

[5] *Al-Shukūk 'alā Baṭlamiyūs*, ed. A. I. Sabra and N. Shehaby, Cairo, 1971.

[6] A. I. Sabra, 'Maqālat al-Ḥasan ibn al-Haytham fī ḥall shukūk ḥarakat al-iltifāf', *Journal for the History of Arabic Science*, 3.2, 1979, pp. 183–212, 388–92.

[7] *Les Mathématiques infinitésimales du IXᵉ au XIᵉ siècle*. Vol. II: *Ibn al-Haytham*, London, 1993, pp. 1–17, 490–1 and 511–38. English translation: *Ibn al-Haytham and Analytical Mathematics*. A History of Arabic Sciences and Mathematics, vol. 2, Culture and Civilization in the Middle East, London, 2012, pp. 11–26, 364–81.

astronomy, possibly because the contradiction was obscured by the fog of cosmological considerations that separates these scholars from the reality of Ibn al-Haytham's astronomy.

It is accordingly more than ever necessary to establish texts of the corpus of Ibn al-Haytham's writings on astronomy and write scholarly commentaries on them. At the moment this must remain a plan for the future, and our present purpose is a more limited one: we shall consider only his treatises on mathematical astronomy and ancillary subjects, to allow us to examine the interactions between astronomy and mathematics, specifically infinitesimal geometry and spherical geometry. We shall show that Ibn al-Haytham's work in astronomy takes place in at least two phases. In the first, where he is criticizing Ptolemy's astronomy, he also makes investigations in several subsidiary fields and already raises some new problems. This as it were preparatory phase is followed by a different phase, in which Ibn al-Haytham works out his new astronomy. It is at this time that, contrary to normal practice, the problem of the height of the heavenly body in the course of its motion becomes the leading problem in astronomical research. Working out this astronomy required a new investigation in infinitesimal geometry. Ibn al-Haytham considers variations of magnitudes and ratios, makes use of calculations of finite differences, and so on. This new research is described in a substantial treatise – one of Ibn al-Haytham's last works – whose importance is comparable to that of his *Book on Optics*: *The Configuration of the Motions of the Seven Wandering Stars*.

The present volume contains the *editio princeps* of the part of this treatise that has come down to us. To put it in context and give a measure of the distance Ibn al-Haytham had travelled since the first phase, we have also given the *editio princeps* of his treatise *On the Variety of Heights of the Wandering Stars*, which Ibn al-Haytham himself later declared to be out of date.

Among Ibn al-Haytham's works in ancillary fields there is a treatise *On the Hour Lines*, which brings to completion the tradition of work on this subject, a tradition that was initiated by Thābit ibn Qurra, who was followed by Ibrāhīm ibn Sinān and then al-Sijzī. Ibn al-Haytham also wrote a book on *Horizontal Sundials*, addressed to artisans, as well as a treatise on *Compasses for Large Circles*.

This volume does not contain all of Ibn al-Haytham's works on astronomy, but it has three purposes: to complete the preceding volumes by making it clear what Ibn al-Haytham's astronomical researches contributed to mathematics; to show how much more advanced his work on hour lines was in comparison with that of his predecessors; and, above all, to examine his new concept of astronomy. Thus we shall be able to observe both of the two directions taken by his research following his critique of Ptolemy: one path led to the construction of other configurations, which avoid the contradictions in Ptolemy, such as those of Naṣīr al-Dīn al-Ṭūsī and then of Ibn al-Shāṭir and those of their successors; the other path led to proposing a celestial kinematics expressed entirely in geometrical terms, which is Ibn al-Haytham's distinctive contribution to astronomy.

Throughout the years in which this volume was in preparation, I have benefited from the unfailing support of Christian Houzel, Emeritus Director of Research at the Centre National de la Recherche Scientifique. I should like to express here my sincere gratitude for the encouragement he gave me, and thank him also for his critical reading of my text, and the corrections he made to my historical and mathematical commentary. I also express my warmest thanks to Badawi El Mabsout, Professor Emeritus at the University of Paris VI, who revised the text of *The Configuration of the Motions of the Seven Wandering Stars* and proposed several improvements to it. In addition, I am grateful to the Reverend Father Régis Morelon, Director of Research at the Centre National de la Recherche Scientifique, for his friendly support and his constant willingness to discuss the ideas I put forward. I also thank Professors Boris Rosenfeld, Mariam Rozhanskaya and Sergei Demidov, who facilitated my access to the manuscript of *The Configuration of the Motions of the Seven Wandering Stars*. Last but not least, I express my immense gratitude to Aline Auger, Ingénieur d'Études at the Centre National de la Recherche Scientifique, who prepared this book for the press and constructed the indexes.

Roshdi Rashed
Bourg-la-Reine, June 2006

NOTE

< > Text between these brackets is an addition to the Arabic text that is necessary for understanding the English text.

[ ] Text between these brackets is an addition to the French text that is necessary for understanding the English text.

In the mathematical commentaries, we have used the following abbreviations: meas.: measure of the angle subtended by an arc; sect.: sector; tr.: triangle.

PART I

CHAPTER I

THE CELESTIAL KINEMATICS OF IBN AL-HAYTHAM

1.1. INTRODUCTION

1.1.1. *The astronomical work of Ibn al-Haytham*

From Pierre Duhem onwards, at least, historians of astronomy have been agreed on the importance of Ibn al-Haytham's contribution to the study of celestial kinematics. Some have paid particular attention to his criticisms of Ptolemy, criticisms that gave rise to the construction by his successors of new planetary models. But Ibn al-Haytham is not seen as having participated in this work, merely as having made the criticisms. Other historians have seen his contribution as synthesizing the *Almagest* with an Aristotelian cosmology. But a careful historical reading of Ibn al-Haytham's writings, including some new texts that have not previously been taken into consideration, shows that these two pictures of him are inaccurate. We find that Ibn al-Haytham tried to carry out a reform of astronomy, excluding any consideration of cosmology and developing the study of celestial kinematics.

However, such a reading requires us to consider Ibn al-Haytham's astronomical work as a whole, so as to define the limits of his concerns and to exclude the writings that have been incorrectly ascribed to him, which distort any assessment of his contribution.

The early bio-bibliographers – al-Qifṭī, Ibn Abī Uṣaybiʿa and an anonymous predecessor – tell us that Ibn al-Haytham wrote twenty-five astronomical works,[1] which means that a quarter of the eminent mathematician's works were concerned with astronomy. Further, that is to say that he wrote twice as many works on this subject as he did on optics, the field with which his name will always be associated. The number of writings alone indicates the huge size of the task accomplished by Ibn al-Haytham and the importance of astronomy in his life work.

---

[1] The first critical examination of what is known about Ibn al-Haytham and his writings is given in R. Rashed, *Ibn al-Haytham and Analytical Mathematics*. A History of Arabic Sciences and Mathematics, vol. 2, Culture and Civilization in the Middle East, London, 2013, together with a summary in the form of a table listing all his works, including those on astronomy (pp. 392–427).

From the writings that have come down to us it becomes clear that, even if the author's primary concerns are theoretical and mathematical, there was no part of astronomy that he neglected. Several treatises relate to technical applications of astronomy, others to methods of astronomical calculation, others again to procedures for making astronomical observations, and so on. One can nevertheless divide his writings into four groups, on the basis of surviving texts or, for lost texts, from titles mentioned in the books of early bibliographers.

The first group consists of about ten treatises in which Ibn al-Haytham is concerned with technical problems: *On the Hour Lines (Fī khuṭūṭ al-sā'āt)*, *On the Horizontal Sundials (Fī al-rukhāmāt al-ufuqiyya)*,[2] *<The Determination> of the Azimuth of the* Qibla *by Calculation (Fī samt al-qibla bi-al-ḥisāb)*,[3] *The Determination of the Height of the Pole with the Greatest Precision (Fī istikhrāj irtifā' al-quṭb 'alā ghāyat al-taḥqīq)*, *The Determination of the Meridian with the Greatest Precision (Fī istikhrāj khaṭṭ niṣf al-nahār 'alā ghāyat al-taḥqīq)*, *The Correction of Astrological Operations (Fī taṣḥīḥ al-a'māl al-nujūmiyya)*,[4] and so on.

The second group is made up of two treatises on astronomical observation: conditions for making observations, the errors that may occur in observation, and so on.

The third group of writings is concerned with various questions and ranges of problems such as those relating to parallaxes, to the Milky Way and so on.

The fourth group is concerned with astronomical theory and can in turn be divided into three subgroups:

In the writings in the first of these, Ibn al-Haytham discusses the work of Ptolemy. We have three books, which are of great historical and theoretical interest:

    1. *Doubts concerning Ptolemy (Fī al-shukūk 'alā Baṭlamiyūs)*[5]

    2. *Corrections to the Almagest (Fī tahdhīb al-Majisṭī)*

    3. *Resolution of Doubts concerning the Almagest (Fī ḥall shukūk fī kitāb al-Majisṭī)*.

Of these three books, only the first and the third have come down to us.

---

[2] See below, Part II, Chap. I and II.

[3] See A. Dallal, 'Ibn al-Haytham's universal solution for finding the direction of the Qibla by calculation', *Arabic Sciences and Philosophy*, 5.2, 1995, pp. 145–93.

[4] See below, Appendix.

[5] *Al-Shukūk 'alā Baṭlamiyūs (Dubitationes in Ptolemaeum)*, ed. A. I. Sabra and N. Shehaby, Cairo, 1971.

In the writings in the second subgroup Ibn al-Haytham examines individual celestial motions:

1. *The Winding Motion* (*Fī ḥarakat al-iltifāf*)

2. *Resolution of Doubts concerning the Winding Motion* (*Fī ḥall shukūk ḥarakat al-iltifāf*)[6]

3. *The Motion of the Moon* (*Fī ḥarakat al-qamar*)

In this subgroup, only the last two texts survive.

The third subgroup includes four titles:

1. *On the Variety that Appears in the Heights of the Wandering Stars* (*Fī mā ya'riḍ min al-ikhtilāf fī irtifā'āt al-kawākib*)

2. *The Ratios of Hourly Arcs to their Heights* (*Fī nisab al-qusiyy al-zamāniyya ilā irtifā'ātihā*)

3. *The Configuration*[7] *of the Motions of Each of the Seven Wandering Stars* (*Fī hay'at ḥarakāt kull wāḥid min al-kawākib al-sab'a*)

4. *The Configuration of the Universe* (*Fī hay'at al-'ālam*)

The first of these texts has come down to us, while the second has been lost. A part of the third survives;[8] the fourth is not to be identified with the apocryphal text of the same title.[9]

This simple summary shows very clearly that this major body of astronomical work is far from being well known, apart from *The Configuration of the Universe* (whose authenticity is doubtful), the treatise *On the Variety that Appears in the Heights of the Wandering Stars*, and *The Configuration of the Motions of Each of the Seven Wandering Stars*.

We also notice that in the three books in which Ibn al-Haytham mentions Ptolemy or the *Almagest* he does so in order to criticize the work. He indeed speaks of 'Doubts', of 'Corrections', of 'Resolution of doubts'. If to that we add the criticism of Ptolemy put forward in *The Resolution of Doubts concerning the Winding Motion*, it is no exaggeration to describe Ibn al-Haytham's researches as explicitly and deliberately designed as criticism and projects for reform. It remains to be seen when this project of reform was actually conceived, and what its outcome was. Here our task is made harder by the fact that some treatises have been lost, and because it is

---

[6] A. I. Sabra, 'Maqālat al-Ḥasan ibn al-Haytham fī ḥall shukūk ḥarakat al-iltifāf', *Journal for the History of Arabic Science*, 3.2, 1979, pp. 183–212, 388–92.

[7] The Arabic *hay'a* could be translated equally by 'configuration' or 'model'.

[8] See below, Part I.

[9] See R. Rashed, '*The Configuration of the Universe*: a Book by al-Ḥasan ibn al-Haytham?', *Revue d'histoire des sciences*, t. 60, no. 1, 2007, pp. 47–63 and *Ibn al-Haytham and Analytical Mathematics*, pp. 362–81.

difficult to date the writings that have survived. We know that *The Doubts concerning Ptolemy* was promised at the end of *The Resolution of Doubts concerning the Winding Motion*. We also know that *The Resolution of Doubts concerning the Almagest* was completed after August 1028, the date when Ibn al-Haytham finished *The Halo and the Rainbow*, which he cites.[10] Lastly, we know that these four books must have been composed at different times. So the order of composition is: *The Winding Motion*, *The Resolution of Doubts concerning the Winding Motion* and, finally, *The Doubts concerning Ptolemy*. Like *The Resolution of Doubts concerning the Almagest*, these three treatises were all composed before 1038, as we learn from the list of Ibn al-Haytham's writings up to that date. So it seems that around 1028, and certainly before 1038, Ibn al-Haytham was actively engaged with astronomy.

Although we cannot speak for the content of the *Corrections to the Almagest*, because the text is lost, the titles of these works make it obvious that Ibn al-Haytham took a critical stance. It is clear that this critical attitude is common to all the titles we have mentioned so far. Even in his book *The Motion of the Moon*, also composed before 1038, where he makes a point of explaining the difficulties in Ptolemy as the result of a first reading, Ibn al-Haytham does not completely abstain from making criticisms. That is to say that his criticisms, far from being merely incidental, are an expression of dissatisfaction with Ptolemy's astronomy. To get a measure of how radical these criticisms of Ptolemy are, by way of example we shall look at what Ibn al-Haytham says in reply to an anonymous scholar who had criticized his treatise *The Winding Motion*:

> From the statements made by the noble Shaykh, it is clear that he believes in Ptolemy's words in everything he says, without relying on a demonstration or calling on a proof, but by pure imitation (*taqlīd*); that is how experts in the prophetic tradition have faith in Prophets, may the blessing of God be upon them. But it is not the way that mathematicians have faith in specialists in the demonstrative sciences. And I have taken note that it gives him (*i.e.* noble Shaykh) pain that I have contradicted Ptolemy, and that he finds it distasteful; his statements suggest that error is foreign to Ptolemy. Now there are many errors in Ptolemy, in many passages in his books. Among others, what he says in the *Almagest*: if one examines it carefully one finds many contradictions. He (*i.e.* Ptolemy) has indeed laid down principles for the models he considers, then he proposes models for the

---

[10] In fact, Ibn al-Haytham himself transcribed his book *The Halo and the Rainbow* (*Fī al-hāla wa-qaws quzaḥ*) in the month of Rajab 419 of the Hegira, that is at the beginning of August 1028. Ibn al-Haytham refers to this book and to his *Optics* in his *Resolution of Doubts concerning the Almagest* (*Fī ḥall shukūk fī kitāb al-Majisṭī*); see ms. Aligarh, 'Abd al-Ḥayy no. 21, fol. 12^r and ms. Istanbul, Beyazit, 2304, fol. 8^v.

motions that are contrary to the principles he has laid down. And this not only in one place but in many passages. If he (*i.e.* noble Shaykh) wishes me to specify them and point them out, I shall do so.

I resolved to write a book to establish the truth in the science of astronomy; in it I show the contradictory passages in the *Almagest*, then the correct passages, and I show how to correct the [faulty] passages.

He made many mistakes in the *Book on Optics*, one of which was a mistake in the proof concerning the shape of mirrors, which shows how uncertain his grasp was.

As for his *Book on Hypotheses*, if one examines the notions he propounded in the second chapter and the models he put forward using spheres and parts of spheres, the demonstration [of the models] is immediately seen to be refuted and flawed. In my reply I have shown his error in regard to the two parts of the sphere, which he postulated for the epicycle, and I have explained with an irrefutable demonstration; and I have shown that, whatever cases one postulates for the [two] parts of spheres, one obtains an indefensible impossibility.[11]

This radical critique has led many historians to believe that Ibn al-Haytham's purpose was merely the limited one of criticism, or as it is

---

[11] Ms. St. Petersburg, B1030/1, fol. 19ᵛ:

وقد تبين لي من تضاعيف كلام مولاي الشيخ أنه يصدق قول بطلميوس في جميع ما يقوله من غير استناد إلى برهان ولا تعويل على حجة ، بل تقليداً محضاً ؛ وهذا هو اعتقاد أصحاب الحديث في الأنبياء ، صلوات الله عليهم . وليس هذا اعتقاد أصحاب التعاليم في أصحاب العلوم البرهانية . ووجدته أيضاً يصعب عليه تغليطي لبطلميوس ويتعض منه ، ويظهر من كلامه أن بطلميوس لا يجوز عليه الغلط . ولبطلميوس أغلاط كثيرة في مواضع كثيرة من كتبه . فمنها أن كلامه في المجسطي ، إذا حقق فيه النظر ، وجد فيه أشياء متناقضة ، وذلك أنه قرر أصولاً للهيئات التي يذكرها ، ثم أتى بهيئات للحركات مناقضة للأصول التي قررها . وليست موضعًا واحداً ، بل مواضع كثيرة . فإن أحب أن أكشفها وأبينها ، فعلت .

وقد كنت عزمت أن أعمل كتابًا في تحقيق الحق من علم الهيئة ، وأبين فيه أولاً المواضع المتناقضة من كتاب المجسطي ، ثم أبين المواضع الصحيحة منه ، ثم أبين كيف تحقق المواضع المتناقضة . وله أغلاط في الكتاب المناظر ؛ فمنها غلط في البرهان في شكل من المرايا يدل على ضعف تصوره .

فأما كتاب الاقتصاص فإن المعاني التي ذكرها في المقالة الثانية والهيئات التي قررها بالأكر والمنشورات إذا حقق النظر فيها بطل أكثرها واضمحل وفي عاجل الحال . قد بينت غلطه في هذا الجواب في المنشورين اللذين فرضهما لفلك التدوير ، وأوضحته بالبرهان الذي لا شك فيه ، وبينت أنه على أي وضع فرض المنشوران عرض منهما المحال الذي لا عذر فيه .

sometimes called 'aporetic'.[12] However, this is not so. During this same period, that is before 1038, Ibn al-Haytham had done some work on a problem that was later to prove fundamental: the heights of planets in the course of their motion. Moreover, in all his other critical writings apart from *Doubts concerning Ptolemy*, Ibn al-Haytham tries to resolve particular problems encountered in the *Almagest*, notably those that are not connected with the work's theoretical structure. In other words, even at this stage, the criticism is also a heuristic strategy. This will become still more apparent when we look at the consequences. It is in the course of these researches, and after carrying out further work to bring them to maturity, that Ibn al-Haytham conceived the idea of writing his monumental book *The Configuration of the Motions of Each of the Seven Wandering Stars*, in which he sets out the details of his new astronomy. That is to say that this last book – in which he again takes up the problem of heights – is the ultimate result of his critical and inventive researches carried out during at least two decades before 1038, and which was very probably not published until shortly after that date.

Now, by an ironic chance, there has recently been a confident attribution to our mathematician, al-Ḥasan ibn al-Haytham, of a Commentary on the *Almagest*, written in strictly Ptolemaic terms, and

---

[12] Because of this intention to criticize, which is clearly stated, some historians have followed S. Pines in believing that Ibn al-Haytham can be seen as belonging to an ancient aporetic tradition. Thus we find the mathematician placed in the same category as the eminent physician al-Rāzī, the author of the famous *Doubts concerning Galen*. This is to overlook an important difference that specifically separates Ibn al-Haytham, al-Rāzī and many others in a very wide range of disciplines, from this so-called aporetic tradition. Indeed, it is one thing to raise difficulties and criticize solutions, quite another to criticize for constructive purposes. For innovative research, of whatever kind, criticism is an integral part of the heuristic procedure. For instance, Ibn al-Haytham's doubts and criticisms were not put forward as arguments for a principle, but as statements the mathematician strove to prove mathematically and with the help of disciplined observations. More importantly still, these doubts and criticisms cannot be understood except in the light of what is in some sense Ibn al-Haytham's final work: *The Configuration of the Motions of Each of the Seven Wandering Stars*. It is thanks to his endeavours to provide a firmer footing for Ptolemy's astronomy by ridding it of its internal inconsistencies that Ibn al-Haytham discovers that to prepare the way for this reformulation he needs to separate an account of the motions – that is celestial kinematics – from cosmology. In short, in the case of Ibn al-Haytham, it is not possible to separate doubts and criticisms from a conscious aim to make fundamental reforms. See S. Pines, 'Ibn al-Haytham's critique of Ptolemy', in *Actes du dixième Congrès international d'histoire des sciences*, 1, no. 10, Paris, 1964, pp. 547–50 and *id.*, 'What was original in Arabic science', in A. C. Crombie (ed.), *Scientific Change*, Leiden, 1963, pp. 181–205.

composed by someone of the same name, a philosopher with an interest in the sciences, but not himself a mathematician, called Muḥammad ibn al-Haytham.[13] Confusion naturally reaches a peak when this text is cited by way of introduction to a deliberately critical book such as the *Doubts*. Such confusion necessarily creates a false impression and makes it impossible to understand al-Ḥasan ibn al-Haytham's astronomy.

But, as we have already seen, Ibn al-Haytham is the subject of another misapprehension, on the part of historians of astronomy. For centuries he has been supposed to be the author of the book called *On the Configuration of the Universe* (*Fī hay'at al-ʿālam*). This book, which is cited by early bio-bibliographers, was translated into Hebrew and into Latin.

Y. T. Langermann, who edited and translated the text, says about it: 'Many of the sharp criticisms of Ptolemy which are developed in the *Doubts* can, in fact, be directed equally well at *On the Configuration*, which faithfully mirrors the astronomical theory of the *Almagest*'.[14] I have added some further observations that cast doubt on the attribution of such a work to Ibn al-Haytham.[15]

To escape from so flagrant a contradiction, one is tempted to claim that this is an early work. But there is no evidence to support such a conjecture. On the contrary. In fact, even in regard to much less significant matters, when Ibn al-Haytham returns to a subject he has treated before, he is in the habit of referring back to his first treatment and warning the reader that it is now superseded by the present one.[16] One would therefore, *a fortiori*, expect a similar gesture here, particularly since he would be in the process of criticizing the theses defended in the first treatment. But it does not happen.

---

[13] In the introduction to the printed edition of *al-Shukūk* (note 5), A. Sabra believes it is possible to shed light on the critical text of this book by calling upon the Commentary on the *Almagest* of Muḥammad ibn al-Haytham, a book which follows Ptolemy to the letter. This strange enterprise stems from the long-standing confusion between Muḥammad ibn al-Haytham and al-Ḥasan ibn al-Haytham. On this matter, see R. Rashed, *Ibn al-Haytham and Analytical Mathematics*, vol. 2, pp. 11–25; *Ibn al-Haytham's Theory of Conics, Geometrical Constructions and Practical Geometry. A History of Arabic Sciences and Mathematics*, vol. 3, Culture and Civilization in the Middle East, London, 2013, pp. 729–34 and *Les Mathématiques infinitésimales du IXᵉ au XIᵉ siècle*, vol. IV: *Méthodes géométriques, transformations ponctuelles et philosophie des mathématiques*, London, 2002, pp. 957–9.

[14] Y. T. Langermann, *Ibn al-Haytham's On the Configuration of the World*, New York/London, 1990, p. 8.

[15] See R. Rashed, *Ibn al-Haytham and Analytical Mathematics*, pp. 362–81.

[16] See for example Ibn al-Haytham, *The Exhaustive Treatise on the Figures of Lunes*, in R. Rashed, *Ibn al-Haytham and Analytical Mathematics*, p. 107; also below, p. 260.

So our present knowledge of Ibn al-Haytham's astronomical work is: some people see no difficulty in attributing to him a thoroughly traditional Commentary on Ptolemy, or a treatise that conforms strictly to Ptolemy, and ignore the contradiction with Ibn al-Haytham's *Doubts* and his criticisms. Others, with good reason, note the contradiction, but stop there. Others still, of much earlier date, concentrate on the *Doubts* and express regret that Ibn al-Haytham was satisfied merely to criticize Ptolemy, without proposing another 'astronomy'. Thus the astronomer al-'Urḍī (who died in 1266) writes:

> No one came after him (Ptolemy) to bring that art (astronomy) to completion in a correct manner; no modern scholar has added anything at all to his work or subtracted anything from it, instead, all have followed him. Some among them have raised doubts, but without contributing more than the expression of doubts, such as Ibn al-Haytham and Ibn Aflaḥ of Andalusia.[17]

If we simply take them at face value, these words of al-'Urḍī are surprising for several reasons. They would seem to ignore what was achieved by Thābit ibn Qurra (826–901) and likewise all the other contributions which were made in the course of three centuries of mathematical astronomy; they would seem to place very little value on the secure observational results obtained by astronomers since the beginning of the ninth century, and they likewise seem to pass over work on instruments; they would also seem to reflect a mistaken outlook, one that had become more extreme in our time, according to which there existed an independent tradition of mathematical astronomy dedicated to criticizing errors in Ptolemy; finally, they would seem to indicate that al-'Urḍī knew no other astronomical text by Ibn al-Haytham apart from the *Doubts concerning Ptolemy*. Now, all this is very improbable, coming as it does from an astronomer like al-'Urḍī, the more so since his future 'boss' at Marāgha, Naṣīr al-Dīn al-Ṭūsī knew, at least, Ibn al-Haytham's book *The Winding Motion*, in which Ibn al-Haytham proposes a model of this motion that combines kinematics with some

---

[17] *Mu'ayyad al-Dīn al-'Urḍī: Kitāb al-Hay'ah*, edition with English and Arabic introductions by G. Saliba, Tārīkh al-'ulūm 'ind al-'Arab 2, Beirut, 1990, p. 214:

ولم يأت من بعده من يكمل هذه الصناعة على الوجه الصواب، ولم يزد أحد من المتأخرين، ولم ينقص شيئًا على ما عمله، لكن تابعوه بأجمعهم، ومنهم من شكك ولم يأت بشيء غير ذكر الشك فقط كأبي علي بن الهيثم وابن الأفلح المغربي.

cosmology.[18] Everything points to the explanation being that al-ʿUrḍī wanted to emphasize that Ibn al-Haytham had not proposed a model of the universe based jointly on the two traditions – that of the *Almagest* and that of the *Planetary Hypotheses* – a model in which celestial kinematics and a cosmology are combined in such a way that the resulting planetary theory is coherent and capable of making predictions that are as accurate as possible; in other words a configuration/model (*hayʾa*), like the one that al-ʿUrḍī thought he had constructed in his own book.[19]

And in fact al-ʿUrḍī's criticism, which in one sense misses the point, is in another sense justified. Ibn al-Haytham did indeed write an *Astronomy*, which will be discussed below. In this *Astronomy*, Ibn al-Haytham has understood that a genuine reform does not consist of constructing a model in the sense in which this was understood by al-ʿUrḍī, but in establishing a kinematic system, on a solid mathematical basis, before thinking about any kind of dynamics.

---

[18] According to what is reported by Naṣīr al-Dīn al-Ṭūsī, on the basis of a text by Ibn al-Haytham that is now lost (see F. J. Ragep, *Naṣīr al-Dīn al-Ṭūsī: Memoir on Astronomy – al-Tadhkira fī ʿilm al-hayʾa*, 2 vols, New York, 1993, vol. 1, pp. 215–17), the matter concerned is the deviation of the apogee and perigee of the epicycle as well as the two points on the epicycle at mean distance. Ibn al-Haytham seems to intend to construct a model using solid orbs as the mechanism for the motion. In this model, Ibn al-Haytham adds three solid orbs for the epicycles of the superior planets and five solid orbs for the inferior planets, so as to take account of the various deviations noticed by observers.

[19] Later, Ibn al-Shāṭir expressed a more qualified opinion than that of al-ʿUrḍī. This can be found in *The New Zīj* (*al-Zīj al-jadīd*, ms. Oxford, Bodleian Library, Arch. Seld. A30, fol. 2ʳ):

وجدت أفاضل المتأخرين مثل المجريطي وأبي الوليد المغربي وابن الهيثم والنصير الطوسي والمؤيد العرضي والقطب الشيرازي وابن شكر المغربي وغيرهم قد أوردوا على هيئة أفلاك الكواكب المشهورة، وهو مذهب بطلميوس، فيها شكوك يقينية مخالفة لما تقرر من الأصول الهندسية والطبيعية، ثم اجتهدوا في وضع أصول تفي بالحركات الطولية والعرضية من غير مخالفة لما تقتضيه الأصول، فلم يوفقوا على ذلك واعترفوا بذلك في كتبهم.

'I have noticed that eminent modern scholars, such as al-Majrīṭī, Abū al-Walīd al-Maghribī [Averroes], Ibn al-Haytham, Naṣīr al-Ṭūsī, Muʾayyid al-Dīn al-ʿUrḍī, Quṭb al-Shīrāzī and Ibn Shukr al-Maghribī, have expressed doubts about the model of the orbs of the planets, that is, the system of Ptolemy, doubts that contest the geometrical and physical principles [he] established, and they [the scholars] have then proceeded to work to put in place principles adequate to [explain] the motions in longitude and in latitude, among those that do not contest what these principles demand. They have not succeeded in this and have admitted as much in their books'.

1.1.2. *The Configuration of the Motions of Each of the Seven Wandering Stars*

Ibn al-Haytham's *Configuration of the Motions of Each of the Seven Wandering Stars* is a monumental achievement.[20] It deals with the 'model', or the 'structure' (*hay'a*), that is to say with a new astronomy or a new theory of the planets. The book, whose mathematical content is at the cutting edge of the science of its time, and which describes work that is both innovative and important, has come down to us in a single manuscript, which is in a pitiful state: a substantial part of it has been cut away, the leaves are out of order, damp has made some parts illegible and the hand-writing is hard to decipher.[21]

The Configuration of the Motions of Each of the Seven Wandering Stars (henceforth *The Configuration of the Motions*) was originally in three books: the first was mathematical astronomy, in which Ibn al-Haytham gives details of his planetary theory; the second was devoted to astronomical calculation or, as he writes, 'all the operations of calculation'; and the third was concerned with an astronomical instrument, one that was easy to manipulate, and designed for precise calculation of the heights of the sun and the planets. Of this complete volume, only the planetary theory has come down to us. The bulk of this first section is a reminder of the original size of the work before so much of it was lost, and allows us to grasp something of the magnitude of the task Ibn al-Haytham undertook. It is very probable that he wanted this book to take in all parts of astronomy, just as his *Book on Optics* had taken in all parts of that subject. But, equally, it also shows us that, at this time, a book about the configuration/model (*hay'a*) would cover several areas of investigation, not only one: a planetary theory, a study of the procedures used in the astronomical calculations needed for compiling tables showing the parameters required for calculating positions of planets (the *zījs*); and research on astronomical instruments.

The first book, which has come down to us, is on the theory of planetary motion; it also includes the introduction to the work as a whole, in which Ibn al-Haytham explains the organisation of the work and the style of presentation. In this introduction Ibn al-Haytham states that the style is that of demonstration, and that *The Configuration of the Motions* supersedes all the works he has previously written on the same subjects. This introduction is followed by mathematical text that takes up slightly less than half the section. This work deals with fifteen propositions which feature as lemmas in the construction of the planetary theory. This latter working appears in the

---

[20] See below.

[21] Ms. St. Petersburg, 600 (formerly Kuibychev, Library V.I. Lenin).

last part of the surviving text. We note that in the first part Ibn al-Haytham breaks new ground in the mathematics of infinitesimals since he is explicitly concerned with variations, variations of elements of a figure as a function of other elements, variations of ratios and variations of trigonometrical relationships. In this new area of research Ibn al-Haytham employs infinitesimal geometry and compares finite differences. This work on variable quantities, set in train by the needs of astronomy, made them a part of the geometry of infinitesimals.

Once he has completed this mathematics, Ibn al-Haytham is in a position to construct his planetary theory. But the length of the treatment and the deep nature of the mathematics in this part of the work are indicative of one of the motives that drive Ibn al-Haytham in his astronomical research: he wants to make planetary theory even more, and much more systematically, mathematical. Here, as in other disciplines, Ibn al-Haytham is following the procedure laid down by his predecessors, from Thābit ibn Qurra onwards, but going more deeply into things and more widely and pushing on further. If we forget this purpose we shall not understand *The Configuration of the Motions*.

But for this further mathematization to be possible, in a framework that continues to be geometric, and so that it can take place without running up against the inconsistencies in Ptolemy that have already been censured in the *Doubts*, Ibn al-Haytham is compelled to rethink the fundamental tenets of Ptolemaic astronomy. Thus in his eyes this systematic mathematization, far from being a merely instrumental or linguistic task, was an undertaking that truly engaged with theory. That is how Ibn al-Haytham came to devise a new planetary theory – one that does not concentrate on anomalies – by starting by separating kinematics from cosmology.

In the *Doubts*, Ibn al-Haytham comes to the conclusion that 'the configuration (*hay'a*) Ptolemy assumes for the motions of the five planets is a false one'.[22] A few lines further on he continues: 'The order in which Ptolemy had placed the motions of the five planets conflicts with the theory <that he proposes>'.[23] A little later he states:

> The configurations that Ptolemy assumed for the <motions of> the five planets are false ones; he decided on them knowing they were false, because he was unable <to propose> other ones. For the motions of the

---

[22] *Al-Shukūk ʿalā Baṭlamiyūs*, ed. Sabra and Shehaby, p. 34:

فقد تبين من جميع ما ذكرناه أن الهيئة التي قررها بطلميوس لحركات الكواكب الخمسة هي هيئة باطلة.

[23] *Ibid.*, pp. 33–4:

فالترتيب الذي رتبه بطلميوس لحركات الكواكب الخمسة خارج عن القياس.

planets there is a true configuration to be found in (*i.e.* from) the actual bodies, a configuration Ptolemy did not obtain and which he did not arrive at.[24]

After making remarks such as these, and many others like them in several places in his writings, a mathematician of Ibn al-Haytham's stature, one who had infinite respect for Ptolemy, as is proved by other comments, had no option but to construct a planetary theory of his own, on a solid mathematical basis, and free of the internal contradictions found in his predecessor. It was precisely the realization of this programme for which Ibn al-Haytham intended his treatise *The Configuration of the Motions*.

Most of the serious contradictions that Ibn al-Haytham censures set the *Almagest* against the *Planetary Hypotheses*. If we wish to describe the irreducible inconsistencies that, according to Ibn al-Haytham, vitiate Ptolemy's astronomy, we may say that they arise from the lack of fit between a mathematical theory of the planets and a cosmology. Ibn al-Haytham had experience of similar, though of course not identical, situations when, in optics, he encountered the inconsistency between geometrical optics and physical optics as understood by philosophers. In reforming optics he as it were adopted 'positivism' (before the term was invented): we do not go beyond experience, and we cannot be content to use pure concepts in investigating natural phenomena. Understanding of these cannot be acquired without mathematics. Thus, once he has assumed light is a material substance, Ibn al-Haytham does not discuss its nature further, but confines himself to considering its propagation and diffusion. In his optics 'the smallest parts of light', as he calls them, retain only properties that can be treated by geometry and verified by experiment; they lack all sensible qualities except energy. That is to say, we begin by insisting on making optics geometrical, or on reforming geometrical optics, leaving aside the 'why' questions that have to do with teleological physics, but prepared to introduce them later when we come to physical optics. It can be shown that this imposition of geometry led Ibn al-Haytham to study the propagation of light in kinematic – mechanical – terms.[25] Ibn al-Haytham adopts a similar approach in astronomy: in *The Configuration of the Motions* he deals with

---

[24] *Al-Shukūk ʿalā Baṭlamiyūs*, ed. Sabra and Shehaby, p. 42:

فالهيئات التي فرضها بطلميوس للكواكب الخمسة هي هيئة باطلة، وقررها على علم منه بأنها باطلة، لأنه لم يقدر على غيرها. ولحركات الكواكب هيئة صحيحة في أجسام لم يقف عليها بطلميوس ولا وصل إليها.

[25] R. Rashed, 'Optique géométrique et doctrine optique chez Ibn al-Haytham', *Archive for History of Exact Sciences*, 6.4, 1970, pp. 271–98; repr. *Optique et Mathématiques: Recherches sur l'histoire de la pensée scientifique en arabe*, Variorum reprints, Aldershot, 1992, II.

the apparent motions of the planets, without ever raising the question of the physical explanation of these motions in terms of dynamics. It is not the causes of celestial motions that interest Ibn al-Haytham, but only the observed motions in space and time. Thus, to proceed with the systematic mathematical treatment, and to avoid the obstacles encountered by Ptolemy, he first needed to break away from any kind of cosmology. And, in fact, in this treatise Ibn al-Haytham does not call upon the theory of material spheres, which had appeared in his *Resolution of Doubts concerning the Winding Motion* and in the *Doubts concerning Ptolemy*. Thus the purpose of Ibn al-Haytham's *Configuration of the Motions* is clear: to construct a geometrical kinematics.

Ibn al-Haytham's second intention is implied by the first one: to avoid the difficulties found in Ptolemy's astronomy. In the *Resolution of Doubts concerning the Almagest*, he states that 'in the *Almagest* as a whole there are doubts (*aporias*) too numerous for one to list them'.[26] All the same, in the *Doubts concerning Ptolemy* he distinguishes between doubts that can be resolved without modifying the structure of the theory and those whose elimination requires the theory to be subjected to radical reform.[27] One of the best examples of the latter type is the concept of the equant, exposed as an error in the *Doubts* and banished from *The Configuration of the Motions*. Ibn al-Haytham rejects the idea because one cannot, at the same time, suppose that a sphere rotates uniformly on its axis and suppose that this same rotation takes place about a line that is not a diameter of the sphere. In rejecting the equant, Ibn al-Haytham is already distancing himself very considerably from Ptolemy.

[26] *Fī ḥall shukūk al-Majisṭī*, ms. Istanbul, Beyazit 2304, fol. 195ʳ:

وفي جميع المجسطي شكوك أكثر من أن تحصى .

[27] *Al-Shukūk ʿalā Baṭlamiyūs*, ed. Sabra and Shehaby, p. 5:

ولسنا نذكر في هذه المقالة جميع الشكوك التي في كتبه، وإنما نذكر المواضع المتناقضة والأغلاط التي لا تأوّل فيها فقط، التي متى لم يخرج لها وجوه صحيحة وهيئات مطردة انتقضت المعاني التي قررها وحركات الكواكب التي حصلها . فأما بقية الشكوك فإنها غير مناقضة للأصول المقررة، وهي تنحل من غير أن ينتقض شيء، من الأصول ولا يتغير .

'We shall not mention in this book all the doubts contained in his works, but shall only mention the passages that contradict one another and the mistakes that cannot be rectified; the ideas he has put in place, and the motions of the planets he has arrived at, collapse if we cannot obtain true methods or uniform models or <correcting> these passages and these errors. As for the remaining doubts, they do not impute error to the established principles and they can be resolved without any of these principles being overturned or altered'.

As the author of two books on astronomical observation and the errors to which it is subject, and moreover as one who is well informed regarding the wealth of observations built up over two centuries, Ibn al-Haytham's third intention in writing *The Configuration of the Motions* is to construct a planetary theory which explains these observations.

These three intentions: mathematization, avoiding Ptolemy's contradictions and accounting for the observations, work together to fulfil Ibn al-Haytham's overall purpose for *The Configuration of the Motions*, that is, to set up a completely geometrical celestial kinematics. But in order to achieve this, he needed to find a way of measuring time. To this end, he introduced a new concept, that of 'required time', that is, a period of time measured by an arc.

A close examination of the way he organizes his exposition of his planetary theory shows that Ibn al-Haytham begins by proposing simple models – that is simple descriptive models – of the motions of each of the seven planets. As the exposition progresses, he makes the models more complicated and increasingly subordinates them to the discipline of mathematics. This increasingly mathematical formulation led him to regroup the motions of several planets as a single model. And it is precisely the mathematical nature of the model which makes this regrouping possible, specifically starting from Proposition 24. This step obviously has the effect of privileging a property that is common to several motions. In this way Ibn al-Haytham opens up the way to achieving his principal objective: to establish a system of celestial kinematics, this without as yet formulating the concept of instantaneous speed, but with the help of the concept of mean speed, represented by a ratio of arcs.

Here we shall explain the principal results Ibn al-Haytham obtained.[28]

1.2. THE STRUCTURE OF *THE CONFIGURATION OF THE MOTIONS*

The first section of *The Configuration of the Motions*, the section that has come down to us, divides into two parts. The first, which is mathematical and chiefly devoted to the study of variable quantities, comprises 15 propositions. The second part deals with planetary theory.

---

[28] See below, Part I.

1.2.1. *Studies of variations*

The fifteen propositions with which the section begins may be separated into several groups. The first consists of the first four propositions, which deal with the variation of trigonometrical functions such as $\frac{\sin x}{x}$. Ibn al-Haytham gives rigorous proofs of the following propositions:

1. If the measures in radians of the arcs α and α_1 of a circle are such that $\alpha + \alpha_1 \leq \frac{\pi}{2}$ and $\alpha > \alpha_1$, then

$$\frac{\alpha}{\alpha_1} > \frac{\sin \alpha}{\sin \alpha_1} \quad \text{and} \quad \frac{\alpha + \alpha_1}{\alpha_1} > \frac{\sin(\alpha + \alpha_1)}{\sin \alpha_1}.$$

2. If the measures in radians of the arcs α and α_1 of a circle and of the arcs β and β_1 of a different circle are such that

$$\beta + \beta_1 < \alpha + \alpha_1 < \frac{\pi}{2} \quad \text{and} \quad \frac{\alpha}{\alpha_1} = \frac{\beta}{\beta_1} = \frac{1}{k} \quad \text{(where } k < 1\text{)},$$

then

$$\frac{\sin \alpha}{\sin \alpha_1} < \frac{\sin \beta}{\sin \beta_1} \quad \text{or} \quad \frac{\sin \alpha}{\sin k\alpha} < \frac{\sin \beta}{\sin k\beta}.$$

As a corollary to this proposition, Ibn al-Haytham proves that

$$\frac{\sin(\alpha + \alpha_1)}{\sin \alpha_1} < \frac{\sin(\beta + \beta_1)}{\sin \beta_1} \quad \text{or} \quad \frac{\sin(1 + k)\alpha}{\sin k\alpha} < \frac{\sin(1 + k)\beta}{\sin k\beta}.$$

Ibn al-Haytham had proved this proposition in his treatise *On the Hour Lines*.[29]

3. If the measures in radians of the arcs α and α_1 of a circle and of the arcs β and β_1 of a different circle are such that

$$\beta + \beta_1 < \alpha + \alpha_1 \leq \frac{\pi}{2} \quad \text{and} \quad \frac{\sin(\beta + \beta_1)}{\sin \beta_1} = \frac{\sin(\alpha + \alpha_1)}{\sin \alpha_1},$$

then

[29] See below, Part II.

$$\beta < \alpha \quad \text{and} \quad \frac{\beta}{\beta_1} < \frac{\alpha}{\alpha_1}.$$

4. If the measures in radians of the arcs α and α_1 of a circle and of the arcs β and β_1 of a different circle are such that

$$\beta + \beta_1 < \alpha + \alpha_1 \leq \frac{\pi}{2}, \ \alpha_1 < \alpha, \ \beta_1 < \beta \ \text{and} \ \frac{\sin\beta}{\sin\beta_1} \leq \frac{\sin\alpha}{\sin\alpha_1},$$

then

$$\frac{\alpha}{\alpha_1} > \frac{\beta}{\beta_1}.$$

And if

$$\frac{\sin(\beta + \beta_1)}{\sin\beta_1} \leq \frac{\sin(\alpha + \alpha_1)}{\sin\alpha_1},$$

then

$$\frac{\beta + \beta_1}{\beta_1} < \frac{\alpha + \alpha_1}{\alpha_1} \quad \text{and} \quad \frac{\beta}{\beta_1} < \frac{\alpha}{\alpha_1}.$$

The second group is made up of the following three Propositions – 5, 6 and 7 – which also deal with variations of quantities and of ratios. In the first two – 5 and 6 – Ibn al-Haytham considers variations in the angular position of a point on a quadrant of a circle. In Proposition 7, he examines variations in right ascension. In the course of these propositions he compares finite differences, calls upon ideas about the geometry of infinitesimals and makes use of the sine rule (which was known to mathematicians of the time such as Abū al-Wafā' al-Būzjānī and Ibn 'Irāq).[30]

In Propositions 5 and 6, Ibn al-Haytham considers a sphere with centre ω on which positions are described with respect to a great circle ABC of diameter AC, its pole K and the great circle KC orthogonal to ABC (Fig. 1.1).

A great circle of diameter AC cuts the arc KB in the point D. With any point, such as H, on the arc CD there is associated a great circle KH that cuts the arc CB in the point P, and a circle through H parallel to (ABC) which cuts the arc KC in the point U, we have $\overset{\frown}{PH} = \overset{\frown}{CU}$. The arcs PH and

---

[30] M.-Th. Debarnot, *Al-Bīrūnī: Kitāb maqālīd 'ilm al-hay'a. La Trigonométrie sphérique chez les Arabes de l'Est à la fin du Xe siècle*, Institut Français de Damas, Damascus, 1985.

CP are, respectively, the *inclination* (the declination if the reference circle is the equator) and the *right ascension* of the point H with respect to the circle ABC.

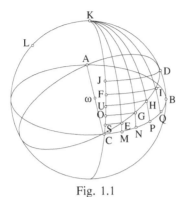

Fig. 1.1

First of all, Ibn al-Haytham considers how the *inclination* of $\widehat{PH}$ varies when the point H describes the arc CD.

Let the (rectilinear) dihedral angle between the planes ABC and ADC be α, we have $B\hat{\omega}D = \alpha$, so $\widehat{BD} = \alpha$. Let us put $\widehat{CH} = x$ and $\widehat{PH} = \widehat{CU} = y$, we have $0 \le x \le \dfrac{\pi}{2}$, $0 \le y \le \alpha$.

The proposition is in two parts which can be summarized as follows (Fig. 1.2):

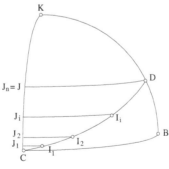

Fig. 1.2

a) The arc CD is divided into n equal parts at the points with spherical abscissae x_i, $0 \le i \le n$, $x_0 = 0$ and $x_n = \dfrac{\pi}{2}$. For $\Delta x = x_i - x_{i-1} = \dfrac{\pi}{2n}$ we have $\Delta y = y_i - y_{i-1}$. We show that Δy decreases when i increases from 1 to n. In other words, y is a concave function of x.

b) We consider two equal arcs with a common endpoint, with $x_i < x_j < x_k$ and $x_j - x_i = x_k - x_j$.

We show that from (a), we have $y_j - y_i > y_k - y_j$. This result may be expressed in the form

$$\frac{x_k - x_j}{x_j - x_i} > \frac{y_k - y_j}{y_j - y_i},$$

or as

$$\frac{y_k - y_j}{x_k - x_j} < \frac{y_j - y_i}{x_j - x_i},$$

which is to say that the gradient of the graph of y as a function of x decreases as x increases.

Proposition 6 extends this result to unequal arcs, such as arcs IJ and JK, where $x_i < x_j < x_k$ and $x_j - x_i \neq x_k - x_j$.

• If the two arcs in question that have an endpoint in common are commensurable, the result follows from a) and b).

• In the case where the same two arcs are incommensurable, Ibn al-Haytham gives a *reductio ad absurdum* argument to show that it is impossible to have

$$\frac{x_k - x_j}{x_j - x_i} \leq \frac{y_k - y_j}{y_j - y_i}.$$

We note that after proving the required inequality holds when the magnitudes are commensurable, Ibn al-Haytham proves the general case by 'extension by continuity' giving a rigorous abductive (apagogic) proof, and by applying his extension of Proposition 1 of *Elements* X.

So we have an argument based on infinitesimals for extending by continuity an inequality for which we have, as yet, no earlier example. We also note that Ibn al-Haytham is treating arcs and angles as magnitudes to which one can apply the theory of proportions.

Let us now return to his discussion of the variation of the inclination and show that his results are correct:

Let us put $y = \widehat{PH}$ as a function of $\widehat{CP} = x$. We have $y = f(x)$.

In the spherical triangle CHP, the arcs PH and PC are orthogonal, so $\hat{P} = \frac{\pi}{2}$, and the angle between arcs CP and CH is that between their tangents and it is equal to $B\hat{\omega}D$, so we have $\hat{C} = \alpha$.

The relation

$$\frac{\sin \widehat{CH}}{\sin \hat{P}} = \frac{\sin \widehat{PH}}{\sin \hat{C}}$$

therefore gives

$$\sin x = \frac{\sin y}{\sin \alpha};$$

so y as a function of x is given by

$$\sin y = \sin \alpha \cdot \sin x, \qquad y = \text{Arc sin} \cdot (\sin \alpha \cdot \sin x),$$

we have

$$\cos y \, dy = \sin \alpha \cdot \cos x \, dx,$$

that is

$$y'_x(x) = \frac{dy}{dx} = \frac{\sin \alpha \cdot \cos x}{\sqrt{1 - \sin^2 \alpha \cdot \sin^2 x}};$$

from which it follows that

$$y'' = -\frac{\sin \alpha \cdot \cos^2 \alpha \cdot \sin x}{\left(1 - \sin^2 \alpha \cdot \sin^2 x\right)^{\frac{3}{2}}}.$$

So for $0 < x < \frac{\pi}{2}$, we have $y' > 0$ and $y'' < 0$, $y = \widehat{PH} = f(x)$ increases from 0 to α.

But $f'(x)$ decreases over the interval $\left[0, \frac{\pi}{2}\right]$ and the function f is thus concave; so

$$\frac{x_m - x_k}{x_j - x_i} > \frac{y_m - y_k}{y_j - y_i}.$$

If in this expression we take:

- $x_m - x_k = x_j - x_i = \dfrac{\pi}{2n}$, we recover the result for a).
- $x_m - x_k = x_j - x_i$, we recover the result for case b), for equal arcs.
- $x_m - x_k \neq x_j - x_i$, we recover the result for case c), for unequal arcs.
  If $x_j = x_k$, the arcs concerned are contiguous.
  If $x_j < x_k$, the arcs concerned are disjoint.

In the seventh proposition (Fig. 1.3), Ibn al-Haytham considers the right ascension $\overset{\frown}{CP}$ when the point H describes the arc CD.

We put $\overset{\frown}{CH} = x$ and $\overset{\frown}{CP} = z$ for $0 \leq x \leq \dfrac{\pi}{2}, 0 \leq z \leq \dfrac{\pi}{2}$, we have:

a) As in considering the inclination, we divide the arc CD into n equal parts at points with spherical abscissa x_i. For the increment $\Delta x = x_i - x_{i-1}$ the corresponding increment in the right ascension, $\Delta z = z_i - z_{i-1}$, and using Menelaus' theorem for the arcs of great circles, we show that Δz increases when i increases from 1 to n.

b) Ibn al-Haytham next says that, as in the treatment of the inclination, one can generalize this result by considering two arcs lying on the arc CD, arcs that are equal to one another or unequal, contiguous or disjoint, and commensurable or incommensurable. Thus, for arcs I_iI_j and I_kI_m with $x_i < x_j \leq x_k < x_m$, one will have

$$\frac{x_m - x_k}{x_j - x_i} < \frac{z_m - z_k}{z_j - z_i}.$$

In other words, z is a convex function of x.

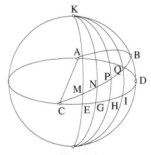

Fig. 1.3

Let us return to his discussion of the right ascension.

Considering $z = \overset{\frown}{CP}$ as a function of $x = \overset{\frown}{CH}$, when H describes the arc CD, $z = g(x)$.

The four circles involved are all great circles, and Menelaus' theorem gives

$$\frac{\sin \overset{\frown}{CH}}{\sin \overset{\frown}{HD}} = \frac{\sin \overset{\frown}{CP}}{\sin \overset{\frown}{PB}} \cdot \frac{\sin \overset{\frown}{KB}}{\sin \overset{\frown}{KD}}$$

$$\overset{\frown}{CH} = x, \quad \overset{\frown}{HD} = \frac{\pi}{2} - x, \quad \overset{\frown}{CP} = z, \quad \overset{\frown}{PB} = \frac{\pi}{2} - z, \quad \overset{\frown}{DB} = \alpha, \quad \overset{\frown}{KD} = \frac{\pi}{2} - \alpha.$$

So we have

$$\frac{\sin x}{\cos x} = \frac{\sin z}{\cos z} \cdot \frac{1}{\cos \alpha},$$

which gives

$$\tan z = \cos \alpha \cdot \tan x.$$

$$z = \text{Arc tan} (\cos \alpha \cdot \tan x) = g(x).$$

So we have

$$(1 + \tan^2 z)\, dz = \cos \alpha \cdot (1 + \tan^2 x)\, dx,$$

$$z' = g'(x) = \frac{\cos \alpha \left(1 + \tan^2 x\right)}{1 + \cos^2 \alpha \cdot \tan^2 x} = \frac{\cos \alpha}{\cos^2 x + \cos^2 \alpha \sin^2 x};$$

from which it follows that

$$z'' = \frac{\sin 2x \cos \alpha \,\sin^2 \alpha}{\left(\cos^2 x + \cos^2 \alpha \sin^2 x\right)^2}.$$

So for $0 < x < \dfrac{\pi}{2}$, we have $z' > 0$, z increases from 0 to $\dfrac{\pi}{2}$. We also have $z'' > 0$, $z' = g'(x)$ increases from 0 to $\dfrac{1}{\cos \alpha}$, hence the result Ibn al-Haytham obtained for the increment Δz.

As in the discussion of the inclination, Ibn al-Haytham indicates that his result can be extended to give an inequality involving differences of the right ascensions for unequal arcs, first in the case where these arcs are commensurable, then in the general case by using an argument of extension by continuity.

The third group is made up of Propositions 8 and 9. Ibn al-Haytham considers a circle (D, DC), that is, with centre D and radius DC, and a point E on DC, as well as the equal arcs AB, BH, HI such that we have the chord $AB < EC$ and he shows that $A\hat{E}B < B\hat{E}H < H\hat{E}I$ (Fig. 1.4).

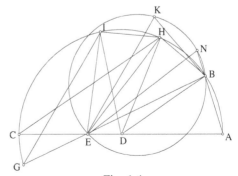

Fig. 1.4

If we put $A\hat{D}B = \theta$, where $\theta \in [0, \pi]$ and $A\hat{E}B = \varphi$, we see that Ibn al-Haytham is considering how φ varies as a function of θ.

In Proposition 9 he considers the sense of its variation.

The fourth group is concerned with the variation of ratios, in cases that become more and more complicated. This work is done in Propositions 10, 11, 12, 14 and 15. Proposition 13 is a lemma to do with technique. In this group, although Proposition 10 does not raise the complicated question of the range of the variations, Propositions 11 and 12, on the one hand, and Propositions 14 and 15, on the other, all require a long discussion, which is given in our commentary.[31]

In Proposition 10, Ibn al-Haytham considers two perpendicular planes $\mathscr{P}$ and $\mathscr{Q}$, two points A and C on their line of intersection, a semicircle of diameter AC lying in the plane $\mathscr{P}$, and a circular arc whose chord is AC, an arc smaller than a semicircle in the plane $\mathscr{Q}$ (Fig. 1.5).

We are trying to prove that there exists a point D such that $DE \perp AC$ and $EB \perp AC$ (where B lies on the semicircle) and such that we have $\dfrac{DB}{DC} > k > 1$, which is the given ratio.

We first show that there exists a unique point K on AC such that $\dfrac{KA}{CK} = k^2$.

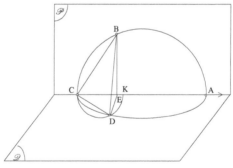

Fig. 1.5

We then draw a circle of diameter CK in the plane $\mathscr{Q}$ and we show that any point D on the circle yields the ratio.

In Propositions 11 and 12, we consider the meridian circle ABC for a given place G, the celestial poles A and C, a circle with centre O parallel to the horizon for G and which cuts the meridian circle in D and E, a circle of

---

[31] See below, Part I.

centre Q parallel to the equator and which cuts the meridian circle in H, the horizontal circle in L and the plane of the circle with centre Q cuts DE in X (Fig. 1.6).

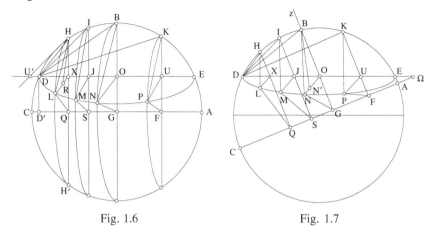

Fig. 1.6   Fig. 1.7

Ibn al-Haytham shows that when the point X moves along DE from D towards E, the point L describes the parallel circle with centre O and the ratio $\dfrac{HL}{HD}$ decreases and tends to 0.

In Proposition 12, we assume that the pole A is above the horizon, and that GOz the vertical at G; we have $D\hat{X}H = D\hat{O}z$, angle independent of the position of X (Fig. 1.7). Ibn al-Haytham shows that: when X moves along DE, the arc HE decreases, $\sin H\hat{D}X$ decreases and $\dfrac{HX}{DH} = \dfrac{\sin H\hat{D}X}{\sin D\hat{X}H}$ consequently also decreases from D to E.

Finally, Propositions 14 and 15 involve the celestial sphere for a given place, its axis, the two poles π and π', the meridian and horizontal planes for the place – the pole π is assumed to lie on or above the horizon.

In Proposition 14, Ibn al-Haytham considers ADB the meridian for an arbitrary place, and ABC a horizontal circle; two circles parallel to the equator cut the meridian in E and D, the circle ABC in I and C and a great circle with diameter $\pi\pi'$ in I and K (Fig. 1.8). Ibn al-Haytham proves that:

if $\widehat{BE} < \widehat{BD} \le \dfrac{1}{2}\widehat{ADB}$, then $\dfrac{\widehat{IE}}{\widehat{EB}} > \dfrac{\widehat{CD}}{\widehat{DB}} > \dfrac{\widehat{CK}}{\widehat{KI}}$.

We are in fact concerned with how $\dfrac{\widehat{IE}}{\widehat{EB}}$ varies as a function of $\widehat{BE}$; that is to say, we want to show that this ratio decreases when E moves from B

towards *F* along the chord of the meridian (where *F* is the midpoint of the arc *AB*).

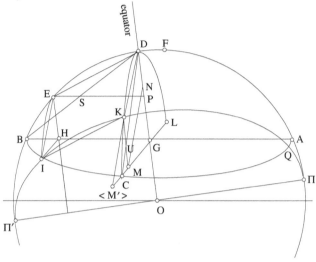

Fig. 1.8

Proposition 15 is a generalization of the preceding one. The two propositions show that Ibn al-Haytham, using the geometrical means at his disposal, investigated the variation of certain trigonometrical ratios; an investigation he could not complete but which sets some new mathematical research in train, as will be shown in our commentary on the translation.

1.2.2. *The planetary theory*

Once he has proved these fifteen mathematical propositions, Ibn al-Haytham immediately moves on to consider the apparent motions of the seven planets. He deals with the apparent motion on the celestial sphere, as seen from a given place, of a planet that is carried round by the diurnal rotation of the world about its axis, in the case where the planet in question has rising and setting points on the horizon of the given place of observation. Throughout his investigation, the place in question is in the northern hemisphere. From the first propositions, and starting from the results Ptolemy obtained for the orbs of the planets and for the different motions of the planets, Ibn al-Haytham shows that the observed trajectory of each of the planets, as seen on the celestial sphere, is different from the hour circle passing through a point of that trajectory, that is to say, it is different from the circle parallel to the equator swept out by a star whose position coincides, at a given moment, with that of the planet.[32] He deals in turn with the moon, the sun and the five planets, and, for the motion of these last along their orbs,[33] he distinguishes direct motion, retrograde motion and the planet's stations.

From this investigation, Ibn al-Haytham draws out and defines two new concepts: 'the required time' (al-zamān al-muhaṣṣal), and 'the inclination proper to the required time' (al-mayl alladhī yakhuṣṣu al-zamān al-muhaṣṣal). The 'required time' corresponds to two known positions of the planet in the course of a motion of known duration. It is measured by an arc of the hour circle, and it is equal to the difference of the right ascensions of the two observed positions. The inclination proper to the 'required time' is equal to the difference of their inclinations. We may note that, since the celestial sphere rotates uniformly, so that physical time can be represented by an arc of the hour circle, this concept of 'required time' is essentially a geometrical one. It is precisely in this way that Ibn al-Haytham represents physical time, and this has the further effect of permitting him to call upon the theory of proportions when time is involved.

Ibn al-Haytham then shows that, in all possible configurations, there exists a ratio greater than the ratio of the required time to the inclination for that time. Thanks to this property he proves that, for each of the planets observed from a given place, the planetary position whose height above the local horizon is a maximum does not correspond to the point of the planet's meridian transit, which is unlike the situation for a star. For a planet, the maximum height is greater than that of its meridian transit and, depending on the position of the planet in its trajectory, maximum altitude will be

---

[32] In his treatise on *The Variety that Appears in the Heights of the Wandering Stars*, composed earlier, Ibn al-Haytham writes as if the trajectory of this apparent motion could be identified with an hour circle (see below, Part I, Text no. 2).

[33] In Arabic astronomy, the word *falak* designates the orb as defined in the *Almagest*, *i.e.* the spherical shell within which the planet moves. Every planet has its own orb. For example, Thābit ibn Qurra (d. 901) wrote in his *Almagest simplified*: 'The orb in which the moon moves is the nearest orb to the earth and it is thick (*lahu sumk*). The moon moves sometimes in its upper part, sometimes in its lower part, and sometimes between them. The same happens for all the other planets' (Thābit ibn Qurra, *Œuvres d'astronomie*, ed. R. Morelon, Paris, 1987, p. 5, for Arabic text with French translation). This was the conventional meaning of the word in the Arabic tradition of Ptolemy, and it is also the sense in which Ibn al-Haytham employed the word in the works he wrote before *The Configuration of the Motions*. In this last book, Ibn al-Haytham used the word *falak* in a new – and unconventional – meaning, indeed so unconventional that the word 'orb' seems in places inappropriate. The right translation, as we shall see later, would be 'trajectory', 'path', or even simply 'orbit'. But, as Ibn al-Haytham himself continued to use the same word, though with a new meaning, we have no choice but to follow his example.

reached either before meridian transit, and thus to the east of the meridian, or after meridian transit, to the west of the meridian.

The investigation of the apparent motion of a planet, when it is above the horizon, ends with a discussion of the case where the geographical latitude of the place from which observations are made is equal to the complement of the maximum declination of the observed axis, or is very close to it. Ibn al-Haytham shows that, for places like these, the planet may set in the east and then rise in the east, or rise in the west and then set in the west.

In the course of this work, whose main lines have been sketched here, we encounter a concept of astronomy that is new in several respects. Ibn al-Haytham sets himself the task of describing the motions of the planets exactly in accordance with their paths on the celestial sphere. He is not trying to save the appearances, that is, to explain the irregularities in the assumed motion by means of artifices such as the equant – a notion he criticizes in his *Doubts concerning Ptolemy* – nor is he willing to account for the observed motions by appealing to underlying mechanisms whose nature is hidden. He wants to give a rigorously exact description of the observed motions in terms of geometry. The only mechanical device involved in the description of the motions of the planets (other than the sun and the moon) is an epicycle, which is employed to account for retrograde motions and the variable speeds near apogee and perigee. Ibn al-Haytham doubtless knew that using an epicycle and deferent was equivalent to using an eccentric circle, and also knew the precise conditions for this equivalence.

What Ibn al-Haytham proposes – and the proposal takes him a little further away from the Ptolemaic tradition – is thus to give a description of the motion in two dimensions on the celestial sphere. He considers the motion appears to be composed of two elementary motions along great circles of the celestial sphere. The free parameters are the speeds of the elementary motions, considered to be independent of one another. But for planets whose trajectory has a variable inclination to the ecliptic, Ibn al-Haytham nevertheless calls upon an epicycle to account for the variation in the inclination, thus (for the moment) returning to a model in a three-dimensional space. That indeed puts him in the Ptolemaic tradition, but without recourse to an equant.

So the guiding principle of Ibn al-Haytham's description is clear: to use Ptolemy's mechanisms as sparingly as possible. In considering the apparent motions of the planets on the celestial sphere, always with respect to the horizon, Ibn al-Haytham picks out four reference points: those of rising, meridian transit, setting and maximum altitude. He shows that this last point is unique and may lie to the east or to the west of the point of meridian transit.

The new astronomy no longer aims at constructing a model of the Universe, but only at describing the apparent motion of each planet, a motion composed of elementary motions, and, for the inferior planets, also of an epicycle. Ibn al-Haytham considers various properties of this apparent motion: localisation and kinematic properties of the variations in speed. In the last part of *The Configuration of the Motions*, he considers the apparent motion of the planet on the celestial sphere in the course of a day and proves that the planet attains its maximum height once and only once, and that any height less than the maximum is reached twice, once on each side of the maximum height. For heights greater than that of meridian transit, the two points where such a height is reached are on the same side of the meridian. Taken together, these discussions make twenty-one propositions.

In this new astronomy, as in the old one, every observed motion is circular and uniform, or composed of motions that are circular and uniform. Ibn al-Haytham considers three basic motions: diurnal motion parallel to the equator; motion of the oblique orb relative to the axis (the line joining the two poles of the ecliptic); and motion of the nodes of the proper orb. The observed motion of a planet is composed of these three motions plus, for the five planets (excluding the sun and moon), a motion on an epicycle. For the sun, only the first two basic motions are involved. To find these motions Ibn al-Haytham makes use of various systems of spherical coordinates: equatorial coordinates – required time and proper inclination for it – which are the first coordinates; horizon coordinates – altitude and azimuth; and ecliptic coordinates.

The use of equatorial coordinates marks a break with Hellenistic astronomy. In the latter, the motion along the orbs was measured against the ecliptic, and all coordinates were referred to the ecliptic (latitude and longitude). Thus analyzing the motion of the planets from their apparent motions changes the reference system for the data; we are now dealing with right ascension and declination. This book by Ibn al-Haytham takes us into a different system of analysis.

Ibn al-Haytham then considers how the speed of change in the inclination varies for any planet, measuring it as the mean speed over an interval that is itself variable. He looks at the change in height of the planets between their rising and setting. These investigations are all carried out rigorously, using the mathematical propositions proved in the first part and rely upon considerations involving infinitesimals (which make repeated appearances). The geometrical proofs that are employed assume only that the motion of the planet is from east to west, and that it is monotonic along the north-south axis.

When the geometry is conceptualized in this way, the question of a possible motion of the earth does not arise, because we are concerned only with the motion of the planet on the celestial sphere as it appears to a terrestrial observer. In other words, we have a kind of phenomenological description of the motions of the planets, which however can be given only in terms of spherical geometry, infinitesimal geometry and trigonometry. There is nothing surprising about this since Ibn al-Haytham is concerned to ensure that his description employs only minimal hypotheses about the properties that characterize the motions: variation of speed and day by day variation of height.

Let us briefly summarize the various chapters of this astronomical part.

I. *The apparent motion of the planets*

In the first part of the section devoted to astronomy, Ibn al-Haytham starts from results Ptolemy proved for each of the seven planets (the three fundamental motions) and introduces definitions of the 'required time', the inclination of the motion of the planet and the inclination of the ascending node. He investigates in turn: 1. The motion of a planet between rising and meridian transit; 2. Motion of known duration between two points of known position.

1.1 *The apparent motion of the moon between rising and meridian transit*

Ibn al-Haytham begins by citing the results proved by Ptolemy in relation to the inclined orb of the moon, and the position of this orb in relation to the circle of the ecliptic and to the nodes, that is to the points of intersection of these two orbs. Ibn al-Haytham considers the dihedral angle between the plane of the inclined orb and the plane of the ecliptic to be fixed. In fact, this angle is almost constant and remains close to 5°. The orb of the moon would thus lie within the Zodiac.

Ibn al-Haytham then reminds us that the motion of the moon on its orb is in the direction of the signs of the Zodiac (direct motion, motion in consequence) and that each of the nodes has a uniform motion round the ecliptic in the direction opposite to that of the signs of the Zodiac (retrograde motion, motion in precedence). Thus the north pole of the orb of the moon, X, describes on the celestial sphere a circle centred on the pole of the ecliptic, P, and each point of the orb of the moon describes a circle parallel to the ecliptic (in retrograde motion). Now, the angle between the circle of the ecliptic and the circle of the equator is constant; but because the nodes move, the inclination of the orb of the moon to the equator of the celestial sphere will vary. The inclination will be equal to the arc of the great circle HX, where H is the north pole of the celestial equator.

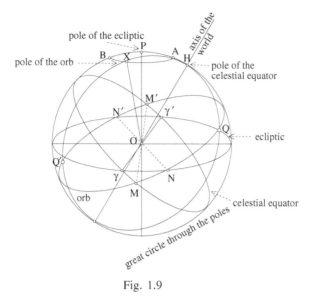

Fig. 1.9

Ibn al-Haytham investigates in minute detail how this arc varies as the node N makes a complete circuit around the ecliptic. In this preliminary investigation, he ignores the precession of the equinoxes (in his terms, the retrograde motion of the equinoxes), which is very slow. He treats the planes of the celestial equator, of the ecliptic and of the circle through the poles of the celestial sphere as if they were all fixed with respect to one another. He does the same thing later on, making it explicit, when he is considering the extreme (maximum and minimum) inclinations of the orb of each of the seven planets to the equator. He behaves as if he were deliberately first constructing a simple model and intended to make it more complicated later.

Ibn al-Haytham then defines the most northerly and the most southerly points of the lunar orb with respect to the equator. These points are the midpoints of the semicircles into which the orb is divided by the diameter, that is, its line of intersection with the plane of the equator. They accordingly lie on the great circle HX that passes through the poles of the Lunar orb and those of the equator; their inclination to the equator is equal to HX and is thus variable (Fig. 1.9).

Ibn al-Haytham then investigates the apparent motion of the moon, in relation to a horizon $ABCD$, between its rising at B and its meridian transit at a point N (Fig. 1.10, where ABC is the east half of the circle of the horizon); he first considers the case in which the motion along its orb is from north to south, then the case in which it is from south to north. He points out that his argument does not involve the horizon, and is consequently applicable to the motion of the moon between any point B on its trajectory (where B lies to the east of the meridian) and the point of its meridian transit. This is the moment at which Ibn al-Haytham introduces the following three definitions:

Required time: the time a fixed star takes to travel from a point B to a point
 I on the meridian; this is the arc BI. It is also the difference of the two
 right ascensions, $\delta(B, N)$, the difference between the right ascension of
 the moon's initial position, B, and the right ascension of its final position,
 N. This arc BI will also be called the right ascension of the motion.
Inclination of the motion of the moon: $\widehat{IN} = \Delta(B,N)$, the difference
 between the inclinations of the initial position, B, and the final position,
 N.
Inclination of the motion of the ascending node: $\widehat{IQ}$.

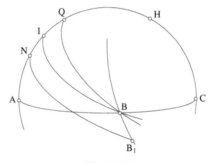

Fig. 1.10

This investigation of the apparent motion of the moon from its rising to its meridian transit is interrupted by a discussion of the relative positions of two circles through B, whose poles are the pole of the equator and the pole of the ecliptic. Finally, Ibn al-Haytham considers the motion of the moon between its meridian transit and its setting, for which he makes use of the concepts he has already defined.

  We note that, in this geometrico-kinematic model, Ibn al-Haytham does not introduce an epicycle since, as he says, 'the epicycle of the moon does not depart from the plane of the orb, so the centre of the moon does not depart from the plane of the inclined orb'.[34]

---

[34] See below, p. 334; Arabic text in Rashed, *Les mathématiques infinitésimales*, V, p. 429, 23–25.

1.2. *The apparent motion of the sun between its rising and its meridian transit*

Ibn al-Haytham works through the same stages as in the previous investigation: he begins by reminding us of what is known about the orb of the sun – the ecliptic – and the sun's proper direct motion through the signs of the Zodiac. He defines the points of the orb that are called equinoxes and solstices. He then deals with two examples, referred to a horizon *ABCD*, concerned with the apparent motion of the sun between its rising at *B* and its meridian transit. In the first case, the motion of the sun along its orb is from north to south, in relation to the equator; and in the second case from south to north. In each example, Ibn al-Haytham finds the arcs that represent 'the required time' and the inclination of the motion of the sun.

This investigation is simpler than the one he carried out for the motion of the moon, which required one to take account of the motion of the node along the ecliptic.

1.3. *The apparent motion of each of the five planets between rising and meridian transit*

Here, as in previous cases, Ibn al-Haytham begins by reminding us of what was established by Ptolemy. He also tells us that his investigation will not take account of the motion of the node, since, he writes, it is 'a slow motion that does not become perceptible'.[35] We should recollect that Ibn al-Haytham has always maintained that, unlike in mathematics, where reasoning is exact, in physics we always allow a certain amount of approximation. And, here, the inclination of the plane of the epicycle to that of the orb is variable. Accordingly, its variation must be taken into account when investigating the motion of each of the five planets towards the meridian circle. Ibn al-Haytham does exactly this when he considers the motion of a planet between its rising, at a point *B* on the horizon, and its meridian transit. He distinguishes three cases: when the planet's motion is direct; when it is retrograde; and finally when the planet is stationary. Ibn al-Haytham's investigation ends with a conclusion on the planets as a whole, concerning the 'required time' and the 'inclination of the motion'.

2. In the previous part of the work, the two positions considered for each of the seven planets were rising, at the point *B*, and meridian transit, at the point *N*; or sometimes the motion considered was from point *N* to setting. In this part of his work, Ibn al-Haytham investigates, for each of the seven planets, an apparent motion of known duration, between two points *A*

---

[35] See below, p. 334; Arabic text in Rashed, *Les mathématiques infinitésimales*, V, p. 429, 2.

and *B*, whose position on the celestial sphere is known. He shows that the 'required time' and the 'inclination of the motion' are then known.

Ibn al-Haytham begins by dealing briskly with the case of the sun, which is simple because his model does not take account of the precession of the equinoxes. Thus if *A* and *B* are respectively the starting and end points of the motion, we at once have: 'required time': $\delta(A, B)$, the difference of the right ascensions of the two points *A* and *B*; 'inclination of the motion': $\Delta(A, B)$, the difference of the declinations of the two points *A* and *B*, that is, the difference of their inclinations to the equator.

The investigation of the motion of the moon must, however, take account of the motion of the ecliptic and the motion of the node of the orb of the moon.[36] Here, as in the case of the sun, the motion is described in equatorial coordinates: 'required time' and 'proper inclination'.

For each of the inferior planets (Venus and Mercury), the ecliptic coordinates – ecliptic latitude and longitude – depend on the inclination of the epicycle to the orb.[37] All the same, if at some known time the ecliptic coordinates are known, we use them to find the equatorial coordinates. Ibn al-Haytham continues his investigation in the same way as he did in the case of the moon.

For the superior planets (Mars, Jupiter and Saturn), the motion of the nodes is very slow, and insensible over the course of a day. As a result, the arc that corresponds to arc *KG*, which is parallel to the ecliptic in the case of the moon, is insensibly small; the point *G* merges with the point *K* and thus lies on the hour circle *AD* (Fig. 1.11).

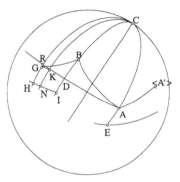

Fig. 1.11

<hr />

[36] See below, Mathematical commentary on Proposition 20.

[37] *I.e.* the circle on which the epicycle moves (*circulus deferens*). We shall find this formulation more than once.

Ibn al-Haytham concludes by taking the five planets together: if the motion of the planet along its orb is direct, 'the required time' $\delta\,(A, B)$ is less than the known time, as happens for the sun and the moon; and if the motion of the planet is retrograde, the 'required time' is then greater than the known time.

II. *The inclination of the planets to the equator*

Ibn al-Haytham begins by discussing the sun, next the moon and then the five planets. As ever, he first of all reminds us of Ptolemy's results. Here, Ibn al-Haytham further determines, in each case, the ecliptic coordinates of the point *I*, the most southerly point of the orb with respect to the equator.

In the case of the sun, the dihedral angle α between the plane of its path (the ecliptic), and the plane of the equator is constant ($\alpha = 23°27'$). This angle α is the maximum inclination of points on the ecliptic with respect to the equator, and corresponds to the solstices. The two points of maximum inclination are thus the first point of Cancer to the north of the equator, and the first point of Capricorn to the south.

In the case of the moon, the dihedral angle β between the orb of the moon and the circle of the ecliptic is constant, but the orb of the moon rotates about the axis of the ecliptic. Accordingly, the dihedral angle δ between the orb of the moon and the plane of the equator is variable; it depends on the position of the ascending node. If the ascending node is at the point γ (Spring equinox) we have $\delta = \alpha + \beta$. But if the descending node is at the point γ, the ascending node is then at the point γ', the Autumn equinox, and we have $\delta = \alpha - \beta$. In either case, the positions of the extreme north and south points of the inclined orb are known.

In the case where the ascending node is not at an equinoctial point, Ibn al-Haytham embarks on a very detailed investigation using spherical trigonometry, in which he applies Menelaus' theorem four times, and shows that, if we know the position of a node on the ecliptic, we can calculate the maximum inclination of the inclined orb with respect to the equator, and find the position, with respect to the ecliptic, of the most northerly or most southerly point of the inclined orb with respect to the equator.

For the superior planets, the procedure is the same as for the moon, since the inclinations of their orbs to the plane of the ecliptic are more or less constant: for Mars, $1°51'$, for Jupiter $1°19'$ and for Saturn $2°30'$. On the other hand, the inclination of the orb of each of the inferior planets to the ecliptic is variable. Ibn al-Haytham accordingly devotes many pages to the investigation of this problem.

He begins by examining the inclination as a function of the position of the planet in its orb, a position for which there is a corresponding point on the eccentric. He shows that this inclination is known at any known time.

He goes on to investigate the case where the nodes occur at the equinoctial points. The most northerly and most southerly points of the orb relative to the equator, and when related to the ecliptic, are the points of the solstices. We calculate the inclination relative to the equator as we did for the moon. Ibn al-Haytham next considers the case where the nodes are not the points of the equinoxes. The positions of the extreme north and south points relative to the equator are found from the extreme north and south points relative to the ecliptic, and the same method is then employed as before.

Ibn al-Haytham goes on to describe – still for the inferior planets – the oscillating motion of the plane of the inclined orb about the line of nodes. The motion of the line of nodes is very slow and for the purposes of this calculation the line is accordingly assumed to be fixed. So any point I of the orb describes a circle with the nodes as its poles, and the point will have a to-and-fro motion along an arc of the orb. With this point I there is associated a point L that represents its position in regard to the ecliptic; this point L will also have a to-and-fro motion along an arc of the ecliptic. In his investigation of the motion of the points I and L, Ibn al-Haytham takes the point I as lying, successively, on each of the four arcs into which the orb is divided by the nodes and the extreme northern and southern points. He assumes that the initial position of the orb is when its inclination to the ecliptic is at a maximum, and he calls the two points in question I and L.[38] He first describes the motion of the points I and L. Next he shows that the circular arc described by the point I in a known time is known; finally, he shows that the arc of the ecliptic described by the point L in a known time is known.

III. From Proposition 24 to the end of the book, Ibn al-Haytham proposes general models for the planets, models that are constructed with the help of the mathematical propositions he has already proved. His work, explicitly analytical and employing infinitesimals, concerns itself with some kinematic properties of the motion. This time, we cannot follow Ibn al-Haytham's procedure without examining his demonstration in detail, which we do in the commentary of his text.[39] Here we shall merely present a general outline.

In the first four propositions – 24 to 27 – Ibn al-Haytham investigates

---

[38] The great circle through the pole cuts the orb in I and the ecliptic in L. The points I and L have the same ecliptic longitude.

[39] See below, Part I.

the variation of the mean speed of a planet. He expresses the mean speed as the inverse ratio $\dfrac{\delta(X,Y)}{\varDelta(X,Y)}$, where X and Y are two general known positions of a planet in its orb, $\delta(X, Y)$ is the 'required time' and $\varDelta(X, Y)$ is the difference between the inclinations of the points X and Y with respect to the equator. Ibn al-Haytham proves that, if we consider the four arcs into which the orb is divided by the diameter, that is, the line of intersection of the planes of the orb and the equator, and the extreme northern and southern points of the path with respect to the equator, and if we take two positions X and Y on one of these arcs, then there always exists a ratio k such that

$$k > \frac{\delta(X,Y)}{\varDelta(X,Y)}.$$

We may note that the known time is a real interval that can be measured by the motion of the planet. Ibn al-Haytham's idea of comparing 'required time', an equatorial coordinate, to this known time, looks like the beginnings of a kinematic description of the motion.

In the following group of propositions, Ibn al-Haytham investigates the apparent motion of a planet above the horizon of a given place. The observed motion depends on the place and on the date of the observation. In the course of this investigation Ibn al-Haytham makes use of the planet's equatorial coordinates, and consequently of its position on its trajectory, of the inclination of the orb to the equator and of the inclination of the equator to the horizon; that is to say, he uses the geographical latitude of the place where the observation is made. Throughout this investigation, Ibn al-Haytham assumes that the celestial sphere is inclined to the south; the observation site thus has a northern latitude. The case of the *sphaera recta*, that is to say, the case where the observer is on the terrestrial equator, appears as a special case. Ibn al-Haytham assumes that the planet's meridian transit takes place between the zenith and the southern horizon, which means that the geographical latitude of the place where the observation is made must be greater than the declination of the planet for the date in question. He also assumes that the latitude of the observation site is smaller than the complement of the declination. Ibn al-Haytham makes a detailed study of the part played by the latitude, which leads him to consider the cases in which meridian transit occurs at the zenith or north of the zenith, and finally the case of places whose latitude is equal to the complement of the maximum declination of the planet.

So, in two propositions, 28 and 29, Ibn al-Haytham investigates heights of a planet above the horizon. Let us suppose that the planet rises at the point B and crosses the meridian at D. The arc BD which it describes is to the east of the meridian plane. Let the height of the planet above the horizon be h (Fig. 1.12). Ibn al-Haytham shows that on arc BD there exist

• points of height $h > h_D$ (the height of point D). Let M be one of these points;

• at least one point X on the arc BM such that $h_X = h_D$;

• at least two points with the same height h with $h_D < h < h_M$, one on the arc XM and the other on the arc MD.

He also shows that, after it crosses the meridian at D, the planet continues its motion towards the western horizon and its height h decreases from h_D to 0. Any height $h < h_D$ is thus reached at least once.

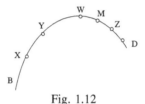

Fig. 1.12

Ibn al-Haytham also shows that, if h_m is the maximum height, the planet will reach this height only once, say at a point W; and that height h_D will be reached once and only once at a point $X \neq D$, such that $X \in \widehat{BW}$.

In Proposition 29, Ibn al-Haytham investigates the movement of the planet from the most southerly to the most northerly points of its trajectory. The planet crosses the meridian at G and sets at D. The arc GD that it describes is to the west of the meridian (Fig. 1.13).

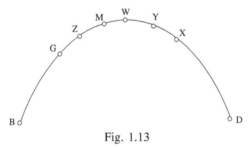

Fig. 1.13

Ibn al-Haytham shows that there exist on the arc GD:

• points with height $h > h_G$, let M be one of them;

• at least one point X, on the arc MD, such that $h_X = h_G$;

• two points with the same height h, where $h_G < h < h_M$, one of which is on the arc XM and the other on the arc MG.

He also shows that, between the planet's rising in the east, at B, and when it crosses the meridian at G, the height h increases from 0 to h_G and that any height $h < h_G$ is reached at least once.

Later on Ibn al-Haytham returns to this investigation to calculate the heights reached by the planet to the west of the meridian. He shows that, if h_m is the maximum height, the planet reaches that height only once – let it be at point W; and that the height h_G which is that of the planet's meridian transit is reached once and only once at a point other than G – let it be at point X on the arc WD; that any height $h < h_G$ is reached once and only once, at a point between X and D; and that any height h such that $h_G < h < h_M$ is reached at two points and at only two points – one on the arc GW and the other on the arc WX.

In Proposition 30, Ibn al-Haytham proves that the point at which maximum height is reached is unique; he then, in Proposition 31, returns to the investigation of heights to the east of the meridian. In these two propositions, Ibn al-Haytham once again introduces innovations in infinitesimal geometry. He is in fact developing a new and entirely original method of working in spherical geometry: he considers infinitesimal curvilinear triangles on the sphere (triangles whose sides are not necessarily arcs of great circle) – constructs a sequence of such triangles whose sides tend to zero – and he handles these triangles as if they were infinitesimal rectilinear triangles. What we are encountering here is in effect a geometry of infinitesimals like what will be used later in the differential geometry.

In order to sum up some results Ibn al-Haytham established in his investigation of the point D where the planet crosses the meridian (that is, in this group of Propositions 28 to 36 where he is investigating the heights of a planet), let us consider the meridian plane, with pole Z, and the equator, whose north pole is N (Fig. 1.14).

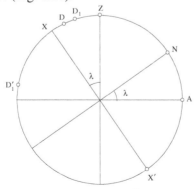

Fig. 1.14: $\alpha < \lambda < \dfrac{\pi}{2} - \alpha$

Let the latitude of the place be λ, the declination of the planet at meridian transit δ and the maximum value of the declination α; we have

$$\widehat{AN} = \widehat{XZ} = \lambda, \qquad \widehat{XD} = \delta, \qquad \widehat{XD_1} = \widehat{XD_1'} = \alpha.$$

We consider only places in the northern hemisphere, and we use the sun as our example, so $\alpha = 23°27'$.

We may summarize the investigation of the position of D, as a function of the geographical latitude λ and the date, in the form of a table. Let us assume that $\alpha < \lambda < \dfrac{\pi}{2} - \alpha$.

latitude	date	position of D
$\lambda = 0$ terrestrial equator	• Spring and Autumn equinoxes • Summer solstice • Winter solstice • Spring and Summer • Autumn and Winter	$D = Z = X$ $D = D_1$ to the north of Z $D = D_1'$ to the south of Z D between Z and D_1, north of Z D between Z and D_1', south of Z
$\lambda = \alpha$ tropic of Cancer	• day of the Summer solstice $\delta = \lambda$ • any other day $\delta < \lambda$	$D = D_1 = Z$ D lies on the arc ZD_1' south of Z
$0 < \lambda < \alpha$ northern tropical zone	• day of the Summer solstice $\delta = \alpha$, so $\delta > \lambda$. The declination $\delta = \lambda$ will be reached once in the Spring and once in the Summer on these two dates between these two dates for any other day of the year	D at D_1 to the north of Z D is at Z D lies on the arc ZD_1 north of Z D lies on the arc ZD_1' south of Z
$\lambda > \alpha$	for any day of the year	D lies south of Z

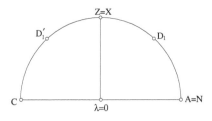

Fig. 1.15.1

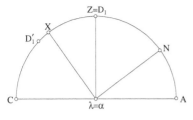

Fig. 1.15.2

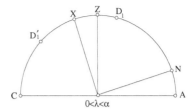

Fig. 1.15.3

In the case of the *sphaera recta*, whether meridian transit is north or south of the zenith Z, we can apply the method employed in Proposition 28 or Proposition 29 and show that the planet will have equal heights h at pairs of points (for $h > h_D$) either to the east of the meridian, or to the west of it.

In the case where the point D, the point of meridian transit, is at the zenith Z,[40] the maximum height of the planet is h_D, and any height $h < h_D$ will be reached once and only once to the east, and the same will apply in the west.

So far, Ibn al-Haytham has considered places north of the equator with latitude $\lambda < \dfrac{\pi}{2} - \alpha$, α being the inclination of the orb to the equator; to complete his investigation of the trajectory of a planet seen above the horizon, he considers places with northern latitude $\lambda = \dfrac{\pi}{2} - \alpha$ or $\lambda \cong \dfrac{\pi}{2} - \alpha$ and shows that in such places, and on particular dates, the planet in question may set in the east and rise in the east and that, on other dates, it may set in the west and rise in the west.

Let *BHID* (Fig. 1.16) be the meridian plane for some place, *BD* the diameter of the horizon, *EG* the diameter of the equator, *H* the pole of the equator $\overset{\frown}{BH} = \lambda = \dfrac{\pi}{2} - \alpha$, $\overset{\frown}{HZ} = \alpha$. If we draw $BI \parallel EG$ and $DI' \parallel EG$, we have $\overset{\frown}{BG} = \overset{\frown}{EI} = \overset{\frown}{ED} = \overset{\frown}{GI'} = \alpha$, so the circles with diameters BI and DI' touch the horizon of the place in question at B and D respectively. The trajectories

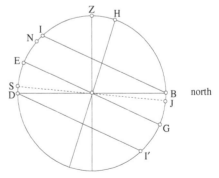

Fig. 1.16

of the planet's diurnal motion therefore lie between these two circles.

[40] See below, p. 249.

We have assumed that: the planet reaches point B, the north cardinal point of the horizon in question, $ABCD$, at the time it gets to the most northerly point of its trajectory, that is to say at the moment when its declination is a maximum and equal to α. So after that the declination decreases and the trajectory of the planet moves away from the circle BI and begins to cut the meridian again at the point N above the horizon.

Ibn al-Haytham then defines:

• a point L that lies on this trajectory and is above the horizon $ABCD$ and to the east of B;

• a horizon circle with diameter JS, at latitude $\lambda + \varepsilon$ which shares the same meridian and is such that the point L is above the horizon.

But the points B and N are above this horizon JS, so when the planet moves from B towards L it sets at a point on the eastern part of this horizon, and when it moves from L towards N it rises at a point which is likewise on the eastern part of this horizon.

At the other extreme, the planet is assumed to be at the most southerly point of its trajectory, at the point B of the hour circle BQI (Fig. 1.17), and this point B is the south cardinal point of the horizon in question, which has latitude $\lambda = \dfrac{\pi}{2} - \alpha$. So after that the declination increases and the trajectory of the planet moves away from the circle BQI and begins to cut the meridian again at a point N above the horizon. The method is accordingly the same as in the previous part. Ibn al-Haytham defines:

• a point L that lies on this trajectory and is above the horizon $ABCD$ and to the west of B;

• a horizon circle $AJCS$ at latitude $\lambda - \varepsilon$ which shares the same meridian and is such that the point L is above it.

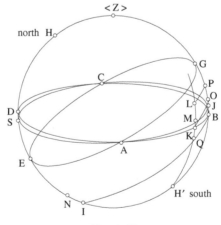

Fig. 1.17

When the planet moves from B towards L it rises at a point on the western part of this horizon, and when it moves from L towards N it sets at a point on the western part of this horizon.

Ibn al-Haytham has thus shown that on the day that a planet reaches its maximum northern declination, α, there exist places in the northern hemisphere, with latitude $\lambda = \dfrac{\pi}{2} - \alpha + \varepsilon$, on whose horizons the planet sets and rises in the east, and that on the day when the planet reaches its maximum southern declination, α, there exist places in the northern hemisphere, with latitude $\lambda = \dfrac{\pi}{2} - \alpha - \varepsilon$, on whose horizons the planet rises and sets in the west.

In both cases, the points at which the planet rises and sets are very close to one another.

We have sketched the principal results that Ibn al-Haytham obtains in his *Configuration of the Motions*. Our aim was not so much to expound all the results in detail, which we do later, but rather to give an overview of what he was trying to do in his book. All the way through *The Configuration of the Motions* he directs his efforts to constructing a descriptive phenomenological theory of the celestial motions, as they are seen by an observer on the earth. This theory, as one can easily assure oneself, does not incorporate any idea of a teleological physics, though it does not conflict with what Aristotle calls the most physical parts of mathematics, which here is geometrical optics, a subject reformed by Ibn al-Haytham himself. When Ibn al-Haytham is constructing his astronomy his obvious concern is, as we have noted, to adopt at each stage the least possible number of hypotheses.

Thus his theory for the motion of the planets calls upon no more than observation and conceptual constructs susceptible of explaining the data, such as the eccentric circle and in some cases the epicycle. However, this theory does not aim to describe anything beyond observation and these concepts, and in no way is it concerned to propose a causal explanation of the motions. In this respect, *The Configuration of the Motions* is both in the astronomical tradition that Ibn al-Haytham inherited and in a tradition that continues after Ibn al-Haytham as far as Kepler. To sum it up, in *The Configuration of the Motions* Ibn al-Haytham's purpose is purely kinematic; more precisely, Ibn al-Haytham wanted to lay the foundations of a completely geometrical kinematic tradition.

Carrying out such a project involves first of all developing some branches of geometry required for solving new problems that arise from this kinematic treatment: Ibn al-Haytham took a huge step forward in spherical geometry as also in plane and spherical trigonometry. To get a measure of how far he has advanced beyond the Greeks, one need only compare *The Configuration of the Motions* with Chapters 9 to 16 of the first book of Ptolemy's *Almagest*; and to appreciate the distance that separates him from his contemporaries one may compare *The Configuration of the Motions* with, for example, the *Almagest* of Abū al-Wafā' al-Būzjānī. As we have seen, Ibn al-Haytham considers the changes in infinitesimal magnitudes that necessarily arise in astronomical research.

In astronomy, there are two major processes that are jointly involved in carrying through this project: freeing celestial kinematics from cosmological connections, that is, from all considerations of dynamics, in the ancient sense of the term; and to reduce physical entities to geometrical ones. The centres of the motions are geometrical points with no physical significance; the centres to which speeds are referred are also geometrical points with no physical significance; even more radically, all that remains of physical time is the 'required time', that is, a geometrical magnitude. In short, in this new kinematics, we are concerned with nothing that identifies celestial bodies as physical bodies. All in all, though it is not yet that of Kepler, this new kinematics is no longer that of Ptolemy nor of any of Ibn al-Haytham's predecessors; it is *sui generis*, half way between Ptolemy and Kepler. It shares two important ideas with ancient kinematics: every celestial motion is composed of elementary uniform circular motions, and the centre for observations is the same as the centre of the Universe. On the other hand, it has in common with modern kinematics the fact that the physical centres of motions and speeds are replaced by geometrical centres.

There remains a major question, that of the relation of this kinematics to the celestial dynamics of the day, that is to say, to cosmology. The question is relevant here only if we come across evidence that Ibn al-Haytham had intended to write on cosmology once he had completed *The Configuration of the Motions*. In that case, one would expect a new cosmology to go with the new kinematics. In fact none of the titles that have come down to us, none of the manuscripts of Ibn al-Haytham's undoubtedly authentic astronomical works, gives grounds for affirming that such a cosmology, based on the new kinematics, ever existed. The only cosmology text known to have been composed by Ibn al-Haytham (that is of well-attested authenticity) is earlier than *The Configuration of the Motions* since it forms part of his treatise on the winding motion. When, in his *Resolution of Doubts concerning the Winding Motion*, he himself mentions this work, which is now lost, he writes:

> The winding motion to which Ptolemy referred, and from which arise the motions in latitude of the five planets, can only be according to the configuration that I demonstrated and according to the account that I gave.

It is a configuration that is not subject to any impossibility or any absurdity. From this motion is generated a motion of the planet which, by the motion of its centre, produces a curve imagined as if the planet were wound round on the body of the small sphere which moves the body of the planet. It is because of the winding of this curve round the body of the epicycle that this motion has been called the winding motion, and for no other reason.[41]

That leaves no room for doubt: Ibn al-Haytham had indeed, in his treatise on the winding motion, proposed a model for the motions in latitude of the epicycles of the five planets, a model in which he considered the physical 'small spheres' that moved the celestial bodies; in other words, he had proposed a cosmology. Many other passages of the treatise confirm this.

Now, from the order of composition of Ibn al-Haytham's writings that we have already established, we know that, of these writings, the two books on the winding motion were composed before the *Doubts concerning Ptolemy*. Moreover, while in the first two books he makes use of the idea of an equant, in the last one he criticizes it, and eventually ends up completely excluding it from *The Configuration of the Motions*. Furthermore, since Ibn al-Haytham emphasizes in the introduction to *The Configuration of the Motions* that the results described in this work supersede any different ones to be found in all his other writings, we may safely conclude that *The Configuration of the Motions* was written after the *Doubts concerning Ptolemy* and, *a fortiori*, after the two books about the winding motion. Thus Ibn al-Haytham's contribution to cosmology is (as it were) local, since it relates only to a particular motion and antedates the *Doubts* and *The Configuration of the Motions*. We proved elsewhere that *The Configuration of the Motions* is also later than his treatise on *The Variety that Appears in the Heights of the Wandering Stars*.[42]

Another argument in favour of this historical and conceptual sequence is drawn from the language used in *The Configuration of the Motions*. The book not only contains new concepts such as 'required time' and 'proper inclination for the required time', but also terms from ancient astronomy whose meaning has changed. For example let us consider a concept central to traditional astronomy, that of *falak*. It is well known that in traditional

---

[41] *Fī ḥall shukūk ḥarakat al-iltifāf*, ms. St. Petersburg, B1030/1, fols 15ᵛ–16ʳ:

وليس يصح أن تكون حركة الالتفاف التي أشار إليها بطلميوس التي تكون منها حركات العرض للكواكب الخمسة إلا على الهيئة التي بينتها وللتفصيل الذي فصلته، وهي هذه لا يعرض فيها شيء، من المحالات فلا يلزمها شيء، من الشناعات، وتتولد منها للكواكب حركة يحدث بها من حركة مركز خط متخيل كأنه ملتف على جسم الكرة الصغرى المحركة لجرم الكوكب. والالتفاف هذا الخط على جسم فلك التدوير سُميت هذه الحركة حركة الالتفاف، لا لعلة أخرى.

[42] See below, Part I.

astronomy this term signifies 'orb'. It refers to the various solid bodies attached to a specific planet. These solid bodies move with uniform circular motions, and the sum of these motions constitutes the apparent motion of the planet concerned, as seen from the earth, which is at the centre of the world. In this system, a planet does not have a motion of its own, it is moved by something else, and one cannot speak of the motion of a planet along its particular orbit, but only of its apparent motion resulting from the composition of the motions of its various spheres. This same word *falak* is also used in the same context to designate the (plane) circles that are the lines on the sky that correspond to the solid bodies in question.

In fact, Ibn al-Haytham uses this term *falak* in these senses in all the works we have cited above, except in *The Variety that Appears in the Heights of the Wandering Stars*, where he has no need of it. On the other hand, in *The Configuration of the Motions*, the term *falak* no longer has the same meaning. In this book it refers mainly to the apparent trajectory of a particular planet across the celestial vault, and everything else derives from the analysis of this apparent motion, without reference to solid bodies that might move the planet in question. This semantic difference, taken together with the new concepts, shows that *The Configuration of the Motions* was composed after the books we referred to earlier. This difference alone also shows that this treatise cannot be placed within a purely Ptolemaic tradition. One might almost translate the term as the 'orbit' of a planet[43] since the apparatus of the orb, in the sense in which the term was conventionally understood, no longer comes into it.

In the *Doubts* we have seen a turning point in Ibn al-Haytham's thoughts about astronomy. There is every indication that *The Configuration of the Motions* is the most substantial result produced by this change. The book gives us a new astronomy even though it retains a geocentric framework within which all motions are circular and uniform. We have a break with tradition despite the background of continuity.

We need to know the reasons for such a change. On this matter the available texts are silent. We may, however, offer the following hypothesis. In the absence of a theory of gravitation, the mathematician-astronomer was faced with two alternatives: either abide by the traditional principle whereby the motion of each planet is due to a cause specific to that body, and thus construct a cosmology of material spheres; or accept the necessity of abandoning that route and instead start by constructing a kinematic account, thus acknowledging the primacy of kinematics in any investigation of dynamics. In many of his astronomical writings, Ibn al-Haytham had been

---

[43] However, we do not do this in the translation, preferring to keep to period usage. We simply need to alert the reader here.

tempted by the first alternative. But, once he had engaged upon mathematizing astronomy and had noted not only the internal contradictions in Ptolemy, but doubtless also the difficulty of constructing a self-consistent mathematical theory of material spheres using an Aristotelian physics, he turned to the second alternative, that of giving a completely geometrized kinematic account. His experience in optics perhaps helped him to take this step: here kinematics and cosmology are entirely separated to effect a reform of astronomy, just as in optics work on the propagation of light is entirely separated from work on vision to effect a reform of optics; in the one case as in the other Ibn al-Haytham arrived at a new idea of the science concerned.

CHAPTER II

MATHEMATICAL COMMENTARY

2.1. GEOMETRY AND TRIGONOMETRY, PLANE AND SPHERICAL

2.1.1. *Plane trigonometry*

Proposition 1. — Let there be three points A, B, C that lie on a circle such that

$$\widehat{AB} > \widehat{BC} \text{ and } \widehat{AB} + \widehat{BC} \leq \pi \Rightarrow \frac{\widehat{AB}}{\widehat{BC}} > \frac{AB}{BC}.$$

By hypothesis $B\hat{A}C + B\hat{C}A \leq \frac{\pi}{2}$, so $A\hat{B}C \geq \frac{\pi}{2}$. Let D be such that $C\hat{B}D = B\hat{A}C$, $C\hat{B}D < C\hat{B}A$, so D lies on $[CA]$. Triangles CBD and CAB are similar, we have

$$\frac{AB}{BC} = \frac{BD}{DC}; \ C\hat{D}B \geq \frac{\pi}{2} \text{ and } CD < CB.$$

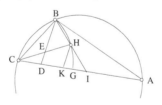

Fig. 2.1

Moreover

$$\frac{B\hat{C}A}{B\hat{A}C} = \frac{\widehat{AB}}{\widehat{BC}},$$

so

$$\frac{B\hat{C}D}{C\hat{B}D} = \frac{\widehat{AB}}{\widehat{BC}}.$$

The circle (C, CB) cuts CA in G, so $CD < CG < CA$.

Let E be a point on BD such that $E\hat{C}D = C\hat{B}D$, the straight line CE cuts the arc BG in H, so

$$\frac{\overarc{GB}}{\overarc{GH}} = \frac{B\hat{C}D}{D\hat{C}E} = \frac{B\hat{C}D}{C\hat{B}D} = \frac{\overarc{AB}}{\overarc{BC}}.$$

Let us draw $HK \parallel BD$; triangles CBD, CED and CHK are similar and $CB = CH$, so

$$\frac{BD}{DC} = \frac{CD}{DE} = \frac{BC}{CE} = \frac{CH}{CE} = \frac{HK}{DE};$$

therefore $HK = CD$.

The straight line BH cuts AC in I, we have

$$\frac{BI}{IH} = \frac{BD}{HK} = \frac{BD}{DC} = \frac{AB}{BC}.$$

We also have

$$\frac{BI}{IH} = \frac{\text{tr.}(CBI)}{\text{tr.}(CHI)}$$

and

$$\frac{\overarc{AB}}{\overarc{BC}} = \frac{\overarc{GB}}{\overarc{GH}} = \frac{\text{sect.}(BCG)}{\text{sect.}(HCG)}.$$

But

$$\text{sect.}(CBH) > \text{tr.}(CBH),$$
$$\text{sect.}(CHG) < \text{tr.}(CHI);$$

so

$$\frac{\text{sect.}(CBH)}{\text{sect.}(CHG)} > \frac{\text{tr.}(CBH)}{\text{tr.}(CHI)};$$

which gives us

$$\frac{\text{sect.}(CBG)}{\text{sect.}(CHG)} > \frac{\text{tr.}(CBI)}{\text{tr.}(CHI)}$$

and consequently

$$\frac{\overarc{AB}}{\overarc{BC}} > \frac{BI}{HI},$$

hence

$$\frac{\widehat{AB}}{\widehat{BC}} > \frac{AB}{BC}.$$

Note: If we put $\widehat{AB} = 2\alpha$, $\widehat{BC} = 2\alpha_1$ where $\alpha > \alpha_1$, $\frac{\pi}{2} > \alpha + \alpha_1$, we have
$AB = 2R \sin \alpha$, $BC = 2R \sin \alpha_1$, if R is the radius of the circle ABC.

$$\frac{\widehat{AB}}{\widehat{BC}} > \frac{AB}{BC} \iff \frac{\alpha}{\alpha_1} > \frac{\sin \alpha}{\sin \alpha_1}$$

$$\alpha > \alpha_1, \ \frac{\pi}{2} > \alpha + \alpha_1, \implies \frac{\alpha}{\alpha_1} > \frac{\sin \alpha}{\sin \alpha_1}.$$

In other words, $\frac{\sin \alpha}{\alpha} < \frac{\sin \alpha_1}{\alpha_1}$, that is, the function $\alpha \mapsto \frac{\sin \alpha}{\alpha}$ decreases
over the interval $0 < \alpha \leq \frac{\pi}{2}$.

Corollary:

$$\frac{\widehat{ABC}}{\widehat{CB}} > \frac{AC}{CB}.$$

We know that $\frac{\widehat{AB}}{\widehat{BC}} > \frac{AB}{BC}$, which gives us

$$\frac{\widehat{ABC}}{\widehat{CB}} > \frac{AB + BC}{BC} > \frac{AC}{BC}.$$

Note: Using the lettering of the previous note, if $\alpha > \alpha_1$ and $\alpha + \alpha_1 < \frac{\pi}{2}$

$$\frac{\alpha + \alpha_1}{\alpha_1} > \frac{\sin(\alpha + \alpha_1)}{\sin \alpha_1}$$

or

$$\frac{\sin(\alpha + \alpha_1)}{\alpha + \alpha_1} < \frac{\sin \alpha_1}{\alpha_1}.$$

The arcs in question belong to two circles, which may be equal or unequal. The assumptions concerning these arcs can be expressed by equalities or inequalities involving their measures in radians.

For example, in Proposition 2, the text has: 'the arc ABC is greater than the arc similar to the arc DEG'. We can write

$$\text{meas. } \overset{\frown}{ABC} > \text{meas. } \overset{\frown}{DEG} \text{ or } \overset{\frown}{ABC} > \overset{\frown}{DEG},$$

where 'meas.' indicates the measure of the angle that subtends the arc.

Proposition 2. — Let there be two arcs ABC and DEG belonging to the same circle or to different circles such that

$$\pi \geq \overset{\frown}{ABC} > \overset{\frown}{DEG}, \ \overset{\frown}{AB} > \overset{\frown}{BC} \text{ and } \overset{\frown}{DE} > \overset{\frown}{EG};$$

if

$$\frac{\overset{\frown}{AB}}{\overset{\frown}{BC}} = \frac{\overset{\frown}{DE}}{\overset{\frown}{EG}} > 1,$$

then

$$\frac{\overset{\frown}{DE}}{\overset{\frown}{EG}} > \frac{\overset{\frown}{AB}}{\overset{\frown}{BC}}.$$

Let AIB be an arc similar to the arc DE, it lies inside the segment defined by the arc AB and its chord.

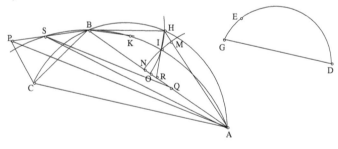

Fig. 2.2

The circle (B, BC) cuts the given arc AB in H, the arc similar to DE in I and the straight line AB in O. The straight line HI cuts AB in R.

We have $BH = BI$ and $B\hat{H}I = B\hat{I}H < \dfrac{\pi}{2}$, hence $B\hat{I}R > \dfrac{\pi}{2}$ and $A\hat{R}I > B\hat{I}R$. The circle (A, AI) cuts AH in M and AB in N. We have

$$\text{sect.}(BHI) > \text{tr.}(BHI)$$

and

$$\text{sect.}(BIO) < \text{tr.}(BIR);$$

so

$$\frac{\text{sect.}(BHI)}{\text{sect.}(BIO)} > \frac{\text{tr.}(BHI)}{\text{tr.}(BIR)};$$

therefore

$$\frac{\text{sect.}(BHO)}{\text{sect.}(BIO)} > \frac{\text{tr.}(BHR)}{\text{tr.}(BIR)},$$

hence

(1) $$\frac{H\hat{B}A}{I\hat{B}A} > \frac{HR}{RI}.$$

Moreover tr.$(AHI) > $ sect.(AMI) and tr.$(AIR) < $ sect.(AIN); so we have

$$\frac{\text{tr.}(AHI)}{\text{tr.}(AIR)} > \frac{\text{sect.}(AMI)}{\text{sect.}(AIN)};$$

therefore

$$\frac{\text{tr.}(AHR)}{\text{tr.}(AIR)} > \frac{\text{sect.}(AMN)}{\text{sect.}(AIN)},$$

hence

(2) $$\frac{HR}{IR} > \frac{H\hat{A}B}{I\hat{A}B}.$$

From (1) and (2), we get

$$\frac{H\hat{B}A}{I\hat{B}A} > \frac{H\hat{A}B}{I\hat{A}B}$$

or

$$\frac{H\hat{B}A}{H\hat{A}B} > \frac{I\hat{B}A}{I\hat{A}B},$$

hence

$$\frac{\widehat{HA}}{\widehat{HB}} > \frac{\widehat{AI}}{\widehat{IB}}.$$

From which we get

$$\frac{\widehat{AHB}}{\widehat{HB}} > \frac{\widehat{AIB}}{\widehat{IB}}.$$

So there exists a point K on the arc BI, $\widehat{BK} < \widehat{BI}$, such that

$$\frac{\widehat{AIB}}{\widehat{BK}} = \frac{\widehat{AHB}}{\widehat{HB}} \qquad \text{(fourth proportional)}.$$

On the circle (AIB) we take the point S such that $\widehat{BS} = \widehat{BK}$, then we have

$$\frac{\widehat{AIB}}{\widehat{BS}} = \frac{\widehat{AHB}}{\widehat{HB}} = \frac{\widehat{AHB}}{\widehat{BC}} = \frac{\widehat{DE}}{\widehat{EG}}.$$

But $\widehat{AIB}$ is similar to $\widehat{DE}$, so $\widehat{BS}$ is similar to $\widehat{EG}$ and $\widehat{ABS}$ is similar to $\widehat{DEG}$. So we have

$$\frac{AB}{BS} = \frac{DE}{EG};$$

but $BS = BK < BI = BC$, so

$$\frac{AB}{BC} < \frac{DE}{EG}.$$

If we put $\widehat{AB} = 2R\alpha$, $\widehat{BC} = 2R\alpha_1$, $\widehat{DE} = 2r\beta$, $\widehat{EG} = 2r\beta_1$, we have by hypothesis $\beta + \beta_1 < \alpha + \alpha_1 < \dfrac{\pi}{2}$ and $\dfrac{\beta}{\beta_1} = \dfrac{\alpha}{\alpha_1}$; the conclusion can be written

$$\frac{\sin\beta}{\sin\beta_1} > \frac{\sin\alpha}{\sin\alpha_1};$$

or, putting $\lambda = \dfrac{\alpha_1}{\alpha}$,

$$\frac{\sin\lambda\alpha}{\sin\alpha} > \frac{\sin\lambda\beta}{\sin\beta}$$

for $\beta < \alpha < \dfrac{\pi}{2}$ and $0 < \lambda < 1$.

In other words, the proposition means that the function $\alpha \mapsto \dfrac{\sin \lambda \alpha}{\sin \alpha}$ increases over the interval $0 < \alpha < \dfrac{\pi}{2}$. Indeed, its derivative can be written

$$\frac{\lambda \sin \alpha \cos \lambda \alpha - \sin \lambda \alpha \cos \alpha}{\sin^2 \alpha}$$

and it is positive if we have the inequality $\lambda \tan \alpha > \tan \lambda \alpha$, that is, if the function $\alpha \mapsto \dfrac{\tan \alpha}{\alpha}$ increases over the same interval.[1]

Corollary: With the same hypotheses

$$\frac{\widehat{AB}}{\widehat{BC}} = \frac{\widehat{DE}}{\widehat{EG}} \Rightarrow \frac{DG}{EG} > \frac{AC}{CB}.$$

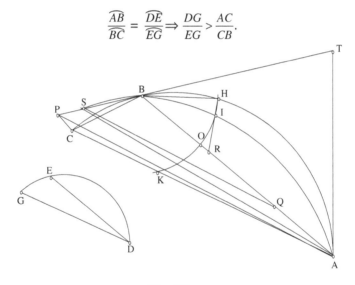

Fig. 2.3

We produce BS to the point P such that $BP = BC$. We draw $QS \parallel AP$, where Q lies on AB, we have

$$B\hat{P}C = B\hat{C}P \text{ and } B\hat{P}C > A\hat{P}C \Rightarrow B\hat{C}P > A\hat{P}C;$$

but $A\hat{C}P > B\hat{C}P$, hence $A\hat{C}P > A\hat{P}C$ and it follows that $AP > AC$, hence

[1] See R. Rashed, *Geometry and Dioptrics in Classical Islam*, London, 2005, pp. 1045–7 and pp. 1051–2. See also Supplementary note [1].

$$\frac{AP}{BP} > \frac{AC}{BC}.$$

Moreover, $\widehat{ABS} < \frac{1}{2}$ circle, so $A\hat{B}S$ is obtuse. Angle AQS is an exterior angle of triangle QBS, hence $A\hat{Q}S > A\hat{B}S$, so $A\hat{Q}S$ is obtuse. Thus $AS > QS$, hence

$$\frac{AS}{SB} > \frac{QS}{SB}.$$

But

$$\frac{QS}{SB} = \frac{AP}{BP},$$

so

$$\frac{AS}{SB} > \frac{AP}{BP} > \frac{AC}{CB}.$$

Moreover

$$\frac{AS}{SB} = \frac{DG}{GE},$$

so

$$\frac{DG}{GE} > \frac{AC}{CB}.$$

Note: Using the lettering of the previous note, the conclusion can be written

$$\frac{\sin(\beta + \beta_1)}{\sin \beta_1} > \frac{\sin(\alpha + \alpha_1)}{\sin \alpha_1},$$

that is, that $\alpha \mapsto \dfrac{\sin(1 + \lambda)\alpha}{\sin \lambda \alpha}$ decreases over the interval $0 < \alpha < \dfrac{\pi}{2(1 + \lambda)}$.

If $\alpha' = (1 + \lambda)\alpha$ and $\mu = \dfrac{\lambda}{1 + \lambda}$, this statement is equivalent to $\alpha' \mapsto \dfrac{\sin \mu \alpha'}{\sin \alpha'}$ increasing over the interval $0 < \alpha' < \dfrac{\pi}{2}$, where $0 < \mu = \dfrac{\lambda}{1 + \lambda} < \dfrac{1}{2}$.

Proposition 3. — If $\pi \geq \widehat{ABC} > \widehat{DEG}$ and $\dfrac{AC}{CB} = \dfrac{DG}{GE}$, then $\widehat{AB} > \widehat{DE}$ and

$$\frac{\overline{AB}}{\overline{BC}} > \frac{\overline{DE}}{\overline{EG}}.$$

B is a point on the arc AC, so $\widehat{BC} < \widehat{AC}$, hence $BC < AC$.

The arc AIC, similar to DEG, lies inside the segment defined by the arc AC (as in Proposition 2). The circle (C, CB) cuts AC in H and the arc AIC in I. The circle (A, AI) cuts the straight line AC in K and the straight line AB in N. The straight line CI cuts the arc ABC in M and the straight line BI cuts AC in L.

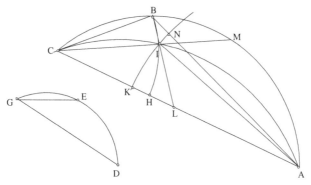

Fig. 2.4

We have

$$\frac{AC}{CI} = \frac{AC}{CB} = \frac{DG}{GE},$$

and since $\widehat{AIC}$ is similar to $\widehat{ABC}$, the arcs AI and IC are similar to the arcs DE and EG respectively. But the arcs AI and AM are similar, so the arc AM is similar to the arc DE, and $\widehat{DE} < \widehat{AB}$.

We have $CB = CI$, so $B\hat{I}C < 1$ right angle and $B\hat{I}M > 1$ right angle, and consequently $A\hat{I}B > 1$ right angle, $L\hat{I}C > 1$ right angle and $A\hat{L}I > 1$ right angle; therefore $BA > AI > AL$. We have

$$\text{sect.}(CBI) > \text{tr.}(CBI), \ \text{sect.}(CIH) < \text{tr.}(CIL),$$

so

$$\frac{\text{sect.}(CBI)}{\text{sect.}(CIH)} > \frac{\text{tr.}(CBI)}{\text{tr.}(CIL)},$$

hence

$$\frac{\text{sect.}(CBH)}{\text{sect.}(CIH)} > \frac{\text{tr.}(CBL)}{\text{tr.}(CIL)};$$

from which we obtain

(1)
$$\frac{A\hat{C}B}{A\hat{C}I} > \frac{BL}{IL}.$$

In the same way, we have

$$\text{tr.}(ABI) > \text{sect.}(AIN), \text{ tr. }(AIL) < \text{sect.}(AIK),$$

hence

$$\frac{\text{tr.}(ABI)}{\text{tr.}(AIL)} > \frac{\text{sect.}(AIN)}{\text{sect.}(AIK)},$$

hence

$$\frac{\text{tr.}(BAL)}{\text{tr.}(AIL)} > \frac{\text{sect.}(NAK)}{\text{sect.}(IAK)}$$

and we obtain

(2)
$$\frac{BL}{IL} > \frac{C\hat{A}B}{C\hat{A}I}.$$

From (1) and (2) we get

$$\frac{A\hat{C}B}{A\hat{C}I} > \frac{C\hat{A}B}{C\hat{A}I},$$

hence

$$\frac{A\hat{C}B}{C\hat{A}B} > \frac{A\hat{C}I}{C\hat{A}I}$$

and consequently

$$\frac{\widehat{AB}}{\widehat{BC}} > \frac{\widehat{AI}}{\widehat{CI}}.$$

But

$$\frac{\widehat{AI}}{\widehat{CI}} = \frac{\widehat{DE}}{\widehat{EG}},$$

hence the conclusion

$$\frac{\widehat{AB}}{\widehat{BC}} > \frac{\widehat{DE}}{\widehat{EG}}.$$

If the arc ABC is greater than the arc similar to the arc DEG, their measures satisfy the inequality $\overset{\frown}{ABC} > \overset{\frown}{DEG}$.

Note: Using the same lettering as before, the assumption made for this proposition can be written

$$\beta + \beta_1 < \alpha + \alpha_1 < \frac{\pi}{2} \text{ and } \frac{\sin(\alpha + \alpha_1)}{\sin \alpha_1} = \frac{\sin(\beta + \beta_1)}{\sin \beta_1}.$$

The conclusion is then that $\alpha > \beta$ and $\dfrac{\alpha}{\alpha_1} > \dfrac{\beta}{\beta_1}$.

Putting $\sin(\alpha + \alpha_1) = z$, $\sin \alpha_1 = x$, $\sin(\beta + \beta_1) = u$ and $\sin \beta_1 = y$, we have

$$\frac{z}{x} = \frac{u}{y} = \lambda,$$

let $z = \lambda x$, $u = \lambda y$ and we wish to prove that

$$\text{Arc } \sin z - \text{Arc } \sin x > \text{Arc } \sin u - \text{Arc } \sin y$$

and

$$\frac{\text{Arc } \sin z - \text{Arc } \sin x}{\text{Arc } \sin x} > \frac{\text{Arc } \sin u - \text{Arc } \sin y}{\text{Arc } \sin y},$$

making the assumption that $y < x$ and $z = \lambda x$, $u = \lambda y$ ($\lambda > 1$).

So this conclusion indicates that $x \mapsto \text{Arc } \sin \lambda x - \text{Arc } \sin x$ and $x \mapsto \dfrac{\text{Arc } \sin \lambda x}{\text{Arc } \sin x}$ are increasing functions over the interval $0 < x < \dfrac{1}{\lambda}$. The derivative of the first of these functions is $\dfrac{\lambda}{\sqrt{1 - \lambda^2 x^2}} - \dfrac{1}{\sqrt{1 - x^2}}$, which is clearly positive since $\lambda > 1$. The derivative of the second function has the same sign as $\lambda\sqrt{1 - x^2} \cdot \text{Arc } \sin x - \sqrt{1 - \lambda^2 x^2}\,\text{Arc } \sin \lambda x$. This last expression is positive because it is zero at 0 and its derivative $\lambda x \left(\dfrac{\lambda}{\sqrt{1 - \lambda^2 x^2}} \text{Arc } \sin \lambda x - \dfrac{1}{\sqrt{1 - x^2}} \text{Arc } \sin x \right)$ remains positive. In fact, if we put $\alpha = \text{Arc } \sin x$ and $\gamma = \text{Arc } \sin \lambda x$, the expression in parentheses comes to $\dfrac{\gamma \sin \gamma}{\sin \alpha \cos \gamma} - \dfrac{\alpha}{\cos \alpha} = \dfrac{1}{\sin \alpha}(\gamma \tan \gamma - \alpha \tan \alpha)$ where $\gamma > \alpha$ and the function $\alpha \mapsto \alpha \tan \alpha$ is obviously increasing.

Proposition 4. — Let there be two arcs ABC and DEG such that

$$\pi > \widehat{ABC} > \widehat{DEG}.$$

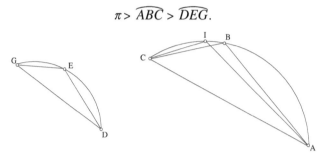

Fig. 2.5

<a> $AB > BC$, $DE > EG$ and $\dfrac{AB}{BC} = \dfrac{DE}{EG} \Rightarrow \dfrac{\widehat{AB}}{\widehat{BC}} > \dfrac{\widehat{DE}}{\widehat{EG}}$.

Argument by *reductio ad absurdum*, using Proposition 2.

• If $\dfrac{\widehat{AB}}{\widehat{BC}} = \dfrac{\widehat{DE}}{\widehat{EG}}$, then $\dfrac{DE}{EG} > \dfrac{AB}{BC}$, from Proposition 2; now $\dfrac{DE}{EG} = \dfrac{AB}{BC}$.

• If $\dfrac{\widehat{AB}}{\widehat{BC}} < \dfrac{\widehat{DE}}{\widehat{EG}}$, then $\dfrac{\widehat{ABC}}{\widehat{BC}} < \dfrac{\widehat{DEG}}{\widehat{EG}}$; so there exists a point I on the arc

BC such that $\dfrac{\widehat{AIC}}{\widehat{CI}} = \dfrac{\widehat{DEG}}{\widehat{EG}}$, hence $\dfrac{\widehat{AI}}{\widehat{CI}} = \dfrac{\widehat{DE}}{\widehat{EG}}$; then, from Proposition 2,

$\dfrac{DE}{EG} > \dfrac{AI}{CI}$.

But $AI > AB$ and $CI < CB$, so $\dfrac{AI}{CI} > \dfrac{AB}{CB}$ and consequently $\dfrac{DE}{EG} > \dfrac{AB}{CB}$;

which is contrary to the original hypothesis.

Conclusion:

$$\dfrac{\widehat{AB}}{\widehat{BC}} > \dfrac{\widehat{DE}}{\widehat{EG}}.$$

<b> $\dfrac{AB}{BC} > \dfrac{DE}{EG} \Rightarrow \dfrac{\widehat{AB}}{\widehat{BC}} > \dfrac{\widehat{DE}}{\widehat{EG}}$.

There exists a point I' between A and B such that

$$\dfrac{AI'}{I'C} = \dfrac{DE}{EG}$$

(since the ratio $\dfrac{AJ}{JC}$ increases from 0 to $\dfrac{AB}{BC}$ as the point J moves along the arc AB from A to B), and from 4<a>, we then have

$$\frac{\widehat{AI'}}{\widehat{I'C}} > \frac{\widehat{DE}}{\widehat{EG}};$$

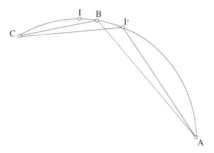

Fig. 2.6

but

$$\frac{\widehat{AB}}{\widehat{BC}} > \frac{\widehat{AI'}}{\widehat{I'C}},$$

so

$$\frac{\widehat{AB}}{\widehat{BC}} > \frac{\widehat{DE}}{\widehat{EG}}.$$

Note: If we keep the same lettering as in Proposition 2, the hypotheses can be written $\beta + \beta_1 < \alpha + \alpha_1 \le \dfrac{\pi}{2}$, $\alpha_1 < \alpha$, $\beta_1 < \beta$ and $\dfrac{\sin \alpha}{\sin \alpha_1} \ge \dfrac{\sin \beta}{\sin \beta_1}$.

The conclusion from <a> and <b> means that $\dfrac{\alpha}{\alpha_1} > \dfrac{\beta}{\beta_1}$.

<c> $\dfrac{AC}{CB} = \dfrac{DG}{EG} \Rightarrow \dfrac{\widehat{ABC}}{\widehat{CB}} > \dfrac{\widehat{DEG}}{\widehat{EG}}.$

• If $\dfrac{\widehat{ABC}}{\widehat{CB}} = \dfrac{\widehat{DEG}}{\widehat{EG}}$, then $\dfrac{DG}{GE} > \dfrac{AC}{CB}$, from the corollary to Proposition 2; which is contrary to the original hypothesis.

• $\dfrac{\widehat{ABC}}{\widehat{CB}} < \dfrac{\widehat{DEG}}{\widehat{EG}}$. In this case, there exists a point I such that

$$\overset{\frown}{CI} < \overset{\frown}{BC} \quad \text{and} \quad \frac{\overset{\frown}{ABC}}{\overset{\frown}{CI}} = \frac{\overset{\frown}{DEG}}{\overset{\frown}{EG}}$$

and in this case $\dfrac{DG}{EG} > \dfrac{AC}{CI}$, from the corollary to Proposition 2, so $\dfrac{DG}{EG} > \dfrac{AC}{BC}$; which is contrary to the original hypothesis.

Conclusion: $\dfrac{\overset{\frown}{ABC}}{\overset{\frown}{CB}} > \dfrac{\overset{\frown}{DEG}}{\overset{\frown}{EG}}$.

<d> $\dfrac{AC}{CB} > \dfrac{DG}{EG} \Rightarrow \dfrac{\overset{\frown}{ABC}}{\overset{\frown}{CB}} > \dfrac{\overset{\frown}{DEG}}{\overset{\frown}{EG}}$ is proved in the same way as <b>.

Ibn al-Haytham summarizes the two paragraphs <c> and <d>.

If $\pi > \overset{\frown}{ABC} > \overset{\frown}{DEG}$,

$$\frac{AC}{CB} \geq \frac{DG}{EG} \Rightarrow \frac{\overset{\frown}{ABC}}{\overset{\frown}{BC}} > \frac{\overset{\frown}{DEG}}{\overset{\frown}{GE}}.$$

Note: Using the same lettering, the hypotheses for <c> and <d> are: $\beta + \beta_1$ $< \alpha + \alpha_1 \leq \dfrac{\pi}{2}$, $\alpha_1 < \alpha$, $\beta_1 < \beta$ and $\dfrac{\sin(\alpha + \alpha_1)}{\sin \alpha_1} \geq \dfrac{\sin(\beta + \beta_1)}{\sin \beta_1}$. The conclusion is $\dfrac{\alpha + \alpha_1}{\alpha_1} > \dfrac{\beta + \beta_1}{\beta_1}$ or $\dfrac{\alpha}{\alpha_1} > \dfrac{\beta}{\beta_1}$.

<e> If the arcs *ABC* and *DEG* are similar, we know that their measures $\overset{\frown}{ABC} = \overset{\frown}{DEG}$, where $\pi > \overset{\frown}{ABC} = \overset{\frown}{DEG}$, and we obtain

$$\frac{AC}{CB} > \frac{DG}{GE} \Rightarrow \frac{\overset{\frown}{ABC}}{\overset{\frown}{CB}} > \frac{\overset{\frown}{DEG}}{\overset{\frown}{EG}}.$$

If $\dfrac{\overset{\frown}{ABC}}{\overset{\frown}{CB}} = \dfrac{\overset{\frown}{DEG}}{\overset{\frown}{EG}}$, the arcs *CB* and *EG* are proportional to the arcs *ABC* and *DEG*, so they are similar and we have in this case

$$\frac{AC}{CB} = \frac{DG}{GE};$$

which is contrary to the original hypothesis.

If $\dfrac{AC}{CB} > \dfrac{DG}{GE}$, then $CB < CI$, if I is the point on the arc AC such that $\dfrac{\overset{\frown}{ABC}}{\overset{\frown}{CI}} = \dfrac{\overset{\frown}{DEG}}{\overset{\frown}{EG}}$, that is, such that the arcs CI and EG are proportional to the arcs ABC and DEG.

But $CB < CI$ implies $\overset{\frown}{CB} < \overset{\frown}{CI}$, and consequently $\dfrac{\overset{\frown}{ABC}}{\overset{\frown}{CB}} > \dfrac{\overset{\frown}{DEG}}{\overset{\frown}{EG}}$.

Note: The case <e> is a special case of <c>.

Corollary of 4 <a> and <b>:

$\overset{\frown}{ABC} > \pi, \ \overset{\frown}{DEG} > \pi$

$\overset{\frown}{AB} > \overset{\frown}{BC}, \ \overset{\frown}{DE} > \overset{\frown}{EG}, \ \overset{\frown}{DE} \leq \pi$ and $\overset{\frown}{AB} \leq \pi$

$\overset{\frown}{AB} \geq$ arc similar to $\overset{\frown}{DE}$.

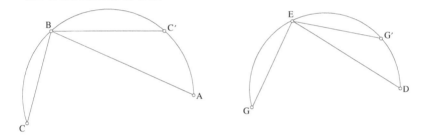

Fig. 2.7

If $\dfrac{AB}{BC} > \dfrac{DE}{EG}$, then $\dfrac{\overset{\frown}{AB}}{\overset{\frown}{BC}} > \dfrac{\overset{\frown}{DE}}{\overset{\frown}{EG}}$.

Let C' be a point such that $\overset{\frown}{BC} = \overset{\frown}{BC'}$ and G' such that $\overset{\frown}{EG} = \overset{\frown}{EG'}$, then we have

$$\frac{AB}{BC} = \frac{AB}{BC'} \text{ and } \frac{DE}{EG} = \frac{DE}{EG'},$$

so

$$\frac{AB}{BC'} > \frac{DE}{EG'}.$$

So, from Proposition 4<b>, we have $\dfrac{\overset{\frown}{AB}}{\overset{\frown}{BC'}} > \dfrac{\overset{\frown}{DE}}{\overset{\frown}{EG'}}$, hence $\dfrac{\overset{\frown}{AB}}{\overset{\frown}{BC}} > \dfrac{\overset{\frown}{DE}}{\overset{\frown}{EG}}$.

Notes:

1) If $\dfrac{AB}{BC} = \dfrac{DE}{DG}$, by using 4<a>, we show that $\dfrac{\widehat{AB}}{\overline{BC}} > \dfrac{\widehat{DE}}{\overline{EG}}$, so that $\dfrac{AB}{BC} \geq \dfrac{DE}{DG} \Rightarrow \dfrac{\widehat{AB}}{\overline{BC}} > \dfrac{\widehat{DE}}{\overline{EG}}$.

2) The hypotheses are $\alpha + \alpha_1, \beta + \beta_1 > \dfrac{\pi}{2}; \dfrac{\pi}{2} \geq \alpha > \alpha_1; \dfrac{\pi}{2} \geq \beta > \beta_1;$ $\alpha \geq \beta$ and $\dfrac{\sin \alpha}{\sin \alpha_1} \geq \dfrac{\sin \beta}{\sin \beta_1}$. The conclusion is $\dfrac{\alpha}{\alpha_1} > \dfrac{\beta}{\beta_1}$.

The first four propositions are matters of plane trigonometry. In them, Ibn al-Haytham compares inequalities between ratios of arcs to the corresponding inequalities between the ratios of chords. The properties he establishes in this way can be expressed by considering the changing values of functions such as $\dfrac{\sin \alpha}{\alpha}$ or $\dfrac{\sin \lambda \alpha}{\alpha}$.

2.1.2. *Spherical geometry and trigonometry*

Proposition 5. — *Inclination*: Let there be on a sphere two great circles *ABC* and *ADC* with common diameter *AC*. Let *K* and *L* be their respective poles where $\widehat{KL} < \dfrac{1}{4}$ circle and $\widehat{AB} = \widehat{BC} = \widehat{AD} = \widehat{DC} = \dfrac{1}{4}$ circle.

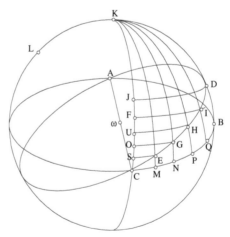

Fig. 2.8

a) We divide the arc CD into equal parts; let E, G, H, I be the points of division. The great circles KE, KG, KH, KI, KD cut the circle ABC in M, N, P, Q and B.

Through the points of division of the circles we draw parallels to the circle ABC that cut the circle KC in S, O, U, F, J; so we have $\widehat{ME} = \widehat{CS}$, $\widehat{NG} = \widehat{CO}$, $\widehat{PH} = \widehat{CU}$, $\widehat{QI} = \widehat{CF}$ and $\widehat{BD} = \widehat{CJ}$.

We may note that if the circle CAB is the equator, the arcs BD, QI, PH, NG and ME are respectively the inclinations of the points D, I, H, G and E.[2]

If we put $\Delta(D, I)$ = inclination of D – inclination of I, we have

$$\Delta(D, I) = \widehat{JF};$$

similarly

$$\Delta(I, H) = \widehat{FU}, \Delta(H, G) = \widehat{UO}, \Delta(G, E) = \widehat{OS} \text{ and } \Delta(E, C) = \widehat{SC}.$$

We prove that if $\widehat{DI} = \widehat{IH} = \widehat{HG} = \widehat{GE} = \widehat{EC}$, then

$$\Delta(D, I) < \Delta(I, H) < \Delta(H, G) < \Delta(G, E) < \Delta(E, C).$$

All the circles we consider, apart from the ones that are parallels, are great circles. The arcs KM and KB are equal and are orthogonal to the arc CMB; we thus have

(1) $$\frac{\sin \widehat{EC}}{\sin \widehat{CD}} = \frac{\sin \widehat{EM}}{\sin \widehat{DB}}$$ (by the sine rule),[3]

---

[2] When the circle ABC is the celestial equator, the arcs BD, QI are called *declinations*. But in the part concerned with astronomy the circle ABC can be the equator, the ecliptic or the horizon. It is accordingly preferable to keep to the term inclination.

[3] In a triangle ABC formed by arcs of great circles on a unit sphere with centre O, we have

$$\frac{\sin B\hat{O}C}{\sin A} = \frac{\sin C\hat{O}A}{\sin B} = \frac{\sin A\hat{O}B}{\sin C}$$

or again

$$\frac{\sin \widehat{BC}}{\sin A} = \frac{\sin \widehat{CA}}{\sin B} = \frac{\sin \widehat{AB}}{\sin C}.$$

In the curvilinear triangles ECM and BCD (see Fig. 2.8), we thus have

$$\frac{\sin \widehat{EC}}{\sin M} = \frac{\sin \widehat{EM}}{\sin C} \text{ and } \frac{\sin \widehat{CD}}{\sin B} = \frac{\sin \widehat{BD}}{\sin C};$$

now $\sin M = \sin B = 1$, hence the equality (1).

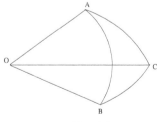

Fig. 2.9

so

$$\frac{\sin \widehat{EC}}{\sin \widehat{CD}} = \frac{\sin \widehat{CS}}{\sin \widehat{CJ}}$$

because $\widehat{EM} = \widehat{CS}$ and $\widehat{DB} = \widehat{CJ}$.
    Similarly

$$\frac{\sin \widehat{CG}}{\sin \widehat{CD}} = \frac{\sin \widehat{CO}}{\sin \widehat{CJ}} \qquad \text{(by the sine rule),}$$

therefore

(2) $\qquad\qquad \dfrac{\sin \widehat{CG}}{\sin \widehat{CE}} = \dfrac{\sin \widehat{CO}}{\sin \widehat{CS}} \qquad$ (by the sine rule).

    We have $\frac{\pi}{2} = \widehat{CD} > \widehat{CE}$, so $\sin \widehat{CD} > \sin \widehat{CE}$. From (1), we then deduce that $\widehat{EC} > \widehat{EM}$, because $\widehat{CD} > \widehat{DB}$ since $\widehat{CD}$ is a quarter of a circle, so $\widehat{EC} > \widehat{CS}$. In the same way we have $\widehat{CG} > \widehat{CO}$.

    Moreover $\widehat{CG} > \widehat{CE} > \widehat{CS}$ (because $\widehat{CG} = 2\widehat{CE}$). From (2) we get $\widehat{CO} > \widehat{CS}$; if we apply Proposition 3 to the triangles formed by the arcs that are double $\widehat{CE}$, $\widehat{CG}$ and $\widehat{CS}$, $\widehat{CO}$, we obtain

$$\widehat{CG} - \widehat{CE} > \widehat{CO} - \widehat{CS}$$

and

$$\frac{\widehat{CG} - \widehat{CE}}{\widehat{CE}} > \frac{\widehat{CO} - \widehat{CS}}{\widehat{CS}},$$

hence

$$\frac{\widehat{GE}}{\widehat{CE}} > \frac{\widehat{OS}}{\widehat{CS}}.$$

But $\widehat{GE} = \widehat{CE}$, hence $\widehat{OS} < \widehat{CS}$, so $\Delta(G, E) < \Delta(E, C)$.

    In the second part, Ibn al-Haytham constructs a second figure (Fig. 2.10) connected with the first one (Fig. 2.8), but with different lettering.
    Let there be a semicircle ACB with the same diameter as the given sphere, and let $\widehat{AC} = \widehat{CB}$; we divide the arc AC into five equal parts at the points E, G, H, K. The points A, E, G, H, K, C in this figure are homologous to the points D, I, H, G, E, C in Fig. 2.8.

On AB we draw an arc ADB similar to double the arc CJ of the first figure; it is implicit that we suppose the arc CJ is less than a quarter of a circle. The circle (B, BG) cuts the arc ADB at the point D, so $BG = BD$.

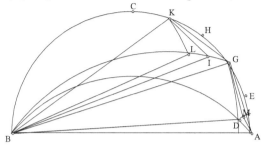

Fig. 2.10

We have $\widehat{AG} = 2\widehat{AE}$ and $\widehat{ACB} = 2\widehat{AC}$, so $\widehat{BCG} = 2\widehat{CE}$ and consequently

$$\frac{AB}{BG} = \frac{AB}{BD} = \frac{\sin\widehat{AC}}{\sin\widehat{CE}} \qquad \text{(Fig. 2.10)};$$

and if we turn back to the arcs in Fig. 2.8, the ratio is equal to

$$\frac{\sin\widehat{CD}}{\sin\widehat{CI}} = \frac{\sin\widehat{DB}}{\sin\widehat{QI}} = \frac{\sin\widehat{CJ}}{\sin\widehat{CF}} \qquad \text{(by the sine rule),}$$

from the analogue to (1) for the point I; so

$$\frac{AB}{BD} = \frac{\sin\widehat{CJ}}{\sin\widehat{CF}}.$$

Now, by hypothesis, $\widehat{ADB}$ is similar to $2\widehat{CJ}$, so $\widehat{BD}$ is similar to $2\widehat{CF}$; there remains $\widehat{AD}$ in Fig. 2.10 which is similar to $2\widehat{JF}$ of Fig. 2.8.

On BG we draw an arc equal to the arc DB and we take on $\widehat{BG}$ the point I such that $\widehat{GI} = \widehat{AD}$. We then have

$$\widehat{BCK} = \widehat{BCG} - \widehat{GK} = \widehat{BCG} - \widehat{AG},$$

hence

(1) $$B\hat{G}K = B\hat{A}G - A\hat{B}G.$$

Moreover

$$\overparen{BIG} = \overparen{BD} \text{ and } \overparen{GI} = \overparen{AD},$$

hence

$$\overparen{BI} = \overparen{BD} - \overparen{AD};$$

thus we have

(2) $I\hat{G}B = B\hat{A}D - A\hat{B}D.$

From (1) and (2), it follows

$$K\hat{G}I = G\hat{A}D - G\hat{B}D.$$

Let M be such that $AD = AM$ and $D\hat{A}M = G\hat{B}D$, then we have $G\hat{A}M = K\hat{G}I$. The point M lies inside the triangle AGD, so $A\hat{G}M < A\hat{G}D$.

Triangles GAM and GKI are equal, because $\overparen{GK} = \overparen{GA}$, so $GK = GA$, $\overparen{AD} = \overparen{GI}$ and $GI = AD = AM$; it follows that $A\hat{G}M = G\hat{K}I$, so $G\hat{K}I < A\hat{G}D$. We have $\overparen{BCG} < \overparen{BCA}$, hence $B\hat{K}G > B\hat{G}A$; then

$$B\hat{K}G - G\hat{K}I > B\hat{G}A - A\hat{G}D,$$

hence $B\hat{K}I > B\hat{G}D$.

Moreover, we have $\overparen{AG} = \overparen{GK}$ and $\overparen{AD} = \overparen{GI}$, hence

$$A\hat{B}G = K\hat{B}G \text{ and } A\hat{B}D = G\hat{B}I,$$

and therefore $D\hat{B}G = K\hat{B}I$.

In triangle BKI, we have

$$B\hat{K}I + K\hat{B}I + K\hat{I}B = \pi,$$

and in triangle DBG, we have

$$B\hat{G}D + D\hat{B}G + B\hat{D}G = \pi.$$

From this we get $B\hat{D}G > K\hat{I}B$, hence $B\hat{G}D > K\hat{I}B$ (isosceles triangle BGD); and consequently $B\hat{K}I > K\hat{I}B$, hence $BI > BK$.

So the circle (B, BK) cuts the arc BIG; say at point L, which will be such that $\widehat{GL} > \widehat{GI}$; so $\widehat{GL} > \widehat{AD}$.

We know that $\widehat{BCG} = 2\widehat{CE}$ and $\widehat{KG} = 2\widehat{GE}$, so $\widehat{BCK} = 2\widehat{CG}$; we then have

$$\frac{BG}{BL} = \frac{BG}{BK} = \frac{\sin \widehat{CE}}{\sin \widehat{CG}}$$

and, if we return to the arcs of the first figure, the previous ratio can be written

$$\frac{\sin \widehat{CI}}{\sin \widehat{CH}} = \frac{\sin \widehat{CF}}{\sin \widehat{CU}} \qquad \text{(the sine rule)}.$$

But the arc BLG (equal to the arc BD) is similar to the arc $2CF$; so the arc BL is similar to the arc $2CU$ and the arc LG is similar to the arc $2UF$.

We have seen that $\widehat{AD}$ is similar to $2\widehat{JF}$ and that $\widehat{AD} < \widehat{LG}$, so $\widehat{JF} < \widehat{FU}$ or $\Delta(D, I) < \Delta(I, H)$.

Using the same method, we can prove that

$$\widehat{FU} < \widehat{UO} \quad \text{and} \quad \widehat{UO} < \widehat{OS}.$$

So we have the stated result:

$$\Delta(D, I) < \Delta(I, H) < \Delta(H, G) < \Delta(G, E) < \Delta(E, C).$$

Comment: We locate each point P of the sphere (Fig. 2.8) by drawing through P a circle parallel to the equator UPH; the coordinates of P are then the inclination $\widehat{CU}$ (arc of the great circle KC) and the right ascension $\widehat{CH}$ (arc of the great circle ADC).

If we divide the arc CD into n equal parts at the points $I_1, I_2, \ldots I_i \ldots I_n = D$, where $I_0 = C$, to each point I_i of the arc CD there corresponds a point J_i of the arc CK such that $\widehat{CJ_i}$ is the inclination of the point I_i; let us put $\widehat{CI_i} = x_i$ and $\widehat{CJ_i} = y_i$.

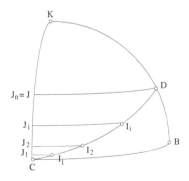

Fig. 2.11

For $x \in [0, \frac{\pi}{2}]$, y increases from $y_0 = 0$ which corresponds to the point C up to $y_n = \widehat{CJ}$ which corresponds to the point D; y is a monotonically increasing function of x.

By hypothesis, for all i from 1 to n, the difference $x_i - x_{i-1} = \Delta x$ is constant, but $y_1 - y_0 > y_2 - y_1 \ldots y_i - y_{i-1} > y_{i+1} - y_i$, so $(\Delta y)_i = y_i - y_{i-1}$ decreases as x_i increases.

b) The proof for two successive arcs – such as $\widehat{CE}$ and $\widehat{EG}$ in the first part, or $\widehat{DI}$ and $\widehat{IH}$ in the second, then $\widehat{IH}$ and $\widehat{HG}$ – involves only equality of arcs taken two at a time, but does not involve either the fact that each of them is equal to $\dfrac{CD}{5}$ or more generally to $\dfrac{CD}{n}$, nor the fact that one or the other of these arcs has an endpoint at C or D.

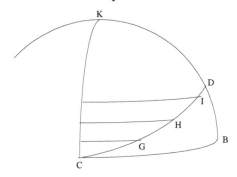

Fig. 2.12

The preceding proof can consequently be generalized to consider two successive arcs that are equal, whether or not each of them is commensurable with a quarter of the circle.

Let us assume we have G, H, I such that $\widehat{GH} = \widehat{HI}$. If $x_G < x_H < x_I$, we have

$$\Delta x = x_H - x_G = x_I - x_H \text{ and } \Delta(G, H) > \Delta(H, I),$$

so

$$\frac{\widehat{HI}}{\widehat{GH}} > \frac{\Delta(H,I)}{\Delta(G,H)}.$$

Using an alternative notation $x_H = x$, $x_G = x - h$ and $x_I = x + h$, we have $y_I - y_H < y_H - y_G$. If we put $y = f(x)$, then

$$\frac{f(x-h)+f(x+h)}{2} < f(x);$$

we know that this inequality and the fact that the function f is continuous imply that the function is concave for all x in the interval and for all $h > 0$. In what follows, Ibn al-Haytham gives a proof that uses the method of abduction to establish a property of exactly this kind.

We may note that this investigation can essentially be interpreted, in other terms, as one of considering the variation of functions by comparing finite differences.

This proposition is equivalent to Proposition 5 of Theodosius' *Sphaerica*, but with a more elegant proof that uses the sine rule.

Proposition 6. — More generally, let there be two general arcs, equal or unequal, contiguous or disjunct; the arcs can be commensurable or incommensurable.

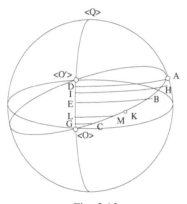

Fig. 2.13

Let O be the point of intersection of the given great circles; let A, B, C be three points such that $\widehat{OA} > \widehat{OB} > \widehat{OC}$; let D, E, G be the points of the arc OQ such that $\widehat{OD}$, $\widehat{OE}$ and $\widehat{OG}$ are the inclinations of A, B and C respectively.

a) $\widehat{AB}$ and $\widehat{BC}$ commensurable.

They can be divided into equal parts, p parts in $\widehat{AB}$ and q parts in $\widehat{BC}$. We then have p unequal parts in $\widehat{DE}$ and q unequal parts in $\widehat{EG}$; these parts become smaller and smaller the greater their distance from G (from Proposition 5)

$$\frac{p}{q} = \frac{\overset{\frown}{AB}}{\overset{\frown}{BC}};$$

but

$$\frac{p}{q} > \frac{\overset{\frown}{DE}}{\overset{\frown}{EG}},$$

so

$$\frac{\overset{\frown}{AB}}{\overset{\frown}{BC}} > \frac{\overset{\frown}{DE}}{\overset{\frown}{EG}} \quad \text{or} \quad \frac{\overset{\frown}{AB}}{\overset{\frown}{BC}} > \frac{\varDelta(A,B)}{\varDelta(B,C)}.$$

So if $x_A > x_B > x_C$, we have

$$\frac{\overset{\frown}{AB}}{\overset{\frown}{BC}} > \frac{\varDelta(A,B)}{\varDelta(B,C)}.$$

b) $\overset{\frown}{AB}$ and $\overset{\frown}{BC}$ incommensurable.

Ibn al-Haytham gives an argument by *reductio ad absurdum* to prove that we again have

$$\frac{\overset{\frown}{AB}}{\overset{\frown}{BC}} > \frac{\varDelta(A,B)}{\varDelta(B,C)}.$$

He first proves that the hypothesis $\dfrac{\overset{\frown}{AB}}{\overset{\frown}{BC}} = \dfrac{\overset{\frown}{DE}}{\overset{\frown}{EG}}$ is absurd.

Let us suppose that $\overset{\frown}{AB}$ is divided into equal p parts, $\overset{\frown}{AB} = p\ \alpha$; and let H be a point on $\overset{\frown}{AB}$ such that $\overset{\frown}{HB} = q\ \alpha\ (q < p)$ and let us suppose $\overset{\frown}{HB} < \overset{\frown}{BC}$.[4] Let I be the point associated with H, the difference of the inclinations of the points H and B is the arc IE and we have $\overset{\frown}{IE} < \overset{\frown}{EG}$. From the above, we have

$$\frac{\overset{\frown}{AH}}{\overset{\frown}{HB}} > \frac{\overset{\frown}{DI}}{\overset{\frown}{IE}},$$

hence

$$\frac{\overset{\frown}{AB}}{\overset{\frown}{BH}} > \frac{\overset{\frown}{DE}}{\overset{\frown}{IE}} \quad \text{or} \quad \frac{\overset{\frown}{IE}}{\overset{\frown}{DE}} > \frac{\overset{\frown}{BH}}{\overset{\frown}{AB}}.$$

---

[4] The division can be carried out with $p = 2^k$, so it can be constructed; the condition $\overset{\frown}{HB} < \overset{\frown}{BC}$ can be met thanks to Lemma X.1 of the *Elements*, as it was reformulated by Ibn al-Haytham (see R. Rashed, *Ibn al-Haytham and Analytical Mathematics. A History of Arabic Sciences and Mathematics*, vol. 2, Culture and Civilization in the Middle East, London, 2013, pp. 235–8, 382–90).

But by hypothesis $\dfrac{\widehat{DE}}{\widehat{EG}} = \dfrac{\widehat{AB}}{\widehat{BC}}$, so we have

$$\frac{\widehat{IE}}{\widehat{EG}} > \frac{\widehat{BH}}{\widehat{BC}}.$$

Let K be such that

$$\frac{\widehat{IE}}{\widehat{EG}} = \frac{\widehat{HB}}{\widehat{BK}},$$

so $\widehat{BC} > \widehat{BK} > \widehat{BH}$. Then we have $q\alpha < \widehat{BK} < \widehat{BC}$. There are then two possible cases:

1) $\alpha < \widehat{KC}$. In this case, there exists an integer m such that

$$(q + m)\, \alpha < \widehat{BC} < (q + m + 1)\, \alpha;$$

and if we put $\widehat{BM} = (q + m)\, \alpha$, we have $\widehat{MC} < \alpha$.

2) $\alpha > \widehat{KC}$, we divide the arc α into equal parts, making them smaller and smaller until we obtain a part $\alpha' < \widehat{KC}$, and we then have a point M that lies between K and C and is such that $\widehat{BM} = n\, \alpha'$ and $\widehat{MC} < \alpha'$.

So in both cases we have found a point M such that the arcs HB and BM are commensurable; we associate with the point M the point L (L lies between E and G) which has the corresponding inclination; then we have

$$\frac{\widehat{HB}}{\widehat{BM}} > \frac{\widehat{IE}}{\widehat{EL}}.$$

But we had

$$\frac{\widehat{IE}}{\widehat{EG}} = \frac{\widehat{BH}}{\widehat{BK}} > \frac{\widehat{BH}}{\widehat{BM}},$$

so

$$\frac{\widehat{IE}}{\widehat{EG}} > \frac{\widehat{IE}}{\widehat{EL}},$$

hence $\widehat{EG} < \widehat{EL}$; which is impossible.

Note: The text does not specify which of the arcs AB and BC is the greater. If we have $\overarc{AB} > \overarc{BC}$ and q such that $q\,\alpha < \overarc{BC} < (q+1)\,\alpha$, we put $\overarc{BH} = q\,\alpha$ and we define K as before; K lies between B and C, so $\overarc{KC} < \alpha$; which corresponds to case 2.

Ibn al-Haytham then proves that the hypothesis

(1) $$\frac{\overarc{AB}}{\overarc{BC}} < \frac{\overarc{DE}}{\overarc{EG}}$$

is also absurd.

Taking H and I as in the previous paragraph, we have

(2) $$\frac{\overarc{IE}}{\overarc{ED}} > \frac{\overarc{BH}}{\overarc{BA}}.$$

From (1) and (2), we get

$$\frac{\overarc{IE}}{\overarc{EG}} > \frac{\overarc{HB}}{\overarc{BC}}.$$

We then define the point K lying between B and C by

$$\frac{\overarc{BH}}{\overarc{BK}} = \frac{\overarc{IE}}{\overarc{EG}}$$

and we complete the proof as in the first case.

Note: So we could prove that $\dfrac{\overarc{AB}}{\overarc{BC}} \leq \dfrac{\overarc{DE}}{\overarc{EG}}$ is absurd without distinguishing the two cases.

In Propositions 5 and 6, Ibn al-Haytham investigated the inclination of points lying on a quadrant of a great circle, the inclination being measured with respect to a great circle with pole K.

In Proposition 7, he investigates the right ascension of points on the quadrant of the first great circle with respect to the second one, which acts as the equator.

The proposition thus follows immediately from Proposition 5 in the case where the arcs AB and BC are commensurable; Ibn al-Haytham moves on to the general case by means of an argument involving infinitesimals that uses Archimedes' axiom and a *reductio ad absurdum*.

We may interpret the result as establishing that declination is a concave function of the position of a point on the circle ACO measured from C.

Proposition 7. — *Right ascension*: The great circle ADC with diameter AC and pole K represents the equator. Let there be a great circle ABC, with diameter AC, whose plane makes an angle α with the plane ADC. We assume $\widehat{AB} = \widehat{BC} = \widehat{AD} = \widehat{DC}$.

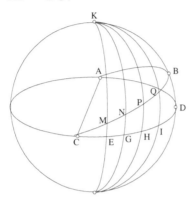

Fig. 2.14

a) We divide $\widehat{BC}$ into equal parts; let M, N, P, Q be the points of division. The great circles KM, KN, KP, KQ cut the arc CD at the points E, G, H, I. The arcs CE, CG, CH, CI and CD are the respective right ascensions of the points M, N, P, Q, B.

If we put, for example,

$$\delta(M, N) = \text{right ascension of } N - \text{right ascension of } M,$$

we have

$$\delta(C, M) = \widehat{CE}, \ \delta(M, N) = \widehat{EG}, \ \delta(N, P) = \widehat{GH},$$
$$\delta(P, Q) = \widehat{HI} \text{ and } \delta(Q, B) = \widehat{ID}.$$

If $\widehat{CM} = \widehat{MN} = \widehat{NP} = \widehat{PQ} = \widehat{QB}$, then

$$\delta(C, M) < \delta(M, N) < \delta(N, P) < \delta(P, Q) < \delta(Q, B).$$

The circles in question are all great circles and the great circles that pass through K are orthogonal to the circle CDA.

From Menelaus' theorem we have

$$\frac{\sin \widehat{CM}}{\sin \widehat{MN}} = \frac{\sin \widehat{CE}}{\sin \widehat{EG}} \cdot \frac{\sin \widehat{KG}}{\sin \widehat{KN}};$$

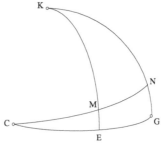

Fig. 2.15

but $\widehat{CM} = \widehat{MN}$ and $\widehat{KG} = \frac{\pi}{2}$; so $\sin \widehat{CE} < \sin \widehat{EG}$, hence $\widehat{CE} < \widehat{EG}$.

We also have

$$\frac{\sin \widehat{CN}}{\sin \widehat{NP}} = \frac{\sin \widehat{CG}}{\sin \widehat{GH}} \cdot \frac{\sin \widehat{KH}}{\sin \widehat{KP}}$$

and

$$\frac{\sin \widehat{CN}}{\sin \widehat{MN}} = \frac{\sin \widehat{CG}}{\sin \widehat{EG}} \cdot \frac{\sin \widehat{KE}}{\sin \widehat{KM}}.$$

But

$$\sin \widehat{NP} = \sin \widehat{MN}, \ \sin \widehat{KH} = \sin \widehat{KE} = 1 \text{ and } \sin \widehat{KP} < \sin KM,$$

so

$$\sin \widehat{GH} > \sin \widehat{EG}$$

and consequently

$$\widehat{EG} < \widehat{GH}.$$

In the same way, we can prove that

$$\widehat{GH} < \widehat{HI} \text{ and } \widehat{HI} < \widehat{ID},$$

hence the conclusion.

The investigation of the right ascension is carried out much more rapidly than that of the inclination, and without recourse to infinitesimals, because Menelaus' theorem can be applied to each pair of great circles passing through K.

Generalizing this to the case of n equal parts on the arc CB is clearly immediate. Let $I_0 = C, I_1 \dots I_i \dots I_n = B$ be the points of division, $K_1, K_i, K_n = D$ the points that define the right ascensions.

If we put $\widehat{CI_i} = x_i$ and $\widehat{CK_i} = z_i$, for $x \in [0, \frac{\pi}{2}]$, z increases from 0 to $\frac{\pi}{2}$ and $z_1 - z_0 < z_2 - z_1 < \dots z_i - z_{i-1} \dots < z_n - z_{n-1}$, so $(\delta z)_i = z_i - z_{i-1}$ increases as x_i increases.

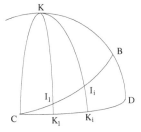

Fig. 2.16

b) Ibn al-Haytham next says that, as in the case of the differences in inclination, we can deduce from the result that has been proved for the differences in right ascensions, as regards arcs equal to $\frac{\pi}{2n}$:

• the result for equal arcs, whether consecutive or not, whether commensurable or not;

• then the result for unequal arcs. So if M, N, P, Q are points lying in that order on the arc CB, we shall have

(*) $$\frac{\widehat{MN}}{\widehat{PQ}} > \frac{\delta(M,N)}{\delta(P,Q)}.$$

So, taking into account Propositions 6 and 7, we have:

$$\frac{\Delta(M,N)}{\Delta(P,Q)} > \frac{\widehat{MN}}{\widehat{PQ}} > \frac{\delta(M,N)}{\delta(P,Q)}.$$

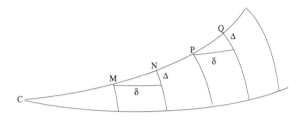

Fig. 2.17

This proposition is equivalent to Proposition 6 of Theodosius' *Sphaerica*. We can express this by saying that the right ascension is a convex function of the position of a point on the circle ABC measured from C. Let us put $x_M = \widehat{CM}$, $x_N = \widehat{CN}$, $x_P = \widehat{CP}$ and $x_Q = \widehat{CQ}$ and let the right ascension of the point R such that $\widehat{CR} = x$ be called $g(x)$. The inequality (*) can be written:

$$\frac{x_N - x_M}{x_Q - x_P} > \frac{g(x_N) - g(x_M)}{g(x_Q) - g(x_P)},$$

that is,

$$\frac{g(x_Q) - g(x_P)}{x_Q - x_P} > \frac{g(x_N) - g(x_M)}{x_N - x_M},$$

which indicates that the function g is convex.

2.1.3. *Plane geometry*

Proposition 8.[5] — Let there be a circle with centre D and diameter AC. We consider a point E on the segment DC, and, on the circle, equal arcs AB, BH, HI; if the chords $AB = BH = HI < EC$, then $A\hat{E}B < B\hat{E}H < H\hat{E}I$.

Triangles ADB, BDH, HDI are isosceles and equal. So we have

---

[5] Proposition 8 should be compared with Theorems 1 and 2 of *Treatise* 4 by Thābit (*Œuvres d'astronomie*, ed. and transl. R. Morelon, Paris, 1987, p. 73); the theorems are summarized in R. Morelon, 'Eastern Arabic astronomy between the eighth and the eleventh century', in R. Rashed (ed.), *Encyclopedia of the History of Arabic Science*, 3 vols, London/New York, 1996, vol. I, pp. 20–57, pp. 34 ff.

$$D\hat{A}B = D\hat{B}A = D\hat{B}H = D\hat{H}B = D\hat{H}I = D\hat{I}H.$$

But $E\hat{H}B > D\hat{H}B$, so

$$E\hat{H}B > E\hat{A}B.$$

The quadrilateral $ABHC$ is convex and inscribed in the circle, so $C\hat{A}B + C\hat{H}B = 2$ right angles; now $C\hat{H}B > E\hat{H}B$, so $E\hat{A}B + E\hat{H}B < 2$ right angles.

Let K be such that $E\hat{B}K = E\hat{A}B$ and $EK = EB$, so we have $E\hat{K}B = E\hat{A}B$, hence $E\hat{K}B + E\hat{H}B < 2$ right angles. We draw the circle EKB.

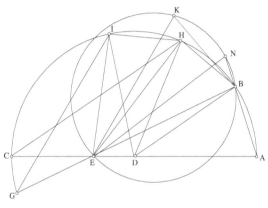

Fig. 2.18

If an inscribed angle that intercepts the arc BKE is called α, we have

$$B\hat{K}E + \alpha = 2 \text{ right angles};$$

but $B\hat{K}E + E\hat{H}B = E\hat{A}B + E\hat{H}B < 2$ right angles, so $\alpha > E\hat{H}B$. Moreover $E\hat{H}B > E\hat{A}B$, so $E\hat{H}B > B\hat{K}E$; the arc of the circle (EKB) intercepted by an angle equal to $E\hat{H}B$ would thus be greater than the arc EB and smaller than the arc EKB; let it be the arc EBN. We have

$$E\hat{N}B = E\hat{K}B = E\hat{B}K \text{ and } E\hat{B}N > E\hat{B}K,$$

hence $E\hat{N}B < E\hat{B}N$ and $EB < EN$.

If we draw the circumcircle of triangle EHB, the chord EB cuts off in this circle a segment similar to the segment EBN cut off by EN in the circle (EKB); now $EB < EN$, so we have

$$\text{circle } (EHB) < \text{circle } (EKB).$$

From the equality $E\hat{A}B = E\hat{K}B$ it follows that the circumcircle of EKB is equal to the circumcircle of EAB (these circles would be symmetrical with respect to the straight line EB). So

$$\text{circle } (EHB) < \text{circle } (EAB).$$

From $EC < EA$ it follows that $EA > EB > EH > EI > EC$. In fact, we have $E\hat{B}A > D\hat{B}A = E\hat{A}B$; $E\hat{H}B > D\hat{H}B = D\hat{B}H > E\hat{B}H$, and so on, and moreover by hypothesis $AB < EC$; so $EA > EB > AB$, hence

$$E\hat{B}A > E\hat{A}B > A\hat{E}B,$$

hence $A\hat{E}B < 1$ right angle. Similarly, $B\hat{E}H < 1$ right angle and $H\hat{E}I < 1$ right angle.

In the circle (EAB), the angle AEB intercepts the arc AB and, in the circle (EHB), the angle BEH intercepts the arc BH; we have

$$\text{circle } (EAB) > \text{circle } (EHB) \text{ and } AB = BH \text{ (chords)},$$

so $A\hat{E}B < B\hat{E}H$.

The straight line BE cuts the circle with centre D again in G. The quadrilateral $BHIG$ inscribed in this circle is convex, so

$$H\hat{B}G + H\hat{I}G = 2 \text{ right angles;}$$

now $H\hat{I}E < H\hat{I}G$, so

$$E\hat{B}H + H\hat{I}E < 2 \text{ right angles.}$$

Moreover

$$E\hat{I}H > D\hat{I}H, \quad D\hat{I}H = D\hat{B}H \text{ and } D\hat{B}H > E\hat{B}H,$$

so

$$E\hat{I}H > E\hat{B}H.$$

The hypotheses for the triangles *BEH* and *HEI* are thus the same as for the triangles *AEB* and *BEH*; so we shall have

$$B\hat{E}H < H\hat{E}I,$$

hence the conclusion

$$A\hat{E}B < B\hat{E}H < H\hat{E}I.$$

Note: We can give a quicker proof by appealing to trigonometry. We have $E\hat{H}B > D\hat{H}B$, hence $E\hat{H}B > E\hat{A}B$.

In the triangles *AEB* and *BEH*, we have (sine rule in the plane)

$$\frac{AB}{\sin A\hat{E}B} = \frac{EB}{\sin E\hat{A}B} \quad \text{and} \quad \frac{BH}{\sin B\hat{E}H} = \frac{EB}{\sin E\hat{H}B};$$

now $AB = BH$ and $\sin E\hat{A}B < \sin E\hat{H}B$, therefore $\sin A\hat{E}B < \sin B\hat{E}H$, hence $A\hat{E}B < B\hat{E}H$.

In triangles *BEH* and *HEI*, we have

$$\frac{BH}{\sin B\hat{E}H} = \frac{EH}{\sin E\hat{B}H} \quad \text{and} \quad \frac{HI}{\sin H\hat{E}I} = \frac{EH}{\sin E\hat{I}H} \quad (\text{where } E\hat{B}H < E\hat{I}H);$$

now $BH = HI$ and $\sin E\hat{B}H < \sin E\hat{I}H$, so $\sin B\hat{E}H < \sin H\hat{E}I$, and consequently $B\hat{E}H < H\hat{E}I$.

This trigonometrical argument, which treats sines as (numerical) functions of angles, does not require the use of several circles, but it is foreign to the mathematics of the period. We can, however, catch a hint of it in the reasoning of Ibn al-Haytham.

Ibn al-Haytham points out that the proof of the inequality of the angles with vertex *E* that intercept the equal arcs *AB*, *BH* and *HI* does not involve the ratio of the arc *AB* to the semicircumference *ABC*. The proof remains valid whether this ratio is rational or not; but it is clear that he is taking it for granted that the sum of the arcs in question is less than the semicircumference.

Notes:

   If we draw the circle (E, EA), the straight lines BE, HE, IE cut it in B', H', I'.

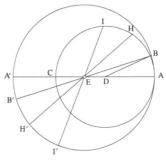

Fig. 2.19

   The result $A\hat{E}B \ < \ B\hat{E}H < H\hat{E}I$ implies $A'\hat{E}B' \ < \ B'\hat{E}H' < H'\hat{E}I'$ and these angles at the centre of the circle (E, EA) intercept unequal arcs

$$\widehat{A'B'} < \widehat{B'H'} < \widehat{H'I'};$$

so

$$\widehat{AB} = \widehat{BH} = \widehat{HI} \Rightarrow \widehat{A'B'} < \widehat{B'H'} < \widehat{H'I'}.$$

   In this proposition, Ibn al-Haytham is investigating changes in certain elements of a figure because of other elements: here angles with vertex E, like angle BEH, changing because of angles with vertex D (the centre of the circle), like angle BDH. Through these geometrical arguments, we are looking at the increase and convexity of the angle AEB considered as a function of the angle ADB.

   Let us establish these two properties analytically:

   Let $A\hat{D}B = \theta \in [0, \pi]$ and $A\hat{E}B = \varphi$. We have

$$\tan \varphi = \frac{r \sin \theta}{r \cos \theta + a},$$

where r is the radius of the circle ABC and $a = DE < r = DA$.

   Differentiation gives,

$$\left(1 + \tan^2 \varphi\right) \frac{d\varphi}{d\theta} = r \frac{r + a \cos \theta}{(r \cos \theta + a)^2} > 0,$$

which shows the increase. Now

$$1 + \tan^2\varphi = \frac{r^2 + 2ar\cos\theta + a^2}{\left(r\cos\theta + a\right)^2};$$

so we have

$$\frac{d\varphi}{d\theta} = r\frac{r + a\cos\theta}{r^2 + 2ar\cos\theta + a^2}$$

a decreasing function of cos θ. As cos θ is a decreasing function of θ, we indeed conclude that φ is convex. Further we verify that

$$\frac{d^2\varphi}{d\theta^2} = \frac{ar\left(r^2 - a^2\right)\sin\theta}{\left(r^2 + 2ar\cos\theta + a^2\right)^2} > 0$$

on $]0, \pi[$.

This is the direction of the increase of the angles *AEB, BEH, HEI*.

Proposition 9. — We return to the circle with centre *D* and diameter *AC*, and a point *E* on the segment *DC*; we draw the circle (*E*, *EA*). Three general straight lines that pass through *E* cut the circle (*D*, *DA*) in the points *B*, *K*, *M* and the circle (*E*, *EA*) in the points *I*, *H*, *N*; we have

$$\frac{\widehat{BK}}{\widehat{KM}} > \frac{\widehat{IH}}{\widehat{HN}}.$$

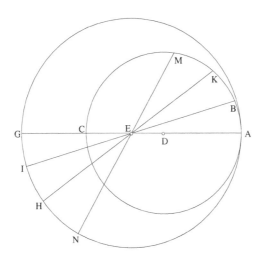

Fig. 2.20

• If $\overparen{BK}$ and $\overparen{KM}$ are commensurable, there exists an arc α such that $\overparen{BK} = p\alpha$ and $\overparen{KM} = q\alpha$. Angles BEK and KEM are then dissected into p angles and q angles respectively, all unequal and increasing in size as we move away from B. The same will be true for angles IEH and HEN which are the angles at the centre that intercept the arcs IH and HN. Each of the arcs IH and HN is then dissected, to become the sum of unequal arcs

$$\overparen{IH} = \sum_{i=1}^{p} \alpha_i, \quad \overparen{HN} = \sum_{i=p+1}^{p+q} \alpha_i \quad \text{where } \alpha_i < \alpha_{i+1}.$$

So we have

$$p\alpha_1 < \overparen{IH} < p\alpha_p \text{ and } q\alpha_{p+1} < \overparen{HN} < q\alpha_{p+q};$$

but $\alpha_p < \alpha_{p+1}$, hence

$$q\alpha_p < \overparen{HN} < q\alpha_{p+q},$$

and consequently

$$\frac{\overparen{IH}}{\overparen{HN}} < \frac{p\alpha_p}{q\alpha_p},$$

so

$$\frac{\overparen{IH}}{\overparen{HN}} < \frac{p}{q} \text{ or } \frac{\overparen{IH}}{\overparen{HN}} < \frac{\overparen{BK}}{\overparen{KM}}.$$

• $\overparen{BK}$ and $\overparen{KM}$ incommensurable.

Ibn al-Haytham says that the proof is carried out as in Proposition 6 (see p. 72). It is proved, by a *reductio ad absurdum*, that we cannot have either

$$\frac{\overparen{IH}}{\overparen{HN}} = \frac{\overparen{BK}}{\overparen{KM}} \text{ or } \frac{\overparen{IH}}{\overparen{HN}} > \frac{\overparen{BK}}{\overparen{KM}}.$$

Notes:

1) Special case: If $\overparen{BK} = \overparen{KM}$, we have seen that $\overparen{IH} < \overparen{HN}$, so we have

$$\frac{\overparen{BK}}{\overparen{KM}} > \frac{\overparen{IH}}{\overparen{HN}}.$$

So, in all cases, we have

$$\frac{\overparen{BK}}{\overparen{KM}} > \frac{\overparen{IH}}{\overparen{HN}}.$$

This proposition may also be expressed in the form: if φ_0, φ_1, φ_2 correspond to three values θ_0, θ_1, θ_2 of θ such that $\theta_0 < \theta_1 < \theta_2$, we have

$$\frac{\varphi_1 - \varphi_0}{\varphi_2 - \varphi_1} < \frac{\theta_1 - \theta_0}{\theta_2 - \theta_1}.$$

This states that φ is a convex function of θ. Ibn al-Haytham proves it by first assuming that the ratio $\dfrac{\theta_1 - \theta_0}{\theta_2 - \theta_1}$ is rational, a case in which he applies Proposition 8; he then extends the inequality by considerations of continuity, as we have seen him do earlier.

It is clear that Ibn al-Haytham is developing a method of analysis whose stages can be expressed as follows:

We want to establish that a function $\varphi = f(\theta)$ increases and is convex. Ibn al-Haytham begins by proving it increases, then proves the inequality:

$$f(\theta + h) - f(\theta) > f(\theta) - f(\theta - h),$$

which is a special case of convexity. In a second stage he deduces from this that

$$\frac{f(\theta_2) - f(\theta_1)}{f(\theta_1) - f(\theta_0)} > \frac{\theta_2 - \theta_1}{\theta_1 - \theta_0}$$

in the case where this ratio is rational; to do this, we divide the intervals $[\theta_0, \theta_1]$ and $[\theta_1, \theta_2]$ into p and q intervals each equal to the same quantity α and we proceed as for the previous stage. Extending this to the case in which the ratio $\dfrac{\theta_2 - \theta_1}{\theta_1 - \theta_0}$ is irrational is accomplished by *reductio ad absurdum* in the classical manner of infinitesimal methods (using Archimedes' axiom). In modern terms, this is an extension by an argument of continuity.

2) We find these inequalities again in the astronomical part of Ibn al-Haytham's work. The circle (E, EA) is taken as the inclined orb of a planet and the circle (D, DA) as its eccentric circle. The preceding inequalities are obtained if we suppose the movement is from the apogee A towards the perigee C.

If the displacement is from C towards A passing through the points P, Q (to which points P_1, Q_1 correspond), we shall have $\hat{GEP_1} > \hat{CDP}$, hence

$$\frac{\overline{CP}}{PQ} < \frac{\overline{GP_1}}{P_1Q_1}, \quad \frac{PQ}{QM} < \frac{\overline{P_1Q_1}}{Q_1M_1}$$

or

$$\frac{\overline{CP}}{\overline{GP_1}} < \frac{\overline{PQ}}{P_1Q_1} > \frac{\overline{QM}}{Q_1M_1}.$$

Proposition 10. — Let there be a segment AC on the line of intersection of two perpendicular planes, a semicircle ABC with diameter AC in one of the planes and a segment of a circle AD that is less than a semicircle in the other plane.

We consider a ratio $k = \dfrac{\overline{HG}}{\overline{HP}} > 1$ (this ratio is defined by a point P on a segment HG).

We want to find, on the arc ADC, a point D such that, if $DE \perp AC$ in the plane ADC and $EB \perp AC$ in the plane ABC, we have

$$\frac{BD}{DC} > \frac{GH}{HP} > 1.$$

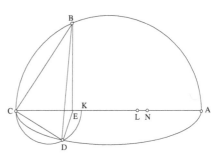

Fig. 2.21

We define the points M and I on the straight line HPG by making $\dfrac{HG}{HP} = \dfrac{HP}{HM}$ (with M between H and P) and $GI = HM$ (I beyond G); we then have $IM = GH$.

Let K be a point on AC such that $\dfrac{AK}{KC} = \dfrac{IM}{MH} = \dfrac{GH}{MH} > 1$, so $AK > KC$ and $\dfrac{AK}{KC} = k^2 > 1$.

The semicircle with diameter CK in the plane ADC cuts the arc ADC at the point D. The point E such that $C\hat{E}D = 1$ right angle satisfies $CE < CK$.

Let there be, on AC, points L and N defined by $AL = KC$ and $AN = CE$, we have $NE > LK$. But

$$\frac{LK}{KC} = \frac{AK - AL}{KC} = \frac{AK - CK}{KC} = \frac{GH - MH}{MH} = \frac{GM}{MH},$$

so

$$\frac{NE}{KC} > \frac{GM}{MH}.$$

We have

$$\frac{NE}{KC} = \frac{NE \cdot EC}{KC \cdot EC};$$

but

$$NE \cdot EC = AE \cdot EC - AN \cdot EC = EB^2 - EC^2, \quad KC \cdot EC = DC^2,$$

so

$$\frac{NE}{KC} = \frac{EB^2 - EC^2}{DC^2} > \frac{GM}{MH}.$$

But

$$EB^2 - EC^2 = EB^2 - (CD^2 - ED^2) = EB^2 + ED^2 - CD^2 = BD^2 - CD^2,$$

so

$$\frac{BD^2 - CD^2}{CD^2} > \frac{GM}{MH},$$

hence

$$\frac{BD^2}{CD^2} > \frac{GH}{MH};$$

but

$$\frac{GH}{MH} = \frac{GH^2}{MH.GH} = \frac{GH^2}{HP^2},$$

so we have

$$\frac{BD}{CD} > \frac{GH}{HP} > 1.$$

Commentary: By hypothesis

$$\frac{\overline{HG}}{\overline{HP}} = k = \frac{\overline{HP}}{\overline{HM}},$$

so

$$\frac{\overline{HG}}{\overline{HM}} = k^2 > 1.$$

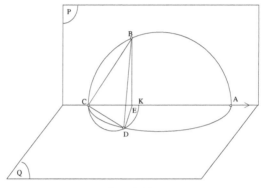

Fig. 2.22

We are given a semicircle of diameter AC in a plane P. On AC there exists a unique point K such that $\dfrac{\overline{KA}}{\overline{CK}} = k^2$.

We draw in the plane Q, perpendicular to P, a semicircle of diameter CK; to any point D on this semicircle there correspond a point E on AC such that $DE \perp AC$ and a point B on the semicircle in the plane P such that $BE \perp AC$. Then we have

$$BE^2 = \overline{CE} \cdot \overline{EA}, \quad ED^2 = \overline{CE} \cdot \overline{EK};$$

from which it follows that

$$BD^2 = BE^2 + ED^2 = \overline{CE} \cdot \left(\overline{EA} + \overline{EK}\right) = \overline{CE} \cdot \left(2\overline{EK} + \overline{KA}\right).$$

Moreover

$$CD^2 = \overline{CE} \cdot \overline{CK},$$

so we have

$$\frac{BD^2}{CD^2} = \frac{\left(2\overline{EK} + \overline{KA}\right)}{\overline{CK}}.$$

Assuming that CA is orientated from C to A, we have $\overline{CK} > 0$, $\overline{KA} > 0$ and $\overline{KA} > \overline{CK}$, $\overline{CK} > \overline{CE} > 0$, hence $\overline{EK} > 0$. So we have

$$\frac{BD^2}{CD^2} > \frac{\overline{KA}}{\overline{CK}} \qquad\qquad \text{with } \frac{\overline{KA}}{\overline{CK}} = k^2,$$

so

$$\frac{BD}{CD} > k.$$

• Any point D taken on the semicircle of diameter CK thus provides a solution to the problem, and to each point D there corresponds, in the plane Q, an arc of the circle ADC that is less than a semicircle.

• If the arc less than a semicircle is given, we have on this arc a unique point D that provides a solution to the problem, the point being constructed in this way, that is, with $\dfrac{KA}{CK} = k^2$.

But, as the condition for the problem is an inequality, all points D on a certain arc of the circle ADC must be suitable. We can find that arc in the following way:

In the plane of the circle ADC, we take coordinate axes CA and CT; the coordinates of the centre Ω are $(\dfrac{d}{2}, h)$ and those of A are $(d, 0)$.

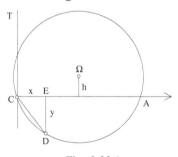

Fig. 2.23.1

If $D = (x, y)$ is a point on the circle, we have

$$\left(x - \frac{d}{2}\right)^2 + (y - h)^2 = \frac{d^2}{4} + h^2,$$

or

$$x(x - d) + y(y - 2h) = 0.$$

Now $BD^2 = BE^2 + ED^2 = x(d - x) + y^2$ and $CD^2 = x^2 + y^2$. The condition for the problem can thus be written

$$y(y - 2h) + y^2 > k^2 (dx + 2hy),$$

or

$$k^2 dx < 2y^2 - 2hy - 2k^2 hy;$$

which means that the point D must lie outside the parabola with equation $k^2 dx = 2y^2 - 2h(1 + k^2)y$, which has axis $y = \dfrac{h}{2} (k^2 + 1)$ and whose vertex has coordinates $\left(-\dfrac{(1 + k^2)^2 h^2}{2k^2 d}, \ \dfrac{h}{2}(k^2 + 1) \right)$.

The point C, the origin of the coordinates, lies on the parabola, while the point $A = (d, 0)$ lies inside it; in fact the straight line with equation $y = 0$ cuts the parabola at C, to the left of A. The parabola cuts the circular arc ADC in a unique point D_0 and the required point D must lie on the arc CD_0. We find D_0 from an equation of degree 3.[6]

Let us consider the same problem with $D\hat{E}C$ acute.
Let D be the point we obtain, and as in the previous case, we associate with it points S and O such that $DS \perp AC$ and $OS \perp AC$, so we have

$$\frac{DO}{DC} > k.$$

Let E be a point on $[CA]$ such that $D\hat{E}C$ is acute ($D\hat{E}C < D\hat{S}C$, hence $CE > CS$) and let B be a point on the given semicircle such that $EB \perp AC$. We want to prove that

$$\frac{BD}{DC} > k.$$

---

[6] $x_0 (x_0 - d) = y_0(y_0 - 2h)$ and $k^2 d \cdot x_0 = 2y_0 [y_0 - (1 + k^2)h]$, hence

$$\frac{x_0 - d}{k^2 d} = \frac{2h - y_0}{2(y_0 - (1 + k^2)h)} \quad \text{and} \quad x_0 = d \frac{y_0(2 - k^2) - 2h}{2(y_0 - h(1 + k^2))};$$

then we have

$$k^2 d^2 (y_0(2 - k^2) - 2h) = 4y_0 (y_0 - h(1 + k^2))^2.$$

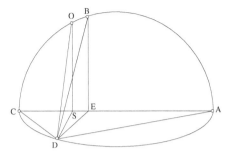

Fig. 2.23.2

Commentary:

Let us put $\overline{CS} = x_0$, $\overline{CE} = x$, $\overline{CA} = d$, $\overline{SD} = a$, with $x_0 < \dfrac{d}{2}$ and $d > x > x_0$. We then have:

(1) $OD^2 = OS^2 + SD^2 = x_0\,(d - x_0) + a^2 = a^2 + dx_0 - x_0^2$

$DE^2 = DS^2 + SE^2 = a^2 + (x - x_0)^2$, $BE^2 = x(d - x)$

$BD^2 = DE^2 + BE^2 = a^2 + (x - x_0)^2 + x(d - x)$

(2) $BD^2 = a^2 + dx - x_0\,(2x - x_0)$.

From (1) and (2)

$$BD^2 > OD^2 \Leftrightarrow d(x - x_0) - 2x_0\,(x - x_0) > 0$$
$$\Leftrightarrow (d - 2x_0)\,(x - x_0) > 0.$$

This inequality is satisfied because $x > x_0$ and $d > 2x_0$. So we have $BD > OD$, hence the conclusion

$$\frac{BD}{CD} > k.$$

Note: We also have $AD > BD$, because we can write

$$AD^2 = a^2 + (d - x_0)^2 = a^2 + x_0^2 + d(d - 2x_0),$$

$$BD^2 = a^2 + x_0^2 + x(d - 2x_0), \quad d - 2x_0 > 0,$$

and we have $x < d$. So, when the point E describes $[SA]$, x increases from x_0 to d, the point B describes the arc OA and the length BD increases; we have $OD < BD < AD$.

Ibn al-Haytham then considers arcs constructed on OD or BD, assuming that each of these arcs is part of a circle equal to the circle ADC; he comes back to the same result from considering the circle ADC and putting into it two chords DO' and DB' such that $DO' = DO$ and $DB' = DB$. We thus have

$$DC < DO' < DB' < DA,$$

hence

$$\widehat{DC} < \widehat{DO'} < \widehat{DB'} < \widehat{DA}$$

and

$$\widehat{DC} + \widehat{DO'} < \widehat{CD} + \widehat{DA} < \frac{1}{2} \text{ circle}$$

and in the same way

$$\widehat{DC} + \widehat{DB'} < \frac{1}{2} \text{ circle}.$$

Hence, from the first proposition

$$\frac{\widehat{DO'}}{\widehat{DC}} > \frac{DO'}{DC} > k, \quad \frac{\widehat{DB'}}{\widehat{DC}} > \frac{DB'}{DC} > k.$$

The conclusion remains the same if we consider arcs DO or DB that are parts of a circle smaller than the circle ADC.

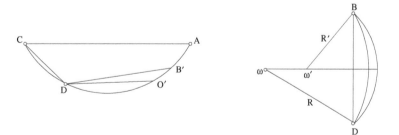

Fig. 2.24

Let there be a chord BD and two circles (ω, R) and (ω', R') that pass through B and D where $R' < R$. The arc BD of the circle centre ω' lies outside the circle centre ω, so the length of the arc is greater than that of the arc BD of the circle centre ω. So we have

$$\frac{\overset{\frown}{DB}}{\overset{\frown}{DC}} > \frac{DB}{DC} > k.$$

Note: In Proposition 10, the axis of the circle ABC is the perpendicular bisector of AC in the plane ADC. The centre of the circle ADC lies on this perpendicular bisector, so it is the centre of a sphere whose surface contains the two circles ABC and ADC. The property established in Proposition 10 is used several times in the part that deals with astronomy (see pp. 381, 393).

Propositions 11 and 12: These two propositions are concerned with a property of points on a parallel on the celestial sphere. Let ABC be a meridian circle, A and C the celestial poles, DNE a circle parallel to the horizon and ED its diameter.

Proposition 11. — We suppose that A and C lie on the horizon (so the place we are considering in this case is on the terrestrial equator). The circle of the equator, which has centre G, cuts the meridian in B and GB cuts DE in O. Three circles parallel to the equator with centres Q, S, F, respectively cut the meridian in H, I, K and the horizontal circle (DNE) in L, M and P, and their planes cut the straight line DE at the points X, J, U. We suppose that $DX < DJ < DO < DU < DE$, thus

$$\frac{LH}{HD} > \frac{MI}{ID} > \frac{NB}{BD} > \frac{PK}{KD}.$$

The plane of the meridian (ABC) is perpendicular to the plane of the equator and to the planes of the circles that are parallels. The inequalities $DX < DJ < DO$ imply $OX > OJ$, hence $LX < MJ$, because $LX^2 = LO^2 - OX^2 = OD^2 - OX^2$ and $MJ^2 = MO^2 - OJ^2 = OD^2 - OJ^2$.

Moreover $QX = SJ$ and $Q\hat{X}L = S\hat{J}M = 1$ right angle, so $X\hat{Q}L < J\hat{S}M$. In the same way we have $J\hat{S}M < O\hat{G}N$. We have $X\hat{Q}L = 2H\hat{L}X$, because, in the circle with centre Q, $X\hat{Q}L$ is an angle at the centre that intercepts the arc LH and $H\hat{L}X$ is an angle at the circumference that intercepts an arc equal to the arc LH (an arc symmetrical with LH with respect to QX). In the same way we have $J\hat{S}M < 2I\hat{M}J$ and $O\hat{G}N = 2O\hat{N}B$, hence

$$H\hat{L}X > I\hat{M}J < O\hat{N}B$$

and consequently

$$L\hat{H}X > M\hat{I}J > N\hat{B}O.$$

Let the point R on the segment LX be such that $X\hat{H}R = M\hat{I}J$; triangles HXR and IJM are right-angled and similar, so we have

$$\frac{RH}{HX} = \frac{MI}{IJ}.$$

But $HL > HR$, so

$$\frac{HL}{HX} > \frac{MI}{IJ}.$$

In the same way we can show that

$$\frac{MI}{IJ} > \frac{NB}{BO}.$$

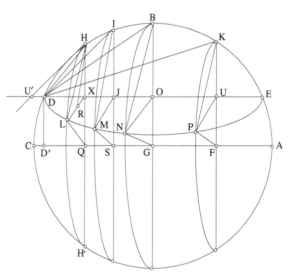

Fig. 2.25

We have $H\hat{D}X > I\hat{D}J > B\hat{D}O$ and consequently $D\hat{H}X < D\hat{I}J < D\hat{B}O$. Let the point U' on the straight line DX be such that $U'\hat{H}X = D\hat{I}J$ ($HU' \parallel ID$); we have

$$\frac{HX}{HU'} = \frac{IJ}{ID};$$

but $HU' > HD$, so we have

$$\frac{HX}{HD} > \frac{JI}{ID};$$

and we have seen that

$$\frac{HL}{HX} > \frac{MI}{IJ},$$

so

$$\frac{HL}{HD} > \frac{MI}{ID},$$

and in the same way we have

$$\frac{MI}{ID} > \frac{NB}{BD}.$$

For the circle (F, FK), we can show in the same way that $KU < BO$, $UP < NO$, $KP < BN$; now $KD > BD$, so

$$\frac{NB}{BD} > \frac{KP}{KD};$$

therefore

$$\frac{HL}{HD} > \frac{MI}{ID} > \frac{NB}{BD} > \frac{KP}{KD}.$$

We may note that $\frac{HX}{DH} = \sin H\hat{D}X$, where the angle HDX, which intercepts the arc HE, decreases as X moves along DE from D to E, with limits of the angle θ between the tangent at D and DE when X is at D; and zero when X is at E.

When X goes from D to O, LX increases; so angle LQX increases since the distance $QX = GO$ remains constant. Therefore $H\hat{L}X = \frac{1}{2} L\hat{Q}X$ also increases and the same is true of $\frac{HX}{HL} = \sin H\hat{L}X$. Thus

$$\frac{HL}{HD} = \frac{HL}{HX} \cdot \frac{HX}{DH}$$

decreases when DX increases from zero to DO.

Note: If DE coincides with AC, $GO = 0$ and angle HLX remains constant and equal to $\frac{\pi}{4}$, then $\frac{HL}{HX}$ also remains constant.

   When X goes from O to E, HL decreases while DH increases and the ratio $\frac{HL}{DH}$ consequently decreases.

Proposition 12. — We suppose that the pole A lies between the horizon and the zenith. The circle (G, GB) is no longer the circle of the equator but, as in Proposition 11, its plane passes through O, the midpoint of DE. In fact, in this case, G is defined as the orthogonal projection of O, the midpoint of DE, on the diameter AC and B is the point of the meridian on GO produced. There are three cases:

   1) AC cuts DE beyond E, at the point Ω;
   2) AC cuts DE at the point E ($\Omega = E = A$);
   3) AC cuts DE between E and O, O being the midpoint of DE.

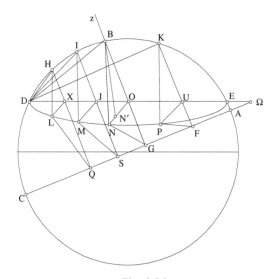

Fig. 2.26

In case 1, as before, we have the straight lines HQ, IS, BG and KF perpendicular to AC, hence $XQ > JS > OG > UF$; and the straight lines LX, MJ, NO, PU perpendicular to the meridian, so perpendicular to DE;

$$D\hat{X}L = D\hat{J}M = D\hat{O}N = D\hat{U}P = 1 \text{ right angle;}$$

$$L\hat{X}H = M\hat{J}I = N\hat{O}B = P\hat{U}K = 1 \text{ right angle.}$$

Moreover, ON is a semi-diameter of the circle $DLMNPE$, so $XL < MJ < NO$ and $NO > UP$. We have $XQ > JS$ and $XL < MJ$, so

$$\frac{XQ}{XL} > \frac{JS}{MJ},$$

that is cotan $L\hat{Q}X > $ cotan $M\hat{S}J$, so $L\hat{Q}X < M\hat{S}J$.

In the circle (Q, QH), the angle at the circumference HLX intercepts an arc equal to the arc HL and the angle at the centre LQH intercepts the arc HL, so

$$H\hat{L}X = \frac{1}{2} L\hat{Q}H;$$

similarly

$$I\hat{M}J = \frac{1}{2} M\hat{S}J \text{ and } O\hat{N}B = \frac{1}{2} N\hat{G}B,$$

so $H\hat{L}X < I\hat{M}J$; and similarly $I\hat{M}J < O\hat{N}B$.

Then we have

$$\sin H\hat{L}X = \frac{HX}{HL} \text{ and } \sin I\hat{M}J = \frac{IJ}{IM}.$$

So we have

$$\frac{HL}{HX} > \frac{IM}{IJ};$$

in the same way, we can prove that

$$\frac{IM}{IJ} > \frac{BN}{BO}.$$

The proof is completed as in Proposition 11. We have

$$\frac{HX}{\sin H\hat{D}X} = \frac{DH}{\sin D\hat{X}H}$$

with an angle $D\hat{X}H = D\hat{O}z$ (where GOz is the vertical of the place) that is independent of the position of X when X moves along DE; the angle HDX intercepts the arc HE which decreases when DX increases, so its sine decreases and the same is true for

$$\frac{HX}{DH} = \frac{\sin H\hat{D}X}{\sin A\hat{O}z}.$$

So for the circles with centres Q, S and G we have the double inequality

$$\frac{HL}{HD} > \frac{MI}{ID} > \frac{NB}{BD}.$$

For these three circles, the proof is the same in cases 1, 2 and 3, because in every version of the figure we have

$$XQ > JS > OG \text{ and } XL < JM < ON,$$

and consequently

$$L\hat{Q}X < M\hat{S}J < N\hat{G}O < \frac{\pi}{2},$$

hence we can conclude that

$$H\hat{L}X < I\hat{M}J < B\hat{N}O < \frac{\pi}{4}.$$

Investigation of a circle such as (F, FK)

In case 1 (Fig. 2.26) or case 2 when E and A are the same point, we have $OG > UF$ and $ON > UP$; we can draw no conclusions about the ratios $\frac{OG}{ON}$ and $\frac{UF}{UP}$. We can have

a) $\dfrac{OG}{ON} < \dfrac{UF}{UP} \Rightarrow O\hat{G}N > U\hat{F}P \Rightarrow O\hat{N}B > U\hat{P}K;$

b) $\dfrac{OG}{ON} = \dfrac{UF}{UP} \Rightarrow O\hat{N}B < U\hat{P}K;$

c) $\dfrac{OG}{ON} > \dfrac{UF}{UP} \Rightarrow O\hat{N}B < U\hat{P}K.$

In case 3, where Ω is between O and E, we can have:

• U between O and Ω; in this case, U lies between K and F and we have $K\hat{P}U < \frac{\pi}{4}$; and there are three possibilities a, b, c.

• U at point Ω; in this case, $U = F = \Omega$ and we have $K\hat{P}U = \frac{\pi}{4}$.

• U between Ω and E; in this case, F lies between K and U, thus we have $K\hat{P}U > \frac{\pi}{4}$, so $K\hat{P}U > B\hat{N}O$ (case c).

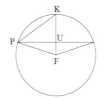 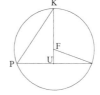

Fig. 2.27

In cases a and b, $K\hat{P}U \leq B\hat{N}O$; the circle (KP) is smaller than the circle (BN), because both of them lie between the equator and the pole A, and the circle (KP) is nearer the pole. So we have $KP < BN$.

Moreover, $KD > BD$, so

$$\frac{BN}{BD} > \frac{PK}{KD}.$$

In case c, we have $K\hat{P}U > B\hat{N}O$ and we have $P\hat{K}U < N\hat{B}O$.

We have a point N' on ON such that $O\hat{B}N' = P\hat{K}U$ and triangles OBN' and PKU are right-angled and similar; we have

$$\frac{KP}{KU} = \frac{BN'}{BO} < \frac{BN}{BO}.$$

Moreover

$$\frac{BO}{BD} > \frac{UK}{KD},$$

so

$$\frac{NB}{BD} > \frac{PK}{KD}.$$

If a circle $(K'P')$ is nearer the pole A than the circle (KP), we have

$$\frac{KP}{KD} > \frac{K'P'}{K'D}.$$

So in all cases we have

$$\frac{LH}{HD} > \frac{MI}{ID} > \frac{BN}{BD} > \frac{KP}{KD}.$$

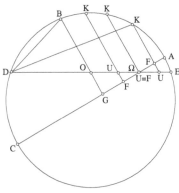

Fig. 2.28

Commentary on Propositions 11 and 12:

Let us now give an analytical proof of Propositions 11 and 12 taken together.

Let us consider a sphere with axis AC and centre K, and a small circle DLE of diameter DE in the meridian plane ABC (Fig. 2.29.1); in the case of Proposition 11, the diameter DE is parallel to the axis AC, and in the case of Proposition 12 meets it in a point Ω. Let F be the midpoint of the arc DE and O the midpoint of the straight line DE.

We suppose that the pole A lies between the horizon (a great circle parallel to DLE) and the zenith (the pole F of the circle DLE). Let HLH' be a variable circle with axis AC; its diameter HH', in the plane ABC, meets DE in X; we want to establish that the ratio $\dfrac{HL}{HD}$ decreases monotonically when X moves along DE from D towards E, that is, when H moves along the arc DE' where $E' = E$ if E is above AC and, if it is not, if E' is the point symmetrical with E with respect to AC.

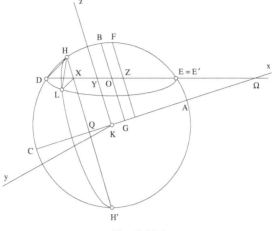

Fig. 2.29.1

Let us use α to denote the colatitude of F, the midpoint of the arc DE, let $\alpha = A\hat{K}F$, and use β to denote the angle EKF, the difference of the colatitudes of E and F. Let θ be the difference of the colatitude of F and that of H: $\theta = H\hat{K}F$, which varies from $-\beta$ (when H is at D) to β if $\alpha \geq \beta$ or to $2\alpha - \beta$ if $\alpha < \beta$ (when H is at E'). In fact, in the latter case, when X is at E, HX the perpendicular to AK is $E'E$, so the point H is at E' (see Fig. 2.29.2).

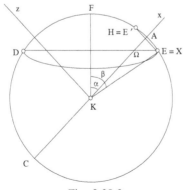

Fig. 2.29.2

We then have

$$\theta = \left(\overrightarrow{KE'}, \overrightarrow{KF}\right) = \left(\overrightarrow{KE'}, \overrightarrow{KA}\right) + \left(\overrightarrow{KA}, \overrightarrow{KE}\right) + \left(\overrightarrow{KE}, \overrightarrow{KF}\right) = 2\left(\overrightarrow{KA}, \overrightarrow{KE}\right) + \beta;$$

now $\left(\overrightarrow{KA}, \overrightarrow{KE}\right) = \left(\overrightarrow{KA}, \overrightarrow{KF}\right) + \left(\overrightarrow{KF}, \overrightarrow{KE}\right) = \alpha - \beta$, so $\theta = 2\alpha - \beta$.

The equation of the diameter DE is:

(1) $$x \cos \alpha + z \sin \alpha = r \cos \beta.$$

By hypothesis $0 < \alpha, \beta \le \dfrac{\pi}{2}$, and x, the abscissa of the point X, is equal to $KQ = r \cos(\alpha - \theta)$, like the abscissa of H, so its z coordinate is

$$z = \frac{r}{\sin \alpha}\left(\cos \beta - \cos \alpha \cos(\alpha - \theta)\right).$$

We have $HL^2 = HX \cdot HH'$ (circle HLH') with

$$HH' = 2r \sin(\alpha - \theta),$$

$$HX = r\sin(\alpha - \theta) - z = \frac{r}{\sin \alpha}\left(\sin \alpha \sin(\alpha - \theta) + \cos \alpha \cos(\alpha - \theta) - \cos \beta\right).$$

(2) $$HX = r\frac{\cos \theta - \cos \beta}{\sin \alpha} = 2r\frac{\sin \dfrac{\beta - \theta}{2}\sin \dfrac{\beta + \theta}{2}}{\sin \alpha}.$$

Further

(3) $$HD = 2r\sin \frac{\beta + \theta}{2}$$

since this chord subtends the angle $HKD = \beta + \theta$ at the centre. Thus

(4) $$\frac{HL^2}{HD^2} = 2r\frac{\sin \dfrac{\beta - \theta}{2}\sin \dfrac{\beta + \theta}{2}}{\sin \alpha} \cdot 2r\sin(\alpha - \theta) \cdot \frac{1}{4r^2 \sin^2 \dfrac{\beta + \theta}{2}}$$

$$= \frac{\sin(\alpha - \theta)\sin \dfrac{\beta - \theta}{2}}{\sin \alpha \sin \dfrac{\beta + \theta}{2}}.$$

We have[7] $(\beta - \alpha)^+ \leq \dfrac{\beta - \theta}{2} \leq \beta$ and $0 \leq \dfrac{\beta + \theta}{2} \leq \inf(\alpha, \beta)$; thus $\sin\dfrac{\beta - \theta}{2}$ is a decreasing function of θ and $\sin\dfrac{\beta + \theta}{2}$ is an increasing function of θ.

The function $\sin(\alpha - \theta)$, where $|\alpha - \beta| \leq \alpha - \theta \leq \alpha + \beta$, increases of θ if $\theta \leq \alpha - \dfrac{\pi}{2}$ and decreases if $\theta \geq \alpha - \dfrac{\pi}{2}$.

Let Y be the point where the axis Kz meets DE; if the point X lies between Y and E, we have $\theta \geq \alpha - \dfrac{\pi}{2}$, so the ratio $\dfrac{HL^2}{HD^2}$ decreases. This conclusion is sufficient if Y lies beyond D outside the sphere, that is, if $\alpha + \beta \leq \dfrac{\pi}{2}$.

If this is not so, that is, where $\alpha + \beta > \dfrac{\pi}{2}$, Y lies between D and E and we need to give another proof for the case where X lies between D and Y. From (2), HX increases for $\theta \leq 0$, that is, when X lies between D and the point Z with the same abscissa ($r \cos \alpha$) as the zenith F (H at F).

We have

$$\frac{HL^2}{HD^2} = \frac{HH'}{HX} \cdot \frac{HX^2}{HD^2},$$

where

$$\frac{HH'}{HX} = \frac{2r\sin(\alpha - \theta)}{r\sin(\alpha - \theta) - z} = 2\left(1 + \frac{z}{r\sin(\alpha - \theta) - z}\right)$$

which decreases between D and Z, because z decreases and $HX = r\sin(\alpha - \theta) - z$ increases.

Further

(5)
$$\frac{HX}{HD} = \frac{\sin\dfrac{\beta - \theta}{2}}{\sin\alpha}$$

which decreases from D to E when X moves along DE from D to E; thus $\dfrac{HL^2}{HD^2}$ continues to decrease from D to Z when X moves along EZ from E to Z. If Z lies beyond E outside the sphere, that is, if $2\alpha \leq \beta$, this conclusion is

---

[7] We use the notation $x^+ = \sup(x, 0)$ the positive part of a number x.

sufficient. If not, since $\alpha - \dfrac{\pi}{2} \leq 0$, Y is to the left of Z and the two proofs are complementary; $\dfrac{HL}{HD}$ decreases from D to E when X moves along DE from D to E.

Notes:

1) Ibn al-Haytham is here undertaking an investigation of the variation of a ratio in a case that is much more complicated than the ones he considered earlier.

As we have shown above, he employs different arguments in two intervals over which X varies: between O and E, then between D and O, where O is the midpoint of DE. We may note that O, which has abscissa $r \cos \alpha \cos \beta$, always lies between Y and Z.

2) In expression (5) we have the angles $\dfrac{\beta - \theta}{2} = H\hat{D}X$ and $\alpha = H\hat{X}D$; the first is the angle at the circumference which subtends the arc HE and the second is constant. Ibn al-Haytham arrives at the same expression for $\dfrac{HX}{HD}$ by using the fact that the sines of the angles are proportional to the opposite sides in triangle HDX.

3) Proposition 11 concerns the case in which $\alpha = \dfrac{\pi}{2}$; this simplifies the calculations because $\sin \alpha = 1$ and $\cos \alpha = 0$, so that $z = r \cos \beta$ is constant and $HX = r (\cos \theta - \cos \beta)$. In this case, the points Y, O and Z coincide.

4) Ibn al-Haytham interprets $\dfrac{HL}{HX}$ as $\dfrac{1}{\sin H\hat{L}X}$; so

$$\frac{HH'}{HX} = \frac{HL^2}{HX^2} = \frac{1}{\sin^2 H\hat{L}X}.$$

Now $H\hat{L}X = \dfrac{1}{2} L\hat{Q}X$ (angle in the circle HLH'); he finds the direction of variation of $\dfrac{HH'}{HX}$ from that of angle $L\hat{Q}X$. Now $L\hat{X}Q$ is a right angle, so

$$\tan L\hat{Q}X = \frac{LX}{XQ}.$$

In the case considered in Proposition 11, $XQ = r \cos \beta$ is constant, so $L\hat{Q}X$ increases or decreases with LX. In the general case considered in

Proposition 12, $XQ = z$ always decreases and LX increases for X between D and O.

5) If Ω lies between O and E, that is, if $\alpha \leq \beta$, Ibn al-Haytham distinguishes the case in which X lies between O and Ω from those in which X lies between Ω and E. The use of analytical notation allows us to ignore that distinction; we simply have $z < 0$ when X lies between Ω and E.

6) Let us calculate the limits of $\dfrac{HL}{HD}$ when X is at D and at E. In expression (4) we can see that $\dfrac{HL^2}{HD^2}$ tends to infinity when X tends to D, and θ becomes equal to $-\beta$.

If $\alpha \geq \beta$, for X at E we have $\theta = \beta$, so $\dfrac{HL^2}{HD^2} = 0$; if on the other hand $\alpha < \beta$, we have $\theta = 2\alpha - \beta$ and $\dfrac{HL^2}{HD^2} = \dfrac{\sin^2(\beta - \alpha)}{\sin^2 \alpha}$, or $\dfrac{HL}{HD} = \dfrac{\sin(\beta - \alpha)}{\sin \alpha}$, a finite limit.

7) We can also investigate the variation of $\dfrac{HL^2}{HD^2}$ by looking at the sign of its derivative; we need only to look at the sign of the numerator of the derivative of

$$\frac{\sin(\alpha - \theta)\sin \dfrac{\beta - \theta}{2}}{\sin \dfrac{\beta + \theta}{2}}.$$

The value of this numerator is:

$$-\cos(\alpha - \theta)\sin \frac{\beta - \theta}{2}\sin \frac{\beta + \theta}{2} - \frac{1}{2}\sin(\alpha - \theta)\cos \frac{\beta - \theta}{2}\sin \frac{\beta + \theta}{2} -$$

$$\frac{1}{2}\sin(\alpha - \theta)\sin \frac{\beta - \theta}{2}\cos \frac{\beta - \theta}{2}$$

$$= -\frac{1}{2}\left[\cos(\alpha - \theta)\cos\theta - \cos(\alpha - \theta)\cos\beta + \sin\beta\sin(\alpha - \theta)\right]$$

$$= \frac{1}{2}\left(\cos(\alpha + \beta - \theta) - \cos(\alpha - \theta)\cos\theta\right).$$

So we need to investigate the sign of $\cos(\alpha + \beta - \theta) - \cos(\alpha - \theta)\cos\theta$, whose derivative is:

(6) $\sin(\alpha + \beta - \theta) - \sin(\alpha - \theta)\cos\theta + \cos(\alpha - \theta)\sin\theta =$

$\sin(\alpha + \beta - \theta) + \sin(2\theta - \alpha) = 2\sin\dfrac{\beta + \theta}{2}\cos\left(\alpha + \dfrac{\beta - 3\theta}{2}\right)$

As we know that $\sin\dfrac{\beta + \theta}{2} \geq 0$, it is enough to determine the sign of $\cos\left(\alpha + \dfrac{\beta - 3\theta}{2}\right)$. We can check that, in all cases, $0 \leq \alpha + \dfrac{\beta - 3\theta}{2} \leq \dfrac{3\pi}{2}$; so the cosine is positive if $\alpha + \dfrac{\beta - 3\theta}{2} \leq \dfrac{\pi}{2}$ and negative if $\alpha + \dfrac{\beta - 3\theta}{2} > \dfrac{\pi}{2}$. Thus the derivative (6) is positive if $\theta \geq \dfrac{2\alpha + \beta - \pi}{3}$ and negative otherwise.

If $\alpha + 2\beta < \dfrac{\pi}{2}$, we always have

$$\theta \geq -\beta > \frac{2\alpha + \beta - \pi}{3},$$

so $\cos(\alpha + \beta - \theta) - \cos(\alpha - \theta)\cos\theta$ increases monotonically up to

$$\cos\alpha - \cos(\alpha - \beta)\cos\beta = -\sin(\alpha - \beta)\sin\beta$$

if $\alpha > \beta$; or

$$\cos(2\beta - \alpha) - \cos(\alpha - \beta)\cos(2\alpha - \beta) =$$
$$-\sin(\beta - \alpha)(2\cos(\beta - \alpha)\sin\alpha + \sin\beta)$$

if $\alpha < \beta$; in both cases, this quantity is negative, so

$$\cos(\alpha + \beta - \theta) - \cos(\alpha - \theta)\cos\theta$$

always remains negative. Thus, in this case, $\dfrac{HL^2}{HD^2}$ decreases.

If $\beta - \alpha > \dfrac{\pi}{4}$, we have

$$\theta \leq 2\alpha - \beta < \frac{2\alpha + \beta - \pi}{3},$$

so $\cos(\alpha + \beta - \theta) - \cos(\alpha - \theta)\cos\theta$ decreases monotonically from

$$\cos(\alpha + 2\beta) - \cos(\alpha + \beta)\cos\beta = -\sin\beta\sin(\alpha + \beta) < 0.$$

We can see that, in this case also, $\dfrac{HL^2}{HD^2}$ decreases monotonically.

Finally, if $\alpha \geq \beta - \dfrac{\pi}{4}$ and $\alpha \geq \dfrac{\pi}{2} - 2\beta$, the expression

$$\cos(\alpha + \beta - \theta) - \cos(\alpha - \theta)\cos\theta$$

goes through a minimum for $\theta = \dfrac{2\alpha + \beta - \pi}{3}$. Its extreme values are:

$$\cos(\alpha + 2\beta) - \cos(\alpha + \beta)\cos\beta = -\sin\beta\sin(\alpha + \beta) \leq 0$$

and

$$\cos\alpha - \cos(\alpha - \beta)\cos\beta = -\sin\beta\sin(\alpha - \beta) \leq 0$$

if $\alpha \geq \beta$, because $\alpha - 2\beta \geq -\alpha$; respectively

$$\cos(2\beta - \alpha) - \cos(\beta - \alpha)\cos(2\alpha - \beta) =$$
$$-\sin(\beta - \alpha)(\sin\beta + \sin\alpha\cos(\beta - \alpha)) \leq 0$$

if $\beta \geq \alpha$.

In all cases $\dfrac{HL^2}{HD^2}$ decreases when X moves along DE from D to E.

Proposition 13. — Let (ABC) be the circle of the horizon, D its pole, (ADC) the meridian and (EH) a circle parallel to the horizon. Two circles parallel to the equator cut the meridian circle in N and L and the circle (EH) in M and G. If the points on the meridian circle are in the order C, E, N, L, D, then the arc LG is greater than the arc similar to the arc NM.

a) The sphere is at right angles to the horizon (and is called 'the right sphere'). That is, the equator passes through D and has A and C as its poles. The plane of the equator is a plane of symmetry for all circles with diameter AC and for the horizontal circle EGH, which it cuts in X, the midpoint of the arc EH.

The circle (CMA) cuts the circle (EGH) again in a point K and cuts the arc LG in the point S between G and L. The arc cut off on LG is similar to the arc MN. So we have $\widehat{MN}$ similar to $\widehat{LS}$ and consequently $\widehat{LG}$ is greater than the arc similar to $\widehat{MN}$.

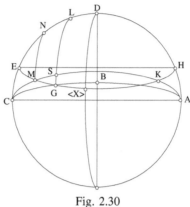

Fig. 2.30

b) The sphere is oblique. Let O be its visible pole; O can lie between A and H, O can be at H or O can lie between H and D.

α) First, let O lie between A and H. We draw through O a great circle to touch (EGH) at K and to cut the horizon at B. We have $OC > OB$. This is so because, since O lies on the meridian circle (which is perpendicular to the horizon), the point O', the projection of O on the plane of the horizon lies on AC and is closer to A than to C; it follows that all the segments $O'B$ joining O' to a point B on the horizon circle are smaller than or equal to $O'C$ and similarly for OB and OC.

So $\overset{\frown}{ODC} > \overset{\frown}{OKB}$ (arcs of great circles).

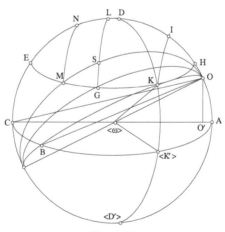

Fig. 2.31

We consider the great circle (DK); it is orthogonal to the circle (ABC) and to the circle (OKB), so its pole is the point B, and the arc BK is a quadrant of a great circle. But the arc DC is also a quadrant of a great circle, so $\widehat{OD} > \widehat{OK}$. We draw through K an arc of a circle parallel to the equator; let it be the arc KI. We have $\widehat{OK} = \widehat{OI}$, so I lies between D and H. The two great circles (OM) and (OG) cut the arc IK in two distinct points.

The point M lies between G and E; so the plane of the great circle OM lies between the plane of the great circle OG and that of the meridian.

Commentary:

We can regard this proposition as a lemma. From the information about the positions of the points C, E, N, L, D given in the statement, the result is obvious:

For all cases, if O is the pole of the equator ($O = A$ or $O \neq A$), the plane of the great circle (OM) lies between the plane of the great circle (OG) and the plane of the meridian; so the arc OM cuts the arc LG in S, the arcs LS and MN are similar; consequently $\widehat{LG}$ is greater than the arc similar to $\widehat{MN}$.

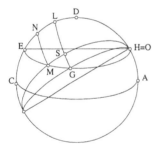 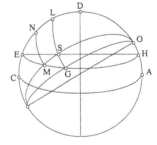

Fig. 2.32	Fig. 2.33
β) The pole is at the point H ($H = 0$)	γ) The pole O lies between H and D

In cases β and γ, the circle (OM) lies between the meridian and the circle (OG).

Proposition 14. — Let there be two circles parallel to the equator cutting the meridian circle in D and E respectively, the circle ABC parallel to the horizon in I and C, and a circle through the axis of the world in I and K.

If $\widehat{BE} < \widehat{BD} \leq \frac{1}{2} \widehat{ADB}$, then $\dfrac{\widehat{IE}}{\widehat{EB}} > \dfrac{\widehat{CD}}{\widehat{DB}} > \dfrac{\widehat{CK}}{\widehat{KI}}$.

If we suppose the plane *ABC* is above that of the horizon and if the visible pole of the circle of the equator is called *Π*, we may have *Π* on the horizon, *Π* between the horizon and the point *A*, *Π* at the point *A*, or *Π* between *A* and the zenith.

In the special case where *ABC* is the horizon and the pole *Π* is at the point *A*, the parts of the circles *EI* and *CD* that are above the horizon are semicircles (the case of the right sphere).

A) *Circle* EI *smaller than the circle* CDL.

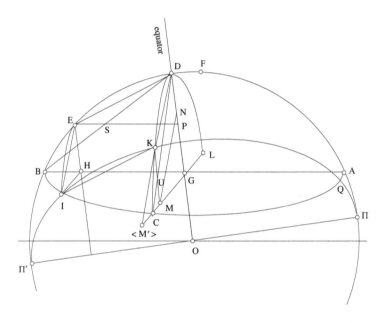

Fig. 2.34: The visible pole *Π* is above the horizon, $\overset{\frown}{BD} \leq \dfrac{1}{2} \overset{\frown}{ADB}$.

D lies on the equator or alternatively *E* and *D* lie between the equator and the point *B*. The point *O*, the centre of circle *CKD*, is alternatively either on the horizon or above the horizon. The centre of the circle *EI* is above the horizon.

We have $DG \perp CG$, $EH \perp HI$. We draw $KM \perp CG$ and $MN \parallel DK$, thus we have $KM = DN$ and $MN = DK$.

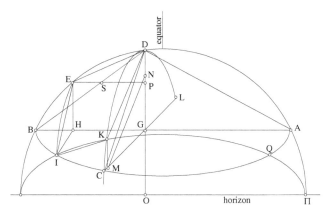

Fig. 2.35: Special case: right sphere.

The pole Π is on the horizon and $\widehat{BD} \leq \frac{1}{2}\widehat{ADB}$

The centres of all the parallel circles thus lie on the horizon.

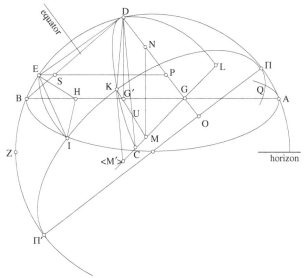

Fig. 2.36: The pole Π is above the horizon, $\widehat{BD} \leq \frac{1}{2}\widehat{ADB}$.

E and D lie on opposite sides of the equator.

The centre of the circle EI is above the horizon.

The centre of the circle CKD is above the horizon.

The arcs *EI* and *DK* are similar, but the circle *CDL* is greater than the circle *EI*, so *DK* > *EI* and *MN* > *EI*. Triangles *EHI* and *NGM* are similar, so *NG* > *EH*.

We draw *EP* ∥ *BG*, so *PG* = *EH*; so *N* lies between *P* and *D*, and *P* lies between *N* and *G*. The straight line *DB* cuts *EP* in *S*, we have

$D\hat{S}P = D\hat{B}H < 1$ right angle.

So $D\hat{S}E$ is obtuse and *DE* > *DS*.

− If the circle *CDL* is the circle of the equator or if it is closer to the hidden pole than the equator is, the arc *CDL* is less than a semicircle.

− If the circle *CKD* lies between the equator and the visible pole, it is closer to the equator than the circle *EI* is.

• If the sphere is right, the part of the circle *CDL* above the horizon is equal to a semicircle, so the arc *CDL* and *a fortiori* the arc *KDL* are less than a semicircle.

• If the sphere is oblique, the part of the circle *CDL* that is above the horizon is greater than a semicircle.

If we denote by *X* the arc of the circle *DKC* that is above the plane *ABC*, the arc *X* is greater than the arc similar to $2\widehat{EI}$; in fact, the arc is greater than the part of the circle *DKC* above the horizon and the part of the circle *EI* above the plane *ABC* is smaller than the part of that circle above the horizon; so we have $X > 2\ \widehat{DK}$.

We have seen that $X > 2\widehat{DK}$, so

$$X + \widehat{CK} > 2\widehat{DK} + \widehat{CK}$$

$$X + \widehat{CK} > \widehat{DK} + \widehat{CD};$$

now $\widehat{CD} = \widehat{DL}$, so

$$X + \widehat{CK} > \widehat{LK}$$

But $X + \widehat{CK} + \widehat{LK}$ is the complete circle, so $\widehat{LK}$ is smaller than a semicircle.

So, in all forms of the figure, we have $\widehat{KCL} < 1$ right angle and $KM \perp CG$ gives a point M between C and G. The straight line DC cuts KM in U. We draw through K the straight line $KM' \parallel DC$, we have $MM' > MC$, so $KM' > KC$, hence

$$\frac{KM'}{KM} > \frac{KC}{KM},$$

so

$$\frac{CU}{UM} > \frac{KC}{KM};$$

but

$$\frac{CU}{UM} = \frac{CD}{DG},$$

so

(α) $$\frac{CD}{DG} > \frac{KC}{KM}.$$

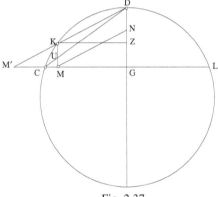

Fig. 2.37

Moreover, $DE > DS$ and $ND < PD$, so

$$\frac{PD}{DS} > \frac{ND}{DE}.$$

But

$$\frac{PD}{DS} = \frac{DG}{DB} \quad \text{and} \quad ND = MK,$$

(β) $\dfrac{DG}{DB} > \dfrac{MK}{DE}$.

From (α) and (β), we obtain

$$\frac{CD}{DB} > \frac{KC}{DE}$$

or again by permutation

$$\frac{CD}{CK} > \frac{DB}{DE} \quad \text{or} \quad \frac{CD}{CK} > \frac{DB}{KI} \qquad \text{(because } DE = KI\text{)}.$$

• If the sphere is *right*, we have $DG \perp AB$; but, by hypothesis, $\widehat{DB} \le \frac{1}{2}\widehat{ADB}$; consequently $AG \ge GB$, so $AG \ge \frac{1}{2}AB$. But $GC \le \frac{1}{2}AB$, so $AG \ge GC$, hence $AD \ge DC$; from which we obtain $D\hat{A}G \le D\hat{C}G$, hence $\widehat{DB}$ is less than or equal to the arc similar to $\widehat{DL}$. Now $\widehat{DL} = \widehat{DKC}$, so the arc similar to $\widehat{DKC}$ is at least equal to $\widehat{DB}$.

• If the sphere is *oblique*, if $DG' \perp AB$, then G' lies between G and B and $G'A \ge \frac{1}{2}AB$:

– If $\widehat{DB} = \frac{1}{2}\widehat{ADB}$, then $G'A = \frac{1}{2}AB$ and $GA < \frac{1}{2}AB$, so $GC < \frac{1}{2}AB$ and $GD > G'D$. Now we have

$$\tan D\hat{C}G = \frac{DG}{GC} \quad \text{and} \quad \tan D\hat{A}B = \frac{G'D}{G'A},$$

hence

$$D\hat{C}G > D\hat{A}B,$$

so $\widehat{DL}$ is greater than or equal to the arc similar to $\widehat{DB}$ or $\widehat{DKC}$ is greater than or equal to the arc similar to $\widehat{DB}$.

– If $\widehat{DB} < \frac{1}{2}\widehat{ADB}$, then $G'A > \frac{1}{2}AB$; but $GC \le \frac{1}{2}AB$, so $G'A > GC$ and $G'D < GD$, we also have $D\hat{C}G > D\hat{A}B$, hence $\widehat{DKC}$ is greater than or equal to the arc similar to $\widehat{DB}$.

So in all cases $\widehat{DKC}$ is greater than or equal to the arc similar to $\widehat{DB}$.

We have proved that $\dfrac{CD}{CK} > \dfrac{BD}{DE}$; so from Proposition 4 (for two different circles) we have:

$$\frac{\overset{\frown}{CKD}}{\overset{\frown}{CK}} > \frac{\overset{\frown}{BED}}{\overset{\frown}{DE}}.$$

By permutation, we have

$$\frac{\overset{\frown}{CKD}}{\overset{\frown}{BED}} > \frac{\overset{\frown}{CK}}{\overset{\frown}{DE}} = \frac{\overset{\frown}{CK}}{\overset{\frown}{KI}};$$

from which we obtain[8]

$$\frac{\overset{\frown}{CKD} - \overset{\frown}{CK}}{\overset{\frown}{BED} - \overset{\frown}{DE}} = \frac{\overset{\frown}{KD}}{\overset{\frown}{EB}} > \frac{\overset{\frown}{CKD}}{\overset{\frown}{BED}} > \frac{\overset{\frown}{CK}}{\overset{\frown}{DE}}.$$

If we take into account that $\overset{\frown}{KD}$ is similar to $\overset{\frown}{EI}$ and $\overset{\frown}{DE} = \overset{\frown}{KI}$, then

$$\frac{\overset{\frown}{EI}}{\overset{\frown}{EB}} > \frac{\overset{\frown}{CD}}{\overset{\frown}{BD}} > \frac{\overset{\frown}{CK}}{\overset{\frown}{KI}}.$$

Here Ibn al-Haytham replaces $\overset{\frown}{KD}$ by the similar arc $\overset{\frown}{EI}$; now, in the case under consideration, we have $\overset{\frown}{EI} < \overset{\frown}{KD}$ since we have assumed that the radius of the circle CKD is greater than that of the circle EI. Since the arcs EI and KD subtend the same angle in the different circles, it seems possible that Ibn al-Haytham was thinking in terms of angles while expressing the argument in terms of arcs, hence the confusion. This method cannot lead to a conclusion. We are left with the inequality

$$\frac{\overset{\frown}{CD}}{\overset{\frown}{BD}} > \frac{\overset{\frown}{CK}}{\overset{\frown}{KI}}.$$

Ibn al-Haytham points out that the proof is the same if the circle ABC is the circle of the horizon.

---

[8] $\dfrac{a}{b} > \dfrac{c}{d} \Leftrightarrow bc < ad \Leftrightarrow ab - bc > ab - ad \Leftrightarrow b(a - c) > a(b - d)$, so $\dfrac{a}{b} > \dfrac{c}{d}$ and

$b > d, \ a > c \Rightarrow \dfrac{a - c}{b - d} > \dfrac{a}{b} > \dfrac{c}{d}.$

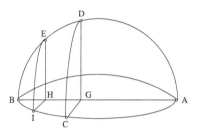

Fig. 2.38

In the special case where the sphere is right, the poles are the points A and B.

A general great circle passing through A and B cuts two circles parallel to the equator and does not cut the circle ABC. The latter circle passes through the points I and C and the point K is then identical with the point C. The arcs IE and CD are similar and by hypothesis we have $\widehat{EB} < \widehat{DB}$; so we have

$$\frac{\widehat{IE}}{\widehat{EB}} > \frac{\widehat{CD}}{\widehat{BD}}.$$

In fact

$$\frac{\widehat{EI}}{\widehat{CD}} = \frac{EH}{DG} = \frac{\sin \widehat{BE}}{\sin \widehat{BD}} > \frac{\widehat{BE}}{\widehat{BD}}.$$

B) *We suppose that the circle* (IE) *is equal to the circle* (CD).

The two circles are thus symmetrical with respect to the plane of the equator.

If, as before, we let X denote the arc of the circle (DC) that is below the plane ABC, we have $X >$ arc similar to $2\widehat{EI}$, so $X > 2\widehat{DK}$, whether the sphere is right or oblique. So we have

$$X + \widehat{CK} > \widehat{LK};$$

and consequently $K\hat{C}G < 1$ right angle and M lies between C and G.

The arcs EI and DK are equal, so $EI = DK$ and $EI \parallel DK$; we also have $MN = EI$ and $MN \parallel EI$. From this we obtain $NG = EH$. So a line through E parallel to AB passes through N, $P = N$, $PD = KM$.

We have $DE > DS$, because angle DSE is obtuse, so

$$\frac{PD}{DS} > \frac{PD}{DE}.$$

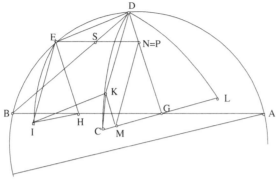

Fig. 2.39

But

$$\frac{PD}{DS} = \frac{DG}{DB}, \quad \frac{PD}{DE} = \frac{KM}{KI} \qquad \text{(because } DE = KI\text{)};$$

so we have

$$\frac{DG}{DB} > \frac{KM}{KI}.$$

But further

$$\frac{CD}{DG} > \frac{CK}{KM} \qquad \left(D\hat{C}G < K\hat{C}M\right),$$

so we have

$$\frac{CD}{DB} > \frac{CK}{KI}.$$

The proof is completed as before and we have

$$\frac{\overset{\frown}{EI}}{\overset{\frown}{EB}} > \frac{\overset{\frown}{CD}}{\overset{\frown}{BD}} > \frac{\overset{\frown}{CK}}{\overset{\frown}{KI}}.$$

Ibn al-Haytham considers two special cases:

• If (ABC) is a horizon and if the sphere is oblique, and we let D' designate the point on the circle CDL diametrically opposite D, then

$$\overset{\frown}{CD'} = \overset{\frown}{IE} = \overset{\frown}{KD}.$$

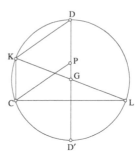

Fig. 2.40

So $KC \parallel DD'$, $K\hat{C}L = 1$ right angle; and $\widehat{LD'C} + \widehat{CK} = \widehat{LDK}$, we then have $C = M$, so $KC = PD$ and

$$\frac{KC}{KI} = \frac{PD}{KI} = \frac{PD}{DE}.$$

But

$$\frac{PD}{DE} < \frac{DG}{DB} < \frac{CD}{DB},$$

so

$$\frac{CD}{DB} > \frac{KC}{KI},$$

and we complete the proof as before.

• If (ABC) is a horizon and if the sphere is right, we have $\widehat{IE} = \widehat{DC}$ and $\widehat{EB} < \widehat{DB}$, so

$$\frac{\widehat{EI}}{\widehat{EB}} > \frac{\widehat{DC}}{\widehat{DB}}.$$

In this case the circle $\Pi\Pi'$ passes through C ($K = C$).

C) *We suppose that the circle* (EI) *is greater than the circle* (CD), and the sphere is inclined towards B. This is possible in three cases:
   • The circle (EI) is the circle of the equator.
   • The two circles lie on opposite sides of the equator with (EI) closer to the equator than (CD).
   • The two circles lie between the equator and the visible pole.

In all three cases we assume as before that $\overset{\frown}{DB} \leq \frac{1}{2} \overset{\frown}{ADB}$.

The straight line DB cuts EH in J; we suppose $\dfrac{EH}{HJ} \geq \dfrac{d_1}{d_2}$, d_1 and d_2 being the diameters of the circles (EI) and (CD) respectively $(d_1 > d_2)$.

• If $\overset{\frown}{LDC} \leq \frac{1}{2}$ circle, then $\overset{\frown}{LDK} < \frac{1}{2}$ circle and we arrive at the conclusion

$$\frac{\overset{\frown}{EI}}{\overset{\frown}{EB}} > \frac{\overset{\frown}{CD}}{\overset{\frown}{BD}} > \frac{\overset{\frown}{CK}}{\overset{\frown}{KI}}.$$

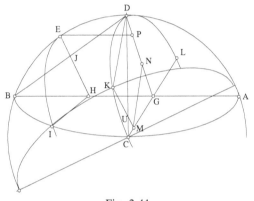

Fig. 2.41

In fact we are returning to the conditions on page 113: $K\hat{C}L$ acute, M between C and G, U between K and M; we have

$$\frac{CU}{UM} > \frac{CK}{KM}$$

with $\dfrac{CU}{UM} = \dfrac{CD}{DG}$ and $KM = ND$, hence

$$\frac{CD}{DG} > \frac{CK}{KM} = \frac{CK}{ND}.$$

• If $\overset{\frown}{LDC} > \frac{1}{2}$ circle, if X designates the part of the circle below ABC, $X < \frac{1}{2}$ circle, if we draw $CS \parallel DG$, we have $\overset{\frown}{DS} = \frac{1}{2}X$ (it is symmetrical with the half of arc X with respect to the centre of the circle CDL).

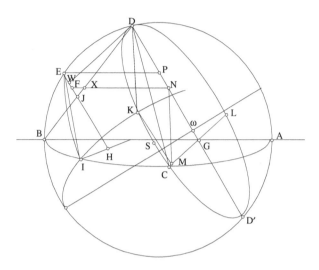

Fig. 2.42

We suppose $\widehat{IE} \leq$ arc similar to X, that is, $\widehat{IE} \leq$ arc similar to $2\widehat{DS}$.[9]

Now $\widehat{IE}$ and $\widehat{DK}$ are similar arcs. So we suppose $\widehat{DK} \leq 2\widehat{DS}$. There are three possible cases:

$$\widehat{EI} < \text{arc similar to } \widehat{DS} \quad \Rightarrow \quad \widehat{DK} < \widehat{DS} \qquad \text{(a)}$$
$$\widehat{EI} = \text{arc similar to } \widehat{DS} \quad \Rightarrow \quad \widehat{DK} = \widehat{DS} \qquad \text{(b)}$$
$$\widehat{EI} > \text{arc similar to } \widehat{DS} \quad \Rightarrow \quad \widehat{DK} > \widehat{DS} \qquad \text{(c)}$$

a) $\widehat{DK} < \widehat{DS}$, so K lies between S and D. $K\hat{C}G$ is acute and M lies between G and C. We have

$$\frac{CD}{DG} > \frac{CK}{KM} \text{ and } KM = ND,$$

hence

$$\frac{CD}{DG} > \frac{CK}{ND}.$$

[9] See Supplementary note [4].

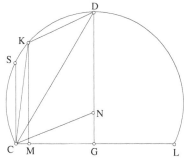

Fig. 2.43

b) $\overset{\frown}{DK} = \overset{\frown}{DS}$, $K = S$, $M = C$, $ND = KM = CK$,

$$\frac{CK}{KM} = \frac{CK}{ND} = 1, \quad \frac{CD}{DG} > \frac{CK}{KM} \text{ or } \frac{CD}{DG} > \frac{CK}{ND}.$$

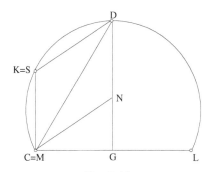

Fig. 2.44

c) $\overset{\frown}{DK} > \overset{\frown}{DS}$, K lies between S and C. But $\overset{\frown}{EI}$ is similar to the arc $\overset{\frown}{DK} \leq 2 \overset{\frown}{DS}$, so $\overset{\frown}{SK} \leq \overset{\frown}{SD}$, $S\hat{C}K \leq S\hat{C}D$, hence $S\hat{C}K \leq C\hat{D}G$.

We draw $KM \perp CG$, then M lies beyond C; we have $MG > CG$ and $KM < GD$. We draw $MN \parallel KD$, we have $KM = ND$ and $M\hat{K}C = K\hat{C}S \leq C\hat{D}G$.

If $K\hat{C}S = C\hat{D}G$, then $\dfrac{CD}{DG} = \dfrac{CK}{KM}$.

If $K\hat{C}S < C\hat{D}G$, then $\dfrac{CD}{DG} > \dfrac{CK}{KM}$.

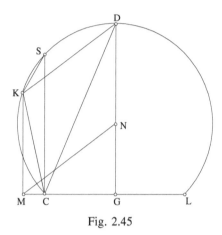

Fig. 2.45

So we have

$$\frac{CD}{DG} \geq \frac{CK}{ND}.$$

Conclusion in all three cases a, b, c: we have

$$CK = ND \ \text{ or } \ \frac{CD}{DG} \geq \frac{CK}{ND}.$$

Arcs DK and EI are similar and the circle EI is greater than the circle DK, so $DK < EI$ and

$$\frac{EI}{DK} = \frac{d_1}{d_2}.$$

We have $MN = DK$, and consequently $MN < EI$. Triangles MNG and EIH are similar, so $NG < EH$.

Moreover $EP \parallel AB$, hence $PG = EH$, so we have $NG < PG$ and consequently $ND > PD$. But

$$\frac{EI}{DK} = \frac{EI}{MN} = \frac{EH}{NG} = \frac{d_1}{d_2};$$

now by hypothesis $\dfrac{EH}{HJ} \geq \dfrac{d_1}{d_2}$, so

$$\frac{EH}{HJ} \geq \frac{EH}{NG},$$

hence $HJ \leq NG$.

α) If $HJ = NG$, then $NJ \parallel GH$ and $\dfrac{ND}{DJ} = \dfrac{DG}{DB}$.

In that case

$$\frac{EH}{HJ} = \frac{d_1}{d_2},$$

hence

$$\frac{EH}{EJ} = \frac{d_1}{d_1 - d_2}.$$

β) If $NG > HJ$, we draw $NF \parallel GH$; we have $NG = HF > HJ$; the straight line NF cuts DJ in the point X between J and D. We have

$$\frac{ND}{DX} = \frac{GD}{DB} \quad \text{and} \quad \frac{EH}{HF} = \frac{d_1}{d_2} \quad \text{and} \quad \frac{EH}{EF} = \frac{d_1}{d_1 - d_2}.$$

We draw $DW \perp EH$, then $d_1 - d_2 = 2EW$ (because DW is parallel to the line through the poles, that is the axis of the circles EI and DC).

In case α, we have $EJ < 2EW$ (because $EH < d_1$).

• If $EH = \dfrac{1}{2}d_1$, then $EJ = EW$, $J = W$, then we have $DJ \perp EH$, so $DJ < DE$.

• If $EH < \dfrac{1}{2}d_1$, then $EJ < EW$, so $D\hat{J}E$ is obtuse and we again have $DJ < DE$.

• If $EH > \dfrac{1}{2}d_1$, then $EJ > EW$, $D\hat{J}E$ is acute; but $EJ < 2EW$, so $WJ < WE$ and $DJ < DE$, so in all cases $DJ < DE$.

Then we have

$$\frac{ND}{DJ} > \frac{ND}{DE};$$

but

$$\frac{ND}{DJ} = \frac{DG}{DB},$$

so

$$\frac{GD}{DB} > \frac{ND}{DE}.$$

In case β, $EH < d_1$ implies $EF < 2EW$, so for EF (as for EJ in case α) there are three possibilities:

$$EF = EW, \ EF < EW, \ 2EW > EF > EW;$$

and in all three cases we can show that $DE > DF$; but $D\hat{X}F$ is obtuse, so $DF > DX$, so we have $DE > DX$ and consequently

$$\frac{ND}{DX} > \frac{ND}{DE}.$$

Now

$$\frac{ND}{DX} = \frac{GD}{DB},$$

so we have

$$\frac{GD}{DB} > \frac{ND}{DE}.$$

So in part C, we have proved that, if $d_1 > d_2$ and $\frac{EH}{HJ} \geq \frac{d_1}{d_2}$, then $CK = ND$ or $\frac{CK}{ND} \leq \frac{CD}{DG}$ and $\frac{ND}{DE} < \frac{GD}{DB}$.

But $DE = KI$, so
- If $CK = ND$, we have $\frac{ND}{DE} = \frac{CK}{KI}$, hence $\frac{GD}{DB} > \frac{CK}{KI}$; now $CD > DG$, so we have $\frac{CD}{DB} > \frac{CK}{KI}$.

- If $\frac{CK}{ND} \leq \frac{CD}{DG}$ and $\frac{ND}{DE} < \frac{GD}{DB}$, then $\frac{CK}{DE} < \frac{CD}{DB}$, so $\frac{CD}{DB} > \frac{CK}{KI}$.

So in all cases we have obtained

$$\frac{CD}{CK} > \frac{DB}{DE},$$

hence

$$\frac{\overline{CD}}{\overline{CK}} > \frac{\overline{DB}}{\overline{DE}} \quad \text{or} \quad \frac{\overline{CD}}{\overline{DB}} > \frac{\overline{CK}}{\overline{DE}};$$

but $\overset{\frown}{DE} = \overset{\frown}{KI}$, hence

$$\frac{\overset{\frown}{CD}}{\overset{\frown}{DB}} > \frac{\overset{\frown}{CK}}{\overset{\frown}{KI}};$$

therefore

$$\frac{\overset{\frown}{CD} - \overset{\frown}{CK}}{\overset{\frown}{BD} - \overset{\frown}{DE}} > \frac{\overset{\frown}{CD}}{\overset{\frown}{DB}}, \quad [10]$$

or

$$\frac{\overset{\frown}{KD}}{\overset{\frown}{EB}} > \frac{\overset{\frown}{CD}}{\overset{\frown}{DB}};$$

but $\overset{\frown}{KD} < \overset{\frown}{EI}$, hence

$$\frac{\overset{\frown}{EI}}{\overset{\frown}{EB}} > \frac{\overset{\frown}{CD}}{\overset{\frown}{DB}} > \frac{\overset{\frown}{CK}}{\overset{\frown}{KI}}.$$

Commentary:

In this proposition Ibn al-Haytham also examines the way a ratio varies in more difficult cases. He is concerned to prove

a) that the ratio $\dfrac{\overset{\frown}{EI}}{\overset{\frown}{EB}}$ decreases when E moves from B towards F (the midpoint of the arc AB) along the meridian circle AEB;

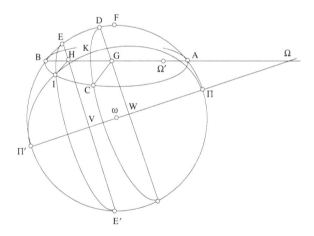

Fig. 2.46

[10] See above, footnote 8.

b) that $\dfrac{\overline{DC}}{\overline{DB}} \ge \dfrac{\overline{CK}}{\overline{KI}}$, where K is the point of intersection of the circle DKC (like EI parallel to the equator) and the meridian circle ΠKI that passes through I (Fig. 2.46).

Putting $E\hat{V}I = \psi$ and $E\hat{\omega}F = \theta$ (measured in the opposite sense to α and β), we have $\Pi\hat{\omega}E = \alpha - \theta$, hence

$$(1) \qquad \cos\psi = \frac{\overline{VH}}{\overline{EV}} = \frac{z}{r\sin(\alpha - \theta)} = \frac{\cos\beta - \cos\alpha\cos(\alpha - \theta)}{\sin\alpha\sin(\alpha - \theta)}$$

where z is the height of H (α the colatitude of F, $\beta = A\hat{\omega}F$, see commentary on Propositions 11 and 12).

If $\alpha = \dfrac{\pi}{2}$, $\cos\psi = \dfrac{\cos\beta}{\cos\theta}$; if $\beta = \dfrac{\pi}{2}$, $\cos\psi = -\dfrac{1}{\tan\alpha\tan(\alpha - \theta)}$.

We have $-\beta \le \theta \le \inf(\beta, 2\alpha - \beta)$, or $|\alpha - \beta| \le \alpha - \theta \le \alpha + \beta$, so there exists on the sphere of unit radius a triangle with sides α, β and $\alpha - \theta$; equation (1) means that ψ is the angle opposite β in the triangle $AB\Gamma$ (see Fig. 2.47).

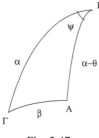

Fig. 2.47

Since $\overset{\frown}{EI} = r\,\psi\sin(\alpha - \theta)$ and $\overset{\frown}{EB} = r(\beta + \theta)$, we have

$$(2) \qquad \frac{\overset{\frown}{EI}}{\overset{\frown}{EB}} = \frac{\psi}{\beta + \theta}\sin(\alpha - \theta) = \xi.$$

Moreover, if $\Psi = D\hat{W}C$ and $\Theta = D\hat{\omega}F$, in the same way we have

$$\widehat{DC} = r\,\Psi \sin(\alpha - \Theta), \ \widehat{DK} = r\,\psi \sin(\alpha - \Theta),$$

hence

$$\widehat{CK} = r(\Psi - \psi)\sin(\alpha - \Theta); \ \ \widehat{KI} = \widehat{DE} = r(\Theta - \theta),$$

so that

$$\frac{\widehat{CK}}{\widehat{KI}} = \frac{\Psi - \psi}{\Theta - \theta}\sin(\alpha - \Theta) \ \text{and} \ \frac{\widehat{DC}}{\widehat{DB}} = \frac{\Psi}{\beta + \Theta}\sin(\alpha - \Theta).$$

So inequality b) can be written

$$\frac{\Psi}{\beta + \Theta} \geq \frac{\Psi - \psi}{\Theta - \theta},$$

which is equivalent to

$$\frac{\Psi}{\beta + \Theta} \leq \frac{\psi}{\beta + \theta},$$

because $\dfrac{a}{b} \geq \dfrac{a-c}{b-d}$ can be written $ab - ad \geq ba - bc$, or $bc \geq ad$ or again $\dfrac{a}{b} \leq \dfrac{c}{d}$. If we put

$$(3) \qquad\qquad \eta = \frac{\psi}{\beta + \theta}$$

where ψ is defined by equation (1), then the statement b) means that η is a decreasing function of θ.

Notes:

1) Ibn al-Haytham considers ratios of arcs of circles (possibly with different radii) rather than ratios of angles. So he cannot formulate b) in terms of the decrease of η. Using his notation, the transformation that is carried out leads to

$$\frac{\widehat{DC}}{\widehat{DB}} \leq \frac{\widehat{DK}}{\widehat{BE}}.$$

2) We have $\xi = \eta\sin(\alpha - \theta)$; if $\alpha + \beta \leq \dfrac{\pi}{2}$, $\sin(\alpha - \theta)$ is a decreasing function of θ, so a) follows from b). However, if $\alpha + \beta > \dfrac{\pi}{2}$, $\sin(\alpha - \theta)$

increases for $-\beta \le \theta \le \alpha - \dfrac{\pi}{2}$, then decreases for $\alpha - \dfrac{\pi}{2} \le \theta \le \inf(\beta, 2\alpha - \beta)$; then b) follows from a) in the first interval and a) follows from b) in the second.

First part: investigation of angle ψ

Let us consider a fixed arc $B\Gamma = \alpha$ on the unit sphere; A, the third vertex of the triangle $AB\Gamma$, is the point of intersection of the circle with centre Γ and radius β and the circle with centre B and radius $\alpha - \theta$, which varies within the interval $[|\alpha - \beta|, \alpha + \beta]$. We need to distinguish two cases, according to whether $\alpha \ge \beta$ or $\alpha < \beta$; in the first case, B lies outside the circle γ with centre Γ and radius β (or on the circle if $\alpha = \beta$), whereas in the second case B lies inside the circle.

1) For $\alpha \ge \beta$: there exists a great circle through B and touching the circle γ at a point A_m. When θ increases from $-\beta$ to β, $\alpha - \theta$ decreases from $\alpha + \beta$ to $\alpha - \beta$ and the angle $\psi = \Gamma \hat{B} A$ increases from 0 to $\psi_m = \Gamma \hat{B} A_m$ (between 0 and $\dfrac{\pi}{2}$), then decreases from ψ_m to 0. Since $B \hat{A}_m \Gamma = \dfrac{\pi}{2}$, we have, from spherical trigonometry: $\dfrac{\sin \psi_m}{\sin \beta} = \dfrac{1}{\sin \alpha}$, or $\sin \psi_m = \dfrac{\sin \beta}{\sin \alpha}$ (see Fig. 2.48) and this value of ψ is reached when $\cos(\alpha - \theta) = \dfrac{\cos \alpha}{\cos \beta} = \cos \mu$, or $\theta = \alpha - \mu$ ($0 \le \mu < \alpha$). If $\alpha = \beta$, $\mu = 0$ and $\psi_m = \dfrac{\pi}{2}$; otherwise, $\alpha - \mu < \beta$ and $\psi_m < \dfrac{\pi}{2}$.

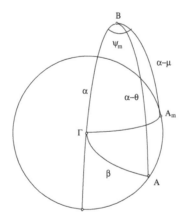

Fig. 2.48

For a given value of $\psi \in [0, \psi_m[$, there are two possible values $\theta_+(\psi)$ and $\theta_-(\psi)$, with $\theta_-(\psi) < \alpha - \mu < \theta_+(\psi)$.

2) For $\alpha < \beta$. When θ increases from $-\beta$ to $2\alpha - \beta$, $\alpha - \theta$ decreases from $\alpha + \beta$ to $\beta - \alpha$ and ψ increases from 0 to π.

For a given value of $\psi \in [0, \pi]$, only one value $\theta_-(\psi) \in [-\beta, 2\alpha - \beta]$ is possible, because $\alpha - \theta \geq 0$ (see Fig. 2.49).

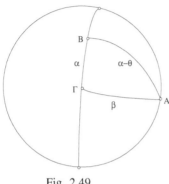

Fig. 2.49

Notes:

1. The abscissa of Ω measured along BA from its midpoint is $r \cos \beta \tan \alpha$; the abscissa is greater than $\frac{1}{2} BA = r \sin \alpha$ in the first case ($\alpha \geq \beta$), which means that Ω lies outside the sphere. $\frac{r}{\sin \alpha}(\cos(\alpha - \theta) - \cos \alpha \cos \beta)$, the abscissa of H, becomes $\frac{r \sin^2 \beta}{\tan \alpha \cos \beta}$ when $\theta = \alpha - \mu$; the corresponding position of H, say Ω', is thus the harmonic conjugate of Ω with respect to A and B (Fig. 2.46).

2. In the second case ($\alpha < \beta$), ψ passes through the value $\frac{\pi}{2}$ when $\theta = \alpha - \mu'$, where $\cos \mu' = \frac{\cos \beta}{\cos \alpha}$, that is, when H is at Ω; we have $\mu' \leq \alpha$ if $\cos \beta \geq \cos^2 \alpha$, that is, if $\sin \alpha \geq \sqrt{2} \sin \frac{\beta}{2}$.

For what follows, we need to investigate the convexity of ψ as a function of θ. Differentiating (1), we find

$$(4) \quad \psi' \sin \psi = \frac{\cos \alpha - \cos \beta \cos(\alpha - \theta)}{\sin \alpha \sin^2(\alpha - \theta)},$$

where $\psi' = \dfrac{d\psi}{d\theta}$.

Notes:

1. If $\alpha = \beta$ (Ω at A), we can simplify (4) by $2\sin^2 \dfrac{\alpha - \theta}{2}$ and we then obtain $\psi' \sin \psi = \dfrac{1}{2\tan\alpha \, \cos^2 \dfrac{\alpha - \theta}{2}}$.

2. For $\theta = -\beta$, $\psi = 0$ and $\psi' \sin \psi = \dfrac{\sin \beta}{\sin \alpha \sin(\alpha + \beta)}$, hence $\psi' = +\infty$. For $\theta = \inf(\beta, 2\alpha - \beta)$, $\alpha \neq \beta$, we have $\psi = 0$ or π and $\psi' \sin \psi = \dfrac{\sin \beta}{\sin \alpha \sin(\beta - \alpha)}$, hence $\psi' = \operatorname{sgn}(\beta - \alpha) \cdot \infty$. For $\alpha = \beta$ and $\theta = \alpha$, $\psi = \dfrac{\pi}{2}$ and $\psi' = \dfrac{1}{2\tan\alpha}$. If $\theta = -\beta + u$, $\psi^2 \approx \dfrac{2u \sin \beta}{\sin \alpha \sin(\alpha + \beta)}$ when u tends to 0. In the same way, if $\theta = \inf(\beta, 2\alpha - \beta) - u$ $(\alpha \neq \beta)$ and $\psi = v$, respectively $\pi - v$ (depending on whether $\alpha > \beta$ or $\alpha < \beta$), we find that $v^2 \approx \dfrac{2u \sin \beta}{\sin \alpha \sin|\alpha - \beta|}$.

3. For $\theta = \alpha - \mu'$, we find that $\psi' \sin \psi = \dfrac{1}{\tan\alpha}$.

By differentiating (4), we obtain

$$(5) \quad \psi'' \sin \psi + \psi'^2 \cos \psi = \frac{2\cos\alpha \, \cos(\alpha - \theta) - \cos \beta \left(1 + \cos^2(\alpha - \theta)\right)}{\sin \alpha \sin^3(\alpha - \theta)}.$$

Let us put $A = \cos \alpha$, $B = \cos \beta$ and $X = \cos(\alpha - \theta)$; we have:

$$(1') \quad \cos \psi = \frac{B - AX}{\sin \alpha \sin(\alpha - \theta)}$$

(4') $\psi' \sin \psi = \dfrac{A - BX}{\sin \alpha \, \sin^2(\alpha - \theta)}$

(5') $\left(\psi'' \sin \psi + \psi'^2 \cos \psi\right) \sin \alpha \sin^3(\alpha - \theta) = 2AX - B\left(1 + X^2\right).$

From (1') we get

$$\frac{\cos \psi}{\sin^2 \psi} = \frac{B - AX}{\left(1 - A^2\right)\left(1 - X^2\right) - (B - AX)^2} \sin \alpha \, \sin(\alpha - \theta)$$

$$= \frac{B - AX}{1 - A^2 - B^2 - X^2 + 2ABX} \sin \alpha \, \sin(\alpha - \theta)$$

and (4') then gives

$$\psi'^2 \cos \psi \, \sin \alpha \, \sin^3(\alpha - \theta) = \frac{(A - BX)^2 (B - AX)}{1 - A^2 - B^2 - X^2 + 2ABX}.$$

Finally

(6) $\psi'' \sin \psi \, \sin \alpha \, \sin^3(\alpha - \theta) =$

$$2AX - B\left(1 + X^2\right) - \frac{(A - BX)^2 (B - AX)}{1 - A^2 - B^2 - X^2 + 2ABX} = \frac{P(X)}{Q(X)}$$

with

(6') $P(X) = BX^4 - A\left(B^2 + 2\right)X^3 + 3A^2BX^2 - A\left(A^2 + 2B^2 - 2\right)X + B^3 - B$

and

$$Q(X) = 1 - A^2 - B^2 - X^2 + 2ABX \geq 0.$$

The sign of ψ'' is that of $P(X)$, which we shall examine for cases where $\cos(\alpha + \beta) \leq X \leq \cos(\alpha - \beta)$. We have

$$P(\cos(\alpha \pm \beta)) = \cos \beta \cos^4(\alpha \pm \beta) - \cos \alpha \cos^2 \beta \cos^3(\alpha \pm \beta) -$$
$$2\cos \alpha \cos^3(\alpha \pm \beta) + 3\cos^2 \alpha \cos \beta \cos^2(\alpha \pm \beta) +$$
$$2\cos \alpha \sin^2 \beta \cos(\alpha \pm \beta) - \cos^3 \alpha \cos(\alpha \pm \beta) - \cos \beta \sin^2 \beta =$$
$$-\sin^2 \beta \sin \alpha \sin^3(\alpha \pm \beta);$$

thus $P(\cos(\alpha + \beta)) < 0$ and $P(\cos(\alpha - \beta))$ has the sign of $\beta - \alpha$ (zero if $\alpha = \beta$).

Calculation gives us

$$P'(X) = 4BX^3 - 3A\left(B^2 + 2\right)X^2 + 6A^2 BX - A\left(A^2 + 2B^2 - 2\right)$$

and

$$P''(X) = 6\left(2BX^2 - A\left(B^2 + 2\right)X + A^2 B\right) = 6(2X - AB)(BX - A);$$

$$P'(\cos(\alpha \pm \beta)) = 4\cos\beta\cos^3(\alpha \pm \beta) - 3\cos\alpha\cos^2\beta\cos^2(\alpha \pm \beta) - $$
$$6\cos\alpha\cos^2(\alpha \pm \beta) + 6\cos^2\alpha\cos\beta\cos(\alpha \pm \beta) + 2\cos\alpha\sin^2\beta - \cos^3\alpha = $$
$$\frac{5}{2}\sin^2\beta\sin(\alpha \pm \beta)\left(\sin(2\alpha \pm \beta) \mp \frac{3}{5}\sin\beta\right).$$

Thus $P'(\cos(\alpha + \beta)) > 0$ is equivalent to $\sin(2\alpha + \beta) > \frac{3}{5}\sin\beta$. If $\alpha > \beta$, $P'(\cos(\alpha - \beta)) > 0$ and if $\alpha = \beta$, $P'(1) = 0$; if $\alpha < \beta$, $P'(\cos(\alpha - \beta)) > 0$ is equivalent to $\sin(\beta - 2\alpha) > \frac{3}{5}\sin\beta$.

Note: $\sin(\beta + 2\alpha) - \sin(\beta - 2\alpha) = 2\cos\beta\sin 2\alpha \geq 0$, so $\sin(\beta - 2\alpha) > \frac{3}{5}\sin\beta$ implies $\sin(\beta + 2\alpha) > \frac{3}{5}\sin\beta$.

The second derivative $P''(X)$ is negative for $\frac{AB}{2} \leq X \leq \frac{A}{B}$, positive elsewhere. We have

$$P'\left(\frac{AB}{2}\right) = \frac{A^3 B^4}{2} - \frac{3A^3 B^4}{4} + \frac{3}{2}A^3 B^2 - 2AB^2 - A^3 + 2A$$

$$= \frac{A}{4}\left(A^2 B^2\left(2 - B^2\right) + 4\left(2 - A^2\right)\left(1 - B^2\right)\right) > 0$$

and

$$P'\left(\frac{A}{B}\right) = \frac{4A^3}{B^2} - 3A^3 - \frac{6A^3}{B^2} + 6A^3 - 2AB^2 - A^3 + 2A = \frac{2A}{B^2}\left(1 - B^2\right)\left(B^2 - A^2\right),$$

with the sign of $\alpha - \beta$ (zero if $\alpha = \beta$). We may note that

$$\frac{AB}{2} = \frac{1}{2}\cos\alpha\cos\beta < \cos(\alpha - \beta) \text{ and } \frac{A}{B} = \frac{\cos\alpha}{\cos\beta} > \cos(\alpha + \beta)$$

since $\cos\alpha - \cos\beta\cos(\alpha + \beta) = \sin\alpha\sin(\alpha + \beta) > 0$.

The inequality

$$\frac{AB}{2} \geq \cos(\alpha + \beta) \text{ (respectively } \frac{A}{B} \leq \cos(\alpha - \beta))$$

can be written

$$\cos\alpha\cos\beta \geq 2\cos(\alpha + \beta),$$

or $3\cos(\alpha + \beta) \leq \cos(\alpha - \beta)$ (respectively $\cos\alpha \leq \cos\beta\cos(\alpha - \beta)$), or $\sin\beta\sin(\alpha - \beta) \geq 0$ where again $\alpha \geq \beta$.

Note: If $\alpha = \frac{\pi}{2}$, $\frac{AB}{2} = \frac{A}{B} = 0$ and elsewhere $P''(X) \geq 0$. If $\alpha = \beta$, $\frac{A}{B} = 1 = \cos(\alpha - \beta)$ and $\frac{AB}{2} \geq \cos(\alpha + \beta) = \cos 2\alpha$ is equivalent to $\cos 2\alpha \leq \frac{1}{3}$, or $\alpha \geq 0.615479709$ (a little less than $\frac{\pi}{5}$). If $\beta = \frac{\pi}{2}$, $\frac{AB}{2} = 0$ and $\frac{A}{B} = +\infty$.

There are four cases:

1) $\alpha \geq \beta$ and $3\cos(\alpha + \beta) \leq \cos(\alpha - \beta)$; then

$$\cos(\alpha + \beta) \leq \frac{AB}{2} < \frac{A}{B} \leq \cos(\alpha - \beta).$$

We have $\frac{A}{B} = \cos\mu$ and we put $\cos\nu = \frac{AB}{2} = \frac{1}{2}\cos\alpha\cos\beta$; we see that $P''(X) \leq 0$ for $\alpha - \nu \leq \theta \leq \alpha - \mu$ and $P''(X) > 0$ elsewhere. We may note that $\nu > \alpha$.

2) $\alpha \geq \beta$ and $3\cos(\alpha + \beta) > \cos(\alpha - \beta)$, then

$$\frac{AB}{2} < \cos(\alpha + \beta) < \frac{A}{B} \leq \cos(\alpha - \beta)$$

and $P''(X) \leq 0$ for $-\beta \leq \theta \leq \alpha - \mu$, that is, H lying between B and Ω'.

3) $\alpha < \beta$ and $3\cos(\alpha + \beta) \leq \cos(\alpha - \beta)$; then

$$\cos(\alpha + \beta) \le \frac{AB}{2} < \cos(\alpha - \beta) < \frac{A}{B}$$

and $P''(X) \le 0$ for $\alpha - v \le \theta \le 2\alpha - \beta$.

4) $\alpha < \beta$ and $3\cos(\alpha + \beta) > \cos(\alpha - \beta)$; then

$$\frac{AB}{2} < \cos(\alpha + \beta) < \cos(\alpha - \beta) < \frac{A}{B}$$

and $P''(X)$ always stays negative.

The tables of variation of $P'(X)$ are as follows, for the four cases:

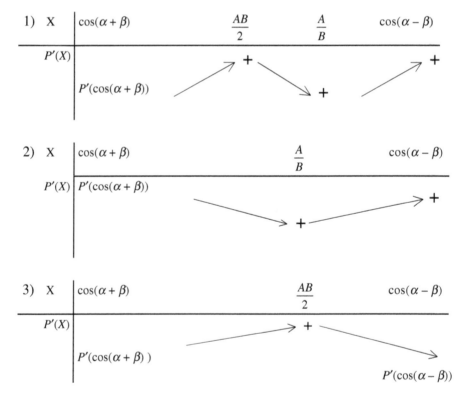

4) X $\cos(\alpha + \beta)$ $\cos(\alpha - \beta)$

$P'(X)$

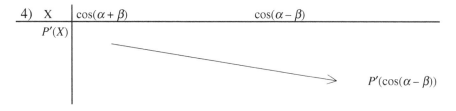

$P'(\cos(\alpha - \beta))$

In case 1, $P'(X)$ stays positive if $P'(\cos(\alpha + \beta)) \geq 0$, that is, if $\sin(2\alpha + \beta) \geq \frac{3}{5}\sin\beta$ and it goes from negative to positive for a value $X_1 = \cos(\alpha - \theta_1) \in \;]\cos(\alpha + \beta), \frac{1}{2}\cos\alpha\cos\beta[$, or $-\beta < \theta_1 < \alpha - v$ if $\sin(2\alpha + \beta) < \frac{3}{5}\sin\beta$.

Note: If $\alpha = \beta$, the inequality $\sin 3\alpha < \frac{3}{5}\sin\alpha$ is equivalent to $\sin\alpha > \sqrt{\frac{3}{5}}$, or $\alpha > 0.886077124$ (between $\frac{\pi}{4}$ and $\frac{\pi}{3}$).

In case 2, $P'(X)$ always stays positive. In case 3, $P'(X)$ stays positive if $\sin(\beta - 2\alpha) \geq \frac{3}{5}\sin\beta$; if $\sin(\beta - 2\alpha) < \frac{3}{5}\sin\beta \leq \sin(\beta + 2\alpha)$, $P'(\cos(\alpha + \beta)) \geq 0$ and $P'(\cos(\alpha - \beta)) < 0$, so $P'(X)$ goes from positive to negative for a value $X_2 = \cos(\alpha - \theta_2)$, $\alpha - v < \theta_2 < 2\alpha - \beta$. Finally, if $\sin(\beta + 2\alpha) < \frac{3}{5}\sin\beta$, $P'(\cos(\alpha \pm \beta)) < 0$, so $P'(X)$ goes from negative to positive at $X_1 = \cos(\alpha - \theta_1)$, $-\beta < \theta_1 < \alpha - v$, then from positive to negative at $X_2 = \cos(\alpha - \theta_2)$, $a - v < \theta_2 < 2\alpha - \beta$.

In case 4, $P'(X)$ stays positive if $\sin(\beta - 2\alpha) \geq \frac{3}{5}\sin\beta$, as in case 3; if $\sin(\beta - 2\alpha) < \frac{3}{5}\sin\beta \leq \sin(\beta + 2\alpha)$, $P'(X)$ goes from positive to negative at $X_2 = \cos(\alpha - \theta_2)$, $\alpha - v < \theta_2 < 2\alpha - \beta$. Finally, if $\sin(\beta + 2\alpha) < \frac{3}{5}\sin\beta$, $P'(X)$ stays negative; but we can check that this does not occur, because it is incompatible with $3\cos(a + \beta) > \cos(\alpha - \beta)$.

In cases 1 and 2, which correspond to $\alpha \geq \beta$, we can see that $P(X)$ increases up to $P(\cos(\alpha - \beta)) \leq 0$, so $P(X)$ stays negative and ψ is a *concave* function of θ.

In cases 3 and 4, corresponding to $\alpha < \beta$, $P(X)$ increases from $P(\cos(\alpha + \beta)) < 0$ to $P(\cos(\alpha - \beta)) > 0$ if $\sin(\beta - 2\alpha) \geq \frac{3}{5}\sin\beta$; so it goes from negative to positive for a value $X_0 = \cos(\alpha - \theta_0)$. Since

$$P\left(\frac{AB}{2}\right) = \frac{A^4 B^5}{16} - \frac{A^4 B^3}{8}\left(B^2 + 2\right) + \frac{3A^4 B^3}{4} - \frac{A^2 B}{2}\left(A^2 + 2B^2 - 2\right) + B^3 - B$$

$$= -\frac{A^4 B^5}{16} - \frac{B}{2}\left(1 - B^2\right)\left(\left(1 - A^2\right)^2 + 1\right) \leq 0,$$

we can see that $\alpha - v \leq \theta_0 < 2\alpha - \beta$. If $\sin(\beta - 2\alpha) < \frac{3}{5}\sin\beta \leq \sin(\beta + 2\alpha)$, $P(X)$ decreases to $P(\cos(\alpha - \beta)) > 0$ after an interval, $-\beta \leq \theta \leq \theta_2$, in which it increases; thus $P(X)$ goes from negative to positive at a value $X_0 = \cos(\alpha - \theta_0)$ as under the previous conditions. We have $\alpha - v \leq \theta_0 \leq \theta_2$. Finally, if $\sin(\beta + 2\alpha) < \frac{3}{5}\sin\beta$, $P(X)$ increases only for $X_1 \leq X \leq X_2$ and goes from negative to positive at $X_0 = \cos(\alpha - \theta_0)$ where $\alpha - v \leq \theta_0 \leq \theta_2$, as above.

To summarize, ψ is concave if $\alpha \geq \beta$; however, if $\alpha < \beta$, ψ is concave for $-\beta \leq \theta \leq \theta_0$ and convex for $\theta_0 \leq \theta \leq 2\alpha - \beta$. The angle θ_0 is determined from the equation $P(\cos(\alpha - \theta_0)) = 0$. We have

$$\cos\mu' = \frac{\cos\beta}{\cos\alpha} = \frac{B}{A} \quad \text{and} \quad P\left(\frac{B}{A}\right) = \frac{B(1 - A^2)(B^2 - A)^2}{A^4} \geq 0;$$

this shows that $\cos\mu' \geq \cos(\alpha - \theta_0)$, or $\theta_0 \leq \alpha - \mu'$.

Note: For $\theta = \alpha - \frac{\pi}{2}$, $X = \cos(\alpha - \theta) = 0$ and $P(X) = B^3 - B \leq 0$, which proves that $\theta_0 \geq \alpha - \frac{\pi}{2}$.

Thus the point of inflexion in the graph of ψ occurs before ψ goes through the value $\frac{\pi}{2}$. If $\beta = \frac{\pi}{2}$, $\theta_0 = \alpha - \frac{\pi}{2}$.

If $\alpha \leq \frac{\beta}{2}$, $\theta_0 < 2\alpha - \beta \leq 0$. If $\alpha > \frac{\beta}{2}$, the inequality $\theta_0 \geq 0$ is equivalent to $P(\cos\alpha) \leq 0$, or

(7) $(3 - B) A^4 - 2(1 + B) A^2 + B(1 + B) \geq 0,$

because
$$P(\cos \alpha) = A^4 B - A^4 (B^2 + 2) + 3A^4 B - A^2(A^2 + 2B^2 - 2) + B^3 - B$$
$$= (B - 1)((3 - B)A^4 - 2(1 + B)A^2 + B (1 + B)).$$

Let us put $\Pi(x) = (3 - B)x^2 - 2(1 + B) x + B(1 + B)$; we have

$$\Pi(1) = (1 - B)^2 > 0, \quad \Pi\left(\cos^2 \frac{\beta}{2}\right) = -\frac{1+B}{4}(1 - B)^2 < 0$$

and

$$\Pi(\cos^2 \beta) = B(1 - B)^2 (1 + B - B^2) > 0,$$

so Π has a root between $\cos^2 \beta$ and $\cos^2 \dfrac{\beta}{2}$ and another root between $\cos^2 \dfrac{\beta}{2}$ and 1. If we assume $\alpha \geq \dfrac{\beta}{2}$, condition (7) is equivalent to

$$A^2 = \cos^2 \alpha \leq \frac{1 + B - (1 - B)\sqrt{1 + B}}{3 - B} = \cos\frac{\beta}{2} \frac{\cos\dfrac{\beta}{2} - \sqrt{2}\sin^2\dfrac{\beta}{2}}{1 + \sin^2\dfrac{\beta}{2}},$$

or $\alpha \geq \alpha_1 (\beta)$ where $\alpha_1 (\beta)$ is defined by

(8) $\quad \cos^2 \alpha_1(\beta) = \cos\dfrac{\beta}{2} \dfrac{\cos\dfrac{\beta}{2} - \sqrt{2}\sin^2\dfrac{\beta}{2}}{1 + \sin^2\dfrac{\beta}{2}}$

or

$$\tan^2 \alpha_1(\beta) = \frac{\sin^2\dfrac{\beta}{2}\left(2 + \sqrt{2}\cos\dfrac{\beta}{2}\right)}{\cos\dfrac{\beta}{2}\left(\cos\dfrac{\beta}{2} - \sqrt{2}\sin^2\dfrac{\beta}{2}\right)} = \frac{\sin\dfrac{\beta}{2}\tan\dfrac{\beta}{2}}{\cos\dfrac{\beta}{2} - \dfrac{\sqrt{2}}{2}},$$

which shows that α_1 is increasing.

We can see that $\alpha_1(0) = 0$; for β close to 0, we have

$$\alpha_1(\beta)^2 \approx \frac{\beta^2}{4\left(1 - \frac{\sqrt{2}}{2}\right)} = \frac{\beta^2}{4}\left(2 + \sqrt{2}\right),$$

that is,

$$\alpha_1(\beta) \approx \frac{\sqrt{2 + \sqrt{2}}}{2}\beta$$

or

$$\beta \approx \alpha_1(\beta)\sqrt{2\left(2 - \sqrt{2}\right)} \quad \left(\frac{\sqrt{2 + \sqrt{2}}}{2} = 0.923879533, \ \sqrt{2\left(2 - \sqrt{2}\right)} = 1.17157288\right).$$

When $\beta = \dfrac{\pi}{2}$, $\alpha_1\left(\dfrac{\pi}{2}\right) = \dfrac{\pi}{2}$; putting $\beta = \dfrac{\pi}{2} - v$, $\alpha_1 = \dfrac{\pi}{2} - u$, we have

$$\frac{1}{\tan^2 u} = \frac{\sin\left(\dfrac{\pi}{4} - \dfrac{v}{2}\right)\tan\left(\dfrac{\pi}{4} - \dfrac{v}{2}\right)}{\cos\left(\dfrac{\pi}{4} - \dfrac{v}{2}\right) - \dfrac{\sqrt{2}}{2}},$$

hence $u^2 \approx \dfrac{v}{2}$ and $\left.\dfrac{d\beta}{d\alpha_1}\right|_{\alpha_1 = \frac{\pi}{2}} = 0$ (see Fig. 2.50).

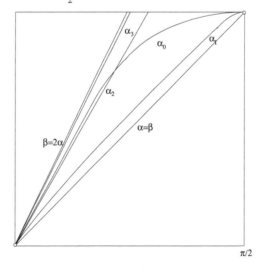

Fig. 2.50

Second part: investigation of $\eta = \dfrac{\psi}{\beta + \theta}$

We have $\eta' = \dfrac{(\beta + \theta)\psi' - \psi}{(\beta + \theta)^2}$, with the sign of $(\beta + \theta)\psi' - \psi$. If $\theta = -\beta +$

u, we know that $\psi \approx C\sqrt{u}$ when u tends to 0, where $C = \sqrt{\dfrac{2\sin\beta}{\sin\alpha\,\sin(\alpha + \beta)}}$;

so $\eta = \dfrac{\psi}{\beta + \theta} \approx \dfrac{C}{\sqrt{u}}$ tends to $+\infty$ when θ tends to $-\beta$, because $\beta + \theta = u$. By

(4) we know that $\psi'\sin\psi$ tends to $\dfrac{\sin\beta}{\sin\alpha\,\sin(\alpha + \beta)} = \dfrac{C^2}{2}$, so

$$\psi' \approx \frac{C}{2\sqrt{u}} \quad \text{and} \quad (\beta + \theta)\psi' - \psi \approx -\frac{C}{2}\sqrt{u} < 0.$$

The derivative of $(\beta + \theta)\psi' - \psi$ is $(\beta + \theta)\psi''$, with the sign of ψ''; so $(\beta + \theta)\psi' - \psi$ decreases and stays negative while $\psi'' \leq 0$. We conclude that η decreases for $-\beta \leq \theta \leq \beta$ in the case where $\alpha \geq \beta$, and for $-\beta \leq \theta \leq \theta_0$ in the case where $\alpha < \beta$.

If $\alpha < \beta$, let us put $\theta = 2\alpha - \beta - u$ and $\psi = \pi - v$; we have seen that $v \approx C'\sqrt{u}$ when u tends to 0, with

$$C' = \sqrt{\frac{2\sin\beta}{\sin\alpha\,\sin(\beta - \alpha)}} \quad \text{(see Note 2, p. 130).}$$

The formula (4) gives

$$\lim \psi'\sin\psi = \frac{\sin\beta}{\sin\alpha\,\sin(\beta - \alpha)} = \frac{C'^2}{2},$$

so $\psi' \approx \dfrac{C'}{2\sqrt{u}}$ and consequently $(\beta + \theta)\psi' - \psi \approx 2\alpha\psi' \approx \dfrac{\alpha\,C'}{\sqrt{u}}$, which tends to

$+\infty$; $\eta' \approx \dfrac{C'}{4\alpha\sqrt{u}}$ also tends to $+\infty$. Thus $(\beta + \theta)\psi' - \psi$, which increases for

$\theta_0 \leq \theta \leq 2\alpha - \beta$, becomes zero and then positive for a value of θ in the interval $\theta_3 \geq \theta_0 \geq \alpha - \dfrac{\pi}{2}$ (see Fig. 2.61). Accordingly, η decreases in the interval $-\beta \leq \theta \leq \theta_3$, then increases in the interval $\theta_3 \leq \theta \leq 2\alpha - \beta$.

We can see that the decrease of η when $\theta \le 0$ requires $\theta_3 \ge 0$. We thus have $2\alpha - \beta > 0$, that is $\alpha > \dfrac{\beta}{2}$ and the value $\beta\psi_0' - \psi_0$ of $(\beta+\theta)\psi' - \psi$ for $\theta = 0$ must be negative. Now

$$\cos\psi_0 = \frac{\cos\beta - \cos^2\alpha}{\sin^2\alpha} = 1 - 2\frac{\sin^2\dfrac{\beta}{2}}{\sin^2\alpha},$$

that is,

$$\sin\frac{\psi_0}{2} = \frac{\sin\dfrac{\beta}{2}}{\sin\alpha} \quad \text{and} \quad \psi_0'\sin\psi_0 = \cos\alpha\frac{1-\cos\beta}{\sin^3\alpha} = \frac{2\cos\alpha\sin^2\dfrac{\beta}{2}}{\sin^3\alpha}.$$

So the condition $\beta\psi_0' \le \psi_0$ can be written

$$\beta\psi_0'\sin\psi_0 = 2\beta\frac{\cos\alpha\sin^2\dfrac{\beta}{2}}{\sin^3\alpha} = 2\beta\frac{\sin^2\dfrac{\psi_0}{2}}{\tan\alpha} \le \psi_0\sin\psi_0,$$

that is,

(9)
$$\frac{\beta}{\tan\alpha} \le \frac{\psi_0}{\tan\dfrac{\psi_0}{2}}.$$

We define a function $\alpha_2(\beta)$ by the implicit equation $\dfrac{\beta}{\tan\alpha_2} = \dfrac{\psi_0}{\tan\dfrac{\psi_0}{2}}$ $\left(\dfrac{\beta}{2} \le \alpha \le \beta\right)$; in fact $\dfrac{\beta}{\tan\alpha}$ is a decreasing function of α so long as $\dfrac{\psi_0}{\tan\dfrac{\psi_0}{2}}$ is an increasing function of α. To put it more precisely, $\dfrac{\beta}{\tan\alpha} - \dfrac{\psi_0}{\tan\dfrac{\psi_0}{2}}$ increases as a function of β and decreases as a function of α; the result is that α_2 increases as a function of β and that the statement b) (that η decreases) is true for $\alpha \ge \alpha_2(\beta)$. When $\beta = 0$, we have $\alpha_2(0) = 0$ because $\alpha_2(\beta) \le \beta$. Further, $\sin\dfrac{\psi_0}{2} \approx \dfrac{\beta}{2\alpha} = \dfrac{1}{2\lambda}$ where $\lambda = \lim\dfrac{\alpha_2}{\beta}$; so

$$\tan\frac{\psi_0}{2} = \frac{1}{\sqrt{4\lambda^2-1}} \quad \text{and} \quad \sqrt{4\lambda^2-1}\,\tan\frac{1}{2\lambda\sqrt{4\lambda^2-1}} = 1,$$

which gives $\lambda = 0.667618851$ or $\left.\dfrac{d\beta}{d\alpha_2}\right|_{\alpha_2=0} = 1.49786064$.

For $\beta = \dfrac{\pi}{2}$, $\sin\dfrac{\psi_0}{2} = \dfrac{1}{\sqrt{2}\sin\alpha}$, hence

$$\tan\frac{\psi_0}{2} = \frac{1}{\sqrt{-\cos 2\alpha}} \quad \text{and} \quad \frac{\pi}{2\tan\alpha_2} = 2\sqrt{-\cos 2\alpha_2}\ \text{Arc}\tan\frac{1}{\sqrt{-\cos 2\alpha_2}},$$

which gives $\alpha_2 = 0.933682485$ – a little less than $\dfrac{3\pi}{10}$ – and

$\dfrac{\alpha_2}{\beta} = \dfrac{2\alpha_2}{\pi} = 0.594400731$.

Note: If we make $\alpha_2 = \lambda\beta$ in (9), $\sin\dfrac{\psi_0}{2} = \dfrac{\sin\dfrac{\beta}{2}}{\sin\lambda\beta}$, that becomes an increasing function of β for a fixed value of $\lambda \geq \dfrac{1}{2}$ (Proposition 2), so

$$\frac{1}{2\lambda} \leq \sin\frac{\psi_0}{2} \leq \frac{1}{\sqrt{2}\sin\dfrac{\pi\lambda}{2}}.$$

Since $\dfrac{\psi_0}{\tan\dfrac{\psi_0}{2}}$ is a decreasing function of ψ_0, we obtain

$$2\sqrt{-\cos\pi\lambda}\ \text{Arc}\tan\frac{1}{\sqrt{-\cos\pi\lambda}} \leq \frac{\psi_0}{\tan\dfrac{\psi_0}{2}} \leq 2\sqrt{4\lambda^2-1}\ \text{Arc}\tan\frac{1}{\sqrt{4\lambda^2-1}};$$

however $\dfrac{\beta}{\tan\lambda\beta}$, a decreasing function of β, is bounded by $\dfrac{\pi}{2\tan\dfrac{\pi\lambda}{2}}$ and $\dfrac{1}{\lambda}$.

So we must have

$$2\sqrt{-\cos\pi\lambda}\ \text{Arc}\tan\frac{1}{\sqrt{-\cos\pi\lambda}} \leq \frac{1}{\lambda}$$

and

$$\frac{\pi}{2\tan\dfrac{\pi\lambda}{2}} \leq 2\sqrt{4\lambda^2-1}\ \text{Arc}\tan\frac{1}{\sqrt{4\lambda^2-1}},$$

inequalities that can be expressed by

$$0.579590352 \le \lambda \le 0.706698767.$$

But we can prove that $\dfrac{\alpha_2}{\beta}$ is a decreasing function of β, so that $0.594 \le$

$\dfrac{\alpha_2}{\beta} \le 0.668$ (see Fig. 2.50).

Third part: investigation of $\xi = \dfrac{\psi}{\beta + \theta} \sin(\alpha - \theta)$

We have

$$\xi = \frac{\widehat{EI}}{EH} \cdot \frac{EH}{\widehat{EB}} = \frac{\psi}{1 - \cos\psi} \cdot \frac{\cos\theta - \cos\beta}{(\beta + \theta)\sin\alpha}.$$

Since

$$\frac{\cos\theta - \cos\beta}{\beta + \theta} = \sin\frac{\beta - \theta}{2} \cdot \frac{2\sin\dfrac{\beta + \theta}{2}}{\beta + \theta}$$

is a decreasing function of θ, it is sufficient to establish that $\dfrac{\psi}{1 - \cos\psi}$ decreases in order to prove that statement a) is true. Now ψ is an increasing function of θ if $\alpha \ge \beta$ and $\theta \le \alpha - \mu$ or if $\alpha < \beta$; so we still need to see whether $\dfrac{\psi}{1 - \cos\psi}$ is a decreasing function of ψ. We have

$$\frac{d}{d\psi} \frac{\psi}{1 - \cos\psi} = \frac{1 - \cos\psi - \psi\sin\psi}{(1 - \cos\psi)^2} = \frac{1}{2\sin^2\dfrac{\psi}{2}}\left(1 - \frac{\psi}{\tan\dfrac{\psi}{2}}\right),$$

so $\dfrac{\psi}{1 - \cos\psi}$ decreasing is equivalent to $\psi \ge \tan\dfrac{\psi}{2}$, that is $\psi \le \delta$ where δ is defined by $\delta = \tan\dfrac{\delta}{2}$ $(0 < \delta \le \pi)$; we find $\delta = 2.33112237$ radians (a little less than $\dfrac{3\pi}{4}$).

In the case where $\alpha \ge \beta$, $\psi \le \psi_m \le \dfrac{\pi}{2} < \delta$, so ξ decreases in the interval $-\beta \le \theta \le \alpha - \mu$. The same is also true in the interval $\alpha - \mu \le \theta \le \beta$ where ψ is a decreasing function of θ; in fact

$$\xi = \frac{\widehat{El}}{El} \cdot \frac{El}{EB} \cdot \frac{EB}{\widehat{EB}} = \frac{\psi}{\sin\frac{\psi}{2}} \cdot \frac{El}{EB} \cdot \frac{2\sin\frac{\beta+\theta}{2}}{\beta+\theta},$$

where the last two ratios decrease, by Propositions 4, 11 and 12, and where the first ratio decreases if $\theta \geq \alpha - \mu$. Thus, in the case where $\alpha \geq \beta$, ξ is a decreasing function of θ throughout the interval $[-\beta, \beta]$.

Let us now consider the case where $\alpha < \beta$: ξ decreases in the interval $-\beta \leq \theta \leq \varepsilon$ where ε is defined by $\dfrac{\cos\beta - \cos\alpha \cos(\alpha-\varepsilon)}{\sin\alpha \sin(\alpha-\varepsilon)} = \cos\delta$.

Since $\delta > \dfrac{\pi}{2}$, ψ goes through the value $\dfrac{\pi}{2}$ before reaching δ and $\alpha - \mu' < \varepsilon$. If $\theta = 2\alpha - \beta - u$,

$$\xi' = \eta' \sin(\alpha - \theta) - \eta \cos(\alpha - \theta) \approx \frac{C'}{4\alpha\sqrt{u}} \sin(\beta - \alpha)$$

when u tends to 0; so we see that ξ' tends to $+\infty$ when θ tends to $2\alpha - \beta$. Thus ξ' must become zero somewhere between ε and $2\alpha - \beta$; if θ_4 is the first value of θ that makes ξ' zero, ξ decreases for $\theta \leq \theta_4$ but increases after that. Since $\theta_3 \geq \alpha - \dfrac{\pi}{2}$ (p. 139), we have $\alpha - \theta \leq \dfrac{\pi}{2}$ for $\theta \geq \theta_3$ and from this $\theta_4 \geq \theta_3$. It can be established that θ_4 is unique and that ξ increases for $\theta_4 \leq \theta \leq 2\alpha - \beta$ (see Fig. 2.61). So statement a) is true only if $\theta_4 \geq 0$, which requires $\alpha \geq \dfrac{\beta}{2}$. Now $\theta_4 \geq 0$ means that $\xi' \leq 0$ for $\theta = 0$, that is

$$\eta_0' \sin\alpha - \eta_0 \cos\alpha = \frac{\beta\psi_0' - \psi_0}{\beta^2} \sin\alpha - \frac{\psi_0}{\beta}\cos\alpha \leq 0$$

or again

$$\beta\frac{\psi_0'}{\psi_0} - 1 \leq \frac{\beta}{\tan\alpha}$$

which can be expresses as

(10) $$\frac{\beta}{\beta+\tan\alpha} \leq \frac{\psi_0}{\tan\dfrac{\psi_0}{2}}.$$

We define a function $\alpha_3 (\beta)$ by

$$\frac{\beta}{\beta + \tan \alpha_3} = \frac{\psi_0}{\tan \dfrac{\psi_0}{2}},$$

where

$$\sin \frac{\psi_0}{2} = \frac{\sin \dfrac{\beta}{2}}{\sin \alpha_3}, \qquad\qquad \text{where } \frac{\beta}{2} \leq \alpha_3 \leq \beta.$$

We see that α_3 increases and that $\alpha_3 \leq \alpha_2$; statement a) is true for $\alpha \geq \alpha_3 (\beta)$.

When $\beta = 0$, $\alpha_3 (0) = 0$; if $\alpha_3 (\beta) \approx \lambda \beta$ for β tending to 0, $\sin \dfrac{\psi_0}{2}$ tends to $\dfrac{1}{2\lambda}$, $\tan \dfrac{\psi_0}{2}$ to $\dfrac{1}{\sqrt{4\lambda^2 - 1}}$ and we have

$$\sqrt{4\lambda^2 - 1} \tan \frac{1}{2(1+\lambda)\sqrt{4\lambda^2 - 1}} = 1,$$

hence $\lambda = 0.5152252767$, or $\dfrac{d\beta}{d\alpha_3}\bigg|_{\alpha_3 = 0} = 1.94089856 .$

For $\beta = \dfrac{\pi}{2}$,

$$\sin \frac{\psi_0}{2} = \frac{1}{\sqrt{2}\sin \alpha}, \qquad \tan \frac{\psi_0}{2} = \frac{1}{\sqrt{-\cos 2\alpha}}$$

and

$$\sqrt{-\cos 2\alpha_3} \tan \frac{1}{\left(2 + \dfrac{4}{\pi}\tan \alpha_3\right)\sqrt{-\cos 2\alpha_3}} = 1,$$

hence

$$\alpha_3\left(\frac{\pi}{2}\right) = 0.8099378632, \text{ between } \frac{\pi}{4} \text{ and } \frac{4\pi}{15}, \text{ and } \frac{\alpha_3}{\beta} = \frac{2\alpha_3}{\pi} = 0.515622458.$$

Note: $\dfrac{\beta}{\beta+\tan\lambda\beta}=\dfrac{1}{1+\dfrac{\tan\lambda\beta}{\beta}}$ is again a decreasing function of β and it is

bounded by $\dfrac{1}{1+\dfrac{2}{\pi}\tan\dfrac{\pi\lambda}{2}}$ and $\dfrac{1}{1+\lambda}$. So if $\alpha_3=\lambda\beta$ in (10), we have

$$2\sqrt{-\cos\pi\lambda}\;\mathrm{Arc\,tan}\,\dfrac{1}{\sqrt{-\cos\pi\lambda}}\le\dfrac{1}{1+\lambda}$$

and

$$\dfrac{1}{1+\dfrac{2}{\pi}\tan\dfrac{\pi\lambda}{2}}\le 2\sqrt{4\lambda^2-1}\;\mathrm{Arc\,tan}\,\dfrac{1}{\sqrt{4\lambda^2-1}}.$$

These inequalities give $0.51\le\lambda\le 0.52$.

But it can be established that $\dfrac{\alpha_3}{\beta}$ is an increasing function of β, so that $0.5152\le\lambda\le 0.516$ (see Fig. 2.50).

Taking axes of coordinates with origin ω, respectively perpendicular to ωF (that is, parallel to BA) and along ωF, the coordinates of the point Ω, the point of intersection of $\Pi'\Pi$ and AB, are $r\cos\beta\tan\alpha$ and $r\cos\beta$. Since the functions $\alpha_j(\beta)$ $(j=1,2,3)$ increase, the conditions $\alpha\ge\alpha_j(\beta)$ mean that Ω lies to the right of the curve with index number j in Fig. 2.51.

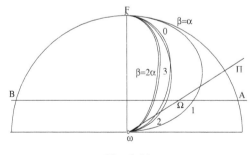

Fig. 2.51

In the position $\omega\Pi$, Proposition 14 is true, but only for the second inequality.

Fourth part: Examples

In all these examples, we have made $\beta = \dfrac{\pi}{3}$; the corresponding values of $\alpha_j(\beta)$ $(j = 1, 2, 3)$ are respectively:

$$\alpha_1\left(\frac{\pi}{3}\right) = 0.932458266, \quad \sin\alpha = \sqrt{2}\sin\frac{\beta}{2} \text{ for } \alpha = \frac{\pi}{4},$$

$$\alpha_2\left(\frac{\pi}{3}\right) = 0.659224289 \text{ and } \alpha_3\left(\frac{\pi}{3}\right) = 0.53978010008.$$

For $\alpha \ge \dfrac{\pi}{3}$, ξ and η decrease in the interval $-\dfrac{\pi}{3} \le \theta \le \dfrac{\pi}{3}$ and ψ is a concave function of θ. We have drawn the graphs of ξ and η for $\alpha = \dfrac{5\pi}{12}$ (Fig. 2.52) and for $\alpha = \dfrac{\pi}{3}$ (Fig. 2.53).

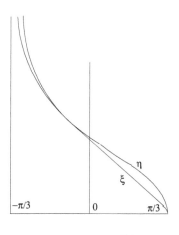

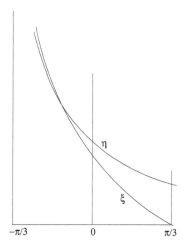

Fig. 2.52: $\alpha = \dfrac{5\pi}{12}$ Fig. 2.53: $\alpha = \dfrac{\pi}{3}$

For $0.932458266 \le \alpha < \dfrac{\pi}{3}$, ψ is concave in the interval $-\dfrac{\pi}{3} \le \theta \le 0$ and ξ and η decrease in that interval (Fig. 2.54, where $\alpha = 1$).

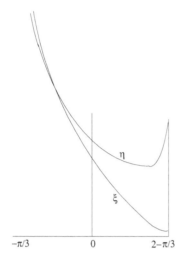

Fig. 2.54: $\alpha = 1$
$\theta_3 = 0.72055$, min of $\eta = 0.935784028$
$\theta_4 = 0.9500790346$, min of $\xi = 0.0700698854$

For $\frac{\pi}{4} \le \alpha < 0.932458266$, ψ ceases to be concave for a value $\theta_0 < 0$ of θ but remains $\le \frac{\pi}{2}$ while $\theta \le 0$; ξ and η decrease in the interval $-\frac{\pi}{3} \le \theta \le 0$ (Fig. 2.55, where $\alpha = 0.8$). For $0.659224289 \le \alpha < \frac{\pi}{4}$, ξ and η still decrease in the interval $-\frac{\pi}{3} \le \theta \le 0$, but ψ ceases to be concave in $\theta_0 < 0$ and reaches the value $\frac{\pi}{2}$ at $\theta = \alpha - \mu' < 0$ (Fig. 2.56, where $\alpha = 0.75$).

For $0.53978010008 \le \alpha < 0.659224289$, ξ decreases in the interval $-\frac{\pi}{3} \le \theta \le 0$, but η goes through a minimum for a value $\theta_3 < 0$ (Fig. 2.57, where $\alpha = \frac{\pi}{5}$ and Fig. 2.58, where $\alpha = 0.55$).

Finally, for $\alpha < 0.53978010008$, ξ and η both have a minimum for negative values of θ, θ_4 and θ_3 respectively (Fig. 2.59, where $\alpha = 0.524$). If $\alpha < \frac{\beta}{2}$, θ always remains negative (Fig. 2.60, where $\alpha = \frac{\pi}{12}$).

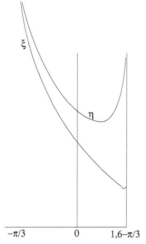

−π/3   0   1,6−π/3

Fig. 2.55: $\alpha = 0.8$
$\theta_3 = 0.24483$,
min of $\eta = 1.40613647$
$\theta_4 = 0.537776499$,
min of $\xi = 0.452137244$

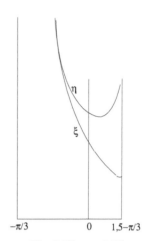

−π/3   0   1,5−π/3

Fig. 2.56: $\alpha = 0.75$
$\theta_3 = 0.15413$,
min of $\eta = 1.5403285$
$\theta_4 = 0.4346042281$,
min of $\xi = 0.576095779$

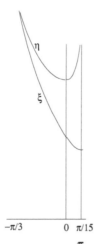

−π/3   0   π/15

Fig. 2.57: $\alpha = \dfrac{\pi}{5}$
$\theta_3 = -0.0503$,
min of $\eta = 1.9362108$
$\theta_4 = 0.1831447956$,
min of $\xi = 0.949376423$

−π/3   1,1−π/3

Fig. 2.58: $\alpha = 0.55$
$\theta_3 = -0.17463$,
min of $\eta = 2.27086432$
$\theta_4 = 0.021138879$,
min. of $\xi = 1.2654303$

In Fig. 2.61, we have drawn the graphs of $\theta = 2\alpha - \beta$, $\theta_4(\alpha)$, $\theta_3(\alpha)$, $\alpha - \mu'$ and $\theta_0(\alpha)$ as a function of $\alpha \in\,]0, \beta]$ for $\beta = \dfrac{\pi}{3}$. The angle θ is constrained to vary from $-\beta$ to $2\alpha - \beta$, so only the area below the diagonal $\theta = 2\alpha - \beta$ is useful. The statement a) is true below the curve θ_4 and the statement b) above the curve θ_3.

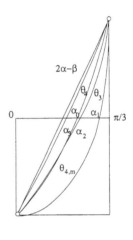

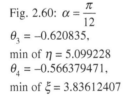

Fig. 2.59: $\alpha = 0.524$	Fig. 2.60: $\alpha = \dfrac{\pi}{12}$	Fig. 2.61
$\theta_3 = -0.21518$,	$\theta_3 = -0.620835$,	
min of $\eta = 2.40218858$	min of $\eta = 5.099228$	
$\theta_4 = -0.03261719$,	$\theta_4 = -0.566379471$,	
min of $\xi = 1.3887698$	min of $\xi = 3.83612407$	

The fact that this long discussion is so complex shows that an exact determination of the conditions for the proposition to be valid was far beyond the scope of the mathematics of Ibn al-Haytham's time, and indeed that of the mathematics of any period before the end of the eighteenth century.

Proposition 15. — Our assumptions about the circles ABC, EI, DC are the same as before: (ABC) is the horizon, (EI) and (DC) are circles parallel to the equator (three cases).

The circle $ADEB$ is the meridian circle and we suppose $\overset{\frown}{BD} \le \dfrac{1}{2}\,\overset{\frown}{BDA}$.

Further, we consider another great circle that passes through the poles; this great circle cuts the circle ABC in O and H, the arc EI in U and the arc DC in L.[11]

With these assumptions, we have

$$\frac{\overset{\frown}{UI}}{\overset{\frown}{UO}} > \frac{\overset{\frown}{CL}}{\overset{\frown}{LO}} > \frac{\overset{\frown}{CK}}{\overset{\frown}{KI}}.$$

We want to prove that the result established for the meridian BED applies to a great circle such as OUL. The position of the meridian BED is special because it is perpendicular to the horizon; we replace it by a general meridian; so this proposition is a more general version of the preceding one.

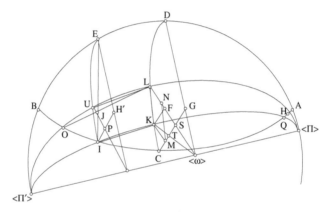

Fig. 2.62

The circle DLC has its centre ω on the line through the poles, the planes of the great circles IKQ and OUL thus cut DLC along two diameters through KT and LS.

We first suppose that the part of DLC above ABC is less than a semicircle, so we have ω below the plane ABC.

We have $D\hat{G}C = 1$ right angle, $L\hat{S}C$ is acute, $K\hat{T}C$ is acute. With the diameter from K, we associate a right-angled triangle $TG\omega$ and with the diameter from L a triangle $SG\omega$ which lies inside the preceding one. So we have $L\hat{S}C > S\hat{T}\omega$, hence $L\hat{S}C > K\hat{T}C$.

We draw $KM \parallel LS$, we have M between T and C and $KM < LS$; we draw $MN \parallel KL$, then we have $KL = MN$ and $KM = NL$.

---

[11] The point L has a significance different from what it had before.

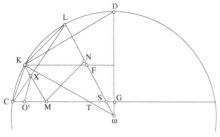

Fig. 2.63.1

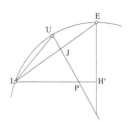

Fig. 2.63.2

The arc KL is similar to the arc IU, so

$$\frac{IU}{KL} = \frac{d_1}{d_2} = \frac{IU}{MN}.$$

Let UP be the diameter from U in the circle EI, we have $LS \parallel UP$, $SC \parallel IP$, hence $L\hat{S}C = U\hat{P}I$.

We draw $KF \parallel CG$, we have $L\hat{K}F = U\hat{I}P$ (because the arcs KL and LD are respectively similar to the arcs IU and UE). Then we have $U\hat{I}P = N\hat{M}S$, so triangles NMS and UIP are similar and consequently

$$\frac{IU}{MN} = \frac{UP}{NS} = \frac{d_1}{d_2}.$$

1) So if $d_1 < d_2$, we have $UP < NS$ and if $d_1 = d_2$, we have $UP = NS$.

By an argument like the one in the previous proposition, we prove that

$$\frac{SL}{LO} > \frac{KM}{KI};$$

now $KM = NL$ and $KI = LU$, so we have

$$\frac{SL}{LO} > \frac{NL}{LU}.$$

2) If $d_1 > d_2$, and if $\dfrac{UP}{PJ} \geq \dfrac{d_1}{d_2}$, we also prove that

$$\frac{SL}{LO} > \frac{NL}{LU} \quad \text{or} \quad \frac{SL}{LO} > \frac{KM}{KI}.$$

We have seen that $K\hat{M}C = L\hat{S}C$ is acute; if from K we drop a perpendicular to CG, its foot lies between M and C. The straight line LC cuts KM at the point X, we draw $XO' \parallel KC$. By hypothesis, $K\hat{C}M$ is acute and $K\hat{C}M = X\hat{O}'M$, so $X\hat{O}'C$ is obtuse, hence $XC > XO'$; consequently

$$\frac{CX}{XM} > \frac{O'X}{XM}.$$

But

$$\frac{CX}{XM} = \frac{CL}{LS} \quad \text{and} \quad \frac{O'X}{XM} = \frac{CK}{KM},$$

so we have

$$\frac{CL}{LS} > \frac{CK}{KM}.$$

Moreover

$$\frac{LS}{LO} > \frac{KM}{KI},$$

so

$$\frac{CL}{LO} > \frac{CK}{KI}$$

and consequently

$$\frac{\widehat{CL}}{\widehat{LO}} > \frac{\widehat{CK}}{\widehat{KI}},$$

if we use the corollary to Proposition 4. Unfortunately, the conditions for this corollary to be applicable are not always satisfied (see below).

If the part of the circle CLD above the plane (ABC) is a semicircle, then $\omega = G$, and $GC = GD$. In this case, the visible pole Π is above the plane ABC. The point G lies on the axis joining the poles, G is both in the plane ABC and in each of the planes HLO and QKI; so the straight lines AB, HO and QI intersect at the point G ($G = S = T$); we have $GK = GL = GC$ as radii of the circle CLD, KC is a chord, so $K\hat{C}G$ is acute.

If we produce CK beyond K, it meets GL, because it meets GD. We draw $KM \parallel LG$; the point M lies between C and G and $KM < LG$. We draw

$MN \parallel KL$; CL cuts KM in X; we draw $XO' \parallel KC$; $X\hat{O}'C$ is obtuse, so $XC > XO'$.

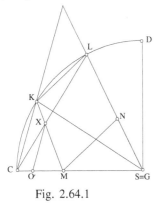

Fig. 2.64.1

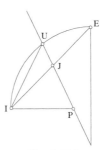

Fig. 2.64.2

But

$$\frac{CL}{LG} = \frac{CX}{XM} \quad \text{and} \quad \frac{XO'}{XM} = \frac{CK}{KM},$$

so

$$\frac{CL}{LG} > \frac{CK}{KM}.$$

Triangle GNM is similar to triangle IUP and

$$\frac{IU}{MN} = \frac{UP}{NG} = \frac{d_1}{d_2}.$$

So if $d_1 \leq d_2$, we have $UP \leq NG$ and as before we have

$$\frac{GL}{LO} > \frac{NL}{LU}.$$

If $d_1 > d_2$ and if $\dfrac{UP}{PJ} \geq \dfrac{d_1}{d_2}$, we again have $\dfrac{GL}{LO} > \dfrac{NL}{LU}$ and consequently

$$\frac{GL}{LO} > \frac{MK}{KI}.$$

Now

$$\frac{CL}{LG} > \frac{CK}{KM},$$

so we have

$$\frac{CL}{LO} > \frac{CK}{KI},$$

and consequently

$$\frac{\overarc{CL}}{\overarc{LO}} > \frac{\overarc{CK}}{\overarc{KI}},$$

as in the first case.

From this result we deduce that

$$\frac{\overarc{CL} - \overarc{CK}}{\overarc{LO} - \overarc{KI}} > \frac{\overarc{CL}}{\overarc{LO}},$$

$\overarc{CL} - \overarc{CK} = \overarc{LK}$ similar to $\overarc{UI}$,[12] $\overarc{LO} - \overarc{KI} = \overarc{LO} - \overarc{LU} = \overarc{UO}$, so we have

$$\frac{\overarc{UI}}{\overarc{UO}} > \frac{\overarc{CL}}{\overarc{LO}} > \frac{\overarc{CK}}{\overarc{KI}}.$$

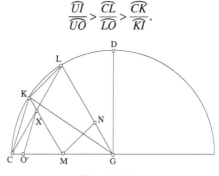

Fig. 2.64.3

We consider another great circle, with diameter $\Pi\Pi'$, which cuts ACB in X and the arc IE in R, and the circle parallel to the circle IE that passes through O; it cuts the arc XR in V.

The argument in the previous section concerning the great circles IKH and OLQ cutting the parallel circles IE and CD can be applied here to the two great circles OLQ and XRZ cutting the parallel circles IE and OV; and we obtain the result

---

[12] Ibn al-Haytham writes here of the equality of the arcs LK and UI; but these similar arcs belong to different circles. So the conclusion is not so general: we cannot draw the conclusion unless $\overarc{LK} \geq \overarc{UI}$, that is unless $d_1 \geq d_2$. Here again, the arcs LK and UI subtend the same angle in two different circles and we might wonder whether Ibn al-Haytham was thinking about the angles while referring to the arcs.

$$\frac{\overline{OV}}{\overline{VX}} > \frac{\overline{IR}}{\overline{RX}} > \frac{\overline{IU}}{\overline{UO}}.$$

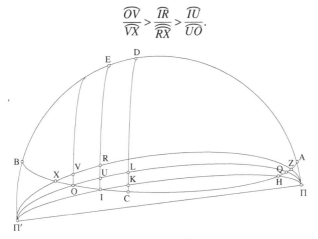

Fig. 2.65[13]

Commentary:

We shall continue to use the notation employed in the commentary on Proposition 14, with an additional variable $\lambda = D\hat{W}L$, which defines the meridian plane OUL, whose equation is $y = z \tan \lambda$. We have $0 \leq \lambda \leq \Psi$, the limiting positions of OUL being the plane BED ($\lambda = 0$) and the meridian plane of C ($\lambda = \Psi$) (Fig. 2.66).

If $\varphi = L\hat{o}O$ and $\Pi\hat{o}O = \alpha - \Theta + \varphi$, the coordinates of O are $x = r \cos (\alpha - \Theta + \varphi)$, $y = r \sin \lambda \sin (\alpha - \Theta + \varphi)$, $z = r \cos \lambda \sin (\alpha - \Theta + \varphi)$. The equation of the plane ABC gives $x \cos \alpha + y \sin \alpha = r \cos \beta$, and the point O lies in the plane ABC, hence

(1) $\cos \alpha \cos (\alpha - \Theta + \varphi) + \cos \lambda \sin \alpha \sin (\alpha - \Theta + \varphi) = \cos \beta,$

an equation that gives two values of $\varphi \in [0, \pi]$, corresponding to the points O and O'; the point O corresponds to the greater of these values of φ.

The angle LWC is equal to $\Psi - \lambda$, so $\overwidehat{CL} = r(\Psi - \lambda) \sin (\alpha - \Theta)$ and we have $\overwidehat{LO} = r\varphi$; thus

---

[13] The letter O has already been used.

$$\frac{\overline{CL}}{\overline{LO}} = \frac{\Psi - \lambda}{\varphi} \sin(\alpha - \Theta).$$

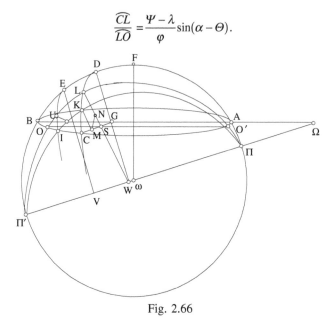

Fig. 2.66

When D is at E, this ratio becomes $\frac{\overline{IU}}{\overline{UO}}$ and we can see that the first inequality means that $\frac{\psi - \lambda}{\varphi} \sin(\alpha - \theta)$ is a decreasing function of θ for a fixed value of λ. When L is at K, the same ratio becomes $\frac{\overline{CK}}{\overline{KI}}$ and the second inequality means that $\frac{\psi - \lambda}{\varphi} \sin(\alpha - \theta)$ is a decreasing function of λ for a fixed value of θ.

First part: investigation of angle φ

Equation (1) gives $\theta - \varphi = \theta_-(\lambda)$, that is $\varphi = \theta - \theta_-(\lambda)$ where θ_- is the function that is the reciprocal of $\theta \mapsto \psi(\theta)$ defined on p. 128. Now θ_- is an increasing function, going from $-\beta$ to $\theta - \mu$ for $0 \leq \lambda \leq \psi_m$ if $\alpha \geq \beta$ and from $-\beta$ to $2\alpha - \beta$ for $0 \leq \lambda \leq \pi$ if $\alpha < \beta$. In the case where $\alpha \geq \beta$, θ_- is convex; in the case where $\alpha < \beta$, θ_- is convex until it reaches the value θ_0 (defined by $P(\cos(\alpha - \theta_0)) = 0$), and then concave (see p. 135). As a result φ is a decreasing function of λ, from $\theta + \beta$ to $\theta - \alpha + \mu$ for $0 \leq \lambda \leq \psi_m$ if $\alpha \geq \beta$ and from $\theta + \beta$ to $\theta - 2\alpha + \beta$ for $0 \leq \lambda \leq \pi$ if $\alpha < \beta$. In the case where $\alpha \geq \beta$, φ is concave; the same is true in the case where $\alpha < \beta$, if we remain within the interval $0 \leq \lambda \leq \lambda_0$ where λ_0 is such that $\theta_-(\lambda_0) = \theta_0$.

We have

$$\cos \lambda - \cos \psi = \frac{\cos \beta - \cos \alpha \cos(\alpha - \theta + \varphi)}{\sin \alpha \sin(\alpha - \theta + \varphi)} - \frac{\cos \beta - \cos \alpha \cos(\alpha - \theta)}{\sin \alpha \sin(\alpha - \theta)} =$$

$$\frac{1}{\sin \alpha \sin(\alpha - \theta)\sin(\alpha - \theta + \varphi)} \times \big(\cos \beta(\sin(\alpha - \theta)) - \sin(\alpha - \theta + \varphi)\big) +$$

$$\cos \alpha \big(\cos(\alpha - \theta)\sin(\alpha - \theta + \varphi) - \sin(\alpha - \theta)\cos(\alpha - \theta + \varphi)\big) =$$

$$\frac{1}{\sin \alpha \sin(\alpha - \theta)\sin(\alpha - \theta + \varphi)} \left(-2\cos \beta \sin \frac{\varphi}{2} \cos\left(\alpha - \theta + \frac{\varphi}{2}\right) + \cos \alpha \sin \varphi\right) =$$

$$\frac{2\sin \dfrac{\varphi}{2}}{\sin \alpha \sin(\alpha - \theta)\sin(\alpha - \theta + \varphi)} \left(\cos \alpha \cos \frac{\varphi}{2} - \cos \beta \cos\left(\alpha - \theta + \frac{\varphi}{2}\right)\right) =$$

$$\frac{2\sin \dfrac{\varphi}{2}}{\sin \alpha \sin(\alpha - \theta)\sin(\alpha - \theta + \varphi)}$$
$$\left(\cos \alpha \cos \frac{\varphi}{2} - \cos \beta\left(\cos(\alpha - \theta + \varphi)\cos \frac{\varphi}{2} + \sin(\alpha - \theta + \varphi)\sin \frac{\varphi}{2}\right)\right) =$$

$$\frac{2\sin \dfrac{\varphi}{2}}{\sin \alpha \sin(\alpha - \theta)} \left(\frac{\cos \alpha - \cos \beta \cos(\alpha - \theta + \varphi)}{\sin(\alpha - \theta + \varphi)}\cos \frac{\varphi}{2} - \cos \beta \sin \frac{\varphi}{2}\right),$$

that is,

(2) $$\cos \lambda - \cos \psi = 2\sin \frac{\varphi}{2} \frac{t\cos \dfrac{\varphi}{2} - \cos \beta \sin \dfrac{\varphi}{2}}{\sin \alpha \sin(\alpha - \theta)}$$

with

(3) $$t = \frac{\cos \alpha - \cos \beta \cos(\alpha - \theta + \varphi)}{\sin(\alpha - \theta + \varphi)}.$$

Now

$$t^2 = \frac{\cos^2\alpha - 2\cos\alpha\,\cos\beta\cos(\alpha-\theta+\varphi) + \cos^2\beta\cos^2(\alpha-\theta+\varphi)}{\sin^2(\alpha-\theta+\varphi)} =$$

$$\cos^2\lambda\sin^2\alpha + \cos^2\alpha - \cos^2\beta = \sin^2\beta - \sin^2\alpha\sin^2\lambda;$$

we can see that $t = 0$ makes $\sin\lambda = \dfrac{\sin\beta}{\sin\alpha}$, which is possible only when $\alpha \geq \beta$; then we have $\lambda = \psi_m$ ($\lambda \leq \psi_m \leq \dfrac{\pi}{2}$), and thus also $\psi = \psi_m$ and $\theta = \alpha - \mu$. So we can see that t does not become zero in the interval $0 \leq \lambda < \psi$; so it keeps the same sign. For $\lambda = 0$, $\alpha - \theta + \varphi = \alpha + \beta$ and $t = \sin\beta > 0$; thus t remains positive and we have

(4) $$t = \sqrt{\sin^2\beta - \sin^2\alpha\sin^2\lambda}\,.$$

We may note that, if we make $\varphi = \theta - \theta_-(\lambda)$ in (3), the numerator becomes $\cos\alpha - \cos\beta\cos(\alpha - \theta_-(\lambda))$, which is always positive if $\alpha < \beta$ and also if $\alpha \geq \beta$, since by definition $\theta_-(\lambda) \leq \alpha - \mu$.

For $\lambda = \psi$, we have $2\sin\dfrac{\varphi}{2}\left(t\cos\dfrac{\varphi}{2} - \cos\beta\sin\dfrac{\varphi}{2}\right) = 0$ and $\varphi = \theta - \theta_-(\lambda)$ with $\theta = \theta_-(\psi)$ if $\alpha < \beta$ or if $\alpha \geq \beta$ and $\theta \leq \alpha - \mu$, but $\theta = \theta_+(\psi)$ if $\alpha \geq \beta$ and $\theta \geq \alpha - \mu$. Thus $\varphi = 0$ if $\alpha < \beta$ or $\alpha \geq \beta$ and $\theta \leq \alpha - \mu$; if $\alpha \geq \beta$ and $\theta > \alpha - \mu$, $\varphi = \theta_+(\psi) - \theta_-(\psi) > 0$, so $\tan\dfrac{\varphi}{2} = \dfrac{t}{\cos\beta}$.

Equation (4) in the commentary on Proposition 14 gives

$$\varphi'_\lambda = \frac{\sin\lambda\sin\alpha\sin^2(\alpha - \theta_-(\lambda))}{\cos\beta\cos(\alpha - \theta_-(\lambda)) - \cos\alpha}, \qquad \text{where } \varphi'_\lambda = \frac{\partial\varphi}{\partial\lambda}.$$

We can check that this quantity is negative. It becomes infinite if $\alpha \geq \beta$ and $\lambda = \psi = \psi_m$, because we then have $\alpha - \theta_-(\psi_m) = \mu$. Let us make $\psi = \psi_m$, let $\theta = \alpha - \mu$ and $\theta_-(\lambda) = \theta - \varphi = \alpha - \mu - \varphi$; we have

$$\varphi'_\lambda = \frac{\sin\lambda\sin\alpha\sin^2(\mu+\varphi)}{\cos\beta\cos(\mu+\varphi) - \cos\alpha}.$$

The denominator

$$\cos\beta\cos(\mu+\varphi)-\cos\alpha = \cos\beta\cos\mu\cos\varphi-\cos\beta\sin\mu\sin\varphi-\cos\alpha =$$

$$-2\sin\frac{\varphi}{2}\left(\cos\beta\sin\mu\cos\frac{\varphi}{2}+\cos\alpha\sin\frac{\varphi}{2}\right)$$

is equivalent to $-\varphi\cos\beta\sin\mu$ when φ tends to 0, while the numerator tends to $\sin\beta\sin^2\mu$; thus

$$\varphi'_\lambda \approx -\frac{\tan\beta\sin\mu}{\varphi}.$$

Now, from (3),

$$t = \frac{\cos\alpha-\cos\beta\cos(\mu+\varphi)}{\sin(\mu+\varphi)} \approx \varphi\cos\beta;$$

if $\lambda = \psi_m - u$, we have

$$t^2 = \sin^2\beta-\sin^2\alpha\left(\sin\psi_m\cos u-\cos\psi_m\sin u\right)^2 =$$

$$\left(2\sin^2\beta-\sin^2\alpha\right)\sin^2 u+\sin\beta\cos\beta\sin\mu\sin 2u,$$

because $\cos\psi_m = \dfrac{\cos\beta}{\sin\alpha}\sin\mu$. Thus

$$t^2 \approx 2\sin\beta\cos\beta\sin\mu\cdot u$$

when u tends to 0; this gives

$$\varphi \approx \sqrt{2u\tan\beta\sin\mu}$$

and finally

(5) $$\varphi'_\lambda \approx -\sqrt{\frac{\tan\beta\sin\mu}{2u}}$$

for $\lambda = \psi_m - u$, u tending to 0.

Second part: investigation of $\eta = \dfrac{\psi - \lambda}{\varphi}$ *as a function of* λ

We have

$$\frac{\partial}{\partial \lambda} \frac{\psi - \lambda}{\varphi} = \frac{-\varphi - (\psi - \lambda)\varphi'_\lambda}{\varphi^2}$$

with a sign opposite to that of $\varphi + (\psi - \lambda)\varphi'_\lambda$ and

$$\frac{\partial}{\partial \lambda}\left(\varphi + (\psi - \lambda)\varphi'_\lambda\right) = (\psi - \lambda)\varphi''_{\lambda^2}$$

with the same sign as φ''_{λ^2} because $\lambda \le \psi$. We have seen that $\varphi''_{\lambda^2} \le 0$ for $\alpha \ge \beta$ or for $\alpha < \beta$ and $\lambda \le \lambda_0$ (where $\theta_-(\lambda_0) = \theta_0$), while $\varphi''_{\lambda^2} > 0$ for $\alpha < \beta$ and $\lambda > \lambda_0$.

So if $\alpha \ge \beta$, $\varphi + (\psi - \lambda)\varphi'_\lambda$ decreases from its initial value $\varphi = \beta + \theta$ (for $\lambda = 0$, where $\varphi'_\lambda = 0$) to its final value $\varphi = 0$ or $2 \operatorname{Arc\,tan}\dfrac{t}{\cos \beta} > 0$, depending on whether $\theta \le \alpha - \mu$ or $\theta > \alpha - \mu$ (for $\lambda = \psi$).

In fact, for $\lambda = \psi$, $(\psi - \lambda)\varphi'_\lambda = 0$ even in the case where φ'_λ is infinite, because if $\lambda = \psi_m - u$, $(\psi_m - \lambda)\varphi'_\lambda \approx -u \cdot \sqrt{\dfrac{\tan \beta \sin \mu}{2u}} = -\sqrt{\dfrac{u \tan \beta \sin \mu}{2}}$ which tends to 0 with u. As a result $\varphi + (\psi - \lambda)\varphi'_\lambda$ remains positive and $\dfrac{\psi - \lambda}{\varphi}$ decreases monotonically. Ibn al-Haytham's second inequality is thus true in this case, and that holds for $-\beta \le \theta \le \beta, 0 \le \lambda \le \psi$.

If $\alpha < \beta$, $\varphi + (\psi - \lambda)\varphi'_\lambda$ decreases from $\beta + \theta$ to a minimum that is reached for $\lambda = \lambda_0$, then increases for $\lambda_0 \le \lambda \le \psi$; the final value is $\varphi = 0$, so the minimum is negative and there exists a unique $\lambda_1 \in]0, \lambda_0[$ where $\varphi + (\psi - \lambda)\varphi'_\lambda$ is zero. We have $\varphi + (\psi - \lambda)\varphi'_\lambda \ge 0$ for $0 \le \lambda \le \lambda_1$ and $\varphi + (\psi - \lambda)\varphi'_\lambda \le 0$ for $\lambda_1 \le \lambda \le \psi$; we put $\theta_1 = \theta_-(\lambda_1)$. We see that $\dfrac{\psi - \lambda}{\varphi}$ decreases as long as $\lambda \le \lambda_1$ but increases for $\lambda > \lambda_1$.

For $\lambda = \lambda_1$, $\varphi = \theta - \theta_1$, so we have $\psi - \lambda_1 = \dfrac{\theta - \theta_1}{\theta'_-(\lambda_1)}$; which means that (θ_1, λ_1) is the point of contact of the tangent to the graph of ψ as a function of θ drawn from the point (θ, ψ) of the graph (see Figs 2.70–71). We have

$\theta_1 \leq \theta_0$ and the fact that ψ is concave for $\theta \leq \theta_0$ proves that θ_1 is a decreasing function of θ; for $\theta = \theta_0$, $\theta_1 = \theta_0$.

Note: Let p denote the value of the derivative $\dfrac{d\lambda_1}{d\theta_1}$ when $\theta = \theta_0$, that is the slope of the tangent at the point of inflexion. Let $\theta = \theta_0 + u$ and $\psi = \lambda_0 + pu + qu^3 + \ldots$ the corresponding value of ψ where u tends to 0 and where q is the value of $\dfrac{1}{6}\dfrac{d^3\lambda_1}{d\theta_1^3}$ at $\theta = \theta_0$. The tangent to the graph of ψ through the point (θ, ψ) touches the curve at a point (θ_1, λ_1) where $\theta_1 = \theta_0 - v$, $\lambda_1 = \lambda_0 - pv - qv^3 - \ldots$ and $\dfrac{d\lambda_1}{d\theta_1} = p + 3qv^2 + \ldots$; we have

$$\lambda_1 + \frac{d\lambda_1}{d\theta_1}(\theta - \theta_1) = \lambda_0 + pu + 3quv^2 + 2qv^3 + \ldots = \psi = \lambda_0 + pu + qu^3 + \ldots$$

So we have $u^3 = 3uv^2 + 2v^3$, which gives $\dfrac{v}{u} = \dfrac{1}{2}$; thus the derivative of λ_1 with respect to θ has the value $-\dfrac{p}{2}$ for $\theta = \theta_0$.

For $\theta = 2\alpha - \beta$, θ_1 has a minimum value $\theta_{1,m}$ and λ_1 has a minimum value $\lambda_{1,m}$. If $\theta = 2\alpha - \beta - u$, we have seen that $\psi = \pi - v$ where $v \approx \sqrt{\dfrac{2u\sin\beta}{\sin\alpha\sin(\beta - \alpha)}}$ when u tends to 0. If to $\theta = 2\alpha - \beta - u$ there corresponds $\theta_1 = \theta_{1,m} + w$, we have

$$\lambda_1 = \lambda_{1,m} + \lambda'_{1,m}w + qw^2 + \ldots \text{ and } \lambda'_1 = \lambda'_{1,m} + 2qw + \ldots,$$

so

$$\psi = \lambda_{1,m} + \lambda'_{1,m}w + qw^2 + \ldots + (\lambda'_{1,m} + 2qw + \ldots)(2\alpha - \beta - u - \theta_{1,m} - w)$$

$$= \lambda_{1,m} + \lambda'_{1,m}(2\alpha - \beta - \theta_{1,m}) - \lambda'_{1,m}u + 2qw(2\alpha - \beta - \theta_{1,m}) - qw(2u + w)$$

$$= \pi - \lambda'_{1,m}u + 2qw(2\alpha - \beta - \theta_{1,m}) + \ldots;$$

thus

$$v \approx \sqrt{\frac{2u\sin\beta}{\sin\alpha\sin(\beta - \alpha)}} \approx -2qw(2\alpha - \beta - \theta_{1,m});$$

we see that $q < 0$ and that

$$\frac{\lambda_1 - \lambda_{1,m}}{-u} \approx -\lambda'_{1,m} \frac{w}{u} \approx \frac{\lambda'_{1,m}}{q(2\alpha - \beta - \theta_{1,m})} \sqrt{\frac{\sin \beta}{2u \sin \alpha \sin(\beta - \alpha)}}$$

tends to $-\infty$ when u tends to 0. So the tangent to the limiting curve is vertical for $\theta = 2\alpha - \beta$.

To ensure the second inequality is satisfied, we may make $\theta_0 \geq 0$ in the case where $\alpha < \beta$. We have seen (in the commentary on Proposition 14) that this is equivalent to $\alpha \geq \alpha_1 (\beta)$ where

$$\cos^2 \alpha_1(\beta) = \cos \frac{\beta}{2} \frac{\cos \frac{\beta}{2} - \sqrt{2} \sin^2 \frac{\beta}{2}}{1 + \sin^2 \frac{\beta}{2}}.$$

If $\alpha < \alpha_1 (\beta)$, the second inequality is satisfied if $\theta \leq \theta_0 < 0$, or if $\theta > \theta_0$ and $\lambda \leq \lambda_1 < \psi$.

Third part: investigation of $\xi = \dfrac{\psi - \lambda}{\varphi} \sin(\alpha - \theta)$ *as a function of* θ

We have

$$\xi = \frac{\overline{CL}}{\overline{LO}} = \frac{\overline{CL}}{\overline{LS}} \cdot \frac{\overline{LS}}{\overline{LO}}$$

where

$$LS = LW - SW = r \sin(\alpha - \theta)\left(1 - \frac{\cos \psi}{\cos \lambda}\right)$$

because $SW \cos \lambda = WG = r \sin(\alpha - \theta) \cos \psi$. Thus

(6) $$\xi = \frac{\psi - \lambda}{\cos \lambda - \cos \psi} \cdot \frac{\cos \lambda - \cos \psi}{\varphi} \sin(\alpha - \theta).$$

The value of the second factor is

$$\frac{2 \sin \frac{\varphi}{2}}{\varphi} \cdot \frac{t \cos \frac{\varphi}{2} - \cos \beta \sin \frac{\varphi}{2}}{\sin \alpha}$$

from equation (2); this is clearly a decreasing function of $\varphi = \theta - \theta_-(\lambda)$, and thus also of θ (bearing in mind that $t \geq 0$). So we now need to look at the first factor $\dfrac{\psi - \lambda}{\cos\lambda - \cos\psi}$; we have

$$\frac{\partial}{\partial\psi}\frac{\psi - \lambda}{\cos\lambda - \cos\psi} = \frac{\cos\lambda - \cos\psi - (\psi - \lambda)\sin\psi}{(\cos\lambda - \cos\psi)^2}.$$

So $\dfrac{\psi - \lambda}{\cos\lambda - \cos\psi}$ is a decreasing function of ψ provided that

(7) $(\psi - \lambda)\sin\psi + \cos\psi - \cos\lambda \geq 0;$

we need to investigate this inequality for $\lambda \leq \psi \leq \pi$ (λ being fixed). Now

$$\frac{\partial}{\partial\psi}\big((\psi - \lambda)\sin\psi + \cos\psi - \cos\lambda\big) = (\psi - \lambda)\cos\psi$$

has the same sign as $\cos\psi$: the first term of (7) increases for $\lambda \leq \psi \leq \dfrac{\pi}{2}$, then decreases for $\dfrac{\pi}{2} \leq \psi \leq \pi$. If $\lambda \geq \dfrac{\pi}{2}$, the conditions of the second case always apply and as the first term of (7) is zero for $\psi = \lambda$, it stays negative: (7) cannot be satisfied and $\dfrac{\psi - \lambda}{\cos\lambda - \cos\psi}$ increases with ψ. On the other hand, if $\lambda < \dfrac{\pi}{2}$, the first term of (7) increases from 0 to a maximum value $\dfrac{\pi}{2} - \lambda - \cos\lambda > 0$, which is reached when $\psi = \dfrac{\pi}{2}$, then decreases to $-1 - \cos\lambda = -2\cos^2\dfrac{\lambda}{2} < 0$, a value reached for $\psi = \pi$. So there exists a (unique) value $f(\lambda) \in [\dfrac{\pi}{2}, \pi[$ of ψ which makes $(\psi - \lambda)\sin\psi + \cos\psi - \cos\lambda$ zero and (7) is satisfied for $\lambda \leq \psi \leq f(\lambda)$, but not for $\psi > f(\lambda)$.

Let us examine the function $f(\lambda)$, defined by

(8) $(f(\lambda) - \lambda)\sin f(\lambda) + \cos f(\lambda) - \cos\lambda = 0,$ $\dfrac{\pi}{2} \leq f(\lambda) < \pi,$

where $0 \le \lambda \le \frac{\pi}{2}$; for $\lambda = 0$, $f(0) \sin f(0) + \cos f(0) - 1 = 0$; let $f(0) = \tan \frac{f(0)}{2}$, which gives $f(0) = \delta = 2.33112237$ radians (see commentary on Proposition 14); for $\lambda = \frac{\pi}{2}$,

$$\left(f\left(\frac{\pi}{2}\right) - \frac{\pi}{2}\right) \sin f\left(\frac{\pi}{2}\right) + \cos f\left(\frac{\pi}{2}\right) = 0,$$

or

$$f\left(\frac{\pi}{2}\right) - \frac{\pi}{2} = \tan\left(f\left(\frac{\pi}{2}\right) - \frac{\pi}{2}\right),$$

which has only one solution in the interval $\left[\frac{\pi}{2}, \pi\right]$, namely $f\left(\frac{\pi}{2}\right) = \frac{\pi}{2}$. Differentiating (8), we have

$$(f'(\lambda) - 1) \sin f(\lambda) + \{(f(\lambda) - \lambda) \cos f(\lambda) - \sin f(\lambda)\} f'(\lambda) + \sin \lambda = 0$$

$$(9) \quad f'(\lambda) = \frac{\sin f(\lambda) - \sin \lambda}{\cos \lambda - \cos f(\lambda)} \tan f(\lambda) = \frac{\tan f(\lambda)}{\tan \dfrac{f(\lambda) + \lambda}{2}},$$

since $f(\lambda) - \lambda = \dfrac{\cos \lambda - \cos f(\lambda)}{\sin f(\lambda)}$.

We can see that $f'(\lambda) \le 0$ so long as $f(\lambda) + \lambda \le \pi$ (because $\tan f(\lambda) \le 0$). For $\lambda = 0$, $f(\lambda) + \lambda = \delta < \pi$ and $f'(\lambda) + 1 = 1 + \dfrac{\tan\delta}{\tan \dfrac{\delta}{2}} = 0.54893386 > 0$. So $f(\lambda) + \lambda$ increases from δ so long as $f'(\lambda) \ge -1$. If there exists a value of λ such that $f'(\lambda) = -1$, we have, for that value,

$$\tan f(\lambda) = -\tan \frac{f(\lambda) + \lambda}{2} = \tan\left(\pi - \frac{f(\lambda) + \lambda}{2}\right),$$

hence $\lambda = 2\pi - 3 f(\lambda)$, which makes $f(\lambda) \le \dfrac{2\pi}{3}$.

Putting $\lambda = 2\pi - 3 f(\lambda)$ into (8), we have

$$(4f(\lambda) - 2\pi) \sin f(\lambda) + \cos f(\lambda) - \cos 3f(\lambda) =$$
$$2\sin f(\lambda) (2f(\lambda) - \pi + \sin 2f(\lambda)) = 0,$$

which gives $f(\lambda) = \dfrac{\pi}{2}$, so $\lambda = \dfrac{\pi}{2}$, a value for which formula (9) is indeterminate. Let us put $\lambda = \dfrac{\pi}{2} + u$, $f(\lambda) = \dfrac{\pi}{2} + v$; we have

$$f'(\lambda) = \frac{\tan \dfrac{u+v}{2}}{\tan v},$$

where $v \approx f'\left(\dfrac{\pi}{2}\right) u$ when u tends to 0. From which it follows, if we look at the limit, that

$$f'\left(\frac{\pi}{2}\right) = \frac{f'\left(\dfrac{\pi}{2}\right) + 1}{2f'\left(\dfrac{\pi}{2}\right)},$$

an equation that gives $f'\left(\dfrac{\pi}{2}\right) = -\dfrac{1}{2}$ (the only negative root). So we conclude that $f'(\lambda) > -1$ and that $f(\lambda) + \lambda$ increases from δ to π and thus has an upper bound π; it follows that $f'(\lambda) \le 0$ and that f decreases from δ to $\dfrac{\pi}{2}$ for $0 \le \lambda \le \dfrac{\pi}{2}$.

We know that ψ is an increasing function of θ if $\alpha < \beta$ or if $\alpha \ge \beta$ and $0 \le \alpha - \mu$; it follows that $\dfrac{\psi - \lambda}{\cos \lambda - \cos \psi}$ is a decreasing function of θ if $\alpha < \beta$, $\lambda \le \dfrac{\pi}{2}$ and $\lambda \le \psi \le f(\lambda)$, or if $\alpha \ge \beta$, $\theta \le \alpha - \mu$, in which case $\lambda \le \psi \le \psi_m \le \dfrac{\pi}{2}$.

In the case where $\alpha \ge \beta$, we can also examine the interval $\alpha - \mu \le \theta \le \beta$ thanks to the expression

$$\xi = \frac{\widehat{CL}}{CL} \cdot \frac{CL}{LO} \cdot \frac{LO}{\widehat{LO}} = \frac{\psi - \lambda}{2\sin \dfrac{\psi - \lambda}{2}} \cdot \frac{\sin \dfrac{\psi - \lambda}{2} \sin(\alpha - \theta)}{\sin \dfrac{\varphi}{2}} \cdot \frac{2\sin \dfrac{\varphi}{2}}{\varphi},$$

in which the last factor decreases as a function of φ, and thus of θ. The first two factors are decreasing functions of θ if the same is true of ψ, that is, if $\alpha \geq \beta$ and $\theta \geq \alpha - \mu$. Thus, when $\alpha \geq \beta$, ξ decreases as a function of $\theta \in [-\beta, \beta]$ for $0 \leq \lambda \leq \psi$, that is, when the point (θ, λ) is below the curve for ψ (Figs 2.67 and 2.68).

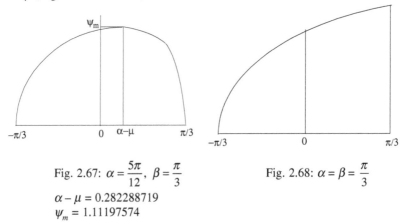

$$\text{Fig. 2.67: } \alpha = \frac{5\pi}{12}, \ \beta = \frac{\pi}{3} \qquad\qquad \text{Fig. 2.68: } \alpha = \beta = \frac{\pi}{3}$$

$$\alpha - \mu = 0.282288719$$
$$\psi_m = 1.11197574$$

In the case where $\alpha < \beta$, ξ will decrease if $\lambda \leq \frac{\pi}{2}$ and $\psi \leq f(\lambda)$, that is, $\theta \leq \theta_-(f(\lambda))$; when $\theta = \theta_-(f(\lambda))$, we have $\frac{d\xi}{d\theta} < 0$, but $\frac{d\xi}{d\theta}$ becomes plus infinity for $\theta = 2\alpha - \beta$. So there exists a value θ_4 of θ between $\theta_-(f(\lambda))$ and $2\alpha - \beta$ for which $\frac{d\xi}{d\theta}$ goes from negative to positive; it can be established that this value is unique and that ξ decreases in the interval $-\beta \leq \theta \leq \theta_4$ and then increases. Similarly, if $\lambda \geq \frac{\pi}{2}$, there exists a value θ_4 between $\theta_-(\lambda)$ and $2\alpha - \beta$ such that ξ decreases as far as θ_4 and then increases. The region of the plane (θ, λ) in which Ibn al-Haytham's statement is true is determined by the inequality

$$\frac{\partial\xi}{\partial\theta} = \frac{\varphi\psi' - \psi + \lambda}{\varphi^2}\sin(\alpha - \theta) - \frac{\psi - \lambda}{\varphi}\cos(\alpha - \theta) \leq 0,$$

or

(10) $((\theta - \theta_-(\lambda))\,\psi' - \psi + \lambda)\sin(\alpha - \theta) - (\theta - \theta_-(\lambda))(\psi - \lambda)\cos(\alpha - \theta) \leq 0$

and θ_4 is determined by the equality in (10).

Note: If $\theta \le \alpha - \mu'$, $\psi \le \dfrac{\pi}{2}$ and we have $\lambda \le \psi \le f(\lambda)$, so $\dfrac{d\xi}{d\theta} \le 0$ and it follows that $\theta_4 \ge \alpha - \mu'$.

We may note that θ_4 is a decreasing function of λ. Its minimum value is $\theta_-(\lambda)$ since $\lambda \le \psi$. Let us write $\theta = \theta_-(\lambda) + u$ in the equality in (10). We have $\psi = \lambda + \psi'_* u + \dfrac{\psi''_*}{2} u^2 + \dots$ and $\psi' = \psi'_* + \psi''_* u + \dots$ using ψ'_*, ψ''_* to denote the values of the derivatives for $\theta = \theta_-(\lambda)$.

Putting these into the first term of (10), we get

$$u^2\left[\frac{1}{2}\psi''_*\sin\big(\alpha - \theta_-(\lambda)\big) - \psi'_*\cos\big(\alpha - \theta_-(\lambda)\big)\right] + \dots$$

The minimum value of θ_4 is thus given by the equation

(11) $\qquad \psi'' \sin(\alpha - \theta) - 2\psi' \cos(\alpha - \theta) = 0.$

If we use equation (6) of the commentary on Proposition 14, this equation can be transformed into

(12) $\qquad \dfrac{P(X)}{Q(X)} = 2X(A - BX)$

or

$$R(X) = 2X(A - BX)\,Q(X) - P(X) = 0.$$

We have

$$R(X) = BX^4 - 3AB^2\,X^3 + BX^2\,(3A^2 + 2B^2 - 2) - A^3 X + B - B^3,$$

where $A = \cos \alpha$, $B = \cos \beta$ and $X = \cos(\alpha - \theta)$. For $\theta = \alpha - \mu'$, let $X = \cos\mu' = \dfrac{B}{A}$, we have

$$R\left(\frac{B}{A}\right) = \frac{1}{A^4} B\left(1 - A^2\right)\left(A^2 - B^2\right)^2 > 0$$

and for $\theta = 2\alpha - \beta$, $X = \cos(\beta - \alpha)$, we get

$$R(\cos(\beta - \alpha)) = -\sin^2\beta \, \sin^3(\beta - \alpha) \sin \alpha < 0.$$

An investigation like that carried out in the commentary on Proposition 14 shows that $R(X)$ decreases from $R\left(\dfrac{B}{A}\right)$ to $R(\cos(\beta - \alpha))$ for $\dfrac{B}{A} \leq X \leq$ $\cos(\beta - \alpha)$. Thus R becomes zero for a determinate value of $X = \cos(\alpha - \theta)$ that corresponds to the minimum value of θ_4, namely, $\theta_{4,m}$, which is the value that was required.

Ibn al-Haytham's statement is true in the case where $\theta_{4,m} \geq 0$, which is equivalent to $R(\cos \alpha) \geq 0$. Now

$$R(\cos \alpha) = R(A) = (1 - B)\,(A^4\,(3B - 1) + B(B + 1)\,(1 - 2A^2));$$

so the required inequality is

$$A^4\,(3B - 1) - 2B(B + 1)\,A^2 + B(B + 1) \geq 0.$$

Let us put $S(x) = (3B - 1)\,x^2 - 2\,B(B + 1)\,x + B(B + 1)$; we have

$$S(1) = -(B - 1)^2 < 0 \text{ and } S(B^2) = B(B - 1)^2\,(3B^2 + 3B + 1) > 0;$$

so S has a root x_0 between B^2 and 1, and $\theta_{4,m} \geq 0$ is equivalent to $x_0 \geq A^2 =$ $\cos^2\alpha_0(\beta)$. The condition is $\alpha \geq \alpha_0\,(\beta)$, with

$$(13) \qquad \cos^2 \alpha_0(\beta) = 2\cos\frac{\beta}{2}\,\frac{\cos\beta\cos\dfrac{\beta}{2} - \sin^2\dfrac{\beta}{2}}{3\cos\beta - 1}\sqrt{2\cos\beta} = \frac{1}{1 + \sqrt{2}\,\dfrac{\sin\dfrac{\beta}{2}\,\tan\dfrac{\beta}{2}}{\sqrt{\cos\beta}}},$$

or

$$\tan^2\alpha_0(\beta) = \sqrt{2}\,\frac{\sin\dfrac{\beta}{2}\,\tan\dfrac{\beta}{2}}{\sqrt{\cos\beta}}.$$

We can see that $\alpha_0(0) = 0$ and $\alpha_0^2 \approx \dfrac{\beta^2\sqrt{2}}{4}$ as β tends to 0; thus

$$\left.\frac{d\alpha_0}{d\beta}\right|_{\beta=0} = \frac{\sqrt[4]{2}}{2} = 0.594603558 \text{ and } \left.\frac{d\beta}{d\alpha_0}\right|_{\alpha_0=0} = \sqrt[4]{8} = 1.68179283.$$

We also have $\alpha_0\left(\dfrac{\pi}{2}\right) = \dfrac{\pi}{2}$ and $\dfrac{d\beta}{d\alpha_0}\bigg|_{\alpha_0 = \frac{\pi}{2}} = \lim_{\alpha_0 \to \frac{\pi}{2}} \dfrac{\cos\beta}{\cos\alpha_0} = 0$.

In Fig. 2.50 we have drawn the graph of α_0; Fig. 2.51 shows the corresponding limit for the point Ω.

The maximum value of θ_4, namely $\theta_{4,m}$, is obtained for $\lambda = 0$; it is the value that was denoted simply by θ_4 in the commentary on Proposition 14.

Fourth part: examples

We have chosen the same numerical values as in the commentary on Proposition 14: $\beta = \dfrac{\pi}{3}$ and $\alpha = \dfrac{5\pi}{12}$, $\dfrac{\pi}{3}$, 1, 0.8, 0.75, $\dfrac{\pi}{5}$, 0.55, 0.524 and $\dfrac{\pi}{12}$ (Figs 2.67 to 2.75).

We have $\alpha_0\left(\dfrac{\pi}{3}\right) = 0.649766287$ which lies between 0.75 and $\dfrac{\pi}{5}$ and $\alpha_1\left(\dfrac{\pi}{3}\right) = 0.932458266$, between 1 and 0.8. In Figs 2.69 to 2.75, we have drawn three curves numbered I, II and III respectively. Curve I is the graph of ψ as a function of θ; it goes from the point $(-\beta, 0)$ to $(2\alpha - \beta, \pi)$, the function being an increasing one. Curve II is the graph of λ_1 as a function of θ; it goes from the point of inflection (θ_0, λ_0) of curve I to the point $(2\alpha - \beta, \lambda_{1,m})$ and it ends in the vertical segment $\theta = 2\alpha - \beta$, $0 \le \lambda \le \lambda_{1,m}$. The tangent to curve I at the point with ordinate λ_1 cuts the curve again at the point with abscissa θ (Figs 2.69 to 2.71); the function λ_1 of θ decreases. Curve III is the locus of points (θ_4, λ); it runs between the points $(\theta_{4,m}, \lambda_{4,m})$ and $(\theta_{4,M}, 0)$ and λ is a decreasing function of θ_4.

Ibn al-Haytham's first inequality holds for points (θ, λ) below curves I and III while his second inequality holds for points (θ, λ) below curves I and II. So the two inequalities both hold in the convex region below the three curves I, II and III.

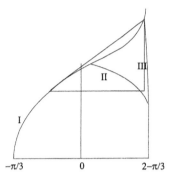

Fig. 2.69: $\alpha = 1$, $\beta = \dfrac{\pi}{3}$; $\theta_0 = 0.220685446$, $\lambda_0 = 1.37400009$

$\theta_{1,m} = -0.833589085$, $\lambda_{1,m} = 0.663153782$
$\theta_{4,m} = 0.876987428$, $\lambda_{4,m} = 1.92921309$
$\theta_{4,M} = 0.9500790346$

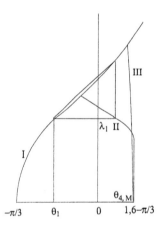

Fig. 2.70: $\alpha = 0.8$, $\beta = \dfrac{\pi}{3}$; $\theta_0 = -0.237427409$, $\lambda_0 = 1.333256925$

$\theta_{1,m} = -0.8722605915$, $\lambda_{1,m} = 0.644350074$
$\theta_{4,m} = 0.349257621$, $\lambda_{4,m} = 1.98972898$
$\theta_{4,M} = 0.537776499$

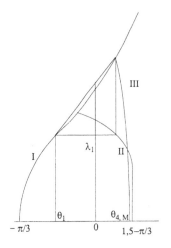

Fig. 2.71: $\alpha = 0.75$, $\beta = \dfrac{\pi}{3}$; $\theta_0 = 0.306153376$

$\lambda_0 = 1.33286849$, $\theta_{1,m} = -0.881787545$

$\lambda_{1,m} = 0.642761272$, $\theta_{4,m} = 0.4346042281$

$\lambda_{4,m} = 1.97974742$, $\theta_{4,M} = 0.229510063$

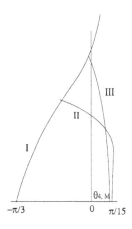

Fig. 2.72: $\alpha = \dfrac{\pi}{5}$, $\beta = \dfrac{\pi}{3}$; $\theta_0 = -0.450887379$, $\lambda_0 = 1.3408219$

$\theta_{1,m} = -0.904851513$, $\lambda_{1,m} = 0.644432659$

$\theta_{4,m} = -0.0472107752$, $\lambda_{4,m} = 1.93620825$

$\theta_{4,M} = 0.1831447956$

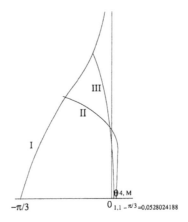

$-\pi/3$ 0 $_{1,1-\pi/3\,=0.0528024188}$

Fig. 2.73: $\alpha = 0.55$, $\beta = \dfrac{\pi}{3}$; $\theta_0 = -0.53311435$, $\lambda_0 = 1.35133667$

$\theta_{1,m} = -0.919752538$, $\lambda_{1,m} = 0.649924588$
$\theta_{4,m} = -0.213335252$, $\lambda_{4,m} = 1.89753524$
$\theta_{4,M} = 0.0211388793$

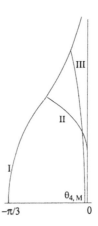

$-\pi/3$ 0

Fig. 2.74: $\alpha = 0.524$, $\beta = \dfrac{\pi}{3}$; $\theta_0 = -0.55910491$, $\lambda_0 = 1.35583723$

$\theta_{1,m} = -0.9247411$, $\lambda_{1,m} = 0.652562409$
$\theta_{4,m} = -0.266150532$, $\lambda_{4,m} = 1.88337596$
$\theta_{4,M} = -0.0326171933$

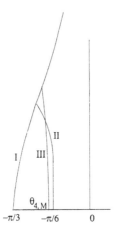

Fig. 2.75: $\alpha = \dfrac{\pi}{12}$, $\beta = \dfrac{\pi}{3}$; $\theta_0 = -0.801761542$, $\lambda_0 = 1.43427582$

$\theta_{1,m} = -0.978448494$, $\lambda_{1,m} = 0.704691461$

$\theta_{4,m} = -0.724212075$, $\lambda_{4,m} = 1.72529646$

$\theta_{4,M} = -0.566379471$

This long analytical investigation, illustrated with examples and figures, shows that Ibn al-Haytham's statements describe the variation of certain rather complicated transcendental functions. The validity of his statements depends on certain conditions that Ibn al-Haytham could not express in explicit terms; formulating these conditions belongs to a kind of mathematics that would not appear until eight centuries later.

All the same, the investigation of the variation of trigonometric functions, as carried out by Ibn al-Haytham, and prompted by research in astronomy, marks the beginning of work in a new area of mathematical research, one in which operations combine methods that might be described as function-oriented and ones that relate to infinitesimals.

2.2. ASTRONOMY

In the second part of his book, Ibn al-Haytham at once sets out upon an investigation of the apparent movement of the seven wandering stars, beginning with the two luminaries.

2.2.1. *The apparent movement of the seven wandering stars*

The movement of the moon

Ibn al-Haytham begins by mentioning some of the results established by Ptolemy, although he does not adopt the models Ptolemy proposes in the *Almagest*. Ibn al-Haytham had indeed criticized these models in his *Doubts concerning Ptolemy*.[14] Let us first of all mention some of these results set out by Ibn al-Haytham.

• In its apparent movement on the celestial sphere, the centre of the moon always lies in the plane of a great circle; this is the oblique orb.

• The oblique orb cuts the circle of the ecliptic along the line of the nodes *NN'* (see Fig. 2.76) and makes an angle with the plane of the ecliptic. Ibn al-Haytham considers this angle as fixed. In fact, it varies very slightly, remaining close to 5°. The oblique orb thus lies within the Zodiac.

The movement of the centre of the moon on its oblique orb is *direct*, that is, it is in the same sense as the signs of the Zodiac (the period is one month).

• The movement of a node along the circle of the ecliptic is *retrograde*, that is, it is in the opposite sense to the signs of the Zodiac (the period is eighteen years and eight months).

• The plane of the oblique orb rotates about the axis defined by the poles of the ecliptic, and every point of the oblique orb of the moon describes a circle about that axis.

• If we refer the oblique orb of the moon to the equator, we can show that the angle the ecliptic makes with the plane of the equator is 24°, according to Ptolemy; 23°33' according to the calculations of ninth-century astronomers, and 23°27' according to later calculations; and the plane of the ecliptic cuts the equator along the diameter $\gamma\gamma'$ (the line of the equinoxes).

• The oblique orb cuts the plane of the equator along a diameter *MM'*.

• Since the nodes *N* and *N'* move along the ecliptic, the inclination of the oblique orb to the equator is variable.

---

[14] *Al-Shukūk 'alā Baṭlamiyūs* (*Doubts concerning Ptolemy*), ed. A. Sabra and N. Shehaby, Cairo, 1971, pp. 15–19.

Once these results have been set out as being secure – unlike all the other results put forward by Ptolemy – and the terminology has been defined, Ibn al-Haytham proceeds to construct his models for the movement of the wandering stars, starting with the moon. But in order to do this he introduces a new concept, that of the 'time required'. This expression is used to denote the time that elapses during the diurnal motion of the moon (or of any wandering star), from the equator to the meridian, a time represented by an arc of a circle. Since all the simple motions involved in constructing the apparent motions are circular and uniform, that makes it possible to give a measure for the time required by an arc of a circle, and the time can thus be handled by means of the theory of proportions.

The apparent motion of the moon is complicated. It is the result of combining three motions: first, motion in the plane of its oblique orb, from north to south and back again, with respect to the equator – this motion is direct, that is, from west to east. The second motion is that of the orb, inclined at an angle, about the axis of the ecliptic – the motion of the node. Finally, there is the diurnal motion.

This composition of motions produces a phenomenon that Ibn al-Haytham finds extremely interesting. Let us suppose that the moon is at the point B in its orb. The point B is a point on the celestial sphere, and thus participates in the diurnal motion, that is it moves round a circle parallel to the circle of the equator. The moon also participates in this motion. The point B is a point on the oblique orb, and is thus subject to the motion of the node along a circle parallel to the ecliptic. The moon is also subject to this motion. But in addition the moon has its own motion along its oblique orb. Thus, after a time t, the point reached by the point B and that reached by the moon cannot be identical. It is the distance between B and the moon that we need to be able to find, and it will be the main subject of Ibn al-Haytham's investigation, as we shall soon see.

Proposition 16. — Let O be the centre of the universe, P the north pole of the ecliptic, H the north pole of the equator. The great circle PH is called the circle of the poles.

If X is the north pole of the oblique orb, the angle POX is equal to the inclination of the orb to the plane of the ecliptic, so $P\hat{O}X \cong 5°$, an angle Ibn al-Haytham considers to be constant. So when the node N moves round the ecliptic, the pole X describes a circle round the axis OP; this circle cuts the circle of the poles in two points, the point A between P and H, and the point B beyond P. Considering arcs of great circles, we have for all positions of X

$$\overset{\frown}{PA} = \overset{\frown}{PB} = \overset{\frown}{PX} \text{ and } \overset{\frown}{HA} < \overset{\frown}{HX} < \overset{\frown}{HB}.$$

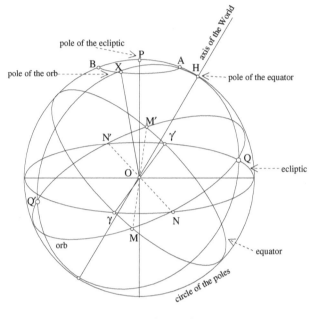

Fig. 2.76

In the course of one revolution, $\widehat{HX}$ increases from $\widehat{HA}$ to $\widehat{HB}$, then decreases from $\widehat{HB}$ to $\widehat{HA}$.

$\widehat{HB} \cong 23°27' + 5°$,

$\widehat{HA} \cong 23°27' - 5°$, with today's values.

The oblique orb cuts the circle of the equator in a diameter MM'; so a semicircle of the oblique orb is north of the equator and a semicircle is south of the equator. The midpoint of the northern semicircle gives the maximum northern inclination of the moon, and the midpoint of the southern semicircle gives the maximum southern inclination. These points, Q and Q', are the points of intersection of the oblique orb and the great circle that passes through the pole H of the equator and the pole X of the oblique orb. So the positions of these points are variable and the inclinations that correspond to them also vary.

The motion of the moon on its orb takes place in the direction of the succession of the signs of the Zodiac, whereas the motion of the node N, like that of a general point in the oblique orb about the axis *OP* of the ecliptic, is in the opposite direction to that of the signs of the Zodiac.[15]

Investigation of the motion of the moon between its rising and its the meridian transit

Let *ABC* be the eastern half of the circle of the horizon, *BED* the half of the orb of the moon that is *below* the horizon and let *H* be the north pole of the equator (Fig. 2.77).

We suppose that the moon starts at the point *B* and that on its orb it moves *from north to south*, from the point *B* towards the point *E*. (Each day the moon describes an arc of about 13° in direct motion.)

Through the point *B* we draw the circle *BIO*[16] with pole *H* (Fig. 2.77).

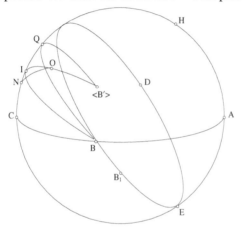

Fig. 2.77

ABC horizon, *AHC* meridian, *BED* oblique orb of the moon
H pole of the equator, *BIO* parallel circle

a) In the course of the *diurnal motion* (which is rapid), the point *B* of the orb describes this circle in retrograde motion and passes across the meridian at the point *I*. The moon participates in the diurnal motion, but does not remain at the point *B* of the orb. When the moon reaches the meridian at the point *N*, the point *B* has passed the point *I* and is at point *O* of the parallel circle *BI*; thus on its orb the moon has traversed the arc BB_1

which has reached the position *ON* to the west of the meridian and *south* of the parallel circle *BIO*.

b) This assumes that the point *B* remains on the parallel circle *BI*, that is that its declination is constant. But *B*, a point on the orb, is carried along by the *motion of the ascending node* on the ecliptic (in a retrograde direction). We draw through *B* an arc of a circle *BQ* whose pole is the pole of the circle of the ecliptic – the circle *BQ* is parallel to the circle of the ecliptic; so *B* moves on the arc *BQ* while still participating in the diurnal motion; it thus reaches a point that *in general is not the same as point O.*
In the figure, we assume the poles *H* and *P* are above the horizon, and that points *I* and *Q* are on the meridian and above the horizon of the point *B*.

Position of the circle BQ *with respect to the circle* BI

• If the great circle from *H*, the north pole of the equator, to the point *B* passes through the pole *P* of the ecliptic, then the circles *BI* and *BQ* touch one another at *B* (Fig. 2.78).
If *P* lies between *H* and *B*, $\widehat{BP} < \widehat{BH}$, then the circle *BQ* is *north* of circle *BI*.
If $\widehat{BP} > \widehat{BH}$, then the circle *BQ* is *south* of the circle *BI*.
In either case, the circle *HPB* is orthogonal to the two circles *BI* and *BQ*.

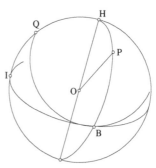

Fig. 2.78: $\widehat{BP} < \widehat{BH}$

• If the great circle from *H* to the point *B* does not pass through the pole *P* of the ecliptic, the circles *BI* and *BQ* intersect at *B* and at a second point *B'* (Fig. 2.79).

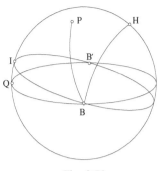

Fig. 2.79

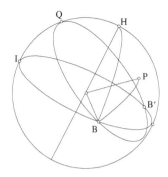

Fig. 2.80

• If the great circle from the pole P to the point B makes an acute angle with the arc BI, that is if $P\hat{B}I$ is acute, then $P\hat{B}I < H\hat{B}I$ (right angle); the arc BQB', which cuts the meridian above the horizon, is thus south of the arc BI (Fig. 2.80).

If the angle $P\hat{B}I$ is obtuse, then $P\hat{B}I > H\hat{B}I$; the arc BQB', which cuts the meridian above the horizon, is thus north of the circle BI.

Let t be the time the moon takes to move from B to N, and let X be the arc corresponding to the motion of the node[17] in this time t. The arc X, which is an arc of the circle BQ, is very small. Let B' be the second point of intersection of the circles BQ and BI.

• If $X = \widehat{BB'}$, then O is at B' and the arc ON is the arc traversed by the moon on the oblique orb in the time t ($\widehat{ON} = \widehat{BB_1}$) (see Fig. 2.77).

• If $X < \widehat{BB'}$, then after time t, B will not have reached B', it will not have returned to the circle BI.

• If $X > \widehat{BB'}$, then B describes the arc BB' of the circle BQ, reaches B' on the circle BI, then goes past B'.

In these last two cases, the position of B after time t is not at point O. Let M be the position of the point B on the circle BQ after time t, that is at the moment when the moon crosses the meridian at N. So in the first case ($O = M = B'$) the point M can be at O on BI and in all other cases it can be north or south of the circle BI.

---

[17] We know that the motion of the node corresponds to an arc of 3′ a day in the direction contrary to that of the signs of the Zodiac (*Fī dhikr al-aflāk*, in *Thābit ibn Qurra, Œuvres d'astronomie*, ed. R. Morelon, p. 21).

If t is the time the moon has taken to move from B to N, t is represented by the arc BS of the circle BI. The point B of the circle BQ has arrived at S on the circle BI and the point B of the orb has arrived at M (Fig. 2.81.1 and 2.81.2).

$\widehat{SM}$ is the arc traversed by the point B of the orb in time t because of the motion of the node.

Moreover, the arc MN is the arc traversed by the moon on its orb in the time t.

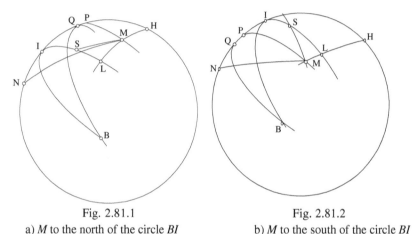

Fig. 2.81.1   Fig. 2.81.2

a) M to the north of the circle BI b) M to the south of the circle BI

Through H, a pole of the equator, we draw the great circle HM which cuts the parallel circle BI at the point L. Through M we draw the parallel circle which cuts the meridian in P.[18] In both cases, $\widehat{NP}$ is the inclination of the arc NM. We have $\widehat{PI} = \widehat{LM}$, the inclination of the arc SM.

We suppose that the moon moves from B towards E and that its motion is *from south to north*; the arc BED of its orb is thus north of the circle BI (Fig. 2.82). The arc BED is, by hypothesis, below the horizon. In the course of the diurnal motion, the point B describes the circle BI. The moon moves on its orb, away from the circle BI and to the north of it. The moon reaches the meridian at the point N, north of I; at that moment the point B arrives at the point O. The arc of the oblique orb described by the moon would then be in the position ON, if we did not have to take into account the motion of the node. We might go on to pursue the investigation as in the first case, taking account of the motion of the node.

---

[18] The letter P does not mean the same as it did before and does not indicate the pole of the ecliptic.

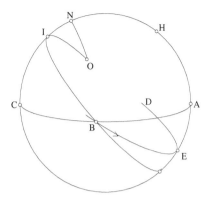

Fig. 2.82

To summarize, if the motion of the moon on its orb is *from north to south*, then *N* will be *south* of the circle *BI* and *M* will be north or south of the circle *BI*. If the motion of the moon on its orb is *from south to north*, then *N* will be *north* of the circle *BI* and *M* will be north or south of this circle. So we can give the following definitions:

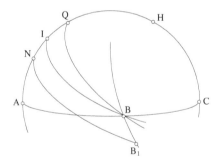

Fig. 2.83

I	the point where *B* crosses the meridian
N	the point where the moon crosses the meridian
Q	point of intersection of the circle *BQ* (motion of the node) and the meridian
$\overset{\frown}{BI}$	time required: time taken by the point *B*, carried by the diurnal motion, to arrive at the meridian
$\overset{\frown}{NI}$	inclination of the motion of the moon
$\overset{\frown}{QI}$	inclination of the motion of the node

Investigation of the motion of the moon between its passage across the meridian and its setting in the west

Horizon *ABCD*, *A* north, *B* east, *C* south, *D* west (Fig. 2.84).

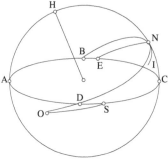

Fig. 2.84

The moon is at *N* on the meridian; let *BND* be the parallel circle of *N*, *NE* an arc of the orb and *NI* the arc described by *N* through the motion of the node.

When the moon reaches the horizon at the point *S*, the point *N* has moved past the point *D* and the arc *NE* described by the moon on its orb is now in the position *SO* (the figure shows the arc *NE* south of the circle *BND*, the moon goes from *N* towards *E*, so the arc *SO* is south of the circle *BND*).

If we take account of the motion of the node, the position of *N* when the moon reaches the horizon is in general not the same as *O*. Let this position be *M*.[19]

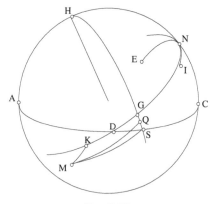

Fig. 2.85

<hr />

[19] The text continues to use the letter *S* in this case.

If we take account of the three motions and if the arc NK is the time the moon takes to travel from N to the point S on the horizon, then the arc NI of the motion of the node has assumed the position KM and the arc NE is in the position SM (Fig. 2.85).

We draw the great circle HS which cuts the circle ND in G and through M we draw an hour circle which cuts the arc HS in Q: the arc ND represents the time required. $\widehat{SG}$ is the inclination of the arc NS described by the moon (or the inclination of the motion of the moon). $\widehat{QG}$ is the inclination of the arc MK (or the inclination of the motion of the node).

The motion of the sun

Proposition 17. — Here, as for the motion of the moon, Ibn al-Haytham begins by defining his terms, setting out principles and determining the various motions that make up the motion of the sun. This time we are concerned with two motions: the diurnal motion and the proper motion along the ecliptic. So the model proposed for the motion of the sun is simpler than that constructed for the motion of the moon.

The sun moves on the ecliptic in the direction of the succession of the signs of the Zodiac, which is to say it has a direct motion around the axis of the ecliptic, if we are looking north.

The circle of the ecliptic cuts the circle of the equator in the equinoctial points γ and γ'. Ibn al-Haytham considers these points γ and γ' as fixed (see Note).

The circle of the ecliptic is divided into four equal arcs by the diameter $\gamma\gamma'$ and the diameter perpendicular to it, $\sigma\sigma'$, which gives the solstices:

σ, north of the equator, is the summer solstice; it is the point of the ecliptic with the maximal northern declination.

σ', south of the equator, is the winter solstice; it is the point of the ecliptic with the maximal southern declination (σ and σ' lie in the plane that contains the poles of the equator and those of the ecliptic).

Note: We may take the plane of the ecliptic as fixed with respect to the stars, but the same is not true for the plane of the equator. Hence the phenomenon of the precession of the equinoxes.

The point γ moves along the ecliptic, in retrograde motion, and makes a complete circuit in 26 000 years, so the equinox occurs earlier.

Thus, after setting out the terms and principles concerned, Ibn al-Haytham constructs a model for the motion of the sun from its rising at a general point on the horizon as far as its passage across the meridian. He considers separately the two following cases.

Let *ABCD* be a horizon, *BKDL* the circle of the ecliptic. First, we consider a point *K* below the horizon and a point *L* above the horizon. The succession of the signs of the Zodiac is in the order *B*, *K*, *D*, *L*.

The equator has a diameter *AC* and its north pole is at the point *H*. We suppose the point *B* is the initial position of the sun. Let the circle *BEM* be the parallel through the point *B* that cuts the meridian in *E*. In the course of the diurnal motion the point *B* describes this parallel circle in a retrograde direction about the axis *ωH* (*ω* being the centre of the sphere of the Universe). In the time that the point *B* takes to describe the arc *BE*, the sun has moved along the ecliptic from *B* to *K*; let B_1 be the position of the sun on the ecliptic when the point *B* reaches the point *E*. So the sun lags behind the point *B* in the diurnal motion; when the sun crosses the meridian at the point *I*, the point *B* has reached the point *G* on the circle *BEM*, *G* being west of the meridian. At that moment the arc BB_1 of the ecliptic has assumed the position *IG*, where *I* is south of *E*. The arc BB_1 of the ecliptic has thus assumed the position *IG* to the west of the meridian circle. So on the celestial sphere the sun has described an arc *BI* lying between the two parallel circles *BE* and B_1I. The arc *BG* represents *the time* in which the sun has moved from the point *B* to the point *I*. The arc *IE* is the *inclination* of the arc *GI* with respect to the parallel circle *BEM*; the arc *IE* is also the *inclination* of the motion of the sun from the point *B* to the point *I*. The arc *BE* is called the *required time*.

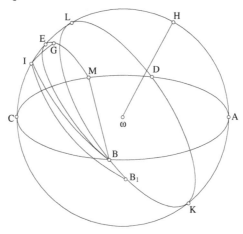

Fig. 2.86

We suppose the semicircle *BKD* is *above* the horizon and the proper motion of the sun is from *B* to *L*. In this case, the succession of the signs of the Zodiac is in the order *B*, *L*, *D*, *K*.

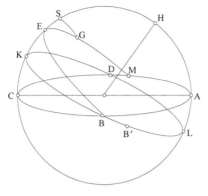

Fig. 2.87

The point B reaches the meridian at the point E, and is at point G when the sun reaches the meridian at the point S. The arc SG which is the arc traversed by the sun on the ecliptic will lie to the *west* of the meridian and to the *north* of the parallel circle BEM. The arc BE is the time required, the arc ES is the inclination of the motion of the sun with respect to the hour circle BEM.

The motion of the planets

Proposition 18. — For Mars, Jupiter and Saturn, the inclination of the orb with respect to the plane of the ecliptic shows no noticeable variation.

For Mercury and Venus, the inclination of the orb with respect to the plane of the ecliptic changes. The plane of the orb oscillates about the line of nodes on either side of the ecliptic. The inclination varies between $0°$ and a limiting angle.[20]

In all cases, that is for the five planets, the inclination – whether variable for Mercury and Venus or taken as constant for Mars, Jupiter and Saturn – is a small fraction of the inclination of the ecliptic to the equator.

As in the case of the moon, the inclination of each of the orbs with respect to the equator is variable, and none of the orbs could be placed in the plane of the equator.

Each of the orbs cuts the plane of the ecliptic along a line of nodes, and has a very slow rotation about the axis of the ecliptic.

---

[20] The maximum of the inclination of the inclined orb with respect to the ecliptic is $7°$ for Mercury and $3°24'$ for Venus. For the superior planets, the inclination is nearly constant, $1°51'$ for Mars, $1°19'$ for Jupiter and $2°30'$ for Saturn.

Motion of each of the five planets on its inclined orb, referred to the circle of the ecliptic

If the motion is *direct*, that is, if it takes place in the direction of the succession of the signs of the Zodiac, the planet moves from west to east, from north to south and from south to north with respect to the poles of the equator, as is the case for the motion of the moon on its orb. But there is a difference between the motion of a planet and that of the moon which arises from the fact that the epicycle of a planet is at an angle to the plane of the orb: because of this, the centre of the planet does not lie in the plane of the orb but to the north or south of it.

If the motion of the planet is retrograde, that is, if its motion with respect to the ecliptic is in the direction contrary to the succession of the signs of the Zodiac, then the planet moves from east to west.

This has no effect on the investigation of the inclination with respect to the equator or with respect to an hour circle.

Halt between retrograde and direct motion (station)

During the halt, we see no motion in longitude, either direct or retrograde, but we may perhaps observe a variation in the latitude caused by the inclination of the epicycle.

If *ABC* is a horizon, *B* a point on the inclined orb, the position of the planet at a given instant, *BED* the hour circle of the point *B*, for all forms of the figure, the planet is moving on its orb towards the meridian circle because of the diurnal motion, and it also has its proper motion on its orb.

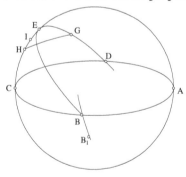

Fig. 2.88

If the planet's motion is *direct* from *B* to B_1, then the point *B* arrives at the meridian circle at the point *E* before the planet. When the point B_1 of the inclined orb reaches the meridian at *H*, the point *B* has reached *G*, the arc BB_1 of the orb has assumed the position HG to the west of the point H, to the north or south of the circle BED.

The position of the planet on the epicycle is given by the arc HI to the north or south of the arc GH. In the time BG, the planet has moved from the point B to the point I; the required time is BE, the inclination of the motion is IE.

If the motion of the planet is *retrograde*, it reaches the meridian before the point B does so.

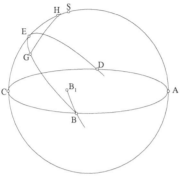

Fig. 2.89

The arc described on the orb is in the position HG, east of the point H and north or south of the circle BED. The position of the epicycle is given by the arc HS, with S north or south of H.

If the point B corresponds to a 'station', a halt between direct and retrograde motion, then for the duration of the halt, the planet departs from the circle BED only by the distance that corresponds to the inclination of the epicycle, a distance that may not be perceptible. If the planet reaches the meridian, it will do so at the point E.

2.2.2. *Required time and inclination*

Let A be the initial position of a wandering star, at a known time t_0; B the position it has reached after a known time t, that is at the instant $t_1 = t_0 + t$. Let CA and CB be the two great circles through the pole C of the equator, C, and let AD be the hour circle through points A and D on the circle CB.

The case of the sun or one of the seven wandering stars

Proposition 19. — It has a proper motion along the ecliptic, and the ecliptic rotates about the axis through the poles *CC'*.

Let *A* be the position of the sun on the circle of the ecliptic at a known time t_0. The point *A* is subject to the diurnal motion and describes an arc *AG* of an hour circle in the known time *t*; this time will be measured by the arc *AG*. The sun, which was at *A*, moves along the ecliptic and in the time *t* it traverses the arc *AA'*; but the ecliptic rotates about the axis *CC'*, and after time *t* the arc *AA'* is in the position *GB*.

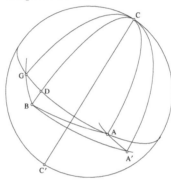

Fig. 2.90

B lies on the parallel passing through *A'* and the angles the arcs *GB* and *AA'* make with the parallel *AD* are equal.

We know the positions of *A*, *G* and *B*; so the arcs *CA*, *CG*, *CB* are known and $\widehat{CA} = \widehat{CG}$, so $\widehat{BD} = \widehat{CB} - \widehat{CA}$ is known.

The point *G* is the position the point *A* reaches after time *t* and the point *D* corresponds to the point *B*, the position reached by the sun; so the arc *GD* is known and represents how far the point *A* is ahead of the sun in the diurnal motion.

On the celestial sphere the sun has traversed an arc *AB* lying between the point *B* and the hour circle *AD* (that is, between the two parallel circles *AD* and *A'B*). The motion along the arc *AB* has two components: the proper motion of the sun on the ecliptic and the diurnal motion.

The arc *AB* has right ascension $\widehat{AD}$[21] and inclination $\widehat{BD}$; these two arcs are known.

---

[21] That is, if δ is the difference between the right ascensions of two points, what we have here is: $\delta(A, B) = \delta(A, D)$ = measure of the arc *AD*.

In fact, if G and B are two known points on the ecliptic, we know $\overset{\frown}{GD}$ and $\overset{\frown}{DB}$ (according to Propositions 5, 6 and 7); as, moreover, we know $\overset{\frown}{AG}$, from which we can find $\overset{\frown}{AD}$.

The case of the moon or one of the five planets

Proposition 20. — Let A be the initial position of the moon; the great circle CA cuts the ecliptic in E; and let B be the position of the moon after time t, the great circle CB cuts the hour circle of A in D and the ecliptic in I. We know the arcs CA, AE, CB, BI; we have $\overset{\frown}{CD} = \overset{\frown}{CA}$, so we know $\overset{\frown}{BD}$ and $\overset{\frown}{DI}$.

In time t, the moon has traversed an arc AA' of its orb, and this arc has reached the position GB. The point G is west of B, and in general is north or south of the circle AD, because of the motion of the node.

The circle CAE has reached the position $CRGH$, where R is on the circle AD; we have $\overset{\frown}{CR} = \overset{\frown}{CA}$ and $\overset{\frown}{GH} = \overset{\frown}{AE}$. The points H and I on the ecliptic are known.

Let $\overset{\frown}{AK}$ be the arc that measures the known time t, the great circle CK cuts the ecliptic in N, $\overset{\frown}{CK} = \overset{\frown}{CA}$.

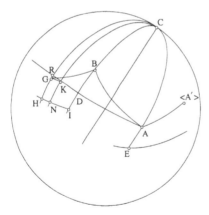

Fig. 2.91: C, the north pole of the equator, and the parallel circle through the point A, the initial position of the wandering star concerned, are fixed elements. The circle of the ecliptic and the orb of the wandering star are both subject to the diurnal motion, the wandering star has a proper motion along its orb, and, in addition, the orb moves with respect to the ecliptic.

If it did not share the motion of the node, after time t the point A would arrive at K; but because of the motion of the node along the ecliptic, A arrives at G, so K and G have the same ecliptic latitude, the arc KG is parallel to the ecliptic ($\widehat{KG} = \widehat{NH}$) and corresponds to the displacement of the node; the point N is known, so $\widehat{NI}$ is known. The arc KD is the right ascension of the arc NI, so $\widehat{KD}$ is known; now $\widehat{AK}$ is known, so $\widehat{AD}$, the required time, is known. And we have seen that $\widehat{BD}$, the inclination of the motion of A to B, is known.

The case of the planets

Proposition 21:

a) *Inferior planets*

For each of these planets, the inclination of the orb with respect to the ecliptic is variable (see above, p. 185), consequently the point G will in most cases lie north or south of the hour circle AD.

The arc AA', which the planet describes on its orb (starting from the point A) in a known time t, is known, and after time t the arc has reached the position BG. If the motion of the planet on its orb is direct, A' is east of A, so G is west of B, and if the motion is retrograde, G is east of B.

The required time and the inclination of the motion are found as in the case of the moon using the initial and final positions A and B. The required time is represented by the arc AD and the inclination is the arc BD.

However, in the case of Mercury and Venus, the position of the body is defined by its latitude and longitude with respect to the ecliptic. Thus this position depends on the inclination of the orb with respect to the ecliptic and the inclination of the epicycle with respect to the orb. When the coordinates of the planet with respect to the ecliptic are known, its coordinates with respect to the equator are also known.

b) *Superior planets*

The motion of the nodes is very slow and has no effect in the course of a day.

The point G lies on the circle AD
– west of B if the motion of the planet is direct
– east of B if the motion is retrograde.

For these three planets, the inclination of the epicycle to the plane of the orb needs to be taken into account; but at any known instant, the planet's latitude with respect to the ecliptic is known; so arcs such as $\widehat{CA}$ and $\widehat{CB}$ are known.

For the five planets:

If the motion of the planet on its orb is direct, then G is west of B, so the required time is less than the known time, that is, it is less than the time that has elapsed during the motion (as in the case of the moon).

If the motion of the planet is retrograde, then G is east of B, so the required time is greater than the known time.

If the planet is at a halt, that is, at station, its position with respect to the ecliptic does not change during the known time, G is at D. The required time is equal to the known time.

Inclination of an orb with respect to the equator[22]

Proposition 22:

Sun: The dihedral angle between the plane of the ecliptic and that of the equator is constant, $\alpha = 23°27'$.

α is the maximum inclination of points of the ecliptic with respect to the equator and corresponds to the solstices:
    – summer solstice to the north (first point of Cancer)
    – winter solstice to the south (first point of Capricorn).

Moon: The dihedral angle β between the orb of the moon and the ecliptic varies very little and Ibn al-Haytham takes it as fixed; it is equal to the maximum inclination, that is, the inclination with respect to the ecliptic of the point furthest north or south. But the points that are furthest north or south change position with respect to the circle of the ecliptic; so their positions with reference to the ecliptic move along the circle of the ecliptic. This change of position is due to the rotation of the orb of the moon about the axis of the ecliptic.

Here, Ibn al-Haytham is repeating the explanations he gave for the moon at the beginning of his treatise (see below, p. 316), explanations relating to the changes in position of the north pole of the orb with respect to the north poles of the equator and of the ecliptic.

The inclination of the orb of the moon with respect to the equator depends on the relative positions of these three poles and the position of the nodes on the ecliptic.

first point of Aries = γ (spring equinox)
first point of Cancer = σ (summer solstice)
first point of Libra = γ' (autumn equinox).

---

[22] See also p. 369.

If the ascending node is at either of the points γ or γ', then the poles H of the equator, P of the ecliptic and X of the orb lie on a great circle that cuts the orb at the points N and S, which are the most northerly and southerly points of both the equator and the ecliptic.

Let α be the inclination of the ecliptic to the equator

β the inclination of the orb to the ecliptic

δ the inclination of the orb to the equator.

• Ascending node at the point γ (first point of Aries).
In this case, $\delta = \alpha + \beta$.

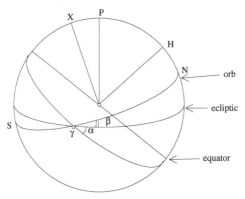

Fig. 2.92

• Descending node at the point γ (so the ascending node is at the point γ' (Libra)).
In this case $\delta = \alpha - \beta$.

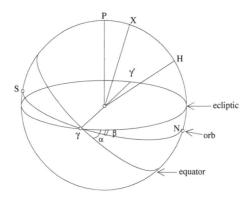

Fig. 2.93

So in these two cases, the positions of the northernmost point N and southernmost point S of the inclined orb with respect to the equator are known.

Investigation of the case in which the ascending node is at neither γ nor γ'

Let ABC be the circle of the ecliptic with pole N, let ADC be the inclined orb with pole E, let B and D be the midpoints of the semicircles with diameter AC. The points N, E, B, D lie on a great circle that cuts the circle of the equator in K. The equator cuts the ecliptic in M and L, the points of the equinoxes. The points A and C are the nodes.

<a> *Let* M *be a point on the arc* BC. The circle MKL, the circle of the equator, cuts the inclined orb in H. Let O be the pole of the equator. The great circle EO cuts the circle of the equator in G, and the arc AD of the inclined orb in I. We have $\widehat{EO} = \widehat{GI}$ and $\widehat{GI}$ is the extreme inclination of the inclined orb with respect to the equator.

If O is the north pole, I is the southernmost point of the orb. If O is the south pole, I is the northernmost point of the orb.

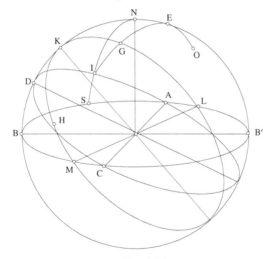

Fig. 2.94

In the figure, we suppose that N, E, O are the north poles of the ecliptic, the orb and the equator; C and A are the descending node and the ascending node respectively; L and M are the spring equinox and the autumn equinox respectively, I is the southernmost point of the orb with respect to the equator.

The points A, C, B, D, L, M are known points; the same is true for the poles E, N, O.

The arc MB of the ecliptic is known, thus the arc MK that corresponds to it on the equator is known and consequently the arc KB of the circle perpendicular to the plane of the ecliptic is also known. Moreover, the arc BD is known, so the arc KD is known too.

Menelaus' theorem gives

$$\frac{\sin \widehat{KD}}{\sin \widehat{DB}} = \frac{\sin \widehat{KH}}{\sin \widehat{HM}} \cdot \frac{\sin \widehat{CM}}{\sin \widehat{CB}};$$

$\widehat{KD}$, $\widehat{DB}$, $\widehat{CM}$ and $\widehat{CB}$ are known arcs, so $\dfrac{\sin \widehat{KH}}{\sin \widehat{HM}}$ is known; so $\widehat{KM} = \widehat{KH} + \widehat{HM}$ is known.

Ibn al-Haytham deduces from this that $\widehat{HM}$ is known.

Justification: $\widehat{KH} = \widehat{KM} - \widehat{HM}$; so we have

$$\frac{\sin\left(\widehat{KM} - \widehat{HM}\right)}{\sin \widehat{HM}} = a, \text{ a known ratio}$$

$$a = \frac{\sin \widehat{KM} \cdot \cos \widehat{HM} - \cos \widehat{KM} \cdot \sin \widehat{HM}}{\sin \widehat{HM}}$$

$$a = \sin \widehat{KM} \cdot \cotan \widehat{HM} - \cos \widehat{KM};$$

$\widehat{KM}$ is known, so $\cotan \widehat{HM}$ is known, so $\widehat{HM}$ is known.

Applying Menelaus' theorem to the arcs of great circles EG, HKG which intersect in G, EK and HI which intersect in D, gives us:

$$\frac{\sin \widehat{EI}}{\sin \widehat{IG}} = \frac{\sin \widehat{ED}}{\sin \widehat{DK}} \cdot \frac{\sin \widehat{KH}}{\sin \widehat{HG}};$$

the ratios of the second term are known; so the first ratio is also known. But $\widehat{EI}$ is a quadrant of a circle, so $\widehat{IG}$ is known and $\widehat{IG}$ is the inclination of the extreme point I with respect to the equator.

We construct a great circle through N and I; it cuts the circle of the ecliptic in S. We have

$$\frac{\sin \widehat{KB}}{\sin \widehat{BD}} = \frac{\sin \widehat{KM}}{\sin \widehat{MH}} \cdot \frac{\sin \widehat{HC}}{\sin \widehat{CD}}.$$

The first two ratios are known and $\widehat{CD}$ is a quadrant of a circle, so $\widehat{HC}$ is known.

The arcs $\widehat{HI}$ on the inclined orb and $\widehat{HG}$ on the equator are quadrants of circles, because G and I lie in the plane of the great circle that passes through the poles E and O.

Now $\widehat{CD}$ is a quadrant of a circle, so $\widehat{CH} = \widehat{DI}$, hence $\widehat{DI} < \dfrac{1}{4}$ circle; I lies between A and D. The great circle through N and I will cut the arc AB in S, which lies between A and B. We have

$$\frac{\sin \widehat{CS}}{\sin \widehat{SB}} = \frac{\sin \widehat{CI}}{\sin \widehat{ID}} \cdot \frac{\sin \widehat{DN}}{\sin \widehat{NB}}.$$

$\widehat{ID} = \widehat{CH}$ is known, so $\widehat{CI}$ is known, $\widehat{DN}$ and $\widehat{NB}$ are also known, so $\dfrac{\sin \widehat{CS}}{\sin \widehat{SB}}$ is known and $\widehat{CB}$ is a quadrant of a circle, $\widehat{CS} = \widehat{CB} + \widehat{BS}$.

Ibn al-Haytham deduces from this that the arc SB is known. In fact $\dfrac{\sin \widehat{CS}}{\sin \widehat{SB}} = \cotan \widehat{SB}$ is known, so SB is known and the point S on the ecliptic is thus known – the position of S gives us the ecliptic longitude of I.

<b> The proof is the same if M, the equinoctial point, lies on the arc AB.

<c> Let us suppose the equinoctial point is at the point B.

The circle of the equator cuts the inclined orb at the point H. Let O be the pole of the equator. The great circle through E and O cuts the equator in G and the orb in I. The arc GI is the maximum inclination of the orb with respect to the equator, and $\widehat{GI} = \widehat{EO}$.

C is the point of a solstice, the great circle NO, the circle through the poles, cuts the equator in K, $\widehat{NC}$ is a quadrant of a circle, $\widehat{KC}$ is the inclination of the ecliptic with respect to the equator; so $\widehat{KN}$ is known.

We have

$$\frac{\sin \widehat{BD}}{\sin \widehat{DN}} = \frac{\sin \widehat{BH}}{\sin \widehat{HK}} \cdot \frac{\sin \widehat{KC}}{\sin \widehat{CN}}.$$

$\widehat{BD}$ and $\widehat{DN}$ are known, so $\dfrac{\sin \widehat{BH}}{\sin \widehat{KH}}$ is known.

We have $\widehat{BK} = \widehat{BH} + \widehat{HK} = \dfrac{1}{4}$ circle, so $\dfrac{\sin \widehat{BH}}{\sin \widehat{KH}} = \tan \widehat{BH}$, so $\widehat{BH}$ and $\widehat{HK}$ are known.

The point H is the pole of the circle EOG, so $\widehat{HG}$ is a quadrant of a circle, hence $\widehat{BG} = \widehat{HK}$ is known.

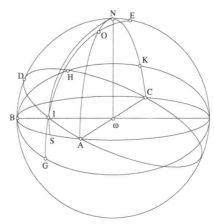

Fig. 2.95
ABC is the ecliptic, with pole N
ADC is the inclined orb, with pole E
BHK is the equator, with pole O

We have

$$\frac{\sin \widehat{EI}}{\sin \widehat{IG}} = \frac{\sin \widehat{ED}}{\sin \widehat{DB}} \cdot \frac{\sin \widehat{BH}}{\sin \widehat{HG}};$$

the last two ratios are known, so $\dfrac{\sin \widehat{EI}}{\sin \widehat{IG}}$ is known and the arc EI is a quadrant of a circle; so $\widehat{IG}$ is known, and is the maximum inclination of the inclined orb to the equator.

In the same way,

$$\frac{\sin \widehat{BN}}{\sin \widehat{ND}} = \frac{\sin \widehat{BK}}{\sin \widehat{HK}} \cdot \frac{\sin \widehat{HC}}{\sin \widehat{HD}},$$

so $\widehat{HC}$ and $\widehat{HD}$ are known and the arc HC is equal to the arc DI. In fact, if we call the centre of the sphere ω, we have $H\omega \perp \omega O$ and $H\omega \perp \omega E$, because H is the point of intersection of the orb with the equator, and E and O are the poles of the equator. So $H\omega$ is perpendicular to the plane $E\omega O$, hence $H\omega \perp \omega I$; so the arc HI is a quadrant of a circle, as is the arc DC. Consequently, $\widehat{DI} = \widehat{HC}$.

Further, the great circle NI cuts the ecliptic in S and we have

$$\frac{\sin \widehat{SC}}{\sin \widehat{SB}} = \frac{\sin \widehat{CI}}{\sin \widehat{ID}} \cdot \frac{\sin \widehat{DN}}{\sin \widehat{NB}},$$

so $\dfrac{\sin \widehat{SC}}{\sin \widehat{SB}}$ is known and the arc CB is a quadrant of a circle; so the arc BS is known and the point S is known.

Proposition 23. — In this proposition Ibn al-Haytham investigates the inclinations of the orbs of the inferior planets: Venus and Mercury.

For each of these planets, the inclination of the orb to the plane of the ecliptic is variable (see Proposition 18).

If the position of the northernmost or southernmost point of the orb with respect to the circle of the ecliptic is known, then the inclination of the orb with respect to the equator can be found by the method given for the moon.

In particular, if the nodes are at the equinoctial points, then the northernmost and southernmost points of the orb (in relation to the equator and referred to the ecliptic) are the points of the solstices. In this case, the values of the inclinations in relation to the equator are found from the inclinations in relation to the ecliptic, using the method given for the moon (see p. 192).

If the nodes are not at the equinoxes, we adopt the same method as for the moon.

A new problem:

The points of intersection of the inclined orb and the equator rotate about the axis defined by the two nodes.

The inclined orb rotates about this axis; so every point of the inclined orb describes an arc of a circle that has the nodes as its poles. For a point on the orb there is an associated point of the ecliptic that has the same longitude (the position referred to the ecliptic; see section 2.1). Ibn al-Haytham gives a very detailed description of the motion of this point. On each orb he considers the four arcs that are the quadrants of circles cut off by the nodes and the northernmost and southernmost points in relation to the ecliptic, taking into account that the apogee moves along the eccentric.

Proof: Let *ABCD* be the circle of the ecliptic and *AECG* the orb of Venus or Mercury. The direction of succession of the signs of the Zodiac is the direction *ABCD* and *H* is the north pole of the ecliptic.

Let E be the southernmost point of the orb of the planet Mercury (if E were the northernmost point of the orb of Venus, it would be necessary to take H' and H as the north and south poles of the ecliptic).

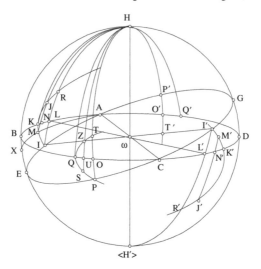

Fig. 2.96

In this figure the points A and C, B and D, E and G are diametrically opposite one another. The nodes are A and C.

a) arc AE and arc CG

Let I be a point on the arc AE and I' a point on the arc CG;[23] the great circle HI cuts the arc AB in L, angle ALI is a right angle and $\widehat{AI} > \widehat{AL}$. The circle with pole A, which passes through I, cuts the arc AB in K, which lies between B and L, and the arc HL in R. The arcs HB and HK are quadrants of circles, since H is the pole of the circle AKB; so the poles of the circles HB and HK lie on the circle AKB and A is the pole of the circle IKR. So the plane AKB is a plane of symmetry for each of the circles HKH' and IKR, which have a common point K; so their common tangent at K is the straight line perpendicular to the plane ABC at K. The point K is the midpoint of the arc IKR and L is the midpoint of the arc ILR.

Let M be a point of the arc KI, the great circle HM cuts the arc KL in N and the arc KR in J. So the two points J and N have the same ecliptic

---

[23] If we take the point I' on the arc CG we may suppose that I and I' are diametrically opposite one another on the circle $AECG$; in that case, the great circle HI cuts the ecliptic in L on the arc AB, in L' on the arc CD and we have L and L' diametrically opposite one another. The same would happen for the points K, K', R, R', etc. which will be defined later.

longitude as the point M, and the point I has the same ecliptic longitude as L.

If the orb rotates about AC until it is superimposed on the ecliptic, the point I describes the arc IMK and the point with the same longitude on the ecliptic describes the arc LNK in the direction of the signs of the Zodiac. If the orb moves on, beyond the circle of the ecliptic, and continues to rotate about AC, the point I describes the arc KJR and the point with the same longitude on the ecliptic describes the arc KNL from K to L, that is, in the direction contrary to that of the signs of the Zodiac.

The arc RL is equal to the arc LI, which was the maximum inclination of the point I south of the ecliptic; so RL will be the maximum inclination north of the ecliptic.

If the inclined orb then rotates so as to return to the ecliptic, the point I describes the arc RK, and the point with the same ecliptic longitude describes the arc LNK in the direction of the succession of the signs of the Zodiac.

If the motion of the inclined orb continues, the point I describes the arc KI and the point with the same ecliptic longitude describes the arc KL in the direction contrary to that of the signs of the Zodiac.

Ibn al-Haytham then gives a summary, for the point I' of the arc CG, of the results he has proved for the point I of the arc AE, and brings in the points H', K', R'.

b) arc EC and arc GA

Let there be a point P on the arc EC and a point P' on the arc AG; we may take P and P' to be at opposite ends of a diameter.

The great circle HP cuts the circle of the ecliptic orthogonally at the point O, so $\overset{\frown}{CP} > \overset{\frown}{CO}$ and $\overset{\frown}{AP} > \overset{\frown}{AO} > \frac{1}{4}$ circle. The circle with pole C that passes through P cuts the circle of the ecliptic in Q and the arc HO between H and O, at the point T. The great circle HQ, orthogonal to the arc BC at the point Q, is tangent to the circle PQT at the point Q. We take a general point S on the arc PQ; the great circle HS cuts the arc QO in U and the arc QT in Z.

Ibn al-Haytham then repeats, for the point P of the arc EC, the work he carried out earlier for the point I of the arc EA, when the inclined orb rotates about AC moving from maximum inclination (southerly) to zero inclination, then from zero inclination to maximum inclination (northerly). When the point P describes the arc PQ then the arc QT, Ibn al-Haytham investigates the change in position of the point of the ecliptic that has the same longitude as P and describes the arc OQ in the direction contrary to that of the signs of the Zodiac, then the arc *QO* in the direction of the signs of the Zodiac.

Ibn al-Haytham then investigates what happens when the orb returns to its initial position: the point *P* then describes first the arc *TQ*, then the arc *QP*.

The same method could be used to investigate the change in position of a point *P'* of the arc *AG*.

Conclusion: The investigation carried out for the point *I* would thus be valid for any point of the orb *AECG*. Ibn al-Haytham carries out the investigation using parallel circles whose poles are the nodes *A* and *C*.

For a point *I*, we consider two motions: the rotation of *I* about the line of the nodes and the movement of its position referred to the ecliptic, that is the movement of the point *L*.

• in the rotation about the axis *AC*, the arc *IM* described by the point *I* in a known time is known; this is a result Ibn al-Haytham proves later.

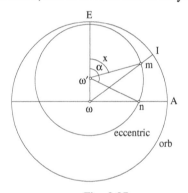

Fig. 2.97

ω' centre of the eccentric	$E\hat{\omega}'m = x$
ω centre of the orb	$E\hat{\omega}'n = \alpha$ $E\hat{\omega}A = 1$ right angle

The *time* taken by the inclined orb to move from maximum inclination to zero inclination is known; it is the time the centre of the epicycle takes to traverse on the eccentric an arc *α* that corresponds to a right angle whose vertex is at *ω*, the centre of the Universe, thus an arc that corresponds to a quadrant of a circle on the inclined orb. This arc *α* is known and the arc *EB* that corresponds to the maximum inclination is known (see Fig. 2.96). While the centre of the epicycle traverses an arc *x* on the eccentric, starting from the apogee, the most northerly or most southerly point of the orb traverses a part $\overarc{EX}$ of the arc *EB* and we have

$$\frac{x}{\alpha} = \frac{\widehat{EX}}{\widehat{EB}}.$$

The motion on the eccentric is uniform; if t_α is the time that corresponds to the arc α, and t_x the time that corresponds to the arc x, we have

$$\frac{x}{\alpha} = \frac{t_x}{t_\alpha}.$$

Moreover if a point such as I reaches the point M when E reaches X, the points A, M, X lie on a great circle that is a possible position of the orb; we have

$$\frac{t_x}{t_\alpha} = \frac{x}{\alpha} = \frac{\widehat{EX}}{\widehat{EB}} = \frac{\widehat{IM}}{\widehat{IK}}.$$

For a known time t_x (where $t_x = 0$ for $x = 0$), the points X of the arc EB and M of the arc IK are known. Consequently, the arcs EX, XB, IM and MK are known.

We have $\widehat{AM} = \widehat{AI}$ which are known arcs; L and N are the points of the ecliptic that have the same longitudes as I and M respectively.

Let us now show that the arc described on the ecliptic by the point L that has the same longitude as I (the ecliptic position of I) in a known time is known.

We return to the preceding figure (Fig. 2.96). The great circle AM cuts the arc EB in X.

Let I be a point of the orb; the arc EB is the inclination of the orb to the circle of the ecliptic. Let I be the position of the point under investigation at a known instant; let m be the position of the centre of the epicycle at that instant (Fig. 2.97) and let x be the arc that separates m from the apogee; $\widehat{Em} = x$ is known.

If m is at the apogee, $x = 0$, and the arc EB is then the maximum inclination i_m which is known.

If m is not at the apogee, then the arc $XB = i$ satisfies

$$\frac{i}{i_m} = \frac{\alpha - x}{\alpha},$$

so the arc $XB = i$ is known.

In a known time, the point I describes the arc IM and the point of the ecliptic that has the same longitude as I describes the arc LN.

To specify the point L, we may write

• $$\frac{\sin \widehat{HB}}{\sin \widehat{BE}} = \frac{\sin \widehat{HL}}{\sin \widehat{LI}} \cdot \frac{\sin \widehat{IA}}{\sin \widehat{AE}} \text{ (triangle } EIH \text{ and circle } ALB\text{)}.$$

$\widehat{HB}$, $\widehat{HL}$ and $\widehat{AE}$ are quadrants of circles, $\widehat{BE}$ and $\widehat{IA}$ are known arcs, so $\widehat{LI}$ is known.

• We also have $\dfrac{\sin \widehat{HE}}{\sin \widehat{EB}} = \dfrac{\sin \widehat{HI}}{\sin \widehat{IL}} \cdot \dfrac{\sin \widehat{AL}}{\sin \widehat{AB}}$ (triangle BHL and circle EIA).

The first two ratios are known, so the third one is also known; now $\widehat{AL} + \widehat{LB} = \widehat{AB} = \dfrac{1}{4}$ circle, so $\widehat{AL}$ is known. So the point L that has the same longitude as I is known.

To specify the point N, we proceed in the same way as before:

• $$\frac{\sin \widehat{HB}}{\sin \widehat{BX}} = \frac{\sin \widehat{HN}}{\sin \widehat{NM}} \cdot \frac{\sin \widehat{MA}}{\sin \widehat{AX}},$$

so the arc NM is known, because all the other arcs are known.

• $$\frac{\sin \widehat{HX}}{\sin \widehat{XB}} = \frac{\sin \widehat{HM}}{\sin \widehat{MN}} \cdot \frac{\sin \widehat{NA}}{\sin \widehat{AB}},$$

so the arc NA is known, the point N, which has the same longitude as M, is thus known.

Conclusion: If a general point, such as I, moves on a circle whose poles are the nodes, so as to traverse the known arc IM in the time t_x and the known arc IK in the time t_o, then the arcs LN and LK of the ecliptic traversed in the same times by the point that has the same longitude are also known. The direction in which these arcs are traversed may be either that of the signs of the Zodiac or the reverse one, as we have seen in our examination of successive points on the arcs AE, EC, CG and GA of the inclined orb.

Proposition 24. — In this proposition Ibn al-Haytham investigates the motion of the planets on their orbs, as well as the upper bounding of the ratio of a required time to the inclination of the part of the motion proper to this required time.

The motion of the sun on its orb from the apogee of the eccentric to its perigee is an *accelerated* motion (the motion of the centre of the epicycle on the eccentric is uniform). The angular speed as seen by an observer on the earth increases. But the motion on the orb from perigee to apogee will be a decelerated motion.

In his proof, Ibn al-Haytham distinguishes four cases for the position of the wandering star. The inclined orb is divided into four arcs by the diameter in which it cuts the plane of the equator and its northernmost and southernmost points in relation to the equator.

Let ABC be the inclined orb, CEG the circle of the equator, whose north pole is D. Let A be the northernmost point (or a point between the northernmost point and the point C). The wandering star is *north of the equator* and moves from A to B, that is, from north to south, and by hypothesis from *apogee to perigee*.

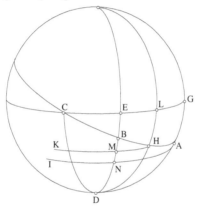

Fig. 2.98.1

The circles CEG, KMH, INA are parallel and the circles DC, DE, DL, DG are orthogonal to them, hence

$$\widehat{EN} = \widehat{AG} = \text{declination of } A$$
$$\widehat{ME} = \widehat{HL} = \text{declination of } H$$
$$\widehat{BE} = \text{declination of } B$$

so

$$\widehat{NB} = \Delta(A, B) \text{ and } \widehat{MB} = \Delta(H, B)\,[24]$$

time $IA = (IA)$ time taken to traverse $\widehat{AB}$

time $KH = (KH)$ time taken to traverse $\widehat{HB}$.

---

[24] Δ indicates the difference of the declinations, that is, of the inclinations with respect to the equator (see p. 65).

Since the wandering star concerned is subject to the diurnal motion, the time required to travel from H to B is $\widehat{KM} = \widehat{KH} - \widehat{MH}$ (Fig. 2.98.2).[25]

Ibn al-Haytham represents times by arcs of hour circles; as we shall see later, (IA) and (KH) appear only in terms of their ratio, so we do not need to know the positions of the points I and K. The argument assumes that the direction from K to H or from I to A is that of the diurnal motion.

Thus far, Ibn al-Haytham was considering motion starting from a position on the equator and the required time was measured by an arc of a great circle. Here, the starting point of the motion is a general point of the sphere and the required time is measured by an arc of the hour circle of that point.

We shall show that

$$\frac{(IA)}{\widehat{NB}} > \frac{(KM)}{\widehat{MB}};$$

the arc MB, part of the arc NB, is called a 'proper' arc for the time (KM).

Let ω be the centre of the orb, the radii ωA, ωH, ωB cut the eccentric in the points A_1, H_1, B_1 respectively; then the times shown as (IA) and (KH) which are the times to traverse the arcs AB and BH of the orb are also the times taken for mean motion on the arcs A_1B_1 and B_1H_1 of the eccentric, so

---

[25] In fact, the wandering star moves from position H to position B on its orb in known time t, let $\widehat{KH} = t$. The arc KH has orientation from K to H in the direction of the diurnal motion. But the points H and B are subject to the diurnal motion. The points H and B of the celestial sphere have moved to H_1 and B_1 respectively after the time t. We have $\widehat{HH_1} \cong \widehat{BB_1} \cong t = \widehat{KH}$ (similar arcs). The wandering star has moved from its initial position H to its final position B; the difference between the right ascensions of these points is $\delta(H, B_1) = \widehat{HM_1} = \widehat{HH_1} - \widehat{H_1M_1}$; but $\widehat{HH_1} = \widehat{KH}$ and $\widehat{H_1M_1} = \widehat{HM}$, so $\delta(H, B_1) = \widehat{KH} - \widehat{HM} = \widehat{KM} = \widehat{HM_1}$.

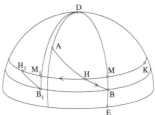

Fig. 2.98.2

So the time KM corresponds to the difference between the right ascensions of the initial and final positions of the wandering star in the motion during time KH, which is the result of the motion on the orb and the diurnal motion. This was the definition of the required time given at the beginning of the part of the work concerned with astronomy.

$$\frac{\text{time}(IA)}{\text{time}(KH)} = \frac{\widehat{A_1 B_1}}{\widehat{B_1 H_1}}$$

because the mean motion is taken as uniform.

But from Propositions 8 and 9

$$\frac{\widehat{A_1 B_1}}{\widehat{B_1 H_1}} > \frac{\widehat{AB}}{\widehat{BH}},$$

so

$$\frac{(IA)}{(KH)} > \frac{\widehat{AB}}{\widehat{BH}};$$

and from Proposition 6

$$\frac{\widehat{AH}}{\widehat{HB}} > \frac{\widehat{NM}}{\widehat{MB}},$$

hence

$$\frac{\widehat{AB}}{\widehat{BH}} > \frac{\widehat{NB}}{\widehat{BM}}.$$

So we have

$$\frac{(IA)}{(KH)} > \frac{\widehat{NB}}{\widehat{BM}}.$$

The ratio of the times (IA) and (KH) is equal to the ratio of the arcs of great circles that are similar to the arcs IA and KH, hence

$$\frac{\text{time}(IA)}{\widehat{NB}} > \frac{\text{time}(KH)}{\widehat{BM}} > \frac{\text{time}(KM)}{\widehat{BM}}.\ ^{26}$$

This result is valid for any point H of the arc AB, a point H with which are associated the required time KM and the arc BM, the difference in inclination proper to the time KM.

$$\frac{\text{time}(IA)}{\Delta(A, B)} > \frac{\text{time}(KM)}{\Delta(H, B)}.$$

---

[26] See previous note.

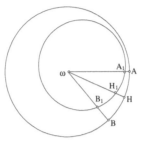

Fig. 2.99

This inequality can be expressed as follows: the mean speed of variation of the inclination over the interval $\overset{\frown}{AB}$ is less than the mean speed of variation of the inclination over the part interval $\overset{\frown}{HB}$. This is a way of stating that the motion of the wandering star over the arc concerned is accelerated.

In the absence of the concept of instantaneous speed, Ibn al-Haytham introduces the mean speed over a variable finite interval.

The conclusion required here is that the speed of variation of the inclination has a lower bound that is a positive quantity.

What we have here is a result in celestial kinematics that represents what Ibn al-Haytham wanted to achieve in his work. The result is expressed in terms of the variation of the mean speed of the motion of a wandering star. The mean speed can then be handled within the theory of proportions, thanks to the fact that the times as well as the distances are represented by arcs of circles.

In fact, given that all the components of the motion are circular and uniform, time can always be represented by arcs of circles and it appears only as a parameter of the motions concerned. Time takes on a quite different meaning when circular and uniform motions are abandoned, as happens in astronomy after Kepler. For Kepler himself, considering elliptical paths, a certain uniformity in the motions is manifested through the area law. So time can be represented geometrically as an area swept out by a radius vector and it does not really enter into the mathematics as a parameter. It is not until Newton that time takes on its full significance as a measure of the motion, with the concepts of instantaneous speed and of acceleration.

Proposition 25. — This proposition deals with the second case for the preceding proposition. We take the wandering star to be south of the equator; it moves southwards, from apogee to perigee.

Let *ABC* be the inclined orb whose southernmost point is *C*, let *AMG* be the circle of the equator whose north pole is *D*.

The planet moves from the point *B* towards the point *H* (in the direction from apogee to perigee); it traverses the arc *BH* in the direction towards the southernmost point.

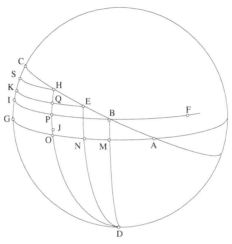

Fig. 2.100

$\widehat{PO} = \widehat{BM} = \Delta\,(A, B)$ (Δ: difference of the inclinations)

$\widehat{QO} = \widehat{EN} = \Delta\,(A, E)$

$\widehat{HO} = \Delta\,(A, H)$

$\widehat{HP} = \Delta\,(B, H)$ —> corresponds to the time (*BI*)

$\widehat{HQ} = \Delta\,(E, H)$ —> corresponds to the time (*EK*)

$(CS) = \Delta\,(H, C)$

$\widehat{BP} = \delta\,(B, H)$ (δ: difference of the right ascensions)

$\widehat{EQ} = \delta\,(E, H),\ \widehat{SH} = \delta\,(H, C).$

$\widehat{AC}$ is a quadrant of a circle, the points *B* and *H* are known, so the arcs *HC*, *HS* and *SC* are known; and the same is true for all the other arcs we have mentioned.

Let us put

$$\frac{\widehat{BP}}{\widehat{PJ}} = \frac{\widehat{HS}}{\widehat{CS}}.$$

This defines the point *J* on *PO*, so the arc *PJ* is a known arc and $\dfrac{\widehat{HP}}{\widehat{PJ}}$ is known.

Let us also put

$$\frac{(FI)}{(IB)} = \frac{\widehat{HP}}{\widehat{PJ}};$$

which defines the point F on $\widehat{IB}$. The time (FI) is then known.

We shall show that

$$\frac{(FI)}{\widehat{PH}} > \frac{(KQ)}{\widehat{QH}}.$$

We have seen above that

$$\frac{(IB)}{(KE)} = \frac{\widehat{BH}}{\widehat{HE}}.$$

but from Proposition 7

$$\frac{\widehat{BH}}{\widehat{HE}} > \frac{\widehat{BP}}{\widehat{EQ}},$$

so

$$\frac{(IB)}{(KE)} > \frac{\widehat{BP}}{\widehat{EQ}},$$

hence

(1) $$\frac{(IB)}{\widehat{BP}} > \frac{(KE)}{\widehat{EQ}}.$$

Moreover, from Propositions 6 and 7 we have

$$\frac{\widehat{HS}}{\widehat{SC}} > \frac{\widehat{EQ}}{\widehat{QH}} \;^{27},$$

so

(2) $$\frac{\widehat{BP}}{\widehat{PJ}} > \frac{\widehat{EQ}}{\widehat{QH}}.$$

From (1) and (2) we get

$$\frac{(IB)}{\widehat{PJ}} > \frac{(KE)}{\widehat{QH}};$$

---

[27] In fact, from Proposition 6, we have $\dfrac{\widehat{HC}}{\widehat{HE}} > \dfrac{\widehat{SC}}{\widehat{HQ}}$ and from Proposition 7 $\dfrac{\widehat{HC}}{\widehat{HE}} < \dfrac{\widehat{HS}}{\widehat{EQ}}$, hence $\dfrac{\widehat{HS}}{\widehat{EQ}} > \dfrac{\widehat{SC}}{\widehat{HQ}}$, so $\dfrac{\widehat{HS}}{\widehat{SC}} > \dfrac{\widehat{EQ}}{\widehat{QH}}$.

but

$$\frac{(IB)}{\overline{PJ}} = \frac{(FI)}{\overline{HP}},$$

so

$$\frac{(FI)}{\overline{HP}} > \frac{(KE)}{\overline{QH}} > \frac{(KQ)}{\overline{QH}};$$

$\frac{(FI)}{\overline{HP}}$, which is a known ratio, is greater than the ratio of the required time (KQ) to the arc that corresponds to it and which is the part $\overline{QH}$ of the arc PH, the difference of the inclinations of the ends of the arc AB that is traversed. There exists a time (FI) such that, for any point E of the arc BH,

$$\frac{\text{time}(FI)}{\Delta(B, H)} > \frac{\text{time}(KQ)}{\Delta(E, H)}.$$

The argument given for the point E, with which is associated the point Q of the arc HP, is valid for any other point of the arc BH.

Ibn al-Haytham again seeks to find the lower bound of the speed of variation of the inclination. But, on the arc in question, we do not know whether the speed of variation of the inclination is increasing, even if we know that the angular speed is increasing. We cannot treat the mean speed on the entire arc as a lower bound. That is why Ibn al-Haytham introduces a suitably defined time (FI).

Ibn al-Haytham then says that the result established in (1) and (2) for an accelerated motion from the northernmost point to the southernmost one, that is, from apogee to perigee, remains true if the motion is from the southernmost point to the northernmost one, that is, from perigee to apogee, and is accelerated.

Ibn al-Haytham notes that we must exclude the two parts close to the northernmost and southernmost points of the orb. At these points the speed we are considering becomes zero. He points out that the upper bound of the ratio of the required time to the difference between the inclinations can be found for any interval that does not include the northernmost and southernmost points of the orb, but that one cannot find an upper bound of this kind for an interval that includes one of these extreme points. It is clear that what he is concerned with here is a local maximum that excludes these two extreme points; the upper bound does not extend over the complete orb.

In other words, Ibn al-Haytham gives a lower bound for the mean speed over any closed intervals in which the speed remains non-zero. This kind of lower bound cannot be extended over the complete orb because the speed does become zero at certain points.

Proposition 26. — Ibn al-Haytham next deals with a third case in which the motion is *from perigee to apogee*. The motion of the centre of the epicycle on the eccentric, which is uniform, is seen on the orb as a decelerated motion; but Ibn al-Haytham assumes that the motion of the wandering star on the epicycle is accelerated.

In the example he considers, the motion is from south to north.

Let *ABC* be the inclined orb and *DEC* the circle of the equator whose north pole is *H*.

The wandering star moves *from perigee to apogee*, and *from south to north*, on the arc of the orb that lies south of the equator.

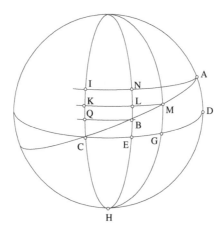

Fig. 2.101.1

The point *A* is the southernmost point of the inclined orb *CBA*. The motion of the planet is from *A* towards *B* on the orb, from perigee towards apogee on the eccentric. In this case, if A_1, M_1 and B_1 are the points of the eccentric that correspond to the points *A*, *M*, *B* on the inclined orb, we have, from Proposition 9,

$$\frac{\overline{B_1 M_1}}{\overline{M_1 A_1}} > \frac{\overline{BM}}{\overline{MA}},$$

hence

$$\frac{\overline{AM}}{\overline{MB}} > \frac{\overline{A_1 M_1}}{\overline{M_1 B_1}}.$$

If we let α and β denote two arcs of the orb (or of a circle equal to it) that are similar to $\overset{\frown}{A_1M_1}$, $\overset{\frown}{M_1B_1}$, we have

$$\frac{\alpha}{\beta} = \frac{\overset{\frown}{A_1M_1}}{\overset{\frown}{M_1B_1}},$$

so

$$\frac{\overset{\frown}{AM}}{\overset{\frown}{MB}} > \frac{\alpha}{\beta}, \quad ^{28}$$

a) $\overset{\frown}{AM} > \alpha$ and $\overset{\frown}{BM} > \beta$ (these inequalities correspond to the case in which A_1, M_1, B_1 lie close to perigee). In this case

$$\frac{\overset{\frown}{AM}}{\overset{\frown}{MB}} > \frac{\alpha}{\beta} \Rightarrow \frac{\overset{\frown}{AM}-\alpha}{\overset{\frown}{BM}-\beta} > \frac{\alpha}{\beta}.$$

We cut off from $\overset{\frown}{AM} - \alpha$ an arc α' such that

$$\frac{\alpha'}{\overset{\frown}{MB}-\beta} = \frac{\alpha}{\beta},$$

hence

$$\frac{\alpha+\alpha'}{\overset{\frown}{MB}} = \frac{\alpha}{\beta} \qquad\qquad \alpha + \alpha' < \overset{\frown}{AM}.$$

---

[28] If we show the orb with centre ω and the eccentric with centre O. If with the point A there is associated the perigee A_1 (Fig. 2.101.2), we have in this case

$$A\hat{\omega}M > A_1\hat{O}M_1, \text{ hence } \overset{\frown}{AM} > \alpha$$

and

$$A\hat{\omega}M < 2A_1\hat{O}M_1, \text{ hence } 2\alpha > \overset{\frown}{AM}.$$

And we have seen (Proposition 9) that for this case of the figure

$$\frac{\overset{\frown}{AM}}{\overset{\frown}{MB}} > \frac{\overset{\frown}{A_1M_1}}{\overset{\frown}{M_1B_1}};$$

that is,

$$\frac{\overset{\frown}{AM}}{\overset{\frown}{BM}} > \frac{\alpha}{\beta}.$$

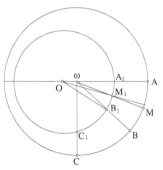

Fig. 2.101.2

We have $\overarc{AM} - \alpha < \alpha,$[29] so

$$\frac{\alpha}{\beta} > \frac{\overarc{AM} - \alpha}{\overarc{MB}},$$

hence

$$\frac{\alpha}{\beta} + \frac{\alpha}{\overarc{MB}} > \frac{\overarc{AM}}{\overarc{MB}},$$

hence

$$\frac{2\alpha}{\beta} > \frac{\overarc{AM}}{\overarc{MB}},$$

so

$$\frac{2\alpha + \beta}{\beta} > \frac{\overarc{AM} + \overarc{MB}}{\overarc{MB}}$$

and *a fortiori*

$$\frac{2(\alpha + \beta)}{\beta} > \frac{\overarc{AB}}{\overarc{BM}}.$$

But

$$\frac{\alpha + \beta}{\beta} = \frac{\overarc{A_1 B_1}}{\overarc{M_1 B_1}} = \frac{(IA)}{(KM)},$$

so

$$\frac{2(IA)}{(KM)} > \frac{\overarc{AB}}{\overarc{BM}}.$$

b) $\overarc{AM} < \alpha$ and $\overarc{BM} < \beta$ (these inequalities correspond to the case in which A_1, M_1, B_1 are close to apogee), with

$$\frac{\overarc{AM}}{\overarc{MB}} > \frac{\alpha}{\beta}.$$

Then we have

$$\frac{\alpha - \overarc{AM}}{\beta - \overarc{BM}} < \frac{\alpha}{\beta} < \frac{\overarc{AM}}{\overarc{MB}}.$$

---

[29] In triangle $\omega O M_1$, we have $O\omega < \omega M_1$, so $\alpha > \hat{M}_1$, $\alpha + \hat{M}_1 < 2\alpha$, so $A\hat{\omega}M < 2\alpha$. So we have $\overarc{AM} < 2\alpha$ or $\overarc{AM} - \alpha < \alpha$.

Let β' be a part of the difference $\beta - \widehat{BM}$ such that

$$\frac{\alpha - \widehat{AM}}{\beta'} = \frac{\widehat{AM}}{\widehat{BM}} \qquad\qquad \beta' < \beta - \widehat{BM}.$$

We know that $\widehat{BM} < \beta$ implies $\beta - \widehat{BM} < \widehat{BM}$,[30] so $\beta' < \widehat{BM} < \beta$. If we add to α an arc α' such that $\dfrac{\alpha'}{\beta'} = \dfrac{\widehat{AM}}{\widehat{BM}}$, we then have $\alpha' < \widehat{AM}$, so $\alpha' < \alpha$.

Assuming that $\widehat{AM} < \alpha$ and $\widehat{BM} < \beta < 2\,\widehat{BM}$ implies $\dfrac{2\alpha}{\beta} > \dfrac{\widehat{AM}}{\widehat{MB}}$, hence

$$\frac{2\alpha + \beta}{\beta} > \frac{\widehat{AM} + \widehat{MB}}{\widehat{MB}}$$

and *a fortiori*

$$\frac{2(\alpha + \beta)}{\beta} > \frac{\widehat{AB}}{\widehat{MB}},$$

so, as in part a), we have

$$\frac{2(IA)}{(KM)} > \frac{\widehat{AB}}{\widehat{MB}}.$$

c) $\widehat{AM} > \alpha$ and $\widehat{BM} < \beta$; this can happen if the points A_1, M_1, B_1 lie about midway between perigee and apogee.

We put $\widehat{JM} = \alpha$, so $\widehat{AJ} < \widehat{JM}$ and $\widehat{AM} < 2\,\widehat{JM}$.

$\widehat{MQ} = \beta$, so $\widehat{BQ} < \widehat{BM}$, because $\widehat{BM} > \dfrac{\beta}{2}$.

Let S be such that $\dfrac{\widehat{SM}}{\widehat{MB}} = \dfrac{\widehat{JM}}{\widehat{MQ}}$, then we have

$$\widehat{JM} > \widehat{SM} \quad \text{and} \quad \frac{\widehat{JS}}{\widehat{BQ}} = \frac{\widehat{JM}}{\widehat{MQ}}.$$

---

[30] We may note that $\beta - \widehat{BM} < \widehat{BM} \Rightarrow \widehat{BM} > \dfrac{\beta}{2} \Rightarrow \dfrac{\widehat{AM}}{\widehat{BM}} < \dfrac{2\widehat{AM}}{\beta} < \dfrac{2\alpha}{\beta}$, which Ibn al-Haytham uses later.

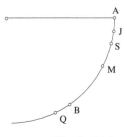

Fig. 2.101.3

We have $\overset{\frown}{AM} < 2\,\overset{\frown}{JM}$, so

$$\frac{\overset{\frown}{AM}}{\overset{\frown}{MB}} < \frac{2\,\overset{\frown}{JM}}{\overset{\frown}{MB}}\,.$$

But $\overset{\frown}{BQ} < \overset{\frown}{BM} \Rightarrow \overset{\frown}{MQ} < 2\,\overset{\frown}{BM}$, so

$$\frac{\overset{\frown}{AM}}{\overset{\frown}{BM}} < \frac{2\,\overset{\frown}{JM}}{\overset{\frown}{BM}} < \frac{4\,\overset{\frown}{JM}}{\overset{\frown}{MQ}}\,.$$

So

$$\frac{4\,\overset{\frown}{JM} + \overset{\frown}{MQ}}{\overset{\frown}{MQ}} > \frac{\overset{\frown}{AB}}{\overset{\frown}{BM}}\,,$$

that is,

$$\frac{4\alpha + \beta}{\beta} > \frac{\overset{\frown}{AB}}{\overset{\frown}{BM}}\,;$$

so *a fortiori*

$$\frac{4(\alpha + \beta)}{\beta} > \frac{\overset{\frown}{AB}}{\overset{\frown}{BM}}\,;$$

but $\dfrac{\alpha + \beta}{\beta} = \dfrac{(IA)}{(KM)}$, the ratio of the times taken to traverse the arcs AB and BM, so

$$\frac{4(IA)}{(KM)} > \frac{\overset{\frown}{AB}}{\overset{\frown}{BM}}\,.$$

So in cases a), b), c), we know how to find a time t such that

$$\frac{t}{(KM)} > \frac{\overset{\frown}{AB}}{\overset{\frown}{BM}}\,.$$

But

$$\frac{\widehat{AB}}{\widehat{BM}} > \frac{\widehat{NB}}{\widehat{BL}}$$

$\widehat{NB} = \Delta\,(A,\,B)$

$\widehat{BL} = \Delta\,(B,\,M)$

$$\frac{t}{(KM)} > \frac{\widehat{NB}}{\widehat{BL}} \quad \text{or} \quad \frac{1}{\widehat{NB}} > \frac{(KM)}{\widehat{BL}},$$

$$\frac{\text{known } t}{\Delta(A,\ B)} > \frac{\text{time } (KM)}{\Delta(B,\ M)}.$$

In this case also, where the motion is decelerated, we find a positive lower bound for the speed of variation of the inclination. This case is even more difficult than the preceding one since the angular speed decreases.

Correction for the epicycle

In investigating the motion of the planet from perigee to apogee, we have considered the arcs α and β of the eccentric associated with the arcs AM and MB of the orb, without taking account of the epicycle.

If the planet is at M_1 on the eccentric where $\widehat{A_1M_1} = \alpha$, the observer, at O, sees it at N on the orb; if the planet is at M_2 on the epicycle, the observer sees it at M.

$\widehat{AN}$ is the arc that has so far been associated with α

$$\widehat{AM} = \widehat{AN} + c,$$

c is the correction which, in the text, it is assumed must be additive.

$\widehat{AN}$ is the arc to be corrected

$$\widehat{AN} = \widehat{AM} - c.$$

$\widehat{AM}$ is the arc obtained after the correction is made.

Similarly, the arc MB will be the arc obtained after the correction is made, the arc to be corrected is $\widehat{MB} - c'$.

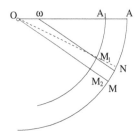

Fig. 2.102.1

Ibn al-Haytham examines three cases:

$$\frac{c}{c'} = \frac{\widehat{AM} - c}{\widehat{BM} - c'}; \quad \frac{c}{c'} < \frac{\widehat{AM} - c}{\widehat{BM} - c'}; \quad \frac{c}{c'} > \frac{\widehat{AM} - c}{\widehat{BM} - c'}.$$

• If $\dfrac{c}{c'} = \dfrac{\widehat{AM} - c}{\widehat{BM} - c'}$, then $\dfrac{c}{c'} = \dfrac{\widehat{AM}}{\widehat{BM}}$.

So we have

$$\frac{\widehat{AM}}{\widehat{BM}} = \frac{\widehat{AM} - c}{\widehat{BM} - c'}.$$

This reduces to the preceding case, because $\widehat{AM} - c$ and $\widehat{BM} - c'$ are the arcs we investigated before.

• If $\dfrac{c}{c'} < \dfrac{\widehat{AM} - c}{\widehat{BM} - c'}$, we then have $\dfrac{c}{c'} < \dfrac{\widehat{AM}}{\widehat{BM}} < \dfrac{\widehat{AM} - c}{\widehat{BM} - c'}$ (we know that $\dfrac{\alpha}{\beta} < \dfrac{\gamma}{\delta}$ implies $\dfrac{\alpha}{\beta} < \dfrac{\alpha + \gamma}{\beta + \delta} < \dfrac{\gamma}{\delta}$).

We should then have, from the above,

$$\frac{\widehat{AM}}{\widehat{BM}} < \frac{\left(\widehat{AM} - c\right) + \left(\widehat{BM} - c'\right)}{\widehat{BM} - c'} < \frac{\text{known time}}{\text{time } (KM)}.$$

• If $\dfrac{c}{c'} > \dfrac{\widehat{AM} - c}{\widehat{BM} - c'}$, we then have $\dfrac{c}{c'} > \dfrac{\widehat{AM}}{\widehat{BM}}$.

Let $\widehat{AS}$ be the correction for the arc AM, so the arc to be corrected is $\widehat{MS}$; and let $\widehat{BO}$ be the correction for the arc MB, the arc to be corrected is then $\widehat{MO}$, so the arc that, after the correction is made, gives $\widehat{AB}$, is $\widehat{SO}$.

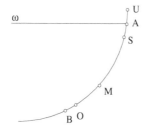

Fig. 2.102.2

We suppose

(1) $\quad \dfrac{\widehat{AS}}{\widehat{BO}} > \dfrac{\widehat{MS}}{\widehat{MO}}.$

We have established that there exists a known time t_c such that

(2) $\quad \dfrac{t_c}{(KM)} > \dfrac{\widehat{SO}}{\widehat{MO}}.$

Let U be such that $\widehat{AU} = \widehat{BO}$, then the arc $\widehat{SU} = \widehat{AS} + \widehat{BO}$ is known and the arc SO is known, so the ratio $\dfrac{\widehat{US}}{\widehat{SO}}$ is known.

Let t be a general time such that $\dfrac{t}{t_c} = \dfrac{\widehat{US}}{\widehat{SO}}$, we have $\dfrac{t + t_c}{t_c} = \dfrac{\widehat{UO}}{\widehat{SO}}$, so from (2) we have

$$\dfrac{t + t_c}{(KM)} > \dfrac{\widehat{UO}}{\widehat{OM}}.$$

But

$$\dfrac{\widehat{UO}}{\widehat{OM}} > \dfrac{\widehat{AO}}{\widehat{OM}} \quad \text{and} \quad \dfrac{\widehat{AO}}{\widehat{OM}} > \dfrac{\widehat{AB}}{\widehat{BM}}^{\,31}$$

and moreover

$$\dfrac{\widehat{AB}}{\widehat{BM}} > \dfrac{\widehat{NB}}{\widehat{BL}},$$

so we have

[31] Since $\dfrac{\widehat{AM}}{\widehat{OM}} > \dfrac{\widehat{AM}}{\widehat{BM}}.$

$$\frac{t+t_c}{(KM)} > \frac{\widehat{NB}}{\overline{BL}}$$

and by permutation

$$\frac{t+t_c}{\widehat{NB}} > \frac{(KM)}{\overline{BL}}.$$

Proposition 27. — This proposition corresponds to the fourth case of Proposition 24.

The circle of the equator is ADG with pole E, and the inclined orb is ABC. The point C is the *northernmost* point on this orb. The planet moves on the arc AC. On the eccentric the motion is *from perigee to apogee*.

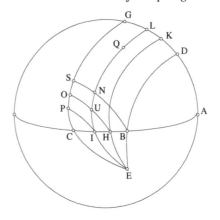

Fig. 2.103

We consider the motion on the arc BI which takes place in a known time, (BS); H is a known point on the arc BI and the arc HI is traversed in a known time, (HO). The times taken to traverse it are represented by arcs of circles parallel to the equator.

The great circles EC, EI, EH and EB cut the circle of the equator in G, L, K and D.

The circles through H and B parallel to the equator cut the arc IL at the points U and N; so from Proposition 7 we have

$$\frac{\widehat{BI}}{\widehat{IH}} > \frac{\widehat{BN}}{\widehat{HU}}.$$

The hour circle through I cuts the arc CG in P. The arc IC is known, so $\widehat{IP}$, the difference between the right ascensions of I and P, and $\widehat{CP}$, the difference between their declinations, are known; and $\dfrac{\widehat{IP}}{\widehat{PC}} > \dfrac{\widehat{HU}}{\widehat{UI}}$, from Proposition 14.

Here we are using Proposition 14 with $\beta = \dfrac{\pi}{2}$ and α close to $23°27'$, or 0.4092797 radians (the obliquity of the ecliptic), that is, satisfying the condition that $\alpha < \dfrac{\beta}{2} = \dfrac{\pi}{4} = 45°$. We know that, for Proposition 14 to hold, the range of θ must be restricted: $0 \le \theta_3 \ (\le \theta_4)$. For $\alpha = 23°27'$, we find $\theta_3 = -0.881811255 = -50°31'27''$.

The angle φ subtended at the centre by the arc HC is given by

$$\cos\varphi = \frac{\cos\alpha\cos\beta - \cos(\alpha-\theta)}{\sin\alpha\sin\beta},$$

here $-\dfrac{\cos(\alpha-\theta)}{\sin\alpha}$ since $\beta = \dfrac{\pi}{2}$.

The condition $\theta \le \theta_3$ can be expressed as

$$\cos\varphi \ge -\frac{\cos(\alpha-\theta_3)}{\sin\alpha} = -0.6937389,$$

which gives $\varphi \le 133°55'36''$. On Fig. 2.104 we have shown the value $\alpha = 23°27'$ and the corresponding value $\theta_3 = -50°31'27''$; we have also indicated the extreme values of α for the case of Mercury, whose orbit is inclined at $7°14''$ to the ecliptic. The corresponding values of θ_3 are $-62°16'18''$ and $-38°48'46''$ respectively, which give maximum values of φ of $133°42'49''$ and $134°18'17''$; the orbits of the other planets have smaller inclinations to the ecliptic than that of Mercury. In any case, when α increases from 0 to $\dfrac{\pi}{2}$, the maximum value of φ increases very slowly from $133°33'51''$ to $180°$.

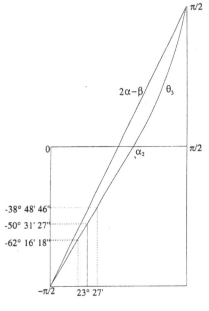

Fig. 2.104

Now, $\overparen{HC} < \overparen{AC}$, which is a quadrant of a circle; so φ is always less than $\frac{\pi}{2} < 133°33'51''$ and Ibn al-Haytham's statement is thus correct.

Let Q be the point on the arc IL defined by $\dfrac{\overparen{BN}}{\overparen{NQ}} > \dfrac{\overparen{IP}}{\overparen{PC}}$, then we have

$$\frac{\overparen{BN}}{\overparen{NQ}} > \frac{\overparen{HU}}{\overparen{UI}},$$

where $\overparen{NQ}$ is a known arc.

Let t_c be the known time defined above, which satisfies

$$\frac{t_c}{(OH)} > \frac{\overparen{BI}}{\overparen{IH}},$$

and let t be the time defined by $\dfrac{t}{t_c} = \dfrac{\overparen{IN}}{\overparen{NQ}}$, we have

(1) $\dfrac{t_c}{\overparen{NQ}} > \dfrac{t}{\overparen{IN}}$

and the time t is known. But

$$\frac{\widehat{BI}}{\widehat{IH}} > \frac{\widehat{BN}}{\widehat{HU}},$$

so

$$\frac{t_c}{(OH)} > \frac{\widehat{BN}}{\widehat{HU}},$$

hence

$$\frac{t_c}{\widehat{BN}} > \frac{(OH)}{\widehat{HU}}.$$

Moreover

$$\frac{\widehat{BN}}{\widehat{NQ}} > \frac{\widehat{UH}}{\widehat{UI}},$$

(subject to the restriction mentioned above for the width of the arc CH), so

(2) $$\frac{t_c}{\widehat{NQ}} > \frac{(OH)}{\widehat{UI}}.$$

From (1) and (2), we obtain

$$\frac{t}{\widehat{NI}} > \frac{(OH)}{\widehat{UI}} > \frac{(OU)}{\widehat{UI}}.$$

OU is the required time associated with the arc UI, the difference between the declinations of the points H and I. The proof is valid for any point H lying between B and I (with the restriction noted above).

So we may use the same method to draw similar conclusions when the motion of the planet is from the southernmost point to the northernmost one, that is from perigee to apogee, and when the motion of the planet is from the northernmost point to the southernmost one, that is from apogee to perigee. In all these cases, we have a lower bound on the mean speed of change of the right ascension.

2.2.3. *Investigation of the heights of a star above the horizon*

The problem is introduced in Propositions 28 and 29. The sphere is assumed to be right or inclined towards the south, that is, the north pole of the equator is on the horizon or above the horizon. Ibn al-Haytham assumes that the star crosses the meridian to the south of the pole of the horizon.

We may note that here Ibn al-Haytham is investigating how the height of the star varies in the course of its motion, that is, as a function of time, in the neighbourhood of M; every height less than that of M is reached at two points of the trajectory. To obtain this result, Ibn al-Haytham appeals to the continuity of the motion: every height intermediate between that of D and that of M is attained once between X and M and once again between M and D. In what follows, he shows that the height of the star decreases after the meridian transit.

The following proposition provides a corresponding investigation for the case where the star moves from the southernmost point to the northernmost one.

In Proposition 30, Ibn al-Haytham proves that the point K at which the body attains its maximum height is unique. He constructs a sequence of points on the trajectory along which the body approaches K and employs arguments from infinitesimal geometry on the sphere to prove that these points all have height less than that of K. The basis for this reasoning is to consider an infinitely small spherical triangle (that is, a triangle such that the diameter of its circumcircle tends to zero) and to suppose it will behave like a plane triangle with the same vertex.

In Proposition 31, Ibn al-Haytham tries to find a global version of the property that, in Proposition 28, he proved held locally: each height less than the maximum one is reached exactly twice. To prove this he makes use of the same methods as in Proposition 30. We may note that he needs to use Proposition 15 in a doubtful case, which slightly constrains the generality of his conclusion.

Proposition 28. — Ibn al-Haytham assumes that, in moving on its orb, the star goes from the northernmost point towards the southernmost one, but without reaching the southernmost point.

For the sun, there is no supplementary hypothesis. For the moon we have an accelerated anomalistic motion or the required additive correction made by the epicycle. For the planets we have accelerated anomalistic motion and/or accelerated motion of the epicycle.

In this case, Ibn al-Haytham investigates the heights of the star

a) between its rising and meridian transit, that is, the *heights in the east*;

b) between its meridian transit and setting, that is, the *heights in the west*.

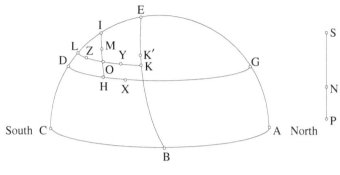

Fig. 2.105

a) Let *ABC* be the horizon with *AC* as a diameter, *ADC* the meridian of the place from which the observation is made; one of the seven planets and luminaries rises at the point *B*.

Through the point *B* we draw the circle *BE* parallel to the equator, this circle cuts the meridian in *E*.[32] If the star had constant declination, in the course of the diurnal motion it would describe this parallel circle. But the star crosses the meridian at the point *D*. Through *D* we draw the circle *DHG* parallel to the horizon and, through a point *H* of that circle, we draw a circle parallel to the circle *BE* to cut the meridian in *I* and such that $\frac{HI}{ID} > \frac{SN}{NP}$, which is a given ratio (Proposition 10).

We have assumed that $\frac{SN}{NP}$ is greater than the ratio of the required time to the inclination of the arc that represents that required time, for every arc the star traverses between the points *B* and *D* (in particular $\frac{SN}{NP} > \frac{\widehat{BE}}{\widehat{ED}}$).

We assume that *I* lies between *E* and *D* (if it does not, we choose a point *I* and from it find another point *H*; Proposition 11 for the right sphere, Proposition 12 for the inclined sphere).

The meridian plane *ADC* passes through the zenith, the pole of the horizon, and the pole of the equator. So it is perpendicular to the plane *DHG* and to the planes of the circles *BE* and *HI*.

---

[32] The argument is valid for a right sphere or an inclined sphere. If the sphere is right, any hour circle (parallel to the equator) is orthogonal to the circle of the horizon.

$$\frac{HI}{ID} > \frac{SN}{NP},$$

so

$$\frac{\widehat{HI}}{\widehat{ID}} > \frac{SN}{NP} > \frac{t \text{ required}}{\widehat{ID}}, \quad {}^{33}$$

so $\widehat{HI}$ > required time corresponding to $\widehat{ID}$.

Let $\widehat{MI}$ be the required time that corresponds to $\widehat{ID}$, so M lies on the arc HI and above the plane DHG. The trajectory of the star cuts the arc HI at the point M so, before it reaches M, the trajectory has cut the circle DHG in a point X that has the same height as D (this height is equal to $\widehat{CD}$). The point X lies between the two parallel circles BE and HI. At the point M, the height of the star is greater than $\widehat{CD}$. Thus, in the eastern part of its motion, the star reaches a point whose height is greater than that of its meridian transit.

---

[33] This assumes that $\dfrac{\widehat{HI}}{\widehat{ID}} > \dfrac{HI}{DI}$. We know that ED is a very small arc and that $\widehat{BC} > \widehat{ED}$, so $\dfrac{SN}{NP} > 1$ and it follows that $\dfrac{HI}{ID} > 1$, $HI > DI$. Arcs DI and HI belong to circles with radii R and R', in general $R > R'$. For example, for the sun we have $R = R'$ on the day of the equinox.

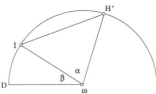

Fig. 2.106

So the chord HI subtends an angle at the centre $\alpha > \beta$, the angle at the centre subtended by chord ID. Let us consider the point H' of the meridian ADC such that $I\hat{\omega}H' = \alpha$. We know, from Proposition 1, that

$$\frac{\widehat{IH'}}{\widehat{ID}} > \frac{IH'}{ID}, \text{ or } \frac{\widehat{IH'}}{IH'} > \frac{\widehat{ID}}{ID};$$

now

$$\frac{\widehat{IH'}}{IH'} = \frac{\widehat{IH}}{IH}.$$

Thus

$$\frac{\widehat{IH}}{IH} > \frac{\widehat{ID}}{ID}, \text{ or } \frac{\widehat{IH}}{\widehat{ID}} > \frac{IH}{ID}.$$

Let O be a general point between M and H on the circle HI; we draw through O a circle LOK parallel to DHG, it cuts the meridian in the point L, which is above D. The point M is above LOK and the points B and D are below, so the star meets the circle LOK twice: a first time at Y when it is moving from B to M and a second time at Z when it is moving from M to D. The points Y and Z have the same height $\widehat{LC}$ $\left(\widehat{LC} > \widehat{DC}\right)$.

The points X, Y, M, Z^{34} of the trajectory all lie to the east of the meridian ADC. Let h be the height $h(X) < h(Y) < h(M)$, $h(X) = h(D)$ and $h(Y) = h(Z)$.

In the same way, to every point O' of the arc HI between O and M there corresponds a horizontal circle $L'O'K'$ that the trajectory of the star meets at Y' and Z' such that $h(Y') = h(Z')$, $h(Y') > h(Y)$, $h(Z') > h(Z)$. All the heights in question are greater than the (common) height of X and D.

b) The motion of the star continues beyond D, that is, from the meridian towards the western horizon. Here the height h of the star is decreasing from $h(D)$ to 0.

The hour circle DQ touches the horizontal circle DHG at D (because the poles of these two circles lie on the meridian of D).

In the diurnal motion the point of the orb that was at the point D describes the arc DQ, but the star has a proper motion that takes it away from the circle DQ and towards the south. Let F be one of its positions, let RFU be the hour circle of F; this circle cuts the meridian in R and $\widehat{CR} < \widehat{CD}$; but if $h(F)$ is the height of F, we have $h(F) < CR < CD$; now $CD = h(D)$, so $h(F) < h(D)$.

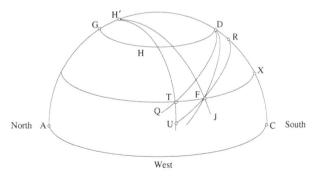

Fig. 2.107

34 X, Y, Z, O', L', K' are not found in the text.

Let *XFT* be the horizontal circle through *F*, it cuts the hour circle *DQ* in a point *T* north of *F*. We construct through *T* and *H′*, the pole of the equator, the great circle *H′T*; it cuts the hour circle *RFU* in a point north of *F*; let this point be *U*. Arcs *RU* and *DT* are similar and $\overparen{RF} < \overparen{RU}$, so the arc *DT* is greater than the arc similar to *RF*.

Arc *FR* lies east of the meridian *H′FJ*. The star does not return to this meridian before setting; since its motion continues to lie south of the circle *RFU*, the star will not meet the arc *FT*, nor the arc *FX*. It will meet the horizontal circle *XF*, only once, at the point *F*. In the same way we may show that, when it moves between the point *D* and setting in the west, the star meets every horizontal circle between the circle *DHG* and the horizon, and meets each of them only once.

Proposition 29. — The celestial sphere in regard to the given horizon is the same as in Proposition 28. Ibn al-Haytham repeats the hypotheses for the moon and the five planets and assumes that, on its orb, the star moves from the *southernmost point* towards the *northernmost point* without reaching the latter.

a) He first investigates the heights of the star between meridian transit and setting.

Let *ABCD* be a horizon, *ANGC* the local meridian.

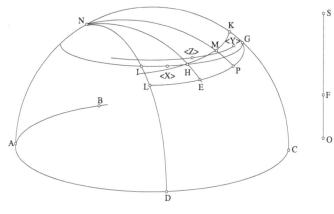

Fig. 2.108

The star rose in the east and crossed the meridian at *G*; we are investigating its motion from the moment of meridian transit until it sets at the point *D*. Let *N* be the pole of the equator and *GHI* the horizontal circle for *G*. The hour circle for *G* cuts the great circle *ND* in *L*. The arc *GL* is the required time and *LD* is the inclination of the motion of the star between the points *G* and *D*.

Let $\dfrac{SF}{FO}$ be a given ratio such that $\dfrac{SF}{FO}$ is greater than the ratio of any required time to the part of the inclination of the motion that is proper to this required time. We choose a point H on the horizontal circle GHI so that the hour circle of H cuts the meridian in the point K, which is such that $\dfrac{\overset{\frown}{HK}}{\overset{\frown}{KG}} > \dfrac{SF}{FO}$. The great circle NH cuts the circle GL in E.

Arcs HK and GE are similar and $\overset{\frown}{KG} = \overset{\frown}{HE}$. So

$$\frac{\overset{\frown}{GE}}{\overset{\frown}{EH}} > \frac{SF}{FO},$$

by an argument like that in note 33 (p. 224).

Let $\overset{\frown}{GP}$ be the required time corresponding to the inclination $\overset{\frown}{EH}$, we have $\overset{\frown}{GP} < \overset{\frown}{GE}$. The great circle NP cuts the arc HK in M; we have $\overset{\frown}{PM} = \overset{\frown}{EH}$, so in time $\overset{\frown}{GP}$ the star has moved from G to M and we have $h(M) > h(G)$. The star sets at the point D, that is, $h(D) = 0$, so the star meets the circle GHI between H and I, because the body is never again south of the arc KH. Let X be the point of transit across the arc HI of the circle GHI, we have $h(X) = h(G)$.

We show as in Proposition 28a that on any horizontal circle of height h such that $h(G) < h < h(M)$, there will be two points where the star crosses it: one, Y, lying between G and M, and the other, Z, between M and X.

b) Investigation of the height of the star between its rising at B and its meridian transit at G.

Let BQ be the hour circle of the point B, let Q be a point on the meridian circle $ANGC$ (we can assume that the point Q lies south of G) and let GL be the hour circle of G.

The hour circle of G touches the horizontal circle GHI at G. So the circle GHI does not cut any of the hour circles that lie between GL and GQ.

Let JUT be a horizontal circle with height $h(J) < h(G)$. As it moves from B to G, the star meets JUT; let it meet it at the point X. Let $H'X$ be the hour circle of X. We have $\overset{\frown}{GU} > \overset{\frown}{H'X}$ (from Proposition 13), so the great circle NX cuts the arc GU, the arc XU lies to the east of X. So, in its motion from X to G, the star will not pass through any point of the arc XU.

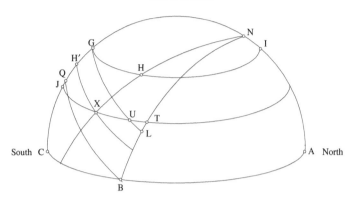

Fig. 2.109

The arc *UT* lies north of the circle *GU* and east of the meridian circle, so the star does not meet this arc either during its motion from *B* to *G*, or after its meridian transit at *G*.

The arc *XJ* lies south of the circle *H'X*, so as the star moves from *X* to *G* it does not return to that arc.

Between rising at *B* and meridian transit, the star has passed through a point of height $h(J)$ once and only once. The same holds for any height $h < h(G)$, so h increases from 0 to $h(G)$.

Proposition 30. — *The point of maximum height is unique.*

Ibn al-Haytham returns to a problem he has already considered (Propositions 28 and 29) and pursues his investigation of the heights the star reaches east of the meridian *DC*, in the part of its trajectory that lies above the horizontal plane through *D*.

We have seen that the star can reach points such as *Y* and *Z*, where $h(Z) = h(Y) > h(D)$ (p. 225).

If h_m is the maximum height, a point of height h_m that is reached by the star cannot lie either on the meridian circle or to the west of it.

Let us suppose that the horizontal circle *OKI* is such that $\widehat{CO} = h_m$. We draw through the pole *N* a circle *NS* that touches the circle *OKI* at *S*.

a) *The star does not meet the arc* SI: let us suppose that the star does cross *SI*, say at *G*; the great circle *NG* cuts the circle *OKI* again in a point *K*. Let $\widehat{GX}$ and $\widehat{KL}$ be the hour arcs of *G* and *K*, then $\widehat{GX}$ is the required time for the arc whose proper inclination is $\widehat{XD}$ as the star moves from *G* to *D*. The arcs *GX* and *KL* are similar, the required time whose proper inclination is $\widehat{LD}$ is part of $\widehat{LK}$; let it be $\widehat{LU}$. In moving from *G* to *D* the star thus meets the arc *KL* in *U*, where $h(U) > h_m$; which is impossible, because h_m is the maximum height.

So the star does not pass through any point of $\overset{\frown}{SI}$.

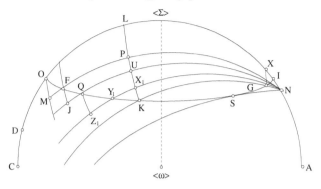

Fig. 2.110

ω : centre of the sphere α : $N\hat{\omega}\Sigma$
Σ : pole of the horizon β : $I\hat{\omega}\Sigma$
N : pole of the equator

Note: To prove that the star does not pass through the point S, we can follow the same line of reasoning as for the point G, considering the hour circle of S.

b) Let us suppose that the point K of the arc SO is a point of meridian transit of the star, we have $h(K) = h_m$.

Let $\overset{\frown}{KL}$ be the hour arc of K where L is a point on the meridian. The arc KL is the required time whose inclination is LD when the star travels from K to D. On the hour circle of O, we consider a point M such that $\overset{\frown}{OM}$ is the required time whose inclination is the arc OD. The great circle NM cuts the arc KL in P and the circle OKI in F, thus the arcs PL and OM are similar and measure the same time. The arc PL is the required time whose inclination is the arc OD, so the arc PK is the time whose inclination is the arc LO; but $\overset{\frown}{LO} = \overset{\frown}{PM}$ and M lies *below* the circle OKI, so $h(M) < h_m$.

In the same way, we draw through the point F an hour circle on which we take a point J such that $\overset{\frown}{FJ}$ is the required time for the inclination $\overset{\frown}{FM}$ ($\overset{\frown}{PM} > \overset{\frown}{PF}$). The great circle NJ cuts OKS in Q and KL in U. Arcs FJ and PU are similar, so $\overset{\frown}{PU}$ is the required time whose inclination is $\overset{\frown}{FM}$, so for the time $\overset{\frown}{UK}$ there is the corresponding inclination $\overset{\frown}{PF}$ which is equal to $\overset{\frown}{UJ}$ and $h(J) < h_m$.

Starting from Q in the same way as we started from F, we draw through the point Q an hour arc QZ_1 equal to the time whose proper inclination is QJ, where Z_1 lies below the circle OKI. The great circle NZ_1 cuts KL in X_1 and KO in Y_1. Thus we have the time KX_1 whose proper inclination is X_1Z_1. Starting from Y_1, which corresponds to Q, we repeat the procedure and obtain the points X_2, Y_2, Z_2, which correspond to X_1, Y_1, Z_1, and so on.

For any i ($i = 1, 2 \ldots n$), there will be a time $\widehat{KX_i}$ whose proper inclination is $\widehat{X_iZ_i}$, with

$$\frac{\widehat{KX_i}}{\widehat{X_iZ_i}} < \frac{\widehat{KX_i}}{\widehat{X_iY_i}}$$

(with $\widehat{X_iZ_i} > \widehat{X_iY_i}$, where Z_i lies below the circle OK).

From Proposition 15, we have

$$\frac{\widehat{KX_{i+1}}}{\widehat{X_{i+1}Y_{i+1}}} < \frac{\widehat{KX_i}}{\widehat{X_iY_i}}.$$

We make use of the second inequality in the proposition, which is always true as long as $\alpha \geq \beta$ – which is the case here since the pole N is assumed to lie outside OKI, the horizontal circle of maximum height. Angle α is $N\hat{\omega}\Sigma$ and angle β, $I\hat{\omega}\Sigma$.

So the ratio $\dfrac{\widehat{KX_i}}{\widehat{X_iY_i}}$ decreases.

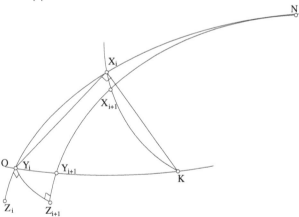

Fig. 2.111

Arcs X_iY_i and KX_i are orthogonal, the angle between their tangents at the point X_i is a right angle.

If we consider the rectilinear triangle KX_iY_i, with vertex X_i and with sides the chords KX_i and X_iY_i; in general the angle KX_iY_i is not a right angle.

It is only after a certain number of iterations n that, for $i > n$, the curvilinear triangle will be so small that the chords X_iK and X_iY_i come to be sufficiently close to the tangents. The curvilinear triangle KX_iY_i is then very close to being a rectilinear and right-angled triangle. In which case,

$$\frac{\widehat{KX_i}}{\widehat{X_iY_i}} \cong \frac{KX_i}{X_iY_i} = \cotan\hat{K}.$$

As the point X_i approaches K, the ratio $\dfrac{\widehat{KX_i}}{\widehat{X_iY_i}}$ decreases and tends to

$\cotan\hat{K}$.

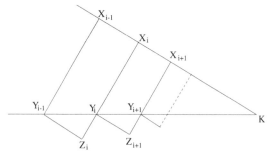

Fig. 2.112

When X_i is close to K, the rectilinear triangle $Y_iZ_{i+1}Y_{i+1}$ is also very close to a rectilinear triangle similar to KX_iY_i, and we have $X_iY_i = X_{i+1}Z_{i+1}$ and $Y_{i+1}Z_{i+1} = X_iY_i - X_{i+1}Y_{i+1}$. As X_i comes close to K, $KX_i \to 0$, so $X_iY_i = KX_i$ $\tan K \to 0$; as a result $Y_{i+1}Z_{i+1} \to 0$, the point Z_{i+1} comes closer and closer to the horizontal circle OKI and approaches the point K.

So we have just proved that while the star passes through the point K, it never at any point crosses the arc KO.

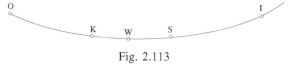

Fig. 2.113

Ibn al-Haytham considers a sequence of points along the trajectory of the star between D and K; this infinite sequence tends to the point K and the height of each of its points is less than the maximum height of the star.

Proposition 31. — *Heights in the east*

This proposition follows on from Proposition 28. We return to the figure for Proposition 28 (p. 223), in which B is the point at which the star rises in the east, D is point of meridian transit and M the point at which the body crosses the hour circle HI with $h(M) > h(D)$. So the star has crossed the horizontal circle DHG at a point that lies between the hour circles BE and HI, let the point be called K. The only points at which the star crosses the horizontal circle DHG are D and K.

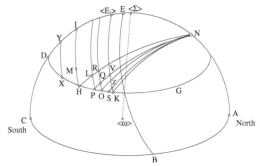

Fig. 2.114

$\theta : E_1\hat{\omega}\Sigma$

λ : angle subtended by the arc E_1V at the centre of the hour circle of K

If X is any point on the arc DH and XY is an arc of an hour circle, where Y lies on DI, then

$$\frac{XY}{YD} > \frac{HI}{ID}$$ (Propositions 11 and 12).

Now $\dfrac{HI}{ID} > \dfrac{\text{required time}}{\text{proper inclination}}$, for any required time when the star is moving from B to D, H being chosen as in Proposition 28, so for any point X on the arc DH, we have

$$\frac{\widehat{XY}}{\widehat{YD}} > \frac{\text{required time}}{\text{proper inclination}}, \qquad \text{since } \frac{\widehat{XY}}{\widehat{YD}} > \frac{XY}{YD}.^{35}$$

• So the star does not at any point cross the arc HD.

• Nor does the star meet the arc KG, since that arc lies to the north of K and the star is moving towards the south.

---

[35] See note 33 (p. 224).

• Ibn al-Haytham goes on to prove that the star cannot meet the arc *KH*.

We imagine a great circle *NH* higher than the point *K*. We know that, as it moves from *K* to *M*, the star passes though the point *M* of the hour arc *HI*; the body meets the great circle *NH* in a point other than *H*, since the body cannot cross the hour circle *HI* at two points *M* and *H*. The point where the body meets the great circle *NH* cannot lie to the south of the hour circle *HI*, since the star could not return from such a southerly point to the point *M*; so the point where it crosses the great circle is a point *L* that lies between *H* and the hour circle of *K*.

The hour circle of *L* cuts *DHG* in a point *P* between *H* and *K*. In its motion from *K* to *L*, the star does not pass through *P*, nor through any other point of *PH*. The same is true for the motion of the body from *L* to *M*, since it moves southwards, it cannot return to lie east of the great circle *NLH*.

In the same way we prove that the star meets the great circle *NP* in a point *Q*. The hour circle of *Q* cuts *DHG* in a point *O* that lies between *P* and *K* and cuts the great circle *NH* at the point *R*. As it moves from *K* to *Q*, the star does not pass through *O*, nor through any other point of the arc *OP*. The arc *QR* is the required time for the motion of the star from the point *Q* to the point *L* and the proper inclination of this required time is $\widehat{LR}$ (where $\widehat{LR} = \widehat{PQ} < \widehat{HR}$).

The hour circle of *K* cuts the great circle *NP* at the point *V*, thus the required time for the star to move from the point *K* to the point *Q* is $\widehat{KV}$ and the inclination proper to it is $\widehat{VQ}$.

So we have proved that if a star passes through *K* it does not pass through any point of the arc *KG* north of *K*. We know that it passes through the points *Q*, *L*, *M* and *D*, but that it does not meet the arcs *DH*, *HP*, *PO*, that is, it does not meet the arc *DO*.

We still need to consider points on the arc *OK*. For this we repeat the previous constructions: the great circle *NO* cuts the arc *KV* in a point V_1, and the star meets the arc OV_1 in a point X_1; the hour circle of X_1 cuts *OK* in Y_1 and *PN* in Z_1. Next let X_2 be the point where the star crosses the great circle NY_1, let Y_2 and Z_2 be the points of intersection of the hour circle of X_2 with the circles *OK* and *ON*, and V_2 the point of intersection of the great circle NY_1 with the arc *KV*. In this way we obtain the points X_i, Y_i, Z_i, V_i ($i = 1, 2...$) such that X_1Z_1 is the required time for the star to move from X_1 to *Q*; the inclination of the motion is Z_1Q, and we have $\widehat{Z_1Q} = \widehat{X_1O}$. In general, $\widehat{X_iZ_i}$ is the required time for the star to move from X_i to X_{i-1}, the inclination of the motion is Z_iX_{i-1} and we have

$$\widehat{Z_iX_{i-1}} = \widehat{X_iY_{i-1}} < \widehat{V_iY_{i-1}}.$$

The star passes through the point X_i, but does not pass through any point of the arc $Y_{i-1}Y_i$.

When the star moves from the point K to the point X_i, the required time is KV_i, its proper inclination is V_iX_i and we have $V_iX_i < V_iY_{i-1}$. The arcs V_iY_{i-1} – and therefore the arcs V_iX_i – are smaller and smaller parts of the arc ED, which is the inclination of the motion of the star from the point B to the point D; this arc ED is itself very small (close to 4′ for the sun, less than 1° for the moon, see footnotes p. 396). So, *a fortiori*, after a certain number of iterations i, the curvilinear triangles KV_iY_{i-1} can be treated as if they were rectilinear triangles, all right-angled at V_i and similar to one another. Then we have, for every i:

$$\frac{\widehat{KV_i}}{\widehat{V_iY_{i-1}}} \approx \frac{KV_i}{V_iY_{i-1}} = \frac{KV}{VP}.$$

The arcs KV_i and KV, the required times, are very small and the arcs V_iX_i and VQ, which are their proper inclinations are also very small and are such that $V_iX_i < V_iY_{i-1}$ and $VQ < VP$; Ibn al-Haytham then draws the conclusion that there is no difference between the ratios $\dfrac{\widehat{KV_i}}{\widehat{V_iX_i}}$ and the ratio $\dfrac{\widehat{KV}}{\widehat{VQ}}$ (see below, p. 397).

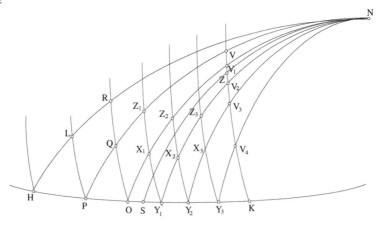

Fig. 2.115.1

Since all the arcs under consideration are very small they can be treated as equivalent to the corresponding chords; the conclusion can then be written, for all values of i,

$$\frac{KV_i}{V_iX_i} = \frac{KV}{VQ}$$

which is the same as saying that the points X_i are very close to the straight line KQ, that is, that the trajectory of the star between the points K and Q can be regarded as a straight line. All the points X_i lie above the horizontal circle KO.

a) We shall show that the star does not pass through any point of the arc KO.

We assume that the arc NH is above the point K.

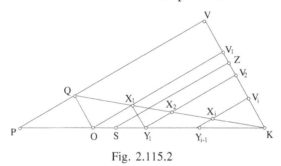

Fig. 2.115.2

Let S be a point of the arc KO, let the great circle NS cut the arc KV at the point Z. The curvilinear triangles KVP and KZS are right-angled, since arcs VP and ZS are orthogonal to arcs VK and ZK respectively. The rectilinear triangles KVP and KZS, which are very small, differ only very slightly from the curvilinear triangles.

• If the difference is negligible, we have $K\hat{V}P = K\hat{Z}S = 1$ right angle and the rectilinear triangles are similar, so

$$\frac{KV}{VP} = \frac{KZ}{ZS} \quad \text{and} \quad \frac{\widehat{KV}}{\widehat{VP}} = \frac{\widehat{KZ}}{\widehat{ZS}}.$$

• In the general case, we have

$$\frac{\widehat{KV}}{\widehat{VP}} > \frac{\widehat{KZ}}{\widehat{ZS}} \qquad \text{(Proposition 15).}$$

We are again using the second inequality of Proposition 15; here the pole N is above the horizontal circle DHG, so that $\alpha < \beta$ ($\alpha = N\hat{\omega}\Sigma$, $\beta = \Sigma\hat{\omega}D$). So the required inequality is satisfied only in very limited circumstances: the point with coordinates (θ, λ) must be below the curve II shown in the figures (Figs 2.21 to 2.27). Since θ is negative, this condition is fulfilled if $\alpha \geq \alpha_1$ (β); this inequality excludes a region close to the earth's north pole.

In every case,

$$\frac{\widehat{KV}}{\widehat{VP}} \geq \frac{\widehat{KZ}}{\widehat{ZS}},$$

so

$$\frac{\widehat{KV}}{\widehat{VQ}} > \frac{\widehat{KZ}}{\widehat{ZS}}.$$

$\widehat{VQ}$ is the proper inclination for the time $\widehat{KV}$ but $\widehat{KZ} < \widehat{KV}$; the proper inclination for the time $\widehat{KZ}$ is smaller than $\widehat{SZ}$; so in its motion from the point K to the point Q the star meets the arc SZ somewhere between S and Z.

The same proof can be given for any point of the arc KO, so the star does not pass through any point of the arc KO.

So if the pole N is above the circle GHD, the star does not meet this circle anywhere except at the points K and D.

Fig. 2.116

b) If the circle NH passes through the point K, the arc KH of the horizontal circle is to the east of K, the star does not return to this arc KH, because it is moving westwards.

c) If the circle NH passes below K, the great circle NK cuts the hour arc HM, the arc KH is to the east of the great circle NK.

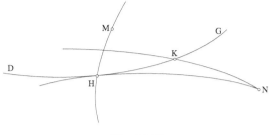

Fig. 2.117

For all versions of the figure the star meets the circle *DHG* at the points *K* and *D* and only at these two points.

• In the east, the star has only a single point of transit across every horizontal circle of height $h < h(D)$.

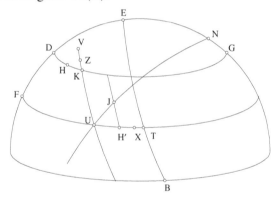

Fig. 2.118

Let *FXT* be a horizontal circle closer to the horizon than the circle *DHG*, and cutting the hour circles *BE* and *VK* in *T* and *U* respectively. In its motion from *B* to *K* the star meets *FXT* at the point *X*, between the points *U* and *T*, and meets the circle *NU* in a point *J* to the north of *U*. So the star does not meet the arc *UF*, since this arc lies south of the arc *KU*, and the body does not meet the arc *XT* which lies north of *X*.

We shall prove that the body does not meet the arc *XU*.

The point *J* is south of *X* and north of *K*, the hour circle of *J* cuts the arc *UT* in *H′* which lies between *U* and *X* (see Fig. 2.114).

Triangle *H′JU* is homologous to triangle *OQP*; we prove that the star does not meet the arc *H′U*, as has been proved for the arc *OP* of the triangle *OQP*; and we prove that the body does not meet the arc *H′X*, as we have proved for the arc *KO*.

• If the great circle *NU* passed through *X* or was below *X*, the arc *XU* would lie east of the great circle *NX*; but the star does not return to lying east of *X*. So the body meets the circle *FXT* only in a single point, the point *X*.

Similarly, in each case there is only one point where the star meets any horizontal circle that is closer to the horizon than the circle *DHG*.

Proposition 32. — *Heights in the east*

In this proposition, Ibn al-Haytham continues his investigation of heights in the east.

If *D* is the point of meridian transit and *K* the highest point the star reaches, then a horizontal circle *FXT* lying between *K* and the horizontal circle *DG* is met by the star twice and only twice.

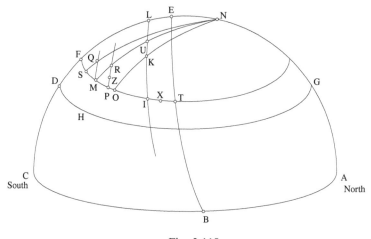

Fig. 2.119

Let *I* and *T* be the points of intersection of the hour circles of *K* and *B* with the horizontal circle *FXT*. As it moves from *B* to *K*, the star meets *FXT* at a point *X* that lies between the circles *LI* and *BE*, so *X* is between *I* and *T*. As it moves from *K* towards *D*, the star meets the circle *FXT* in a point *M* that lies between *I* and *F*.

The star does not meet the circle *FXT* again in a third point.

a) The star does not meet the arc *XT* which lies north of *X*.

b) We can prove that the star does not meet the arc *XI*, in the same way that it was proved in the previous proposition that the body does not meet the arc *KH* of the circle *DHG* (see pp. 232 ff.).

c) Let *O* be the point of intersection of the great circle *NK* and the circle *FXT*. We shall prove that the star does not meet the arc *IM*.

The star does not meet the arc *IO*, since the arc *OI* lies to the east of the circle *NKO*, and the body does not reach *O*.

The great circle *NM* cuts the arc *KI* at the point *U*: $\overset{\frown}{UK}$ is the required time and $\overset{\frown}{UM}$ is the inclination proper to this required time. Let *P* be a point of the arc *OM*, the hour circle of *P* cuts *UM* in *R*. We have

$$\frac{\overset{\frown}{PR}}{\overset{\frown}{RM}} = \frac{\overset{\frown}{IU}}{\overset{\frown}{UM}}$$

(as we have seen already for very small arcs).
But

$$\frac{\overset{\frown}{IU}}{\overset{\frown}{UM}} > \frac{\overset{\frown}{KU}}{\overset{\frown}{UM}},$$

so

$$\frac{\overset{\frown}{PR}}{\overset{\frown}{RM}} > \frac{\overset{\frown}{KU}}{\overset{\frown}{UM}},$$

so the required time whose proper inclination is *RM* is a time $\overset{\frown}{RZ} < \overset{\frown}{RP}$, since $\frac{\overset{\frown}{PR}}{\overset{\frown}{RM}} = \frac{\overset{\frown}{IU}}{\overset{\frown}{UM}}$ in the small triangles *IUM* and *PRM*. The star, in travelling from the point *K* to the point *M*, meets the arc *PR* at the point *Z*, so the body does not pass through the point *P*. The same will be true for any point of the arc *OM*.

d) We shall prove that the star does not meet the arc *MF*.

Let *S* be a point of the arc *MF*, the great circle *NS* cuts the hour circle of *M* at the point *Q*. In the very small triangles *SQM* and *MRP* we have

$$\frac{\overset{\frown}{MQ}}{\overset{\frown}{QS}} = \frac{\overset{\frown}{PR}}{\overset{\frown}{RM}} > \frac{\overset{\frown}{ZR}}{\overset{\frown}{RM}},$$

so the inclination proper to the time *MQ* is greater than *QS*, the star meets the great circle *NQS* in a point below the circle *FXT*, the body does not pass through the point *S* and the same will be true for any point of the arc *MF*.

So the star does not meet the circle *FXT* anywhere except at the two points *X* and *M*.

In this matter Ibn al-Haytham is putting forward an approximate argument that takes account of the fact that the arc *DE* is very small and that the curvilinear triangles concerned are thus close to being rectilinear triangles.

In fact, in this case the first inequality of Proposition 15 gives

$$\frac{\widehat{PR}}{\widehat{RM}} > \frac{\widehat{TU}}{\widehat{UM}}$$

(instead of the equality), which is sufficient. Now this inequality is satisfied if (θ, λ) lies below curves I and III in Figs 2.69–2.75 (see note on Proposition 15); here λ corresponds to the required time $\widehat{LK}$ and θ corresponds to the distance from *L* to the zenith. The fact that $\widehat{DE}$ is small implies that $\widehat{FL}$ is also small, measured by $\beta + \theta$ where β measures the distance from *F* to the zenith. For the inequality to be satisfied, It is sufficient that, in numerical terms, θ remains less than $\theta_{4,m}$.

Conclusion for heights in the east: Let h_D and h_K be the heights of the points *D* and *K*, then any height *h* that satisfies $h_D \le h < h_K$ is reached twice, the height h_K is reached once and any height *h* such that $0 < h < h_D$ is reached once.

Heights in the west (continued)
Proposition 33. — *The maximum height is only attained once.*

Let *OKE* be the horizontal circle corresponding to the maximum height, h_m (where *O* and *E* lie on the meridian); let *NF* be the great circle passing through the pole *N* and touching the circle *OKE* in *F*, and let *OP* be the hour circle of *O*.

We prove, as in the investigation of the heights in the east (Proposition 30), that the star does not pass through any point of the arc *EF*.

We suppose the star meets the circle *OE* in a point *K* to the south of *F*, we have $h_K = h_m$ and we prove that *K* is then the only point that has height h_m.

The great circle *NK* cuts the hour circle *OP* in *P*, and the hour circle *GL* in *J*, where *G* is the point of meridian transit. The arc *GJ* is the required time for the star to move from *G* to *K* and *KJ* is its proper inclination.

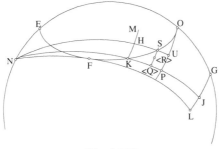

Fig. 2.120

Arcs OP and GJ are similar, they are the measures of the same time, but $\widehat{KP} < \widehat{KJ}$. The required time that corresponds to the inclination KP is smaller than $\widehat{PO}$; let $\widehat{PU}$ be this required time; so, as it moves from G to K the star passes through the point U. Let KM be the hour circle of K, the great circle NU cuts KM at the point H and cuts KO at the point S, we have $\widehat{HS} < \widehat{HU}$; $\widehat{HU}$ is the inclination of the motion of the star from the point U to the point K and $\widehat{HK}$ is the required time, similar to $\widehat{UP}$ (Ibn al-Haytham says 'equal' instead of 'similar', which is not accurate). So we have

$$\frac{\widehat{KH}}{\widehat{HS}} > \frac{\widehat{KH}}{\widehat{HU}}.$$

The hour circle of S cuts KP in a point $<Q>$ such that $\widehat{HS} = \widehat{KQ}$; to the inclination KQ there corresponds a required time $\widehat{QR} < \widehat{QS}$, since the motion continues west of the great circle NSU, so the star passes through the point R, which is below the circle OKE.

We could prove, as in the case of the heights in the east, that all the positions of the star between U and K are below the circle OKE. The star does not meet the arc KO. It does not pass through any point of the arc FE (p. 240). We can prove as in note 131 on page 392 that the star does not pass through any point of the arc KF. So the point where it crosses the circle OKE of height h_m is a unique point, which we have called K.

If the point where the star crosses the circle OE is the point F, it is the only point of height h_m that the body reaches.

Ibn al-Haytham returns to Proposition 29a (Fig. 2.108) to continue his investigation of the heights in the west.

Proposition 34. — Let G be the point of meridian transit and GHI the horizontal circle through G; H and K are the points defined in Proposition 29a.

Assumption: $\dfrac{HK}{KG} > \dfrac{\text{required time}}{\text{proper inclination}}$, for every required time for the star as it moves from the point G to the point D.

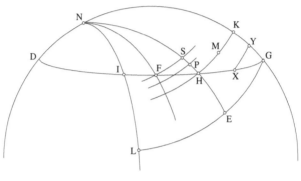

Fig. 2.121

But

$$\frac{\widehat{HK}}{\widehat{KG}} > \frac{HK}{KG},^{36}$$

hence

$$\frac{\widehat{HK}}{\widehat{KG}} > \frac{\text{required time}}{\text{proper inclination}}.$$

Let $\widehat{KM}$ be the required time whose proper inclination is $\widehat{KG}$ $\left(\widehat{KM} < \widehat{KH}\right)$, the star moves from the point G to the point M on the hour circle HK.

If F is the second point at which the star crosses the horizontal circle GHI, then G and F are the only points where it crosses GHI.

Let X be a point of the arc HG and let XY be an arc of an hour circle, then

$$\frac{XY}{YG} > \frac{HK}{KG};^{37}$$

but

---

[36] By an argument analogous to that in note 33 (p. 224).
[37] See Proposition 12.

$$\frac{\widehat{XY}}{\widehat{YG}} > \frac{XY}{YG}, [38]$$

hence

$$\frac{\widehat{XY}}{\widehat{YG}} > \frac{\text{required time}}{\text{proper inclination}}$$

for every required time; so the star does not pass through the point X, it passes through an interior point of the arc XY.

The star does not pass through any point of the arc HG. As it moves from G to M, it is above the arc HG.

Let FS be the hour circle of F. The star moves from M to F; so it cuts the circle NH in a point that cannot be either H, or a point south of H, or the point S, or one north of S. The body cuts the circle NH in a point P lying between H and S. We then prove that the body cannot pass through any point of the arc FH (using the same method of proof as before).

Fig. 2.122.1

We still need to prove that the star does not pass through any point of the arc FI.[39]

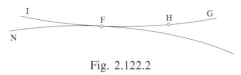

Fig. 2.122.2

We draw the great circle NF. There are three possible cases:

a) The circle NF touches the circle GHI.

b) The circle NF cuts GHI in two points, F being the more northerly one.

In these two cases, the star does not pass through any point of the arc FI.

c) The circle NF cuts GHI in two points, F being the more southerly one. In this case we draw the great circle NXU to touch the circle GHI at X

---

[38] See note 33 (p. 224).

[39] We assume here that the pole N is below the horizontal circle GHI.

(where U lies on the hour circle SF). The hour arc of X cuts the circle NF in R. We have

$$\frac{\overparen{FS}}{\overparen{SH}} = \frac{\overparen{XR}}{\overparen{RF}}$$

(as we have seen in Proposition 31 for very small triangles, F and X being close to one another).

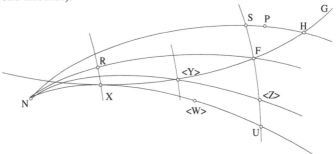

Fig. 2.122.3

But

$$\frac{\overparen{FS}}{\overparen{SP}} > \frac{\overparen{FS}}{\overparen{SH}},$$

so

$$\frac{\overparen{FS}}{\overparen{SP}} > \frac{\overparen{XR}}{\overparen{RF}}.$$

Moreover, if α is the inclination proper to the time XR, we have

$$\frac{\overparen{FS}}{\overparen{SP}} = \frac{\overparen{XR}}{\alpha} > \frac{\overparen{XR}}{\overparen{RF}},$$

(since the required times $\overparen{FS}$ and $\overparen{XR}$ are very small and close to one another), so $\alpha < \overparen{RF}$; but $\overparen{XR}$ is similar to $\overparen{FU}$ (and has a little difference since they are quite close, Ibn al-Haytham considers them equal) and $\overparen{RF} = \overparen{XU}$, so $\alpha < \overparen{XU}$,

$$\frac{\overparen{FS}}{\overparen{SP}} = \frac{\overparen{FU}}{\alpha}.$$

Let W be a point between U and X such that $\overparen{UW} = \alpha$.

The star moves from the point F to the point W between X and U. Any great circle drawn through N and a general point Y on the arc XF cuts the arc FU in Z, the hour arc of Y cuts the arc RF and the arc XU. To the time FZ there corresponds an inclination that is a part of the arc ZY (homologous with WU which is a part of UX); the end of this arc is below the arc XF. So the star does not pass through any point of XF except F. The star does not pass through any point of XI, since the arc XI is to the east of the great circle NU.

Conclusion: The star meets the circle GHI only at the two points G and F.

Let there be a horizontal circle OUT whose height h is less than the height of the point G. The star meets OUT in a point H' north of the hour circle FSU, H' is the only point where the star crosses the circle OUT.

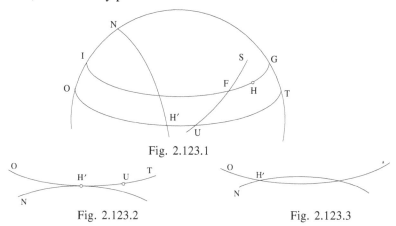

Fig. 2.123.1

Fig. 2.123.2   Fig. 2.123.3

If the great circle NH' touches the circle OUT at H' or if it cuts OUT in a second point south of H', the star does not meet the arc $H'T$, which lies to the south of the point H'.

Fig. 2.124

If the great circle *NH'* cuts the circle *OUT* in *H'* and in a point north of
H', the great circle *NU* cuts the circle *OU* in *U* and in a point north of *H'*;
and it cuts the hour circle of *H'* in a point *J*. The star moves from *F* to the
point *H'*, so it cuts the circle *NU* between the two points *J* and *U*, say at the
point *Y*; the hour circle of *Y* cuts *H'U* in *Z*.

Proceeding as before (p. 243), we prove that the star does not pass
through any point of the arc *H'U*, nor through any point of the arc *H'T*.
Moreover, the body does not pass through any point of the arc *UO*, which
lies to the south of the arc *FU*. So the only point where the body crosses the
circle *OT* is the point *H'*.

In the same way we prove that on every horizontal circle of height
$h < h_G$, there is a single point through which the star passes.

Proposition 35. — We then consider a horizontal circle *OST* whose height
h satisfies the inequality $h_m > h > h_G$, h_m being the maximum height and *U*
the point at which the body reaches this height. The star meets this circle at
two points.

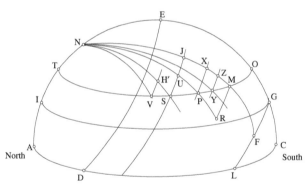

Fig. 2.125

Let *DE* and *GL* be the hour circles of *D* and *G*. The point *U* lies
between these two circles, similarly its hour circle lies between these circles
and cuts the circle *OST* in the point *S*. The great circle *NU* cuts the circle *OT*
in *P*.

The star has moved from the point *G* to the point *U*, so it has cut the
circle *OT* in a point that cannot be either *P*, or *S*, or a point of the arc *SP*,
but which must lie to the south of the arc *UP*. So the point of intersection
lies on the arc *OP*, let it be the point *M*. The great circle *NM* cuts *GL* in *F*,
so the arc *GF* is the required time whose inclination is the arc *FM*.

The star does not pass through any point of the arc MO, which is east of M.

The hour circle of M cuts the circle NU at the point R; the arc MR is the required time whose inclination is RU. The hour circle of U cuts the great circle NM in J; we have $\widehat{JM} = \widehat{UR}$; $\widehat{UJ}$ and $\widehat{RM}$ are similar arcs that measure the same time, so $\widehat{UJ}$ is the required time whose inclination is $\widehat{JM}$; in reality, if we abide by the conventions adopted thus far, the required time is measured by the arc RM not by the arc UJ, but these two arcs, which belong to different circles, subtend the same angle at the centre. We can see that Ibn al-Haytham is maintaining a degree of ambiguity between angles and arcs, as we have already noticed in regard to Propositions 14 and 15. The hour circle of P cuts the circle NJM in X, $\widehat{PX}$ is similar to $\widehat{UJ}$ and $\widehat{JM} > \widehat{XM}$, so

$$\frac{\widehat{PX}}{\widehat{XM}} > \frac{\widehat{UJ}}{\widehat{JM}}.$$

We may prove in the same way as before, by using small triangles similar to triangle PXM, that any hour arc such as YZ drawn through a point of the arc PM to meet the arc XM is such that $\dfrac{\widehat{YZ}}{\widehat{ZM}} = \dfrac{\widehat{PX}}{\widehat{XM}}$ and that $\dfrac{\widehat{YZ}}{\widehat{ZM}}$ is greater than the ratio of $\widehat{YZ}$ to the inclination that is proper to it.

So the star does not pass through any point of the arc PM; so it does not pass through any point of the arc SM.

Let V be a second point where the star crosses the circle OT, V lies between the hour circles SU and DE. The hour circle of V cuts the great circle NS in H'. As it moves from U to V, the star meets the arc $H'S$. We can prove as before that the star does not pass through any point of VS or through any point of VT. The only points at which the body crosses the circle OT are the points M and V.

Every horizontal circle of height h such that $h_G < h < h_m$ is crossed by the star twice and only twice.

Ibn al-Haytham then states a general conclusion relating to the case where the point at which the star reaches its maximum height lies to the west of the meridian circle, that is, the case in which the body is moving from the northernmost point of its orb towards the southernmost one.

From Proposition 28 onwards, the investigation of the heights has been concerned with heavenly bodies whose point of meridian transit is south of the pole of the horizon. We have always considered a sphere inclined to the south, that is, the investigation is conceived for an observer in the northern hemisphere.

Proposition 36. — For places on the terrestrial equator, the sphere is right and the hour circles are orthogonal to the circle of the horizon, since the pole of the equator lies on the horizon.

To every horizontal circle of diameter DD', where DD' is in the meridian plane, with D' to the north and D to the south, there correspond hour arcs that are equal two by two, such as $\overset{\frown}{HI} = \overset{\frown}{H'I'}$ and $\overset{\frown}{ID} = \overset{\frown}{I'D'}$ (symmetry with respect to OZ).

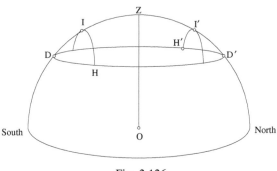

Fig. 2.126

In Proposition 28, the investigation assumed that meridian transit took place at a point such as D south of Z, and we considered an hour arc such as HI that satisfied the inequality $\dfrac{\overset{\frown}{HI}}{\overset{\frown}{ID}} > k$, where k is a given ratio greater than the ratio of the required time to the inclination proper to this required time for every arc traversed by the star between rising and meridian transit.

If the meridian transit occurs at the point D' to the north of Z, we need to consider the arc $H'I'$, symmetrical with HI; this arc has the same property $\dfrac{\overset{\frown}{H'I'}}{\overset{\frown}{I'D'}} > k.$

In the case of the sphere being right, and whether the meridian transit takes place to the north or to the south of the pole of the horizon, we obtain results analogous to those found for a sphere inclined towards the south with the meridian transit south of the pole of the horizon.

If the sphere is inclined to the south, that is, in places in northern latitudes, the planets cannot cross the meridian at the zenith or north of the zenith unless the latitude is small.

The star will cross the meridian at the zenith if, at the moment of transit, the declination of the body is equal to the latitude of the place where the observation is made; and the transit will occur to the north of the zenith if the declination is greater than the latitude of the observer.

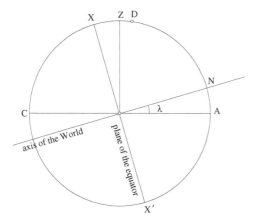

Fig. 2.127

Let Z be the zenith, and XX' the line of intersection of the planes of the equator and the meridian, the latitude is $\lambda = \widehat{AN} = \widehat{XZ}$. If the transit occurs at D, the declination of the star is then $\widehat{XD}$. The point D cannot lie north of the zenith unless $\widehat{XZ} < \widehat{XD}$, that is $\lambda < \widehat{XD}$.

Case where meridian transit occurs at the pole of the horizon

Let ABC be the horizon and D its pole, and let $EGHI$ be a horizontal circle. The hour circle of D cuts this circle in G to the east and in I to the west. A star that crosses the meridian at D would move along the hour circle $BGDI$.

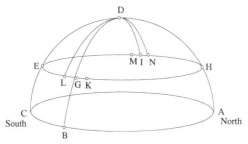

Fig. 2.128

a) The star rises in the east; if it crosses the meridian at D and if its motion on its orb is from north to south, it meets the circle EGH in K, north of G, and between its meridian transit at D and setting the body meets the circle $EGHI$ in M, to the south of I.

b) If the body's motion on its orb is from south to north as it moves between rising and meridian transit at D, the body meets the circle $EGHI$ in L to the south of K and as it moves between D and setting, it cuts the circle again in N to the north of I.

We then prove that the only eastern point where the star meets the horizontal circle $EGHI$ is, in case a), the point K (see end of Proposition 31) and, in case b), the point L (see end of Proposition 29). Similarly, the only western point where the star meets the horizontal circle $EGHI$ is, in case a), the point M (see end of Proposition 28) and, in case b), the point N (see end of Proposition 34).

The same reasoning would apply for every horizontal circle, so on the day that a star crosses the meridian at the point D, the pole of the horizon, the body will not have either two equal heights in the east or two equal heights in the west.

Investigation of horizons for northern latitude λ equal to the complement of the inclination of the orb, and of horizons for latitudes close to λ for which the star has points of rising and setting

1. Let us suppose that the motion of the star on its orb is from the northernmost point towards the equator.

Let $ABCD$ be the horizon of a place for which the height of the north pole of the equator, H, above the horizon – that is, the geographical latitude of the place – is the complement of the inclination of the orb of the star with respect to the equator, that is, a place on the polar circle.

The circle of the equator cuts the horizon in A and C – where C is to the east and A to the west – and cuts the meridian in E and G. The arc DE is the inclination of the orb. We have $\widehat{DE} = \widehat{BG} = \widehat{HZ}$ (Z is the pole of the horizon).

The circle BQI whose pole is H and which passes through B cuts the meridian in I and we have $\widehat{EI} = \widehat{BG}$. If we are considering the sun, BQI is the tropic of Cancer. In the course of the diurnal motion the point B, the northernmost point of the orb, describes the circle BQI, which touches the horizon $ABCD$ at B, $ABCD$ being one of the horizons that have BED as a meridian circle.

Let us suppose that at a given moment the star is at the point B on the horizon $ABCD$; in the course of the diurnal motion, the body will move away from the circle BQI and will come to lie south of that circle (as we have seen in Propositions 16, 17, 18 when the motion of the star on the orb is from the northernmost point towards the southernmost one).

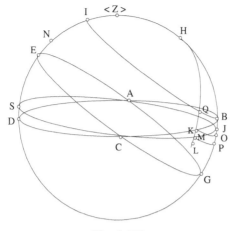

Fig. 2.129

Let K be a point on the horizon and O a point on the meridian such that the hour arc KO satisfies the inequality $\dfrac{\widehat{KO}}{\widehat{OB}} > \dfrac{\text{any required time}}{\text{proper inclination}}$; in the motion of the star, starting from B, the inclination proper to the time $\widehat{KO}$ is $\widehat{BP} > \widehat{BO}$. The hour circle of P cuts the great circle HK in L; we have time $\widehat{KO}$ = time $\widehat{LP}$ = time $\widehat{BQ}$ (the time is in fact the angle subtended by every one of the arcs which are all similar) and $\widehat{QL} = \widehat{BP}$; so $\widehat{QL}$ is the inclination proper to the time $\widehat{BQ}$ and the star thus reaches the point L, below the circle of the horizon $ABCD$.

Let J be a point on the meridian such that $\widehat{BJ} < \widehat{BO}$ and $\widehat{BJ} < \widehat{OP}$; the circle $AJCS$ cuts the arc ED of the meridian in S and the great circle HKL in M, we have $\widehat{KM} < \widehat{BJ} < \widehat{OP}$; but $\widehat{OP} = \widehat{KL}$, hence $\widehat{KM} < \widehat{KL}$; so L is below the circle $AJCS$ and B is above the same circle $AJCS$, which is a horizon that has the circle BED as its meridian; for this horizon the star, which has travelled from the point B to the point L, has thus set at a point of the arc JM, that is, in the east.

Let N be the point of the meridian such that $\widehat{IN}$ is the inclination of the motion of the star, the inclination proper to the time BQI; so the star will travel from the point L to the point N and it will rise at a point of the arc MC, that is, on the eastern part of the horizon $AJCS$.

Notes:

1) We know that, in the case of the sun, the arc IN that measures the decrease in declination over half a day is close to $8'$ (see footnotes p. 396). So the arc BP is much smaller than $8'$, and all the arcs such as OP, BJ, QK

are smaller still. So the figure cannot give an exact account of proportions, but its purpose is to show the points with which the argument is concerned and their relative positions.

It is clear that the arc BK is extremely small and that the same is true of the arc between the points where the star rises and sets on the horizon $AJCS$; the points of rising and setting are practically identical.

2) If λ is the latitude of the place whose horizon is the circle $ABCD$, we have $\overset{\frown}{HB} = \lambda$. But $\overset{\frown}{HJ} > \overset{\frown}{HB}$, so the latitude of the place with horizon $AJCS$ is $\overset{\frown}{HJ} = \lambda + \varepsilon$, where ε is very small, the latitude corresponding to the horizon $AJCS$ is only slightly greater than λ, that is, only slightly greater than the complement of the maximum declination of the star.

If the northern latitude becomes a little bit greater still, the sun will not set but will simply remain below the horizon for the full twenty-four hours.

3) For horizon $ABCD$, with latitude λ, we know by hypothesis that the star reaches the point B when its northern declination is at its maximum and equal to α. Before reaching B its declination was increasing and less than α, its path came closer and closer to the hour circle BQI and, after passing through B, the star will move away from the circle BQI once more; Ibn al-Haytham proves that the star passes through a point L below the horizon $ABCD$, so in the course of its motion from L to N the body meets the horizon $ABCD$ at a point between K and C, that is, in the east.

So, with horizon $ABCD$, the star has set at B, the north cardinal point and has risen in the east.

In what we have said so far, the circle BQI corresponds to the maximum inclination of the star, that is, in the case of the sun its inclination on the day of the summer solstice. On the following day the star meets the meridian in a point X of the arc BG, and through that point there passes an hour circle Γ homologous to the circle BQI; through the point X there passes a great circle Γ' that touches Γ; so Γ' will be the homologue of $ABCD$. Starting from the point X and the circles Γ and Γ', and proceeding as we did when starting from the point B and the circles BQI and $ABCD$, we can find a horizon Γ'', homologous to $AJCS$, for which the star sets in the east and then rises in the east.

Each of the horizons we have considered is for a northern latitude, since the height of the pole above the horizon, that is, the arc HJ or one of its homologues, is less than a quadrant of a circle.

Any point of the circle BQI may be a point of contact of the circle with a great circle that can be taken as a horizon. In this way we obtain the horizons of all points on the Earth with latitude equal to the complement of the maximum inclination of the star concerned, and the preceding argument is valid for all of them.

The star moves from the northernmost point of its orb towards the equator, so its inclination with respect to the equator decreases. On the day when meridian transit below the horizon occurs at a point B' close to G, such that $\widehat{B'G}$ is less than the inclination at meridian transit above the horizon, the transit takes place at a point N' close to the point E and to the south of E.

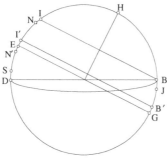

Fig. 2.130

For a horizon $AJCS$ such that the arc ES is greater than the inclination of half a daily revolution, the star will rise in the east at a point of the arc MC or at a point of the arc CS.

2. The motion of the star on its orb is from the southernmost point towards the equator.

Let H and H' be the north and south poles of the equator $AGCE$, and $ABCD$ a horizon, with A to the west and C to the east. If Z is the zenith, we have $\widehat{DE} = \widehat{BG} = \widehat{HZ}$. We suppose the horizon $ABCD$ touches the hour circle BQI (which corresponds to the maximum southern inclination of the star in question) at the point B.

In the case of the sun, BQI is the tropic of Capricorn.

Let K be a point on the horizon and O a point on the meridian such that the hour arc KO satisfies the inequality $\dfrac{\widehat{KO}}{\widehat{OB}} > \dfrac{\text{required time}}{\text{proper inclination}}$ during the motion of the star from B.

The great circle $H'K$ cuts the hour circle of B at the point Q. The time KO is equal to the time BQ (equal angles) and the inclination proper to this time is $\widehat{BP} > \widehat{BO}$. The hour circle of P cuts the circle $H'K$ in L, we have $\widehat{QL} = \widehat{BP}$. So in the time $\widehat{BQ}$, the star travels from the point B to the point L.

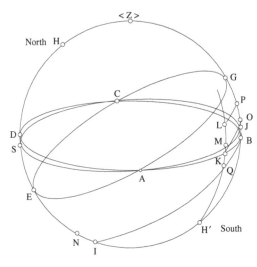

Fig. 2.131

As in the previous proposition we take a point J on the meridian (with $\widehat{BJ} < \widehat{OP}$ and $\widehat{BJ} < \widehat{BO}$) and the great circle *AJCS* cuts the circle *HK* in M; we prove that M lies between K and L, so in its motion in the time BQ the star has travelled from the point B to the point L and has met the circle *AJCS* in a point that lies between B and M; but *AJCS* is the horizon for a place that has meridian *BHD*, B is below the horizon *AJCS* and L is above it, so for this horizon, the star rose in the west at a point between J and M.

The motion of the star takes it beyond L to intersect the arc $H'E$ at the point N, so it meets the arc JA of the horizon *AJCS*, beyond M, and passes below the horizon; so the body sets in the west at a point of the arc *MA*.

Notes:

1) As in the first part, the points of rising and setting are very close to one another.

2) We have $\widehat{HD} = \lambda$, $\widehat{HS} > \widehat{HD}$, $\widehat{HS} = \lambda + \varepsilon$, where ε is very small. So the latitude corresponding to the horizon *AJCS* is only a little greater than the complement of the maximum declination of the star.

If the northern latitude becomes a little greater still, the sun will remain below the horizon for the full twenty-four hours.

3) For the horizon *ABCD*, we could prove as before that the star rises at B, the south cardinal point, and sets in the west at a point very close to B.

In conclusion, we can see that Ibn al-Haytham establishes that the motion of a star on the celestial sphere has the following properties. Between rising and setting, the star passes through a unique point U at which its height above the horizon is a maximum $h_U = h_m$ and any height $h < h_m$ is reached exactly twice. This property depends on the fact that the motion of the star is always from east to west.

For observers who are not near the poles, the star rises in the observer's east and sets in the west; its diurnal course takes the star through a point G of the meridian. The transit at G can take place before or after the body passes through U. Let h_G be the height of the point G; every height $h < h_G$ is reached once between rising and passing through G, and once between G and setting. Any height h such that $h_G < h < h_m$ is reached twice between G and U.

For observers close to the arctic circle or antarctic circle, the rising and setting of the star can both be in the east or both in the west, with no meridian transit in the course of the diurnal motion. Outside the arctic circle or antarctic circle, the star may not rise or (on the contrary) not set.

2.3. HISTORY OF THE TEXT

Of the three books that made up *The Configuration of the Motions of the Seven Wandering Stars*, only the first has survived. We have the book in which Ibn al-Haytham develops his theory of the motions of these wandering stars. This book has come down to us in a single manuscript, which is part of a valuable collection now in the National Library in St. Petersburg, Arabic new series no. 600. A description of this collection can be found in the fourth volume of *Mathématiques infinitésimales*, pp. 24–6, so there is no need for us to repeat it here. Suffice it to recall that the collection was copied towards the middle of the seventeenth century, on thin transparent paper, slightly greyish in colour. So it often happens that the words and figures on the recto of a page show through on the verso. It also happens, because of the deterioration at the edges of the folios, that some areas near the margins are difficult to read. Finally, the rather careless *nasta'līq* script makes some words particularly difficult to read.

The text of *The Configuration* is by a single hand and occupies folios 368^v–420^v. But these folios, like those of other treatises in the same collection – for example *The Properties of Circles* by Ibn al-Haytham – are out of order. Thus the text of *The Configuration* is finally ordered as follows:

368^v, 397^v, 397^r–401^v, 402^v, 402^r, 403^r–408^v, 369^r–396^v, 409^r–420^v.

We may note that, in addition to being out of order, some of the folios are inverted; these accidents most probably occurred when the collection was bound.

This text has been edited according to the rigorous rules that were followed in our other critical editions, rules which we have already explained more than once. There remains, however, the question of how to treat the figures. The figures in the text are, we know, drawn by the last copyist. These figures are indicative, but inaccurate and even ambiguous. In most cases the state of the manuscript renders them illegible. Certain of them are so complicated that it is difficult to distinguish the lines, straight lines and planes. In view of this state of affairs, we have been compelled to reconstruct the figures, partly from the traces in the manuscript, but largely with the help of the written text. It has sometimes happened that we have made one figure into two to make it easier to understand (notably the figures for Propositions 14, 15 and 16). In every case, we have been at pains to make use of the traces of the figures in the manuscript, however illegible they are, in order to stay as close as possible to the original.

TRANSLATED TEXT

Al-Ḥasan ibn al-Haytham

*The Configuration of the Motions of Each
of the Seven Wandering Stars*

In the name of God, the Compassionate the Merciful
From Him comes our help

TREATISE BY SHAYKH ABŪ 'ALĪ IBN AL-HAYTHAM

May God continue his blessing upon him

On the Configuration of the Motions
of Each of the Seven Wandering Stars[1]

And on the proof that each of the seven wandering stars, can have, in
the east,[2] at certain moments of a single day, equal heights at all places on
the Earth where the diurnal <arc> of the star can be divided into two equal
parts; that at certain moments in a single day, in the west,[3] there can be
equal heights at all places on the Earth where the diurnal <arc> of the star
can be divided into two equal parts; that at certain places on the Earth, at
certain moments, a wandering star can set on the eastern horizon[4] and, in
the course of its day rise on the eastern horizon; and that at certain
moments, at these same places on the Earth, a wandering star can in the
course of its day, rise on the western horizon and set on the western horizon. It is omniscient God who gives [us] success.

Since it is accepted that the knowledge of the motions of each of the
wandering stars, together with the differences between them, forms part of
the art of astronomy; that knowledge of the ascendant from the height of
the moon and of the five planets forms part of the art of astronomy; that
what we have stated concerning the setting of the wandering stars in the
east[5] also forms part of the art of astronomy – something about which those
learned in this art must know the truth – we resolved to be even more
especially careful in establishing the truth of the proofs of all the results we
have mentioned, relating to the phenomena that occur among the
wandering stars, so as to be certain that everything we said about them is

---

[1] Lit.: the seven stars; we shall translate *al-kawākib al-sab'a* as 'the seven
wandering stars'; we shall translate *al-kawākib al-khamsa* as 'the five planets'.

[2] Lit.: in the easterly direction.

[3] Lit.: westerly direction.

[4] He means heliacal setting.

[5] Lit.: in the easterly direction and their rising in the westerly direction

indeed as we said; and we are collecting this together in a single book that includes all the proofs [of these results]. We follow it with another book in which we summarize all the calculation procedures that lead [us] to understand the truth of <each of> these results. Then we accomplish what this art requires and we spare the specialists in it the trouble of writing essays on the observation of minutes [of arc] and small parts for the altitude of the sun and of all the wandering stars, by presenting an instrument that is easy to handle[6] and can be understood by everyone, by means of which we find the height of the sun and of each of the [wandering] stars using the minutes and small parts [of the height]. Thanks to this instrument and to the procedures we explain, all the procedures followed by astronomers are shown to be correct and an end is put to all the disputes that arise over principles, because of the fractions that are missed by observers and that they find almost imperceptible, on account of the design of the instruments. It is from God that we ask for help in all things.

Everything we have said in places other than this book, concerning the height of the sun, the heights of the wandering stars and the altitude on the meridian,[7] and where we have not given a precise account of these matters, employed only the method used by the majority of mathematicians, following the conventional principles. However, all we have said concerning heights, in the conventional way, is to be found only in what we have set out in the books [written] before this one, before this idea occurred to us; then, once this idea occurred to us and took a precise form, we wrote this book and in it we have explained these ideas. Thus, if anyone examines this book and our other books and finds a difference in what we have stated about heights, let him be aware that the reason for it is what we have referred to, and that what we have said in this book concerning the heights of the wandering stars is extremely accurate. And what we have said in our other books, written before this one, uses the method conventionally employed by mathematicians.

<1> For two unequal arcs of the same circle, whose sum is not greater than a semicircle, the ratio of the greater arc to the smaller arc is greater than the ratio of the chord of the large arc to the chord of the small arc.

Example: The arcs *AB* and *BC* have as their sum the arc *ABC*, which is not greater than a semicircle, the arc *AB* is greater than the arc *BC* and they are cut off by the two chords *AB* and *BC*.

---

[6] See Appendix.
[7] Lit.: the height <of the arc> of midday. He means the altitude of meridian passage for the bodies concerned.

I say that the ratio of the arc AB to the arc BC is greater than the ratio of the chord AB to the chord BC.

Proof: We draw the straight line AC and we make the angle CBD equal to the angle BAC which is smaller than the angle ABC; the angle BDC is thus equal to the angle ABC and the triangle CBD is similar to the triangle ABC. So the ratio of AB to BC is equal to the ratio of BD to DC. But the straight line AB is greater than the straight line BC, because the arc AB is greater than the arc BC; so the straight line BD is greater than the straight line DC. In the same way, the ratio of the angle BCA to the angle BAC is equal to the ratio of the arc AB to the arc BC. So the ratio of the angle BCD to the angle CBD is equal to the ratio of the arc AB to the arc BC. We take the point C as centre and with the distance CB we draw an arc of a circle, let the arc be BHG; so it cuts the straight line CA; let it cut it at the point G. It is thus clear that the point G lies outside the straight line CD on the same side as the point A, because the straight line GC is equal to the straight line BC and the straight line BC is greater than the straight line CD – and this [is so] because the angle BDC is not smaller than a right angle since it is equal to the angle ABC; and the straight line BC is smaller than the straight line CA, so the point G lies between the two points D and A.

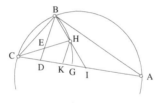

Fig. I.1

We make the angle DCE equal to the angle CBD; so the point E lies between B and D, since the angle BCA is greater than the angle CBD. We draw the straight line CE to meet the arc BG; let it meet it at the point H. Then the ratio of the arc BG to the arc GH is equal to the ratio of the angle BCD to the angle DCE which is equal to the angle CBD. Now we have proved that the ratio of the angle BCD to the angle CBD is equal to the ratio of the arc AB to the arc BC; so the ratio of the arc BG to the arc GH is equal to the ratio of the arc AB to the arc BC. We draw the straight line HK parallel to the straight line DE; since the angle ECD is equal to the angle CBD and the angle CDE is common to the triangles CBD and CED, triangle CED is similar to triangle CBD; thus the ratio of BD to DC is equal to the ratio of CD to DE and is equal to the ratio of BC to CE. But BC is equal to CH; so the ratio of CD to DE is equal to the ratio of CH to CE. But the ratio of CH to CE is equal to the ratio of HK to DE; so the ratio of CD to DE is equal to the ratio of HK to DE, and HK is equal to CD. Now we have proved that BD is greater than CD, so BD is greater than HK. We draw BH and we extend it. It then meets the straight line AC; let it meet it at the point I. So the ratio of BI to IH is equal to the ratio of BD to HK. But the ratio of BD to HK is equal to the ratio of BD to DC which is equal to the ratio of AB to BC; so the ratio of the straight line BI to the straight line IH is equal to the ratio of the straight line AB to the straight line BC. Thus we have proved that the ratio of the arc BG to the arc GH is equal to the ratio of the arc AB to the arc BC. But the ratio of the straight line BI to the straight line IH is equal to the ratio of triangle CBI to triangle CHI, the ratio of the arc BG to the arc GH is equal to the ratio of the sector CBG to the sector CHG, the sector CBH is greater than the triangle CBH and the sector CHG is smaller than the triangle CHI; the ratio of the sector CBH to the sector CHG is thus greater than the ratio of the triangle CBH to the triangle CHI. By composition, the ratio of the sector CBG to the sector CHG is thus greater than the ratio of the triangle CBI to the triangle CHI. So the ratio of the arc BG to the arc GH is greater than the ratio of the straight line BI to the straight line IH. Now we have proved that the ratio of the arc BG to the arc GH is equal to the ratio of the arc AB to the arc BC and the ratio of the straight line BI to the straight line IH is equal to the ratio of the straight line AB to the straight line BC; so the ratio of the arc AB to the arc BC is greater than the ratio of the chord AB to the chord BC. That is what we wanted to prove.

Ptolemy established this result in his book *The Almagest*,[8] but using a method different from this one.

I also say that the ratio of the arc ABC to the arc CB is greater than the ratio of the straight line AC to the straight line CB.

Proof: The ratio of the arc AB to the arc BC is greater than the ratio of the straight line AB to the straight line BC. By composition, the ratio of the arc ABC to the arc CB is thus greater than the ratio of the sum[9] of the straight lines AB and BC to the straight line CB. But the ratio of the sum of AB and BC to the straight line CB is greater than the ratio of the straight line AC to the straight line CB, because the sum of the two straight lines AB and BC is greater than AC. So the ratio of the arc ABC to the arc CB is

---

[8] Book I.10.

[9] We sometimes add the word 'sum' for clarity in the translation.

greater than the ratio of the straight line *AC* to the straight line *CB*. That is what we wanted to prove.

<2> If we have two unequal arcs such that one is greater than the arc[10] similar to the other, [arcs] belonging either to two equal circles or to two unequal circles and such that each of them is no greater than a semicircle, if we divide each of the two arcs into two unequal parts such that the ratio of the greater part of the greater arc to the smaller part of it is equal to the ratio of the greater part of the smaller arc to its smaller part and if we draw the chords of these arcs, then the ratio of the chord of the greater part of the small arc to the chord of its smaller part is greater than the ratio of the chord of the greater part of the large arc to the chord of its smaller part.

Example: Let *ABC* and *DEG* be two unequal arcs; the arc *ABC* is greater than the arc similar to the arc *DEG* and each of them is no greater than a semicircle. We divide the two arcs at the points *B* and *E* such that the arc *AB* is greater than the arc *BC* and the arc *DE* greater than the arc *EG* and such that the ratio of the arc *AB* to the arc *BC* is equal to the ratio of the arc *DE* to the arc *EG*; we draw the chords *AB*, *BC*, *DE* and *EG*.

I say that the ratio of the chord *DE* to the chord *EG* is greater than the ratio of the chord *AB* to the chord *BC*.

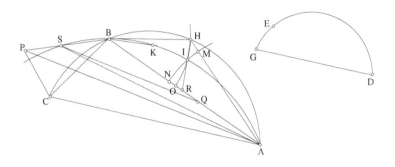

Fig. I.2

Proof: On the chord *AB* we construct an arc similar to the arc *DE*; it falls inside it because the arc *DEG* is smaller than the arc similar to the arc *ABC*.[11] But the ratio of the arc *DE* to the arc *EG* is equal to the ratio of the arc *AB* to the arc *BC*; so the arc *DE* is smaller than the arc similar to the arc *AB*, the angle within which it falls is thus greater than the angle within

---

[10] Lit.: the similar; which we have translated throughout as 'the arc similar'.

[11] See Supplementary note [1].

which the arc AB[12] falls and the arc similar to arc DE thus falls within the arc AB; let it be equal to the arc AIB. We cut off the arc BH equal to the arc BC and we draw the straight lines BH and AH. Since the arc ABC is not greater than a semicircle, the arc AB is smaller than a semicircle, so the angle AHB is obtuse. We take the point B as centre and with the distance BC we draw an arc of a circle. This arc cuts the angle AHB; let this arc be the arc HIO. So the arc HIO cuts the arc AIB; let it cut it at the point I. We draw the straight line BI; it is equal to the straight line BH. We draw the straight line HI and we extend it to R; the angle BHI is thus equal to the angle BIH, the angle BHI is acute, the angle BIR is obtuse and the angle ARI is greater than it, so the angle ARI is obtuse. So the straight line AH is greater than the straight line AI and the straight line AI is greater than the straight line AR. We take the point A as centre and with the distance AI we draw an arc of a circle. This arc cuts the straight line AH in a point between the two points A and H and cuts the straight line AR in a point beyond the point R; let this arc be the arc MIN. Since the sector BHI is greater than the triangle BHI and the triangle BIR is greater than <the sector> BIO, the ratio of the sector BHI to the sector BIO is greater than the ratio of the triangle BHI to the triangle BIR. By composition again, the ratio of the sector BHO to the sector BIO is greater than the ratio of the triangle BHR to the triangle BIR. So the ratio of the angle HBA to the angle IBA is greater than the ratio of the straight line HR to the straight line RI. But also from the fact that the triangle AHI is greater than the sector AMI and the sector AIN is greater than the triangle AIR, the ratio of the triangle AHI to the triangle AIR is greater than the ratio of the sector AMI to the sector AIN. By composition again, the same will hold and we have that the ratio of HR to RI is greater than the ratio of the angle HAB to the angle IAB. But the ratio of the angle HBA to the angle IBA is greater than the ratio of the straight line HR to the straight line RI and the ratio of the straight line HR to the straight line RI is greater than the ratio of the angle HAB to the angle IAB; the ratio of the angle HBA to the angle IBA is thus much greater than the ratio of the angle HAB to the angle IAB. If we permute, the ratio of the angle HBA to the angle HAB is greater then the ratio of the angle IBA to the angle IAB. So the ratio of the arc AH to the arc HB is greater than the ratio of the arc AI to the arc IB. By composition, the ratio of the arc AHB to the arc BH is greater than the ratio of the arc AIB to the arc BI. So the arc to which the arc AIB has a ratio equal to the ratio of the arc AHB to the arc BH is smaller than the arc BI; let it be the arc BK. We draw the straight line BK; it is smaller than the straight line BI, because the arc BI is smaller than a semicircle, so

---

[12] See Supplementary note [1].

the straight line *BK* is smaller than the straight line *BH*. We complete the circle *AIB* and we cut off from it an arc *BS* equal to the arc *BK*. We draw the straight lines *AS* and *BS*. So the ratio of the arc *AIB* to the arc *BS* is equal to the ratio of the arc *AHB* to the arc *BH*, that is, to the arc *BC*; but the ratio of the arc *AHB* to the arc *BC* is equal to the ratio of the arc *DE* to the arc *EG*; so the ratio of the arc *AIB* to the arc *BS* is equal to the ratio of the arc *DE* to the arc *EG*. But the arc *AIB* is similar to the arc *DE*; so the arc *BS* is similar to the arc *EG*, the arc *ABS* is similar to the arc *DEG* and the ratio of the straight line *AB* to the straight line *BS* is equal to the ratio of the straight line *DE* to the straight line *EG*. But *BS* is equal to *BK*, *BK* is smaller than *BH* and *BH* is equal to *BC*; so the straight line *BS* is smaller than the straight line *BC* and the ratio of the straight line *AB* to the straight line *BS* is greater than the ratio of the straight line *AB* to the straight line *BC*. Now we have proved that the ratio of *AB* to *BS* is equal to the ratio of *DE* to *EG*; so the ratio of the straight line *DE* to the straight line *EG* is greater than the ratio of the straight line *AB* to the straight line *BC*. That is what we wanted to prove.

I also say that the ratio of the straight line *DG* to the straight line *GE* is greater than the ratio of the straight line *AC* to the straight line *CB*.

Now we have proved that the straight line *BS* is smaller than the straight line *BC*. We make the straight line *BP* equal to the straight line *BC*, we join *AP* and *PC* and we draw *SQ* parallel to the straight line *AP*. Since *CB* is equal to *BP*, the angle *BPC* is equal to the angle *BCP*. But the angle *BPC* is greater than the angle *APC*; so the angle *BCP* is greater than the angle *APC* and the angle *ACP* is much greater than the angle *APC*. So the straight line *AP* is greater than the straight line *AC* and the ratio of *AP* to *PB* is greater than the ratio of *AC* to *CB*. Moreover the angle *ABS* is obtuse, because the arc *ABS* is smaller than a semicircle, since it is equal to the arc *DEG* which is smaller than the arc similar to the arc *ABC* which is not greater than a semicircle. So the angle *AQS* is obtuse,[13] the straight line *AS* is greater than the straight line *QS* and the ratio of *AS* to *SB* is greater than the ratio of *QS* to *SB*. But the ratio of *QS* to *SB* is equal to the ratio of *AP* to *PB*; so the ratio of *AS* to *SB* is greater than the ratio of *AP* to *PB*. But the ratio of *AP* to *PB* is greater than the ratio of *AC* to *CB*; so the ratio of *AS* to *SB* is much greater than the ratio of *AC* to *CB*. But the ratio of *AS* to *SB* is equal to the ratio of *DG* to *GE*, so the ratio of *DG* to *GE* is greater than the ratio of *AC* to *CB*. That is what we wanted to prove.

---

[13] Angle *AQS* is an exterior angle of the triangle *QBS*, so $A\hat{Q}S > A\hat{B}S$.

<3> If we have two unequal arcs, such that each is not greater than a semicircle and one is greater than the arc similar to the other, if these two arcs belong to two equal circles or to two unequal circles and if in the two arcs we draw two chords such that the ratio of the chord of the large arc to the chord drawn in it is equal to the ratio of the chord of the small arc to the chord drawn in it, then the remaining arc of the large arc is greater than the arc similar to the remaining arc of the small arc and the ratio of the remaining arc of the large arc to the arc cut off by the chord is greater than the ratio of the remaining arc of the small arc to the arc cut off by the chord.

Example: The two arcs *ABC* and *DEG* are unequal and each of them is no greater than a semicircle, the arc *ABC* is greater than the arc similar to the arc *DEG*; we draw in it the two chords *CB* and *EG* and the two chords *AC* and *DG* such that the ratio of the straight line *AC* to the straight line *CB* is equal to the ratio of the straight line *DG* to the straight line *GE*.

I say that the arc *AB* is greater than the arc similar to the arc *DE* and that the ratio of the arc *AB* to the arc *BC* is greater than the ratio of the arc *DE* to the arc *EG*.

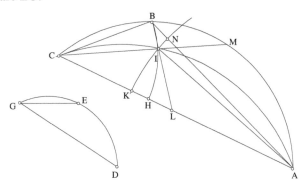

Fig. I.3

Proof: On the straight line *AC* we construct an arc of a circle similar to the arc *DEG*; let the arc be *AIC*. Since the arc *ABC* is not greater than a semicircle, the straight line *AC* is greater than the straight line *CB*. If we take the point *C* as centre and with distance *CB* we draw an arc of a circle, then it cuts the straight line *AC* between the two points *A* and *C*, and if it cuts the straight line *AC* between the two points *A* and *C*, it cuts <the arc> *AIC*; let that arc [the one we have drawn] be the arc *BIH*. We draw the straight line *CI*, we extend it to *M*, we draw the straight line *BI* and we extend it to *L*. We join *AB* and *AI*. So the straight line *CI* is equal to the straight line *CB* and the ratio of the straight line *AC* to the straight line *CI* is equal to the ratio of the straight line AC to the straight line CB. But the ratio of AC to CB was equal to the ratio of the straight line DG to the straight line GE; so the ratio of the straight line AC to the straight line CI is equal to the ratio of the straight line DG to the straight line GE. But the arc AIC is similar to the arc DEG, so the arc AI is similar to the arc DE; the arc IC is similar to the arc EG, the arc AI is similar to the arc AM and the arc AB is greater than the arc AM, so the arc AB is greater than the arc similar to the arc DE. But since the straight line CI is equal to the straight line CB, accordingly the angle BIC is acute and the angle BIM is obtuse, so the angle AIB is obtuse and the angle LIC is obtuse, because it is equal to the angle BIM. But the angle ALI is greater than the angle LIC, so the angle ALI is obtuse, so the straight line BA is greater than the straight line AI and the straight line AI is greater than the straight line AL. We take the point A as centre and with the distance AI we draw an arc of a circle. This arc cuts the straight line AB between the two points A and B and beyond the point M,[14] because the angle AIM is acute. This arc cuts the straight line AL beyond the point L; let this arc be the arc NIK. Since the sector CBI is greater than the triangle CBI and the sector CIH is smaller than the triangle CIL, the ratio of the sector CBI to the sector CIH is greater than the ratio of the triangle CBI to the triangle CIL. By composition, the ratio of the sector CBH to the sector CIH is greater than the ratio of the triangle CBL to the triangle CIL; so the ratio of the angle ACB to the angle ACI is greater than the ratio of the straight line BL to the straight line LI. Similarly, since the triangle ABI is greater than the sector ANI and the triangle AIL is smaller than the sector AIK, the ratio of the triangle ABI to the triangle AIL is greater than the ratio of the sector ANI to the sector AIK. Similarly, also by composition, the ratio of the straight line BL to the straight line LI is greater than the ratio of the angle CAB to the angle CAI. But the ratio of the angle ACB to the angle ACI is greater than the ratio of the straight line BL to the straight line LI and the ratio of the straight line BL to the straight line LI is greater than the ratio of the angle CAB to the angle CAI; so the ratio of the angle ACB to the angle ACI is much greater than the ratio of the angle CAB to the angle CAI. If we permute, the ratio of the angle ACB to the angle CAB is greater than the ratio of the angle ACI to the angle CAI. So the ratio of the arc AB to the arc BC is greater than the ratio of the arc AI to the arc IC; but the ratio of the arc AI to the arc IC is equal to the ratio of the arc DE to the arc EG, because they are similar <two by two>, so the ratio of the arc AB to the arc BC is greater than the ratio of the arc DE to the arc EG. That is what we wanted to prove.

[14] Beyond the straight line IM.

<4> Let there be two unequal arcs such that one is greater than the arc similar to the other and each is no greater than a semicircle. In each of them we draw two unequal chords such that the ratio of the greater chord of the greater arc to the smaller chord of the same arc is equal to the ratio of the greater chord of the smaller arc to the smaller chord of that arc; then the ratio of the greater of the two arcs of the large arc to the smaller of the two arcs of the latter is greater than the ratio of the greater of the two arcs of the smaller arc to the smaller of the two arcs of the latter.

Example: Let the two arcs be *ABC* and *DEG*; the arc *ABC* is greater than the arc similar to the arc *DEG* and each of them is no greater than a semicircle. In these [two arcs] we draw the chords *AB, BC, DE* and *EG* such that the ratio of the chord *AB* to the chord *BC* is equal to the ratio of the chord *DE* to the chord *EG*.[15]

<a> I say that the ratio of the arc *AB* to the arc *BC* is greater than the ratio of the arc *DE* to the arc *EG*.

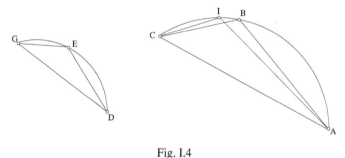

Fig. I.4

Proof: If it were not so, then the ratio of the arc *AB* to the arc *BC* would either be equal to the ratio of the arc *DE* to the arc *EG* or it would be smaller than the ratio of the arc *DE* to the arc *EG*.

If the ratio of the arc *AB* to the arc *BC* is equal to the ratio of the arc *DE* to the arc *EG*, then the ratio of the straight line *DE* to the straight line *EG* will be greater than the ratio of the straight line *AB* to the straight line *BC*, as has been proved in the second proposition. But by hypothesis the ratio of the straight line *AB* to the straight line *BC* is equal to the ratio of the straight line *DE* to the straight line *EG*. So the ratio of the arc *AB* to the arc *BC* is not equal to the ratio of the arc *DE* to the arc *EG*.

If the ratio of the arc *AB* to the arc *BC* is smaller than the ratio of the arc *DE* to the arc *EG*, then, by composition, the ratio of the arc *ABC* to the arc *CB* is smaller than the ratio of the arc *DEG* to the arc *GE*, so the ratio of the arc *DEG* to the arc *GE* will be equal to the ratio of the arc *ABC* to an

[15] From the statement of the proposition, we must assume *AB* > *BC* and *DE* > *EG*.

arc smaller than the arc BC; let it be the arc CI. So the ratio of the arc AI to the arc IC will be equal to the ratio of the arc DE to the arc EG. We draw the straight lines AI and IC. Since the ratio of the arc AI to the arc IC is equal to the ratio of the arc DE to the arc EG and the arc AIC is greater than the arc similar to the arc DEG, the ratio of the straight line DE to the straight line EG is greater than the ratio of the straight line AI to the straight line IC. But the ratio of the straight line AI to the straight line IC is greater than the ratio of the straight line AB to the straight line BC, since the straight line AI is greater than the straight line AB and the straight line IC is smaller than the straight line BC. But if the ratio of the straight line DE to the straight line EG is greater than the ratio of the straight line AI to the straight line IC and if the ratio of the straight line AI to the straight line IC is greater than the ratio of the straight line AB to the straight line BC, the ratio of the straight line DE to the straight line EG is much greater than the ratio of the straight line AB to the straight line BC. But by hypothesis the ratio of the straight line DE to the straight line EG is equal to the ratio of the straight line AB to the straight line BC. So the ratio of the arc AB to the arc BC is not equal to the ratio of the arc DE to the arc EG, nor is it smaller than the ratio of the arc DE to the arc EG; so the ratio of the arc AB to the arc BC is greater than the ratio of the arc DE to the arc EG. That is what we wanted to prove.

<b> It necessarily follows, also, that if the ratio of the straight line AB to the straight line BC is greater than the ratio of the straight line DE to the straight line EG, then the ratio of the arc AB to the arc BC will be greater than the ratio of the arc DE to the arc EG, because the two straight lines whose ratio one to the other is equal to the ratio of the straight line DE to the straight line EG are such that the point they have in common lies between the two points A, B; and this point divides the arc ABC into two parts whose ratio one to the other is greater than the ratio of the arc DE to the arc EG. So the ratio of the arc AB to the arc BC is much greater than the ratio of the arc DE to the arc EG.

<c> I say also that if the ratio of the straight line AC to the straight line CB is equal to the ratio of the straight line DG to the straight line GE, then the ratio of the arc ABC to the arc CB is greater than the ratio of the arc DEG to the arc GE.

Proof: If the ratio of the arc ABC to the arc CB is not greater than the ratio of the arc DEG to the arc GE, then the ratio of the arc ABC to the arc CB is equal to the ratio of the arc DEG to the arc GE or smaller than the ratio of the arc DEG to the arc GE.

If the ratio of the arc *ABC* to the arc *CB* is equal to the ratio of the arc *DEG* to the arc *GE*, then the ratio of the straight line *DG* to the straight line *GE* will be greater than the ratio of the straight line *AC* to the straight line *CB*, as have been proved at the end of the second proposition of this treatise. But by hypothesis the ratio of the straight line *DG* to the straight line *GE* is equal to the ratio of *AC* to *CB*; so the ratio of the arc *ABC* to the arc *CB* is not equal to the ratio of the arc *DEG* to the arc *GE*.

And if the ratio of the arc *ABC* to the arc *CB* is smaller than the ratio of the arc *DEG* to the arc *GE*, then the ratio of the arc *DEG* to the arc *GE* is greater than the ratio of the arc *ABC* to the arc *CB*. So the ratio of the arc *DEG* to the arc *GE* is equal to the ratio of the arc *ABC* to an arc smaller than the arc *CB*; let that arc be the arc *CI*. We join *AI* and *IC*; the ratio of *DG* to *GE* is greater than the ratio of *AC* to *CI*. But the ratio of *AC* to *CI* is greater than the ratio of *AC* to *CB*; so the ratio of *DG* to *GE* is greater than the ratio of *AC* to *CB*. But by hypothesis the ratio of *DG* to *GE* is equal to the ratio of *AC* to *CB*; so the ratio of the arc *ABC* to the arc *CB* is not equal to the ratio of the arc *DEG* to the arc *GE*, nor is it smaller than it; so the ratio of the arc *ABC* to the arc *CB* is greater than the ratio of the arc *DEG* to the arc *GE*. That is what we wanted to prove.

<d> It necessarily follows that if the ratio of the straight line *AC* to the straight line *CB* is greater than the ratio of the straight line *DG* to the straight line *GE*, then the ratio of the arc *ABC* to the arc *CB* will be greater than the ratio of the arc *DEG* to the arc *GE*.

From all this, it necessarily follows that, if we have two unequal arcs one of which is greater than the arc similar to the other and the greater of which is not greater than a semicircle, if we draw two chords in these two arcs, [chords] such that the ratio of the base of the large arc[16] to the chord drawn in it [the arc] is not smaller than the ratio of the base of the small arc to the chord drawn in it, then the ratio of the large arc to what the chord cuts off from it is greater than the ratio of the small arc to what the chord cuts off from it.

<e> I also say that if the two arcs *ABC* and *DEG* are similar and if the ratio of the straight line *AC* to the straight line *CB* is greater than the ratio of the straight line *DG* to the straight line *GE*, then the ratio of the arc *ABC* to the arc *CB* is greater than the ratio of the arc *DEG* to the arc *EG*.

Indeed, if the ratio of the arc *ABC* to the arc *CB* is equal to the ratio of the arc *DEG* to the arc *GE*, then the ratio of the straight line *AC* to the straight line *CB* is equal to the ratio of the straight line *DG* to the straight

---

[16] *I.e.* the chord across the whole arc.

line *GE*. So if the ratio of the straight line *AC* to the straight line *CB* is greater than the ratio of the straight line *DG* to the straight line *GE*, then the straight line *CB* is smaller than the straight line that cuts off the arc proportional to the arc *GE*; so the arc *CB* is smaller than the arc proportional to the arc *GE* and the ratio of the arc *ACB* to the arc *CB* is greater than the ratio of the arc *DEG* to the arc *GE*.

I also say that, if each of the two arcs *ABC* and *DEG* is greater than a semicircle, if each of the two arcs *AB* and *DE* is no greater than a semicircle, if the arc *AB* is no smaller than the arc similar to the arc *DE*, if the arc *AB* is greater than the arc *BC* and the arc *DE* is greater than the arc *EG* and if the ratio of the straight line *AB* to the straight line *BC* is greater than the ratio of the straight line *DE* to the straight line *EG*, then the ratio of the arc *AB* to the arc *BC* is greater than the ratio of the arc *DE* to the arc *EG*.

Proof: If the arc *AB* is greater than the arc *BC* and the arc *DE* is greater than the arc *EG*, then it is possible to cut off from the arc *AB* an arc equal to the arc *BC* and [to cut off] from the arc *DE* an arc equal to the arc *EG* and to cut off their chords; so the ratio of the straight line *AB* to the chord of the part cut off from the arc *AB* will be greater than the ratio of the straight line *DE* to the chord of the part cut off from the arc *DE*. But the arc *AB* is not smaller than the arc similar to the arc *DE* and each of the two arcs *AB* and *DE* is no greater than a semicircle; so the ratio of the arc *AB* to the part cut off from it is greater than the ratio of the arc *DE* to the part cut off from it, as has been shown earlier.[17] But the part cut off from the arc *AB* is equal to the arc *BC* and the part cut off from the arc *DE* is equal to the arc *EG*; so the ratio of the arc *AB* to the arc *BC* is greater than the ratio of the arc *DE* to the arc *EG*. So if each of the two arcs *ABC* and *DEG* is greater than a semicircle, if each of the two arcs *AB* and *DE* is no greater than a semicircle, if the arc *AB* is greater than the arc *BC* and the arc *DE* is greater than the arc *EG*, if the arc *AB* is no smaller than the arc similar to the arc *DE* and if the ratio of the straight line *AB* to the straight line *BC* is greater than the ratio of the straight line *DE* to the straight line *EG*, then the ratio of the arc *AB* to the arc *BC* is greater than the ratio of the arc *DE* to the arc *EG*. That is what we wanted to prove.

<5> Let there be in a sphere two great circles that cut one another, such that the distance between their poles is smaller than a quarter of a circle.

<a> We divide a quarter of one of them into an arbitrary number of equal parts; from the pole of the other circle we draw great circles that pass

---

[17] See <b>, the second part of Proposition 4.

through the points of division of the quarter that has been divided [circles that] end on the other circle; then the differences[18] between the arcs of these [arcs of] circles that are cut off between the first two circles and that are the inclinations of the parts of the quarter that has been divided, are unequal, and those of them that lie on the side towards the point of intersection are greater than the differences between those that are further away from the point of intersection.

Example: Let there be in the sphere two great circles ABC, ADC that cut one another; let the arc CD be a quarter of the circle ADC and the arc AB a quarter of the circle ABC. Let us divide up the quarter [circle] CD into equal parts; let the parts be CE, EG, GH, HI, ID. Let the pole of the circle ABC be the point K and the pole of the circle ADC the point L. We make a great circle pass through the two points L, K, then it passes through the two points D and B, because if the poles of the two circles ABC, ADC lie on the circle LK, then the pole of the circle LK lies on the two circles ABC, ADC; so the point C is the pole of the circle LK and every arc of one of the great circles drawn from the point C to the circle LK is a quarter of a circle. So the circle LK passes through the two points D and B. Since each of the arcs LD, KB is a quarter of a circle, the arc LK is equal to the arc DB and the arc LK is smaller than a quarter of a circle; so the arc DB is smaller than a quarter of a circle. Let us make a great circle pass through the point K and through each of the points C, E, G, H, I, let there be circles KC, KEM, KGN, KHP, KIQ. So the arcs EM, GN, HP, IQ, DB are the inclinations of the arcs CE, CG, CH, CI, CD and the arc DB will be the greatest of the inclinations which is the inclination of the quarter [circle] CD. We draw from the pole K with distances KE, KG, KH, KI and KD arcs of parallel circles that cut the arc KC; let these arcs be the arcs ES, GO, HU, IF and DJ; so the arcs JF, FU, UO, OS are the differences for the inclinations of the arcs CD, CI, CH, CG and CE.[19]

---

[18] The successive differences between the arcs taken two by two.

[19] That is, the successive amounts by which the inclination of each of the arcs CD, CI, CH, CG, CE exceed the inclination of the arc that follows it.

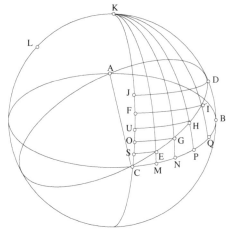

Fig. I.5.1

I say that the arc *JF* is smaller than the arc *FU*, that *FU* is smaller than *UO*, that *UO* is smaller than *OS* and that *OS* is smaller than *SC*. As for *OS* [being] smaller than *SC*, it is proved as we shall describe.

In fact, it has been proved from the sector figure[20] that the ratio of the chord of twice the arc *EC* to the chord of twice the arc *CD* is equal to the ratio of the chord of twice the arc *EM* to the chord of twice the arc *DB*,[21] that is, the ratio of the chord of twice the arc *SC* to the chord of twice the arc *CJ*, because the arc *KM* is equal to the arc *KB*. Also, in the same way, the ratio of the chord of twice the arc *GC* to the chord of twice the arc *CD* is equal to the ratio of the chord of twice the arc *OC* to the chord of twice the arc *CJ*. So we have that the ratio of the chord of twice the arc *GC* to the chord of twice the arc *CE* is equal to the ratio of the chord of twice the arc *OC* to the chord of twice the arc *CS*.

Now, given that the ratio of the chord of twice the arc *CE* to the chord of twice the arc *CD* is equal to the ratio of the chord of twice the arc *EM* to the chord of twice the arc *DB*, the ratio of the chord of twice the arc *CE* to the chord of twice the arc *EM* is equal to the ratio of the chord of twice the arc *CD* to the chord of twice the arc *DB*. But the chord of twice the arc *CD* is greater than the chord of twice the arc *DB*, because the arc *DB* is smaller than a quarter circle; thus the chord of twice the arc *CE* is greater than the chord of twice the arc *EM*, so twice the arc *CE* is greater than twice the arc *EM* and the arc *CE* is greater than the arc *EM*; so the arc *CE* is greater than the arc *CS*. In the same way we prove that the arc *CG* is greater than the arc

---

[20] This is Menelaus' theorem.

[21] See Mathematical commentary.

CO. But the ratio of the chord of twice the arc *GC* to the chord of twice the arc *CE* is equal to the ratio of the chord of twice the arc *OC* to the chord of twice the arc *CS*. But twice the arc *GC* is greater than twice the arc *CO* and twice the arc *CE* is greater than twice the arc *CS*; the amount by which twice the arc *CG* exceeds twice the arc *CE*, which is equal to twice the arc *GE*, is thus greater than the amount by which twice the arc *OC* exceeds twice the arc *CS*, which is equal to twice the arc *OS*; so the arc *GE* is greater than arc *OS*, as has been proved in the third proposition of this treatise, and the ratio of the amount by which twice the arc *GC* exceeds twice the arc *GE* to twice the arc *CE* is greater than the ratio of the amount by which twice the arc *OC* exceeds twice the arc *CS* to twice the arc *CS*, as has also been proved in the third proposition. But the amount by which twice the arc *GC* exceeds twice the arc *CE* is equal to twice the arc *CE*, because these arcs are, by hypothesis equal; so the amount by which twice the arc *OC* exceeds twice the arc *CS* is smaller than twice the arc *CS*; so twice the arc *OS* is smaller than twice the arc *SC* and the arc *OS* is smaller than the arc *SC*.

For the other arcs, the proposition is proved as we shall describe.

We trace a straight line equal to the diameter of the circle *ADC*; let it be *AB*. On it we describe a semicircle, let it be *ACB*; it will be equal to the arc *ADC*. We divide it into two equal parts at the point *C*; so the arc *AC* will be equal to the arc *CD*, the one in the first case of the figure. We also draw, on the straight line *AB*, an arc similar to twice the arc *CJ* which is the greatest inclination, let it be *ADB*; the arc *ADB* will thus be twice the inclination of the arc *AC*.[22] Let the point *C* be the point corresponding to *C* in the first case of the figure, which is the point of intersection; so the point *A* is the point which corresponds to the point *D* in the first case of the figure. Let the two arcs *AE, EG* be equal and let each of them be equal to each of the equal parts into which we have divided the arc *CD*. So the arc *AE* will be equal to the arc *DI* in the first case of the figure and the arc *EG* will be equal to the arc *IH* in the first case of the figure. We join *BG*. Since the arc *ACB* is a semicircle and the arc *AC* is a quarter circle, the straight line *AB* is the chord of twice the arc *AC*. Since the arc *AG* is twice the arc *AE*, the arc *BCG* is twice the arc *CE*; so the straight line *BG* is the chord of twice the arc *CE*; but the straight line *BG* is smaller than the straight line *BA*, since the arc *ACB* is a semicircle. If we take the point *B* as centre and if, with the distance <equal> to the straight line *BG*, we draw an arc of a circle, then it cuts the straight line *AB* between the two points *A* and *B*. If it cuts the straight line *AB* between the two points *A* and *B*, then it cuts the arc

---

[22] That is, twice the inclination of the arc *CD* in the first figure.

ADB; let it cut it at the point *D*. We join *BD*; it will be equal to the straight line *BG*. So the ratio of *AB* to *BG* is equal to the ratio of *AB* to *BD*; but the ratio of *AB* to *BG* is equal to the ratio of the chord of twice the arc *AC* to the chord of twice the arc *CE*, so it is equal to the ratio of the chord of twice the inclination of the arc *AC* to the chord of twice the inclination of the arc *CE* in the second case of the figure; so the ratio of *AB* to *BD*, which is equal to *BG*, is equal to the ratio of the chord of twice the inclination of the arc *AC* to the chord of twice the inclination of the arc *CE*. But for two similar arcs in which two chords are drawn, the ratio of the chord of one of the two arcs to the chord drawn in that arc is equal to the ratio of the chord of the other arc to the chord drawn in that arc. In fact, the two chords drawn in two similar arcs cut off from them two similar arcs. But the arc *BDA* is similar to twice the inclination of the arc *AC* and the ratio of *AB* to *BD* is equal to the ratio of the chord of twice the inclination of the arc *AC* to the chord of twice the inclination of the arc *CE*. So the arc *BD* is similar to twice the inclination of the arc *CE* – that is, twice the arc *CG* for the first case of the figure which is equal to the arc *IG* – which is the inclination of the arc *IC* which is equal to the arc *CE* in the second case of the figure. But the arc *ADB* is similar to twice the arc *CJ* in the first case of the figure and the arc *DB* is similar to twice the arc *CF* in the first case of the figure. So it remains that the arc *AD* is similar to twice the arc *JF* of the first case of the figure; so half of the arc *AD* is similar to the arc *JF* of the first case of the figure.

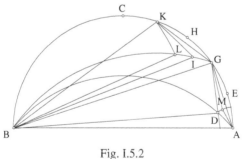

Fig. I.5.2

On the straight line *BG*, which is equal to the straight line *BD*, we draw an arc equal to the arc *BD*; let the arc be *BIG*. Let the two arcs *GH*, *HK* be equal to each of the parts of the arc *CD* of the first figure; so the arc *GK* is equal to the arc *AG*. We draw the straight lines *AG*, *AD*, *DG*, *GK*, *BK*; we cut off from the arc *BIG* the arc *GI* equal to the arc *AD* and we draw the straight lines *BI*, *GI*, *KI*. Since the arc *BCK* is smaller than the arc *BCG* of the arc *GK* which is equal to the arc *AG*, the angle *BGK* is smaller than the angle *BAG* by the angle *ABG*; and since the arc *BIG* is equal to the arc *BD* and the arc *GI* is equal to the arc *AD*, the arc *BI* is smaller than the arc *BD* of the arc *AD*; so the angle *BGI* is smaller than the angle *BAD* by the angle *ABD*. So it remains that the angle *KGI* is smaller than the angle *GAD* by the angle *DBG*. We make the angle *DAM* equal to the angle *DBG*, then it remains that the angle *GAM* is equal to the angle *KGI*. We take the point *A* as centre and with distance *AD* we draw an arc of a circle; that arc lies inside the angle *ADG*, because the angle *ADG* is obtuse. In fact, the straight line *DB* is equal to the straight line *BG*, so the angle *BDG* is acute, and if we extend the straight line *BD* in the direction of *D*, the exterior angle will be obtuse. But the straight line *BD*, if it is extended in the direction of *D*, cuts the angle *ADG*, so the angle *ADG* is obtuse. So the arc drawn taking centre *A* and with distance *AD* lies inside the angle *ADG* and it cuts the straight line *AM*, if we extend *AM*; let that arc cut the straight line *AM* at the point *M*. Since the straight line *AM* lies inside the angle *GAD* and the arc that passes through the two points *D* and *M* lies inside the angle *ADG*, accordingly the point *M* lies inside the triangle *GAD*. We draw the straight line *GM*; then that straight line lies inside the triangle *GAD*, so the angle *AGM* is smaller than the angle *AGD*. Since the arc *AG* is equal to the arc *GK*, the straight line *AG* is equal to the straight line *GK*. But the straight line *AD* is equal to the straight line *AM*, so the straight line *AM* is equal to the straight line *GI*. So the two straight lines *GA*, *AM* are equal to the two straight lines *KG*, *GI* and we have proved that the angle *GAM* is equal to the angle *KGI*; so the straight line *GM* is equal to the straight line *IK* and the triangle *GAM* is equal to the triangle *KGI*; so the angle *AGM* is equal to the angle *GKI*; but the angle *AGM* is smaller than the angle *AGD*, so the angle *GKI* is smaller than the angle *AGD*. But since the arc *BCG* is smaller than the arc *BCA*, the angle *BKG* is greater than the angle *BGA*; but we have proved that the angle *GKI* is smaller than the angle *AGD*, so it remains that the angle *BKI* is greater than the angle *BGD*. But since the arc *AG* is equal to the arc *GK* and the arc *AD* is equal to the arc *GI*, the angle *ABG* is equal to the angle *KBG* and the angle *ABD* is equal to the angle *IBG*; so it remains that the angle *DBG* is equal to angle *KBI*. Now we have proved that the angle *BKI* is greater than the angle *BGD*, so it remains that the angle *GDB* is greater than the angle *KIB*. But the angle *GDB* is equal to the angle *BGD*, because the straight line *BG* is equal to the straight line *BD*, so the angle *BGD* is greater than the angle *KIB*. Now the angle *BKI* is greater than the angle *BGD*, so the angle *BKI* is much greater than the angle *BIK* and the straight line *BI* is greater than the straight line *BK*. If we take the point *B* as centre and if, with distance *BK*, we describe an arc of a circle, then the arc cuts the straight line *BI* between the two points *B*, *I*. But if it cuts the straight line *BI* between the two points *B* and *I*, it will cut the arc *BI* between the two points *B* and *I*; let that arc, whose centre is the point *B*, cut the arc *BI* at the point *L*. So the point *L* lies between the two points *B* and *I*, the arc *GL* is greater than the arc *GI* and the arc *AD* will be smaller than the arc *GL*. We draw the straight line *BL*; it is equal to the straight line *BD*; so the ratio of the straight line *BG* to the straight line *BK* is equal to the ratio of the straight line *BG* to the straight line *BL*. But the straight line *BG* is the chord of twice the arc *CE*, and since the arc *BCG* is twice the arc *CE* and the arc *KG* is twice the arc *GE*, the arc *BCK* is twice the arc *CG*; so the straight line *BK* is the chord of twice the arc *CG*, the ratio of the straight line *BG* to the straight line *BL* is equal to the ratio of the chord of twice the arc *EC* to the chord of twice the arc *CG* and the ratio of the straight line *BG* to the straight line *BL* is equal to the ratio of the chord of twice the inclination of the arc *CE* of the second figure to the chord of twice the inclination of the arc *CG*. But the inclination of the arc *EC* is the arc *CF* in the first figure and the inclination of the arc *CG* is the arc *CU* of the first figure; so the ratio of the straight line *BG* to the straight line *BL* is equal to the ratio of the chord of twice the arc *CF* to the chord of twice the arc *CU*. But the arc *BLG* is similar to twice the arc *CF*, because it is equal to the arc *BD*, so the arc *BL* is similar to twice the arc *CU*; so it remains that the arc *GL* is similar to twice the arc *UF* and half of the arc *GL* is similar to the arc *UF*; but half the arc *AD* is similar to the arc *FJ*; now we have proved that the arc *AD* is smaller than the arc *GL*, so the arc *JF* is smaller than the arc *FU*.

In the same way, we prove that the arc *FU* is smaller than the arc *UO*, if on the straight line *BK* we construct an arc equal to the arc *BL*, if we cut off from the arc *KC* an arc equal to the arc *KG* and if we complete the construction as before.

In the same way, it is also proved that the arc *UO* is smaller than the arc *OS*; but it has been proved that the arc *OS* is smaller than the arc *SC*.

It is clear from what we have proved that the arc *JF* is smaller than the arc *FU*, that the arc *FU* is smaller than the arc that follows it and similarly for all the remaining arcs, that each of them is smaller than the one that follows it.

So if we divide the arc *CD* in the first figure into equal parts, of any number, [and] whether these parts are small or large, then the differences between their inclinations are unequal; and the smallest of them is on the side towards the point *D* which is at the end of the inclination and the largest of them is on the side towards the point *C* which is the point of intersection; and for all the remaining arcs, those that are close to the point *D*, the difference between the inclinations[23] of those that are on the side

---

[23] See Mathematical commentary.

towards the point D is smaller than the difference between the inclinations of those that are further from it. That is what we wanted to prove.

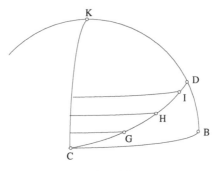

Fig. I.5.3

<b> From what we have proved, it is clear that if we have two consecutive equal arcs cut off on the arc CD, which are parts of the arc CD, commensurable or incommensurable with it, and which are not the two ends of the arc CD, then the difference in the inclination[24] of the one furthest from the point of intersection is smaller than the difference of the inclination of the arc that follows it, which is closer to the point of intersection.

Indeed, the proof for two consecutive equal arcs is the proof that we have set out because, in that proof, it is not necessary for the arc to be commensurable with the whole quarter [circle], and it is not necessary, either, that the endpoint of one of the two arcs should be an endpoint of the quarter [circle]. For any two consecutive equal arcs cut off on a quarter circle, and inclined to another circle, the two differences of the two inclinations of two equal arcs are unequal and the smaller difference is the one that is on the side of the endpoint that has the greater inclination.[25]

<6> Since all this has been proved, we say that if there are two arcs cut off on a quarter circle, inclined to another circle, whether they are commensurable with the quarter circle or incommensurable with it,[26] whether the two arcs are contiguous or disjunct, whether they are equal or unequal, then the ratio of the one that is more distant from the point of intersection of the two circles, one of which is inclined to the other, to the one that is

---

[24] The 'difference in the inclination' means the difference between the inclinations of the two ends of the arc.

[25] See Mathematical commentary, pp. 69–70.

[26] See Supplementary note [2].

closer, is greater than the ratio of the difference in inclination of the more distant one to the difference in inclination of the closer one.

Example: Let the two arcs be *AB*, *BC* cut off on a quarter of a circle, inclined to another circle, the arc *AB* being more distant from the point of intersection than the arc *BC*; the arc *DE* is the difference in the inclination of the arc *AB* and the arc *EG* is the difference in the inclination of the arc *BC*.

I say that the ratio of the arc *AB* to the arc *BC* is greater than the ratio of the arc *DE* to the arc *EG*.

Proof: The two arcs *AB*, *BC* are either commensurable or incommensurable.[27]

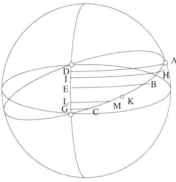

Fig. I.6

If they are commensurable, then they can be divided into equal parts by the part that is their [common] measure. If we cut off the two arcs *DE* and *EG*, by means of the arcs which are the differences of the inclinations of these parts, the parts of the two arcs *DE* and *EG* will be unequal and the number of parts in the arc *DE* will be equal to the number of the equal parts in the arc *AB*; number of parts in the arc *EG* will be equal to the number of equal parts in the arc *BC* and the parts of the arc *DG* that are closer to the point *G* will be smaller than those that are further away from it. So the parts of the arc *DE* are smaller than the parts of the arc *EG*. So the ratio of the number of the parts of the arc *DE* to the number of parts of the arc *EG* is greater than the ratio of the arc *DE* to the arc *EG*, because the ratio of the numbers the one to the other is the ratio of equal parts. But the parts of the arc *DE* are unequal and each of them is smaller than each of the parts of the arc *EG*; so the ratio of the number of parts of the arc *DE* to the number of parts of the arc *EG* is greater than the ratio of the arc *DE* to the arc *EG*. But the ratio of the number of parts of the arc *DE* to the number of parts of the

[27] See Supplementary note [2].

arc *EG* is equal to the ratio of the number of equal parts of the arc *AB* to the number of equal parts of the arc *BC* and the ratio of the number of parts of the arc *AB* to the number of parts of the arc *BC* is the ratio of the arc *AB* to the arc *BC*, because the parts of the two arcs *AB*, *BC* are equal; so the ratio of the arc *AB* to the arc *BC* is greater than the ratio of the arc *DE* to the arc *EG*.

But if these two arcs *AB*, *BC* are incommensurable, we again say that the ratio of the arc *AB* to the arc *BC* is greater than the ratio of the arc *DE* to the arc *EG*.

Proof: If it were not so, then the ratio of the arc *AB* to the arc *BC* would be either equal to the ratio of the arc *DE* to the arc *EG* or smaller than it.

First, let the ratio of the arc *AB* to the arc *BC* be equal to the ratio of the arc *DE* to the arc *EG*. We subdivide up the arc *AB* into any number of equal parts and we take from among these parts one that is of a magnitude smaller than the arc *BC*; let it be the arc *HB*. Let the arc *IE* be the difference of the inclination of the arc *HB*; so the ratio of *AH* to *HB* is greater than the ratio of *DI* to *IE*, as has been proved in the first part of this proposition. Again by composition, the ratio of *AB* to *BH* is greater than the ratio of *DE* to *EI*. So the ratio of *IE* to *ED* is greater than the ratio of *HB* to *BA*. But the ratio of *DE* to *EG* is by hypothesis equal to the ratio of *AB* to *BC*; so the ratio of *IE* to *EG* is greater than the ratio of *HB* to *BC*. Let the ratio of *IE* to *EG* be equal to the ratio of *HB* to *BK*. We cut off from *BC* parts equal [to one another] and equal to the parts contained in the arc *HB* until the parts reach a magnitude greater than the arc *BK* and less than the arc *BC*. If the arc *KC* is smaller than one of the parts that are in *HB*, then we divide up the parts in *HB* into small parts until each of these is smaller than *KC* and we take a quantity of small parts of *BC* <whose sum is> smaller than *BC* and greater than *BK*; let this magnitude [the sum] be the magnitude *BM*. Let *EL* be the difference in inclination of the arc *BM*; so the arc *EL* is smaller than the arc *EG* because *EG* is the difference in inclination of the arc *BC* which is greater then *BM*; the ratio of the arc *HB* to the arc *BM* is greater than the ratio of the arc *IE* to the arc *EL*; but the ratio of the arc *IE* to the arc *EG* is greater than the ratio of the arc *HB* to the arc *BM*, because it is equal to the ratio of *HB* to *BK*; so the ratio of *IE* to *EG* is greater than the ratio of *HB* to *BM*. But the ratio of *HB* to *BM* is greater then the ratio of *IE* to *EL*, so the ratio of *IE* to *EG* is much greater than the ratio of *IE* to *EL* and the arc *EG* is smaller than the arc *EL*; which is impossible. So the ratio of *AB* to *BC* is not equal to the ratio of *DE* to *EG*.

If the ratio of *AB* to *BC* were smaller than the ratio of *DE* to *EG*, the ratio of *IE* to *ED* would be greater than the ratio of *HB* to *BA*. But the ratio of *DE* to *EG* is greater than the ratio of *AB* to *BC*; so the ratio of *IE* to *EG* is much greater than the ratio of *HB* to *BC*. We put the ratio of *HB* to *BK* equal to the ratio of *IE* to *EG* and the proof is completed as before; the arc *EG* would then be smaller than the arc *EL*; which is impossible.

So the ratio of *AB* to *BC* is not equal to the ratio of *DE* to *EG* nor smaller than it; so the ratio of *AB* to *BC* is greater than the ratio of *DE* to *EG*. That is what we wanted to prove.

And by synthesis, the ratio will again have this property.

By this proof we show that, if the two arcs *AB* and *BC* are disjunct, the differences of the inclinations of the arcs that are disjunct are also unequal; and for those of the arcs that are further away from the point of inter-section, <the differences> in the inclinations are smaller.

<7> Similarly, let there be two great circles that cut one another on a sphere; let them be the circles *ABC* and *ADC* and let each of the arcs *CB* and *CD* be a quarter circle. We cause to pass through the two points *B* and *D* a quarter of a great circle; let it be *BDK*. This circle passes through the poles of the circles *ABC* and *ADC*. Let the pole of the circle *ADC* be the point *K*; let us divide the arc *CB* into equal parts; let these be *CM*, *MN*, *NP*, *PQ*, *QB*. We cause great circles to pass through the points of division and through the point *K*; let them be *EMK*, *GNK*, *HPK*, *IQK*; let the points *E*, *G*, *H*, *I* lie on the arc *CD*. If the circle *CDA* is the circle of the equator, the arcs cut off on it between the great circles drawn from the point *K* are the right ascensions[28] for the equal arcs cut off on the arc *CB*.

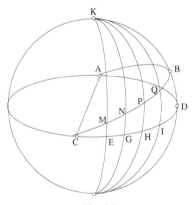

Fig. I.7

---

[28] Lit.: are the ascensions on the right orb.

I say that the arcs *DI*, *IH*, *HG*, *GE*, *EC* are unequal, that *DI* is the largest, that *EC* is the smallest and that those [arcs] that are closer to *DI* are greater than those that are further from it.

Proof: The ratio of the chord of twice the arc *CM* to the chord of twice the arc *MN* is a compound of the ratio of the chord of twice the arc *CE* to the chord of twice the arc *EG* and the ratio of the chord of twice the arc *GK* to the chord of twice the arc *KN*. But the chord of twice the arc *CM* is equal to the chord of twice the arc *MN*, so the ratio compounded of the ratio of the chord of twice the arc *CE* to the chord of twice the arc *EG* and the ratio of the chord of twice the arc *GK* to the chord of twice the arc *KN* is the ratio of equality.[29] But the chord of twice the arc *GK* is greater than the chord of twice the arc *KN* because the chord of twice the arc *GK* is the diameter. If the ratio of equality is compounded of two ratios, one of which is the ratio of the largest to the smallest, then the second ratio is the ratio of the smallest to the largest; so the chord of twice the arc *CE* is smaller than the chord of twice the arc *EG*, twice the arc *CE* is smaller than twice the arc *EG* and the arc *CE* is smaller than the arc *EG*. In the same way, the ratio of the chord of twice the arc *CN* to the chord of twice the arc *NP* is compounded of the ratio of the chord of twice the arc *CG* to the chord of twice the arc *GH* and of the ratio of the chord of twice the arc *HK* to the chord of twice the arc *KP*. But the ratio of the chord of twice the arc *CN* to the chord of twice the arc *NM* is compounded of the ratio of the chord of twice the arc *CG* to the chord of twice the arc *GE* and of the ratio of the chord of twice the arc *EK* to the chord of twice the arc *KM*; but the ratio of the chord of twice the arc *CN* to the chord of twice the arc *NM* is itself the ratio of the chord of twice the arc *CN* to the chord of twice the arc *NP*, because the arc *NM* is equal to the arc *NP*. The ratio compounded of the ratio of the chord of twice the arc *CG* to the chord of twice the arc *GH* and the ratio of the chord of twice the arc *HK* to the chord of twice the arc *KP* is the ratio compounded of the ratio of the chord of twice the arc *CG* to the chord of twice the arc *GE* and the ratio of the chord of twice the arc *EK* to the chord of twice the arc *KM*. But the arc *KP* is smaller than the arc *KM*; so the chord of twice the arc *KP* is smaller than the chord of twice the arc *KM* and the arc *HK* is equal to the arc *EK*; so the ratio of the chord of twice the arc *HK* to the chord of twice the arc *KP* is greater than the ratio of the chord of twice the arc *EK* to the chord of twice the arc *KM*; so it remains that the ratio of the chord of twice the arc *CG* to the chord of twice the arc *GH* is smaller than the ratio of the chord of twice the arc *CG* to the chord

---

[29] We obtain a ratio equal to unity.

of twice the arc *GE*; so the chord of twice the arc *GH* is greater than the chord of twice the arc *GE* and the arc *GH* is greater than the arc *GE*.

In the same way, we prove that the arc *HI* is greater than the arc *HG* and that the arc *ID* is greater than the arc *IH*.

In the same way, we prove that for two equal arcs, [that are] contiguous, [and are] cut off on the arc *BC*, the ascensions[30] are unequal and that the ascension of whichever of the two arcs is closer to the point of intersection is smaller than the ascension of the one that is further away, that each of these two equal arcs will have the magnitude of a quarter circle or it will not be of the magnitude of a quarter circle, whether it is commensurable with the circle or whether it is not commensurable with it. The contiguous equal arcs that have been cut off from the inclined circle have unequal right ascensions and the equal arcs closest to the point of intersection have the smallest ascensions. That is what we wanted to prove.

This having been proved, then for two contiguous arcs cut off on a quarter of a circle inclined to the circle of the equator, [arcs that are] equal or not equal, commensurable with the quarter circle or not commensurable [with it], the ratio of whichever is closer to the point of intersection to the one that is further away from the point of intersection is greater than the ratio of the ascension whichever is the closer to the point of intersection to the ascension of the one that is the further from the point of intersection.

The proof of this is the same as that for the preceding proposition concerning the ratio of two arcs and their inclinations; but in this other one,[31] the ratio of the arc that is further from the point of intersection to the one that is closer to it is greater than the ratio of the difference in inclination of the one that is further away to the difference in inclination of the one that is closer, whereas, in this present proposition, the ratio of whichever of the two arcs is closer to the point of intersection to the one that is further away is greater than the ratio of the ascension of the closer one to the ascension of the further one. And by synthesis, the same will hold for the two ratios.

<8> In any circle we draw one of its diameters, on which we mark a point that is not the centre; from this point we draw straight lines to the circumference of the circle, [lines] that cut off on the circumference of the circle equal arcs each of which has as its chord a straight line smaller than the complement of the diameter; then the angles formed at the chosen point

---

[30] The 'ascension of an arc' means the difference between the ascensions of the two endpoints of the arc.

[31] That is, in the preceding proposition.

are unequal and the angle that lies on the same side as the greater of the two parts of the diameter is the smaller, and the angles that are closer to it are smaller than those that are further from it.

Example: In the circle *ABC* we have drawn a diameter *AC*, the centre is <the point> *D* and we take any point on the diameter, which is the point *E*. From the point *E* we draw the straight lines *EB*, *EH*, *EI* such that the arcs *AB*, *BH*, *HI* are equal and each of their chords is smaller than the straight line *EC*.

I say that the angle *AEB* is smaller than the angle *BEH* and the angle *BEH* is smaller than the angle *HEI*.

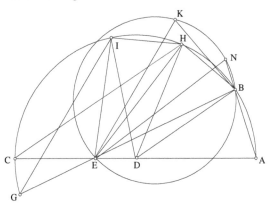

Fig. I.8

Proof: We draw the straight lines *DB*, *DH*, *DI*, *AB*, *BH*, *HI*. Since the arcs *AB*, *BH*, *HI* are equal, the triangles *ADB*, *BDH* and *HDI* are equiangular, so the angles *DAB*, *DBA*, *DHB*, *DBH*, *DHI*, *DIH* are all equal and the angle *EHB* is greater than the angle *DHB*; so the angle *EHB* is greater than the angle *EAB*. We join *HC*; the figure *ABHC* is a quadrilateral inscribed in a circle, so the angle *CAB* plus the angle *CHB* is equal to two right angles, the angle *EAB* plus the angle *EHB* is smaller than two right angles,[32] the angle *EHB* is greater than the angle *EAB* and their sum is smaller than two right angles. On the straight line *EB* we construct an angle *EBK* equal to the angle *EAB*. But the angle *EAB* is acute, so the angle *EBK* is acute. We make the straight line *EK* equal to the straight line *EB*; so the angle *EKB* will be equal to the angle *EBK*. We circumscribe about the triangle *EBK* a circle *BKE*; so the angle *BKE* plus the angle inscribed in the segment [of the circle] *EB*, which is opposite it, is thus equal to two right angles. But the angle *BKE* plus the angle *EHB* is smaller than two right angles, because

___

[32] Because $E\hat{H}B < C\hat{H}B$.

the angle *BKE* is equal to the angle *EAB*; so the angle *EHB* is smaller than the angle inscribed in the segment *EB*. So the segment of the circle *BKE* which is cut off by an angle equal to the angle *EHB* is greater than the segment *EB*; but the angle *EHB* is greater than the angle *EBK*, because it is greater than the angle *EAB*, which is equal to the angle *EBK*. But since the angle *EHB* is greater than the angle *EBK*, the segment of the circle *BKE* in which is inscribed an angle equal to the angle *EHB*, is smaller than the segment *EBK*. But since the segment *EBK* is greater than a semicircle, the segment of the circle *BKE* intercepted by an angle equal to the angle *EHB* is greater than the segment *EB* and smaller than the segment *EBK*. Let there be a segment [of a circle]; let the segment in which is inscribed an angle equal to the angle *EHB* be the segment *EBN*. We join *EN*, *BN*; then the angle *ENB* is equal to the angle *EKB*, because they are [inscribed] in the same segment of a circle. But the angle *EKB* is equal to the angle *EBK*, so the angle *ENB* is equal to the angle *EBK*; but the angle *EBN* is greater than the angle *EBK*, so the angle *EBN* is greater than the angle *ENB* and the straight line *EN* is greater than the straight line *EB*. If about the triangle *EHB* we circumscribe a circle, the segment of this circle cut off by the straight line *EB* is similar to the segment *EBN* and the straight line *EB* which is the base of this segment[33] is smaller than the straight line *EN* which is the base of the segment *EBN*. The circle circumscribed about the triangle *EHB* is thus smaller than the circle *BKE* and the circle *BKE* is equal to the circle circumscribed about the triangle *AEB*, because the straight line *EB* cuts off from the circle circumscribed about the triangle *AEB* a segment similar to the segment *BKE* cut off by the same straight line *EB*. So the circle *BKE* is equal to the circle circumscribed about the triangle *AEB*, so the circle circumscribed about the triangle *AEB* is greater than the circle circumscribed about the triangle *EBH*. But each of the angles at the point *E* is an acute angle, because any straight line drawn from the point *E* to the circumference of the circle is greater than the straight line *EC* and the straight line *EC* is greater than each of the straight lines *AB*, *BH*, *HI*; so the straight line *EB* is greater than the straight line *BA* and the angle *EAB* is greater than the angle *AEB*; but the angle *EAB* is acute, so the angle *AEB* is acute. Similarly, the straight line *EH* is greater than the straight line *EB*, so the angle *EBH* is greater than the angle *BEH*; but the angle *EBH* is acute, so the angle *BEH* is acute. We show this in the same way for all the angles at the point *E*, so all the angles at the point *E* are acute. So the arc of the circle circumscribed about the triangle *AEB* that is intercepted by the angle *AEB* is smaller than a semicircle; similarly the arc of the circle circumscribed about the triangle *EBH* that is intercepted by the angle *BEH*

---

[33] That is, the chord of the arc.

is smaller than a semicircle, so the arc cut off by the straight line AB from the circle circumscribed about the triangle AEB and intercepted by the angle AEB is smaller than a semicircle, and the arc cut off by the straight line BH from the circle circumscribed about the triangle BEH and intercepted by the angle BEH is smaller than a semicircle. But the straight line AB is equal to the straight line BH and the circle circumscribed about the triangle AEB is greater than the circle circumscribed about the triangle BEH; the arc cut off by the straight line AB from the circle circumscribed about the triangle AEB and intercepted by the angle AEB is thus smaller than the arc similar to the arc cut off by the straight line BH from the circle circumscribed about the triangle BEH and intercepted by the angle BEH; so the angle AEB is smaller than the angle BEH.

Similarly, we draw the straight line BE to [meet] the circumference of the circle, let it be the straight line BEG, and we join IG; the figure $BHIG$ is then a quadrilateral inscribed in a circle; so the angle HBG plus the angle HIG add up to two right angles and the angle EBH plus the angle EIH add up to less than two right angles. But the angle EIH is greater than the angle DIH which is equal to the angle DBH and the angle DBH is greater than the angle EBH; so the angle EIH is greater than the angle EBH and, in the triangles BEH and HEI, the angle EIH of the second [triangle] is greater than the angle EBH of the first. But the angle EBH plus the angle EIH adds up to less than two right angles. We then prove for these two triangles what has been proved for the triangles AEB and BEH: that the angle BEH is smaller than the angle HEI. Similarly for all the angles that follow these angles until the equal arcs reach the point C, if the arc AB measures the arc ABC, or until [in sum] the equal arcs add up to an arc smaller than the arc AB on the side towards the point C.

It is clear from the preceding that the angles at the point E are unequal in the way that we have established, that each of the equal arcs cut off on the arc ABC measures the arc ABC or does not measure it, [that is] according to whether it is commensurable or incommensurable with it, and it is also clear by the second proof we set out for the angles BEH, HEI that the angles at the point E are unequal, even if the equal arcs do not begin at the point A.

Let there be a general circle in which we draw a diameter, on it we mark a point that is not the centre and from this point we draw straight lines to the circumference of the circle that cut off on the circumference of the circle equal arcs each of which is subtended by a straight line that is smaller than the complement of the diameter; then the angles formed at the chosen point will be unequal, the smallest is on the side towards the centre and one that is closer to this angle is smaller than one further from it. That is what we wanted to prove.

<9> From here on, it results from what we have proved that if, with centre E, we draw a circle, of the arcs cut off by the straight lines drawn from the point E on the circumference of the circle whose centre is a different point, those [arcs] that are on the side towards the diameter AC are smaller than those that follow them, if the arcs cut off by the straight lines drawn from the point E on the arc ABC are equal.

Let us again draw the circle ABC; let the centre be <the point> D. Let the point E be different from the centre. With centre E let us draw a circle AIG and let us draw from the point E three general straight lines; let them cut the circle ABC at the points B, K, M and them cut the circle AIN at the points I, H, N.

I say that the ratio of the arc BK to the arc KM is greater than the ratio of the arc IH to the arc HN, whether the arcs BK, KM are equal or unequal, whether each of the two arcs is commensurable or incommensurable with the arc BAC; whether the two arcs are contiguous or disjunct.

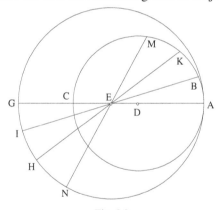

Fig. I.9

Proof: If the two arcs BK and KM are commensurable,[34] we divide them by the magnitude that is their common measure and we draw from the point E to the points of division straight lines that we extend until they meet the circle AIG; these straight lines cut off on the circle AIG unequal arcs such that the one among them that is closest to the point A is smaller than one that is further from it. From this we show that the ratio of the arc BK to the arc KM is greater than the ratio of the arc IH to the arc HN. If the

---

[34] See Supplementary note [2].

arc *BK* is not commensurable with the arc *KM*, then the proof that the ratio
of the arc *BK* to the arc *KM* is greater than the ratio of the arc *IH* to the arc
HN is the same proof as the one used in Proposition 6: the ratio of *BK* to
KM would either be equal to the ratio of *IH* to *HN*, or smaller than it; and
from this impossibility there follows what followed in Proposition 6; the
ratio of the arc *BK* to the arc *KM* is accordingly greater than the ratio of the
arc *IH* to the arc *HN*.

The same will hold if the two arcs *BK* and *KM* are disjunct, not
contiguous, because the arcs cut off on the circle *AIN* are unequal and the
one on the side towards the point *A* is smaller even if they are disjunct.
That is what we wanted to prove.

<10> Similarly, let there be a known circle *ABC* in which a diameter
AC has been drawn, and let there be a segment of the circle smaller than a
semicircle such as the segment *ADC*; let that segment be perpendicular to
the plane of the circle *ABC*, and let the ratio of *GH* to *HP* be known.

We wish to draw in the segment *ADC* a chord such as the chord *CD* so
that, if from its endpoint we draw a perpendicular to the diameter *AC*, such
as the perpendicular *DE*, if from the foot of this perpendicular we draw a
perpendicular to the diameter *AC* in the plane of the circle *ABC*, such as the
perpendicular *EB* and if we join its endpoint to the endpoint of the first
chord with a straight line such as the straight line *BD*, the ratio of *BD* to
DC will be greater than the ratio of *GH* to *HP*.

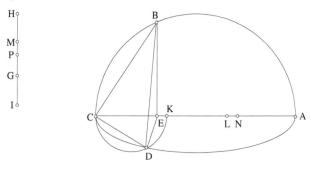

Fig. I.10.1

We take the ratio of *GH* to *HP* equal the ratio of *HP* to *HM*, we extend
HG in the direction of *G* and we take *GI* equal to *HM*. We divide the
straight line *AC* at the point *K* so that the ratio of *AK* to *CK* is equal to the
ratio of *IM* to *MH*. So *AK* is greater than *CK* and *IM* is greater than *MH*; in

fact *IM* is equal to *GH*.[35] On the straight line *KC* we construct a semicircle in the plane of the segment *ADC*; let the semicircle be *KDC*. This semicircle cuts the arc *CDA*. In fact, if from the point *C* we draw a tangent to the segment *CDA*, then it [the tangent] makes an acute angle with the straight line *CA*, because it encloses with it an angle equal to the angle contained in the remainder of the circle *ADC*; but this remainder is greater than a semicircle, so the angle contained in it is an acute angle, the straight line [that is a] tangent to the arc *CDA* at the point *C* makes an acute angle with the straight line *CA* and any straight line drawn from the point *C* that makes an acute angle with the straight line *CA* lies inside the arc *CDK*; so the straight line [that is a] tangent to the arc *CDA* drawn from the point *C* lies between the arc *CDA* and the arc *CDK*, a part of the arc *CDK* lies outside the arc *CDA* and the point *K* of the arc *CDK* lies inside the arc *CDA*; so the semicircle *CDK* cuts the arc *CDA*; let it cut it at the point *D*. We join *CD* and from the point *D* we draw a perpendicular to the straight line *CK*; let it be *DE*. We draw *EB* perpendicular to the straight line *CA* in the plane of the circle *ABC* and we join *BD*.

I say that the ratio of *BD* to *DC* is greater than the ratio of *GH* to *HP*.

Proof: We cut off *AL* equal to *KC*; the ratio of *AK* to *KC* is equal to the ratio of *GH* to *MH* and *KC* is greater than *CE*. We take *AN* equal to *CE*; then *NE* is greater than *LK*. So the ratio of *NE* to *KC* is greater than the ratio of *LK* to *KC*; but the ratio of *LK* to *KC* is equal to the ratio of *GM* to *MH*.[36] So the ratio of *NE* to *KC* is greater than the ratio of *GM* to *MH*; but the ratio of *NE* to *KC* is equal to the ratio of the product of *NE* and *EC* to the product of *KC* and *EC*; so the ratio of the product of *NE* and *EC* to the product of *KC* and *CE* is greater than the ratio of *GH* to *MH*. But the product of *NE* and *EC* is the amount by which the product of *AE* and *EC* exceeds the product of *AN* and *EC*. But the product of *AN* and *EC* is the square of *EC* because *AN* is equal to *EC*, so the product of *NE* and *EC* is the amount by which the product of *AE* and *EC* exceeds the square of *EC*. Now the product of *AE* and *EC* is the square of *EB*, so the product of *NE* and *EC* is the amount by which the square of *BE* exceeds the square of *EC*. But the product of *KC* and *CE* is the square of *CD*, so the ratio of the product of *NE* and *EC* to the product of *KC* and *CE* is the ratio of the amount by which the square of *BE* exceeds the square of *EC*, to the square of *CD*. Now the ratio of the product of *NE* and *EC* to the product of *KC* and *CE* is greater than the ratio of *GM* to *MH*; the ratio of the amount by which

---

[35] So we have $\dfrac{AK}{CK} = \dfrac{GH}{MH}$.

[36] Indeed $\dfrac{LK}{KC} = \dfrac{AK - AL}{KC} = \dfrac{AK - KC}{KC} = \dfrac{GH - MH}{MH} = \dfrac{GM}{MH}$.

the square of EB exceeds the square of EC to the square of CD is accordingly greater than the ratio of GM to MH. But the amount by which the square of BE exceeds the square of EC is the amount by which the square of BD exceeds the square of DC because DE is perpendicular to the plane of the circle ABC; so the ratio of the amount by which the square of BD exceeds the square of DC to the square of DC is greater than the ratio of GM to MH. By composition, the ratio of the square of BD to the square of DC is greater than the ratio of GH to HM which is the ratio of the square of GH to the square of HP; so the ratio of the square of BD to the square of DC is greater than the ratio of the square of GH to the square of HP and the ratio of BD to DC is greater than the ratio of GH to HP. That is what we wanted to construct.

If we wish the straight line DE to make a known acute angle with the diameter AC on the side towards the point C, CDK being the right angle, and the ratio of BD to DC is to be greater than the given ratio, then we carry out the construction[37] as we did before for the right angle.

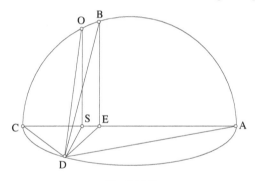

Fig. I.10.2

Let the perpendicular be DS, and the perpendicular drawn from the point S to the circle ABC be the perpendicular SO; we join OD. So the ratio of OD to DC is greater than the ratio given before.[38] From the point D we then draw a straight line, let it be DE, that makes an acute angle with the straight line AC on the side towards the point C. From the point E we draw a perpendicular to the straight line AC; let it be EB. So the perpendicular EB is parallel to the perpendicular SO. We join BD; it will be greater than DO, because <the plane of> the arc ADC is perpendicular to the segment of a circle ABC. But the arc CD is smaller than the arc DA, so the straight line

---

[37] We construct the point D.

[38] That is, the ratio $\dfrac{GH}{HP}$.

BD is greater than the straight line DO[39] and the ratio of *BD* to *DC* is greater than the ratio of *OD* to *DC*. But the ratio of *OD* to *DC* is greater than the given ratio, so the ratio of *BD* to *DC* is greater than the given ratio. That is what we wanted to construct.

If on the straight line *OD* or on the straight line *BD* we draw an arc of a circle equal to the circle *ADC*, the arc that is on the straight line *OD* or on the straight line *BD* is greater than the arc *DC*, because each of the straight lines *OD* and *DB* is greater than the straight line *DC*; the ratio of the arc that is on the straight line *OD* (or *BD*) to the arc *DC* of the circle *ADC* is thus greater than the ratio of the chord *OD* (or *BD*) to the chord *DC*, from what has been proved in the first proposition of this treatise, because the arc that is on the chord *OD* (or *BD*) is smaller than the arc *AD*; so the arc drawn on the chord *OD* (or *BD*), plus the arc *DC*, is smaller than a semicircle. If the arc drawn on the straight line *OD* (or *BD*) is from a circle smaller than the circle *ADC*, it will be greater than the arc similar to the arc of a circle equal to the circle *ADC*; so the ratio of the arc that is on the chord *OD* (or *BD*) to the arc *DC* is much greater than the ratio of the chord *OD* (or *BD*) to the chord *DC*. But the ratio of the chord *OD* (or *BD*) to the chord *DC* is greater than the given ratio, let the arc drawn on the straight line *OD* (or *BD*) be from a circle equal to the circle *ADC* or from a circle smaller than the circle *ADC*. That is what we wanted to prove.

<11> Similarly, if the circle *ABC* is a meridian circle, the diameter *AC* the axis of the sphere and the two points *A* and *C* the two poles of the sphere, if the circle *DNE* is one of the *muqanṭarāt* of altitude – that is, one of the circles parallel to a horizon – which is parallel to a horizon that passes through the points *A* and *C*, and if parallel circles among those circles whose two poles are the points *A* and *C* – such as the circles *HL*, *IM*, *BN*, *KP* – cut this sphere, if among these circles *BN* is the circle of the equator, if the semidiameters of these circles are the straight lines *HQ*, *IS*, *BG*, *KF*,[40] if the straight line *DE* is the [line of] intersection of the *muqanṭara DNE* and the meridian circle, if this straight line cuts the diameters of the parallel circles at the points *X*, *J*, *O*, *U* and if we draw the straight lines *LH*, *MI*, *NB*, *PK*, *HD*, *ID*, *BD*, *KD*, then the ratio of *LH* to *HD* is greater than the ratio of *MI* to *ID*, the ratio of *MI* to *ID* is greater than the ratio of *NB* to *BD* and the ratio of *NB* to *BD* is greater than the ratio of *PK* to *KD*.

---

[39] See Mathematical commentary. Ibn al-Haytham provides no justification for the inequality *BD* > *DO*.

[40] The points *Q*, *S*, *G*, *F* are the centres of the circles.

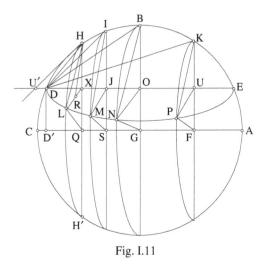

Fig. I.11

Proof: We draw the straight lines *LX*, *MJ*, *NO*, *PU*; these straight lines are the [lines of] intersection of the planes of the parallel circles with the plane of the *muqanṭara*. We draw the straight lines *LQ*, *MS*, *NG*, *PF*. Since the poles of the parallel circles are the points *A* and *C*, their centres lie on the straight line *AC*; since the [lines of] intersection of these parallel circles with the plane of the circle *ABC* are the straight lines *HQ*, *IS*, *BG*, *KF* and since these parallel circles cut the straight line *AC* at the points *Q*, *S*, *G*, *F*, these points are the centres of these circles and the straight lines *LQ*, *HQ*, *MS*, *IS*, *NG*, *GB*, *PF*, *KF* are semidiameters of these circles; since the two points *A* and *C* are the two poles of these circles, the straight line *AC* will be perpendicular to the planes of these circles and since the *muqanṭara* *DNE* is parallel to the horizon that passes through the two points *A* and *C*, the straight line *DE* will be parallel to the straight line *AC* because they are the two [lines of] intersection of the horizon that passes through the points *A* and *C* and the *muqanṭara* *DNE* with the <plane> of the meridian circle; so the straight line *DE* is perpendicular to the planes of the parallel circles. Each of the angles *DXL*, *DXH*, *DJM*, *DJI*, *DON*, *DOB*, *DUP*, *DUK* is a right angle and the straight lines *QX*, *SJ*, *GO*, *FU* are equal and perpendicular to the plane of the *muqanṭara*; in fact they are equal because they are parallel and the straight line *DE* is parallel to the straight line *CA*; they are perpendicular to the plane of the *muqanṭara* because they are the [lines of] intersection of the parallel circles and the meridian circle. Now the parallel circles are perpendicular to the plane of the horizon that passes through the points *A* and *C*, and the meridian circle is also perpendicular to that horizon; so these [lines of] intersection are perpendicular to the plane of that horizon and the plane of the *muqanṭara* is parallel to the plane of that horizon. So these [lines of] intersection are perpendicular to the plane of the *muqanṭara*, each of the angles *QXL*, *SJM*, *GON*, *FUP* is a right angle, the straight line *XL* is smaller than the straight line *JM* and the straight line *JM* is smaller than the straight line *ON*; so the angle *XQL* is smaller than the angle *JSM* and the angle *JSM* is smaller than the angle *OGN*. The points *Q*, *S*, *G* are the centres of the circles; so the arc *LH* is smaller than the arc similar to the arc *MI* and the arc *MI* is smaller than the arc similar to the arc *NB*. If we draw these parallel circles until they cut the *muqanṭara* in the other direction, then each of the arcs cut off on each of them between the *muqanṭara* and the arc *EBD* is equal to the corresponding one among the arcs *LH*, *MI*, *NB*. So the angle *HLX* is smaller than the angle *IMJ*, the angle *IMJ* is smaller than the angle *BNO* and each of the angles *HXL*, *IJM*, *BON* is a right angle. So it remains that the angle *LHX* is greater than the angle *MIJ* and the angle *MIJ* is greater than the angle *NBO*. We draw the angle *XHR* equal to the angle *JIM*; thus the triangle *XHR* will be similar to the triangle *JIM* and the ratio of *RH* to *HX* is equal to the ratio of *MI* to *IJ*. But the straight line *LH* is greater than the straight line *RH* because the angle *LRH* is obtuse; so the ratio of *LH* to *HX* is greater than the ratio of *MI* to *IJ*. In the same way, we prove that the ratio of *MI* to *IJ* is greater than the ratio of *NB* to *BO*.

In the same way, since the angle *HDX* is greater than the angle *IDX* and each of the two angles *HXD* and *IJD* is a right angle, accordingly the angle *DIJ* is greater than the angle *DHX*. We put the angle *XHU′* equal to the angle *DIJ*; then the triangle *XHU′* is similar to the triangle *JID*, so the ratio of *XH* to *HU′* is equal to the ratio of *JI* to *ID*. But *HU′* is greater than *HD*, because the angle *HDU′* is obtuse; so the ratio of *XH* to *HD* is greater than the ratio of *XH* to *HU′* and the ratio of *XH* to *HD* is greater than the ratio of *JI* to *ID*. So the ratio of *LH* to *HX* is greater than the ratio of *MI* to *IJ* and the ratio of *XH* to *HD* is greater than the ratio of *JI* to *ID*, so the ratio of *LH* to *HD* is much greater than the ratio of *MI* to *ID*. In the same way, we prove that the ratio of *MI* to *ID* is greater than the ratio of *NB* to *BD*. In the same way, the perpendicular *KU* is smaller than the perpendicular *BO* and the perpendicular *UP* is smaller than the perpendicular *ON*; so the straight line *KP* is smaller than the straight line *BN*, the straight line *KD* is greater than the straight line *BD* and the ratio of *NB* to *BD* is greater than the ratio of *PK* to *KD*.

We have thus proved that the ratio of the straight line *LH* to the straight line *HD* is greater than the ratio of the straight line *MI* to the straight line *ID*, that the ratio of the straight line *MI* to the straight line *ID* is greater than the ratio of the straight line *NB* to the straight line *BD* and that the ratio of the straight line *NB* to the straight line *BD* is greater than the ratio of the straight line *PK* to the straight line *KD*. That is what we wanted to prove.

<12> In the same way, let us return to the figure; let the pole *A* be higher than the horizon and below the zenith, the pole *C* beneath the horizon, the parallel circles inclined to the plane of the *muqanṭara* and the point *O* the centre of the *muqanṭara*.

I say that the chords that we have mentioned are as they were [before].

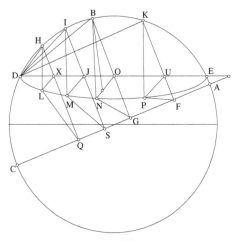

Fig. I.12

Proof: If the pole *A* is above the horizon, the straight line *DE* meets the axis *AC* on the side towards *E* and the centres of the parallel circles are on the axis *AC*; so the straight lines *XQ, JS, OG, UF* are unequal, the longest one is the straight line *XQ* and the shortest is *UF*.[41] Each of the angles *DXL, DJM, DON, DUP, LXH, MJI, NOB, PUK* is a right angle, because the parallel circles are perpendicular to the meridian circle, the *muqanṭara* is perpendicular to the meridian circle and the straight lines *LX, MJ, NO, PU* are the [lines of] intersection of the parallel circles and the *muqanṭara*; so they are perpendicular to the plane of the meridian circle. But the straight line *LX* is smaller than the straight line *MJ* and the straight line *MJ* is smaller than the straight line *NO*, because *NO* is a semidiameter of the *muqanṭara*. Since the straight line *QX* is greater than the straight line *SJ* and the straight line *XL* is smaller than the straight line *JM*, the angle *XQL*

---

[41] In all versions of the figure we have *XQ > JS > OG*, and if the axis *AC* cuts *DE* at the point *E* or beyond *E*, we have *OG > FU*. But if *AC* cuts *DE* between *O* and *E*, we can have *OG ≤ FU*.

will be smaller than the angle *JSM* and the angle *HLX* that intercepts an arc equal to the arc *HL* is half of the angle *XQL*. Similarly, the angle *IMJ* is half of the angle *JSM*, so the angle *HLX* is smaller than the angle *IMJ*. But each of the angles *LXH* and *MJI* is a right angle. We then prove, as was proved in the preceding proposition, that the ratio of *LH* to *HX* is greater than the ratio of *MI* to *IJ* and, in the same way as in the preceding proposition, we also prove that the ratio of *XH* to *HD* is greater than the ratio of *JI* to *ID*; so the ratio of *LH* to *HD* is greater than the ratio of *MI* to *ID*. Similarly, we prove that the ratio of *MI* to *ID* is greater than the ratio of *NB* to *BD*.

The same proof is necessary if the axis *AC* cuts the straight line *DE* inside the meridian circle, because if it cuts it inside the meridian circle, it cannot cut it only between the points *E* and *O* and nothing is changed in the proof, because if the axis *AC* passes through the point *O*, which is the centre of the *muqanṭara*, the pole *A* will be the zenith; now, by hypothesis, it is lower than the zenith. The same proof is also necessary if the point *E* is the upper pole.[42] The ratio of *NB* to *BD* is also greater than the ratio of *PK* to *KD*. Indeed, either the angle *KPU* is smaller than the angle *BNO*, or it is equal to it, or it is greater than it.

If the angle *KPU* is smaller than the angle *BNO*, then the arc *PK* will be smaller than the arc similar to the arc *BN*; but the arc *PK* belongs to a circle that is smaller than the circle *BN*, because the largest of the parallel circles <situated between the pole *A* and the equator> is inclined towards *D*;[43] so the circle *BN* is smaller than the largest of the parallel circles, the circle *KP* is smaller than the circle *BN*, the straight line *KP* is smaller than the straight line *BN* and *KD* is greater than *BD*; so the ratio of *NB* to *BD* is greater than the ratio of *PK* to *KD*.

If the angle *KPU* is equal to the angle *BNO*, then the arc *KP* is similar to the arc *BN*; but *KP* belongs to a smaller circle, so the straight line *KP* is smaller than the straight line *BN*.

If the angle *KPU* is greater than the angle *BNO*, then the angle *PKU* is smaller than the angle *NBO*. If from the angle *NBO* we cut off an angle equal to the angle *PKU*, there is formed inside the triangle *NBO* a triangle similar to the triangle *KPU* and the ratio of the straight line that lies inside the triangle *NBO* to the straight line *BO* is equal to the ratio of *PK* to *KU*; so the ratio of *NB* to *BO* is greater than the ratio of *PK* to *KU*. But the ratio of *OB* to *BD* is greater than the ratio of *UK* to *KD* – since this is proved as we proved it in the preceding proposition. So the ratio of *NB* to *BD* is

---

[42] By implication the one above the horizon.

[43] The largest of the parallel circles we are considering is the one that is closest to *D*, that is, the closest to the equator.

greater than the ratio of *PK* to *KD*. Similarly, we prove that the ratio of *PK* to *KD* is greater than the ratio of any chord corresponding to the straight line *PK*, more inclined towards the point *E* than the straight line *PK*, to the chord that corresponds to the straight line *KD*.

Thus we have proved that the ratio of two chords from among the preceding chords is greater than the ratio of two chords that are further away than the first two, for all the positions of the parallel circles and all the positions of the *muqanṭarāt*. That is what we wanted to prove.

<13> Similarly, if the circle *ABC* is one of the horizons, the circle *ADC* the meridian circle, the point *D* the pole of the horizon, and the circle *EH* one of the *muqanṭarāt* of altitude, if the two arcs *LG* and *NM* belong to hour circles, that is to say to circles parallel to the equator, if the two points *L* and *N* lie on the arc *DE* and if the sphere is right or inclined towards the point *E*, then the arc *LG* is greater than the arc similar to the arc *NM*.

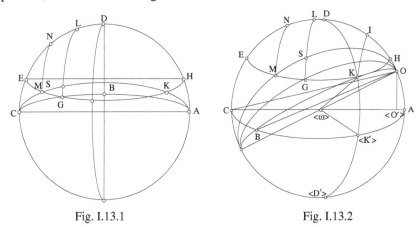

Fig. I.13.1   Fig. I.13.2

Proof: If the sphere is right, then the circle of the equator passes through the point *D*, <its diameter> passes through the centre of the *muqanṭara EGH*, the two points *G* and *M* will lie between the circle of the equator and the point *E* and the two poles of the equator will be the points *A* and *C*; the great circle drawn through the two poles *A*, *C* and which touches the *muqanṭara EGH*, [but] only at the midpoint of the arc *EGH* which is the point of intersection of the *muqanṭara EGH* with the circle of the equator; any great circle drawn through the two poles *A*, *C* and which cuts the *muqanṭara EGH*, cuts off from the *muqanṭara* two equal arcs on either side of the point of contact, and the circle closest to the tangent circle cuts off a smaller arc from the *muqanṭara* on the side towards the point of contact. From this we prove that the two great circles drawn through the points *A* and *C* and which pass through the points *M* and *G* cut the *mu-qanṭara* in such a way that the circle that passes through the point *G* is closer to the tangent circle than the circle that passes through the point *M*, because the point *G* is closer to the midpoint of the arc *EGH* than the point *M*. If this is so, then the great circle that passes through the points *M*, *K* cuts the arc *GL*; if it cuts the arc *GL*, then it cuts off from it an arc similar to the arc *MN*; consequently the arc *GL* is greater than the arc similar to the arc *MN*.

If the sphere is inclined towards *E*, then the visible pole is either below the *muqanṭara EGH* or on the *muqanṭara* itself occupies the position of the point *H*, or [the visible pole] is above the *muqanṭara*.

First, let the pole be below the *muqanṭara*, let it be at the point *O*. We cause to pass through the point *O* a great circle that touches the *muqanṭara EGH*, which is the circle whose inclination to the horizon is equal to the altitude of the *muqanṭara*; let the circle be *OKB*.[44] Since the circle *ADC* is perpendicular to the horizon, accordingly the straight lines drawn from the point *O* to the circumference of the horizon are unequal and the greatest is the chord of the arc *OC*; so the chord of the arc *OC* is greater than the chord of the arc *OB* and the arc *OC* is greater than the arc *OB*. We cause a great circle to pass through the points *D*, *K*; let the circle be *DK*. Since the circle *OKB* touches the circle *EGH* and the circle *DK* is a great circle that passes through the point of contact and through the pole of the circle *EGH*, the circle *DK* passes through the pole of the circle *OKB*; so the pole of the circle *DK* lies on the circumference of the circle *OKB* and, similarly, the pole of the circle *DK* lies on the circumference of the circle *ABC*; so the point *B* is the pole of the circle *DK*, the arc *BK* is a quarter of a circle and the arc *DC* is a quarter of a circle; so it remains that the arc *OD* is greater than the arc *OK*. Through the point *K* we draw the hour arc, let the arc be *KI*; so the point *I* lies between the points *D* and *H*. From the point *O* to the points *G* and *M* we draw two great circles – they cut the hour arc *KI* in two

---

[44] The circle *OKB* cuts the horizon in *B* and touches the horizontal circle *EGH* in *K*. The great circle *DK* cuts the horizon in *K'*, the arc *KK'* indicates the altitude of any point of the circle *EGH*. The tangent to the circle *BKO* at *K* and the tangent to the circle *CBK'* at *K'* are parallel; the straight line *Bω* (where *ω* is the centre of the sphere), the line of intersection of the planes *BKO* and *CBK'*, is thus parallel to them, and consequently *Kω* ⊥ *Bω* and *K'ω* ⊥ *Bω*; the angle *KωK'* is the rectilinear of the dihedral angle formed between the plane *BKO* and the horizontal plane, so the circle *OKB* is 'the circle whose inclination to the horizon is the altitude of the *muqanṭara*'.

different points, because each of them is smaller than a semicircle[45] –
which cut one another before the two points M and G and they also cut the
arc KH of the *muqanṭara*; let the two circles be OG, OM. So the circle OG
is closer to the point of contact than the circle OM, because the point G is
closer to the point of contact than the point M, so the circle OM cuts the arc
GL; let it cut it at the point S. So the arc SL is similar to the arc MN and the
arc GL is thus greater than the arc similar to the arc MN.

If the pole is at the point H, then the great circle that passes through the
two points H, M will be closer to the meridian circle than the great circle
that passes through the two points H and G, so the great circle that passes
through the two points H and M cuts the arc GL.

Similarly, if the pole lies on the arc DH, then the great circle drawn
from the pole to the point M will be closer to the meridian circle than the
great circle drawn from the pole to the point G.

So in all cases the arc GL is greater than the arc similar to the arc MN.
That is what we wanted to prove.

<14> Similarly, let the circle ABC be one of the *muqanṭarāt* parallel to
the horizon. Let the arc AD be part of the meridian circle and let the sphere
be right or inclined towards B. The pole of the equator which is on the side
towards the point A either lies on the circumference of the horizon, or is
above the horizon. Let the arc CDL be the circle of the equator or be
parallel to the circle of the equator. Through the pole of the equator let us
draw a great circle. Let it cut the arc DC and the arc AC; let the circle be
IKQ. Let its intersection with the arc DC be the point K and its intersection
with the arc ACB the point I. Through the point I let us cause to pass an arc
of a circle parallel to the equator, let the arc be IE; let it cut the arc ADB at
the point E and let the arc DB be no greater than half of the arc ADB.

I say that the ratio of the arc IE to the arc EB is greater than the ratio of
the arc CD to the arc DB and that the ratio of the arc CD to the arc DB is
greater than the ratio of the arc CK to the arc KI; I mean by my statement
that 'the ratio of an arc to an arc in two different circles' the ratio of the arc
of a circle equal to the circle, to an arc similar to an arc of the first circle; it
is this that I mean in everything that I shall mention later if I make use of
the ratios of arcs one to another.

---

[45] For each of them the arc above the horizon is smaller than a semicircle.

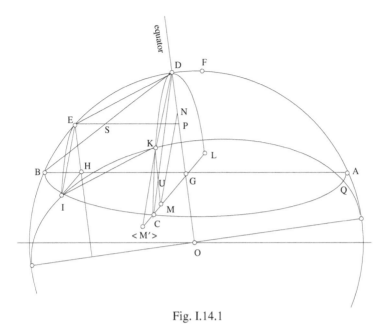

Fig. I.14.1

Proof of what we have referred to: The circle *IE* is either equal to the circle *CDL*, or smaller than it, or greater than it.

<*A*> First, let the circle *IE* be smaller than the circle *CDL*. We draw the straight lines *BE*, *BD*, *DE*, *DC*, *DK*, *KC*, *KI*, *EI*. Let *AB* be the line of intersection of the meridian circle and the plane of the *muqanṭara*, let *CGL* be the line of intersection of the *muqanṭara* and the circle *LDC*, let *DG* be the line of intersection of the meridian circle and the circle *LDC*, let *EH* be the line of intersection of the meridian circle and the circle *EI* and let *HI* be the line of intersection of the circle *EI* and the *muqanṭara*. So we have that *DG* is perpendicular to *GC* because each *muqanṭara* and the circle *DC* are perpendicular to the meridian circle and their line of intersection, which is *CG*, is thus perpendicular to the meridian circle; so the angle *DGC* is a right angle, and similarly for the angle *EHI*. We draw *KM* perpendicular to the straight line *LC*, it will thus be parallel to the straight line *DG*. We draw *MN* parallel to *DK*, we have that *DN* is equal to *KM* and *NM* is equal to *DK*. Since the two arcs *EI* and *DK* are between two of the great circles drawn from their two poles,[46] we have that the arc *DK* is similar to the arc *EI*. But since the circle *DKC* is greater than the circle *EI* and the arc *DK* is similar to the arc *EI*, the straight line *DK* is greater than the straight line *EI*.

---

[46] That is, the poles of the equator.

If from the point K we draw a perpendicular to the straight line DG, a triangle is formed that is similar to the triangle EIH, because the straight line EH is a diameter of the circle EI and the straight line IH is the sine of the arc EI. Similarly, the straight line DG is a diameter of the circle DKC and the perpendicular drawn from the point K to the diameter DG is the sine of the arc DK; but the arc DK is similar to the arc EI, so the triangle formed by the perpendicular drawn from the point K to the straight line DG is similar to the triangle EIH. The triangle NMG is similar to the triangle formed by the perpendicular drawn from the point K, because the straight line MN is parallel to the straight line DK; so the triangle NMG is similar to the triangle EIH. But the straight line MN is equal to the straight line DK and DK is greater than EI; so the straight line NM is greater than the straight line EI. But since the triangle NMG is similar to the triangle EIH, and the straight line NM is greater than the straight line EI, we have that NG is greater than EH.

We draw the straight line ESP parallel to the straight line HG, so PG is equal to EH; but NG is greater than EH, so NG is greater than PG and the point P lies between the two points N and G; so the straight line PD is greater than the straight line ND. But since the circle ABC is a *muqanṭara* parallel to the horizon, the arc ADB is smaller than a semicircle; so the arc AD is much smaller than a semicircle, so the angle DBH is acute; but the angle DSP is equal to the angle DBH; so the angle DSP is acute, so the angle DSE is obtuse and the straight line DE is thus greater than the straight line DS. Since the circle DKC is greater than the circle EI, the circle DKC is either the circle of the equator, or one of the circles that are parallel to it and that is closer to the hidden pole, or one of the circles parallel to the equator and that is closer to the visible pole, or closer to the equator than the circle EI. So if the circle DKC is the circle of the equator, then the part of this circle that is above the *muqanṭara ABC* is smaller than a semicircle, because the part of the equator which is above the horizon is a semicircle, and the arc CDL is thus smaller than a semicircle.[47] Similarly, if the circle DKC is closer to the hidden pole of the equator, then the part of the circle which is above the horizon, if the sphere is right, is a semicircle, and if the sphere is inclined towards B, it is smaller than a semicircle; thus the part of the circle DKC above the *muqanṭara ABC* is smaller than a semicircle and the arc LDK is accordingly much smaller than a semicircle.

If the circle DKC is closer to the visible pole, while still being closer to the equator than the circle EI, then if the sphere is right, the part of the circle DKC above the horizon is a semicircle and thus the part above the

---

[47] This arc has been cut off from the semicircle by the straight line CL.

muqanṭara is smaller than a semicircle; so the arc *LDK* is much smaller than a semicircle. If the sphere is inclined towards *B*, then the part of the circle *DKC* above the horizon will be greater than a semicircle; however the part of this circle that is below the horizon will be greater than the arc similar to the part of the circle *EI* that is above the horizon: in fact, this circle is closer to the equator than the circle *EI*. But the part of the circle *DKC* that is below the *muqanṭara* is greater than the part of this circle that is below the horizon, so that part of the circle *DKC* that is below the *muqanṭara* is much greater than the arc similar to the arc of the circle *EI*, and the part of the circle *EI* that is above the horizon is greater than the part of this circle that is above the *muqanṭara*. The part of the circle *DKC* that is below the *muqanṭara* is thus much greater than the similar arc of the circle *EI* which is above the *muqanṭara*. But the part of the circle *EI* above the *muqanṭara* is twice the arc *EI*, so it is twice the arc similar to the arc *DK*; so the part of the circle *DKC* that is below the *muqanṭara* is greater than twice the arc *DK*. We add to each side the arc *CK*; the part of the circle *DKC* that is below the *muqanṭara*, plus the arc *CK*, is greater than twice the arc *DK*, plus the arc *CK*; but twice the arc *DK*, plus the arc *CK*, is the arc *CD*, plus the arc *DK*, and the arc *CD* is equal to the arc *DL*; so the arc *CD* plus the arc *DK* is equal to the arc *LK*; the part of the circle *DKC* that is below the *muqanṭara*, plus the arc *CK*, is greater than the arc *LK* and the part of the circle *DKC* that is below the *muqanṭara*, plus the arc *CK*, plus the arc *LK*, is the whole circle. So the arc *LK* is smaller than the remainder of the circle, so the arc *LK* is smaller than a semicircle, so in all the cases the arc *LK* will be smaller than a semicircle, and in all the cases the angle *KCL* is an acute angle; so the perpendicular *KM* falls on the straight line *GC*, the point *M* lies between the two points *G* and *C*, and the straight line *DC* cuts the perpendicular *KM*; let it cut it at the point *U*. If from the point *K* we draw a straight line parallel to the straight line *UC*, it meets *MC* outside the triangle and will be greater than *KC*, because the angle *KCM* is acute, so the angle that is adjacent to it is obtuse, so the ratio of the parallel straight line drawn from the point *K* to the straight line *KM* is greater than the ratio of the straight line *CK* to the straight line *KM*. But the ratio of the straight line drawn from the point *K*, parallel to the straight line *UC*, to the straight line *KM* is equal to the ratio of the straight line *CU* to the straight line *UM*, so the ratio of the straight line *CU* to the straight line *UM* is greater than the ratio of *CK* to *KM*. But the ratio of *CU* to *UM* is equal to the ratio of *CD* to *DG*, so the ratio of *CD* to *DG* is greater than the ratio of *CK* to *KM*. But we have also proved that the straight line *ED* is greater than the straight line *DS* and that the straight line *ND* is smaller than the straight line *PD*, so the ratio of *PD* to *DS* is greater than the ratio of *ND* to *DE*. But the ratio of *PD* to *DS* is equal to the ratio of *GD* to *DB*, so the ratio of *GD* to *DB* is greater than the ratio of *ND* to *DE*. But *ND* is equal to *MK*, because the area *KMND* is a parallelogram, and the straight line *DE* is equal to the straight line *KI*, because the arc *DE* is equal to the arc *KI*, since they belong to two equal <great> circles drawn through the pole of the two parallel circles; so the ratio of *ND* to *DE* is equal to the ratio of *MK* to *KI*; and the ratio of *GD* to *DB* is greater than the ratio of *MK* to *KI*. So the ratio of *CD* to *DG* is greater than the ratio of *CK* to *KM* and the ratio of *GD* to *DB* is greater than the ratio of *MK* to *KI*; so the ratio of *CD* to *DB* is much greater than the ratio of *CK* to *KI*; if we permute, the ratio of *DC* to *CK* is greater than the ratio of *BD* to *KI*; but *KI* is equal to *DE*, so the ratio of the straight line *DC* to the straight line *CK* is greater than the ratio of the straight line *BD* to the straight line *DE*.

Similarly, if the sphere is right, then the straight line *DG* is perpendicular to the straight line *AB* and the arc *DB* is not greater than half the arc *ADB*; so the straight line *AG* is not smaller than the straight line *GB*, and the straight line *AB* is a diameter of the *muqanṭara ABC*; so the straight line *AG* is not smaller than the semidiameter of the *muqanṭara ABC* and the straight line *GC* is not greater than the semidiameter of this *muqanṭara*. The straight line *AG* is thus not smaller than the straight line *GC*, and the straight line *GD* is perpendicular to the two straight lines *AG*, *GC*, if the sphere is right. We join the straight line *AD*, it will not be smaller than the straight line *DC*; so the angle *DAG* is not greater than the angle *DCG*. So the arc *DB* is not greater than the arc similar to the arc *DL*. But the arc *DL* is equal to the arc *DKC*, so the arc *DKC* is not smaller than the arc similar to the arc *DB*.

If the sphere is inclined towards *B*, then the perpendicular drawn from the point *D* to the straight line *AB* meets it at a point between the two points *B* and *G* and the part of the straight line *AB* cut off by the perpendicular on the side towards the point *A* is not smaller than the semidiameter. In fact, if the arc *DB* is half the arc *ADB*, then the foot of the perpendicular is at the centre of the *muqanṭara* and the point *G* will lie between the centre and the point *A*, because the straight line *DG* is inclined to the straight line *AB*; so the straight line *GC* is smaller than the semidiameter and the straight line *GD* is greater than the perpendicular. The angle *DCG* is greater than the angle *DAB*, so the arc *LD*, that is, the arc *DC*, is greater than the arc similar to the arc *DB*. But if the arc *DB* is smaller than half the arc *ADB*, then the perpendicular falling from the point *D* onto the straight line *AB*, cuts off from the straight line *AB*, on the side towards *A*, a straight line greater than the semidiameter of the *muqanṭara*; but the straight line *GC* is not greater than the semidiameter of the *muqanṭara*; the straight line cut off by the perpendicular from the straight line *AB* on the side towards the point *A* is thus greater than the straight line *GC* and the perpendicular itself is smaller than the straight line *DG*; so the angle *DCG* is greater than the angle *DAB* and the arc *DKC* is greater than the arc similar to the arc *DB*. Thus in all the cases, the arc *DKC* is not smaller than the arc similar to the arc *DB*.

Now we have proved that the ratio of the straight line *DC* to the straight line *CK* is greater than the ratio of the straight line *BD* to the straight line *DE*; so the ratio of the arc *DKC* to the arc *CK* is greater than the ratio of the arc *BED* to the arc *DE*, as has been shown in Proposition 4 of this treatise. If we permute, the ratio of the arc *CD* to the arc *DB* is greater than the ratio of the arc *CK* to the arc *DE*, that is, to the arc *KI*. But if the ratio of the arc *CD* to the arc *DB* is greater than the ratio of the arc *CK* to the arc *DE*, then the ratio of the remaining arc, which is the arc *DK* – that is, the arc <similar to the arc> *EI* – to the remaining arc, which is the arc *EB*, is greater than the ratio of the arc *CK* to the arc *DE*.[48] The ratio of the arc *IE* to the arc *EB* is greater than the ratio of the arc *CD* to the arc *DB* and the ratio of the arc *CD* to the arc *DB* is greater than the ratio of the arc *CK* to the arc *KI*. That is what we wanted to prove.

If the circle *ABC* is a horizon, if the sphere is inclined, if the circle *ABC* cuts the circle *LDC* and if the circle *LDC* is the circle of the equator or it is closer to the hidden pole than the equator, the part of this circle above the horizon is not greater than a semicircle and all that we have said in connection with the *muqanṭara* necessarily follows. The proof is the same as the one we have given for the *muqanṭara*.

If the circle *LDC* is closer to the visible pole than the equator is, then the arc of the circle which is below the horizon is greater than the arc of the circle *EI* similar to the arc which is above the horizon and the arc of the circle *LDC* which is below the horizon, plus the arc *CK*, is greater than the arc *LK* and the arc *LK* is thus smaller than a semicircle, so the angle *KCG* is acute and we complete the proof, as in what we said before in connection with the *muqanṭara*.

If the sphere is right and if the circle *ABC* is a horizon, then the two poles are the points *A*, *B* and any great circle drawn through the two poles cuts the two circles *DC*, *EI*; it does not cut the arc *ACB* and does not form an arc like the arc *KI*; but the ratio of the arc *IE* to the arc *EB* will be greater than the ratio of the arc *CD* to the arc *DB*.

48 $\dfrac{a}{b} > \dfrac{c}{d} \Leftrightarrow bc < ad \Leftrightarrow ab - bc > ab - ad \Leftrightarrow b(a-c) > a(b-d)$, so $\begin{cases} b > d \\ a > c \end{cases} \Rightarrow \dfrac{a-c}{b-d} > \dfrac{a}{b} > \dfrac{c}{d}$.

In fact, if the sphere is right, then the parts of each of the circles *CD*, *IE*, that are above the horizon are equal to a semicircle. But the arc *EB* is smaller than the arc *DB*, so the ratio of the arc *IE* to the arc *EB* is greater than the ratio of the arc *CD* to the arc *DB*.

<B> Similarly, if the circle *IE* is equal to the circle *CD*, then the two circles *CD*, *IE* are on either side of the equator. But if the two circles *LDC*, *IE* are equal, then the arc of the circle *LDC* that is below the horizon is equal to the arc of the circle *EI* above the horizon, and the arc of the circle *LDC* that is below the *muqanṭara* is greater than the arc of the circle *EI* above the *muqanṭara* – whether the sphere is inclined towards *B* or is right. The arc of the circle *LDC* below the *muqanṭara*, plus the arc *CK*, is greater than the arc *LK*, so the angle *KCG* is acute; the point *M* thus lies between the two points *G* and *C*. If the two circles *LDC*, *IE* are equal, then the arcs *DK*, *EI* are equal and the straight line *DK* will be equal to the straight line *EI*; so the straight line *NM* will be equal to the straight line *EI* and *NG* will be equal to *EH*; so the point *N* will be the point *P* and the straight line *ND* will be the straight line *PD*; so *PD* will be equal to the straight line *MK*. But *DE* is greater than *DS* in all cases, because the angle *DSE* is always obtuse since the angle *DSP* is always acute; in fact it is equal to the acute angle *DBA*, because the arc *AD* is always smaller than a semicircle. But since the straight line *DE* is greater than the straight line *DS*, the ratio of *PD* to *DS* is greater than the ratio of *PD*, which is equal to *MK*, to *DE*. From which it necessarily follows that the ratio of the straight line *CD* to the straight line *DB* is greater than the ratio of the straight line *CK* to the straight line *KI* and we complete the proof as before. Then the ratio of the arc *IE* to the arc *EB* is greater than the ratio of the arc *CD* to the arc *DB* and the ratio of the arc *CD* to the arc *DB* is greater than the ratio of the arc *CK* to the arc *KI*.

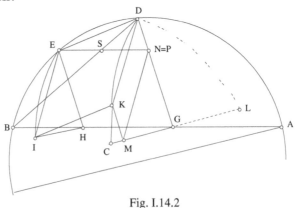

Fig. I.14.2

If the circle *ABC* is a horizon and if the sphere is inclined, then the arc of the circle *LDC* below the horizon is equal to the arc of the circle *EI* that is above the horizon; the arc of the circle *LDC* which is below the horizon, plus the arc *CK*, is thus equal to the arc *LK*. So the angle *KCL* is a right angle, the point *C* is the point *M*, the straight line *KC* is equal to the straight line *PD* and the ratio of *KC* to *KI* is equal to the ratio of *PD* to *KI*; but the ratio of *GD* to *DB* is greater than the ratio of *CK* to *KI*, and the ratio of *CD* to *DB* is much greater than the ratio of *CK* to *KI*, because *CD* is greater than *DG*; and the proof is completed as before.

If the circle *ABC* is a horizon, if the circles *LDC*, *EI* are equal and if the sphere is right, then the two circles *LDC* and *EI* lie on either side of the pole of the horizon, the arc *DC* is equal to the arc *EI* and the arc *EB* is smaller than the arc *DB*; the ratio of the arc *IE* to the arc *EB* is then greater than the ratio of the arc *CD* to the arc *DB*, and the circle drawn through the two poles passes through the point *I* and through the point *C* without cutting the arc *DC*.

<*C*> Let us return to the figure; let the circle *EI* be greater than the circle *LDC* and let the sphere be inclined towards *B*; then either the circle *EI* is the circle of the equator or it is closer to the hidden pole than <the circle of> the equator; so the circle *LDC* is closer than the equator is to the visible pole and the distance from the circle *LDC* to the equator is greater than the distance from the circle *EI* <to the circle of the equator>; or the circles *EI*, *LDC* are both inclined towards the visible pole,[49] and the circle *LDC* is further from the equator than the circle *EI* – because the circle *EI* is not greater than the circle *LDC* while the arc *DB* is not greater than half the arc *ADB*, except if the sphere is inclined towards *B*;[50] so the point *J* is the point in which the straight line *DB* cuts the straight line *EH*.

I say that, if the ratio of the straight line *EH* to the straight line *HJ* is not smaller than the ratio of the semidiameter of the circle *EI* to the semidiameter of the <circle> *DC*[51] and if the arc *LDC* is not greater than a semicircle – this can happen if the circle *ABC* is one of the *muqanṭarāt* parallel to the horizon without being the horizon – then the ratio of the arc *IE* to the arc *EB* is greater than the ratio of the arc *CD* to the arc *DB* and the ratio of the arc *CD* to the arc *DB* is greater than the ratio of the arc *CK* to the arc *KI*; and if the arc *LDC* is greater than a semicircle – this can happen if the circle *ABC* is one of the *muqanṭarāt* parallel to the horizon and is

---

[49] Either *EI* and *LDC* are both, in relation to the equator, on the side towards the visible pole.

[50] See Supplementary note [3].

[51] See Supplementary note [4].

necessary if the circle *ABC* is a horizon – then the ratios of the arcs are as
we have said, if the part of the circle *EI* above the circle *ABC* is not greater
than twice the arc similar to the arc of the circle *LDC* below the circle
ABC.[52] These conditions apply only in this case, I mean if the circle *EI* is
greater than the circle *DC*.

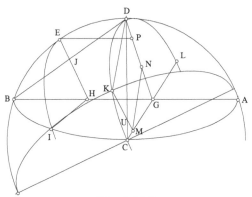

Fig. I.14.3

If the arc *LDC* is not greater than a semicircle, then the arc *LDK* is
smaller than a semicircle; so the angle *KCL* will be acute; and we have the
perpendicular *KM* inside the arc *LDC* and the ratio of *CU* to *UM* is greater
than the ratio of *CK* to *KM*; so the ratio of *CD* to *DG* is greater than the
ratio of *CK* to *KM* which is equal to the ratio of *CK* to *ND*.

If the arc *LDC* is greater than a semicircle and if the part of the circle
EI above the circle *ABC* is not greater than twice the arc similar to the arc
of the circle *LDC* below the circle *ABC*,[53] then we draw the straight line *CS*
parallel to the straight line *GD*; so <the arc> *DS* is equal to half the part of
the circle *LDC* below the circle *ABC*. Since the straight line *DG* is a
diameter of the circle *LDC* and it is perpendicular to the straight line *LGC*,
accordingly it divides the part of the circle *LDC* below the circle *ABC* into
two equal parts. But the arc *EI* is half the part of the circle *EI* above the
circle *ABC*, so the arc *EI* is no greater than twice the arc similar to the arc
DS, and the straight line *CS* is perpendicular to the straight line *CD*,
because it is parallel to the straight line *GD*. But if the arc *EI* is not greater
than twice the arc similar to the arc *DS*, then the arc *EI* is either smaller
than the arc similar to the arc *DS* or equal to the arc similar to the arc *DS* or
greater than the arc similar to the arc *DS*.

---

[52] Here Ibn al-Haytham introduces a supplementary hypothesis (see Supplementary
note [5]).

[53] See Supplementary note [5].

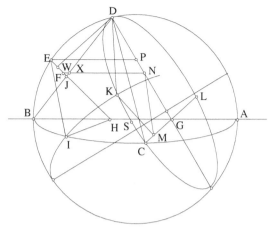

Fig. I.14.4

If the arc *EI* is smaller than the arc similar to the arc *DS*, then the arc *DK* is smaller than the arc *DS* and the point *S* lies between the two points *K* and *C*; then the angle *KCG* will be acute, because the angle *SCG* is a right angle; so the perpendicular *KM* will lie inside the arc *CDL* and the point *M* will lie between the two points *G*, *C*; so the ratio of *CD* to *DG* is greater than the ratio of *CK* to *KM*, as we proved before. But *KM* is equal to *ND*, so the ratio of *CD* to *DG* is greater than the ratio of *CK* to *ND*.

If the arc *EI* is equal to the arc similar to the arc *DS*, then the arc *DK* is the arc *DS*, the point *S* is the point *K* and the straight line *CK* is the straight line *CS*; so the straight line *CK* is perpendicular to the straight line *GC* and the point *M* is the point *C*; the straight line *ND* will be equal to the straight line *CK*.

If the arc *EI* is greater than the arc similar to the arc *DS*, then the arc *DK* will be greater than the arc *DS* and the point *S* will lie between the two points *D*, *K*; but the arc *DK* is not greater than twice the arc *DS*, accordingly the arc *SK* is not greater than the arc *SD* and the angle *SCK* is not greater than the angle *SCD*; but the angle *SCD* is equal to the angle *CDG*, because the straight line *SC* is parallel to the straight line *DG*; so the angle *SCK* is not greater than the angle *CDG* and the perpendicular drawn from the point *K* to the straight line *GC* has its foot at a point outside the point *C*, because it [the perpendicular] will be parallel to the straight line *CS* since the straight line *CS* is perpendicular to the straight line *GC*; the point *M* is outside the straight line *GC* and the perpendicular drawn from the point *K* to the straight line *GC* is smaller than the straight line *GD*, because the straight line *GD* is the sagitta of the arc *LDC* and the sagitta of any arc is the greatest perpendicular from [a point on] the arc to its chord. So the perpendicular drawn from the point K to the straight line GC is smaller than the straight line GD and this perpendicular is equal to the straight line ND, because the straight line MN is drawn from the foot of the perpendicular and it is parallel to the straight line KD. If the perpendicular is drawn from the point K to the straight line GC, then it meets the straight line KC at the point K making an angle equal to the angle KCS, because the perpendicular will be parallel to the straight line CS. But the angle KCS is not greater than the angle CDG, so the angle enclosed by the straight line CK and the perpendicular drawn from the point K to the straight line GC is not greater than the angle CDG. But the right angle at the foot of the perpendicular is equal to the right angle CGD; so if the angle enclosed by the straight line CK and the perpendicular is equal to the angle CDG, then the ratio of CD to DG is equal to the ratio of CK to the perpendicular. But if the angle enclosed by the straight line CK and the perpendicular is smaller than the angle CDG, then the ratio of CD to DG is greater than the ratio of CK to the perpendicular. But the perpendicular drawn from the point K to the straight line GD is equal to the straight line ND. Accordingly, for both cases, the ratio of CD to DG is not smaller than the ratio of CK to ND; and in all cases the straight line CK either will be equal to the straight line ND, or its ratio to the straight line ND will not be greater than the ratio of the straight line CD to the straight line DG.

In the same way, since the arc DK is similar to the arc EI and the circle DC is smaller than the circle EI, the straight line DK will be smaller than the straight line EI; so the straight line MN is smaller than the straight line EI and the triangle NMG is similar to the triangle EIH; so the straight line NG is smaller than the straight line EH, the straight line NG is smaller than the straight line PG, the point N lies between P and G, and the straight line ND is greater than the straight line PD. But since the arc DK is similar to the arc EI, the ratio of the straight line EI to the straight line DK is equal to the ratio of the diameter of the circle EI to the diameter of the circle DC; so the ratio of the straight line EI to the straight line NM is equal to the ratio of the diameter of the circle EI to the diameter of the circle DC; but the ratio of EI to NM is equal to the ratio of EH to NG; so the ratio of EH to NG is equal to the ratio of the diameter of the circle EI to the diameter of the circle DC. But the ratio of EH to HJ is not smaller than the ratio of the diameter of the circle EI to the diameter of the circle DC; so the ratio of EH to HJ is not smaller than the ratio of EH to NG, and the ratio of EH to HJ is either equal to the ratio of EH to NG, or greater than the ratio of EH to NG. If the ratio of EH to HJ is equal to the ratio of EH to NG, then NG is equal to HJ. But if the ratio of EH to HJ is greater than the ratio of EH to GN, then NG is greater than HJ. If NG is equal to HJ, then we join NJ; it will be parallel to *GH* and the ratio of *ND* to *DJ* will be equal to the ratio of *GD* to *DB*. But if *NG* is greater than *JH*, then from the point *N* we draw a straight line parallel to the straight line *GH*; it thus cuts the straight line *EJ*, because *NG* is smaller than *EH* and greater than *JH*; let it cut it at the point *F*. If that parallel straight line cuts the straight line *EJ*, it will cut the straight line *DJ*; then it will cut it at the point *X*; let the parallel straight line be the straight line *NX*, then the ratio of *ND* to *DX* is equal to the ratio of *GD* to *DB* and the ratio of *EH* to *HF* is equal to the ratio of the diameter of the circle *EI* to the diameter of the circle *DC*. Similarly, if the ratio of *EH* to *HJ* is equal to the ratio of the diameter of the circle *EI* to the diameter of the circle *DC*, then the ratio of *EH* to *EJ* is equal to the ratio of the diameter of the circle *EI* to the amount by which it exceeds the diameter of the circle *DC*. But the amount by which the diameter of the circle *EI* exceeds the diameter of the circle *DC* is twice what the perpendicular from the point *D* cuts off on the straight line *EH*, on the side towards the point *E*. But the straight line *EH* is smaller than the diameter of the circle *EI*; so the straight line *EJ* is smaller than twice what the perpendicular to the straight line *EH* from the point *D*, cuts off on the side towards the point *E*. If the straight line *EH* is the semidiameter of the circle *EI*, then the straight line *EJ* will be what the perpendicular from the point *D* cuts off from the straight line *EH*. So the straight line *DJ* will be the perpendicular and the angle *DJE* will be a right angle; the straight line *DE* will then be greater than the straight line *DJ*. If the straight line *EH* is smaller than the semidiameter of the circle *EI*, then the straight line *EJ* will be smaller than the magnitude cut off by the perpendicular on the side towards the point *E*; the angle *DJE* will then be obtuse and the straight line *DE* will be greater than the straight line *DJ*. If the straight line *EH* is greater than the semidiameter of the circle *EI*, then the straight line *EJ* will be greater than the magnitude cut off by the perpendicular. But, in all the cases, the straight line *EH* is smaller than the diameter; so the straight line *EJ* is smaller than twice what the perpendicular cuts off. The perpendicular drawn from the point *D* to the straight line *EJ* thus divides the straight line *EJ* into two unequal parts the greater of which is on the side towards the point *E*; the straight line *DE* will then be greater than the straight line *DJ*. In all the cases, if the ratio of *EH* to *HJ* is equal to the ratio of the diameter of the circle *EI* to the diameter of the circle *DC*, then the straight line *DE* is greater than the straight line *DJ*, and if the ratio of *EH* to *HJ* is equal to the ratio of the diameter of the circle *EI* to the diameter of the circle *DC*, then the straight line *NJ* is parallel to the straight line *GH* and the ratio of *ND* to *DJ* is equal to the ratio of *GD* to *DB*. But the ratio of *ND* to *DJ* is greater than the ratio of *ND* to *DE*, because *DE* is greater than *DJ*; so if the ratio of *EH* to *HJ* is equal to the ratio of the diameter of the circle *EI* to the diameter of the circle *DC*, then the ratio of *GD* to *DB* will be greater than the ratio of *ND* to *DE*. But if the ratio of *EH* to *HJ* is greater than the ratio of the diameter of the circle *EI* to the diameter of the circle *DC*, then the ratio of the diameter of the circle *EI* to the diameter of the circle *DC* is equal to the ratio of *EH* to *HF* and the ratio of *HE* to *EF* is equal to the ratio of the diameter of the circle *EI* to the amount by which it exceeds the diameter of the circle *DC*, so either the straight line *EF* is the magnitude that the perpendicular cuts off, or it is smaller than it, or it is smaller than twice it, as has been proved for the straight line *EJ*; so the straight line *DE* is greater than the straight line from the point *D* to the point *F*; but the straight line from the point *D* to the point *F* is greater than the straight line *DX*, because the angle *DXF* is obtuse, and this [is so] because the angle *DXN* is acute, because it is equal to the angle *DBG*; so the straight line *DE* is much greater than the straight line *DX* and the ratio of *ND* to *DX* is greater than the ratio of *ND* to *DE*; but the ratio of *ND* to *DX* is equal to the ratio of *GD* to *DB*, because the straight line *NX* is parallel to the straight line *GHB*. So the ratio of *GD* to *DB* is greater than the ratio of *ND* to *DE*. So if the ratio of *EH* to *HJ* is not smaller than the ratio of the diameter of the circle *EI* to the diameter of the circle *DC*, then the ratio of *GD* to *DB* is greater than the ratio of *ND* to *DE* in all cases.

Thus we have proved, from what we said, and in accordance with the conditions we introduced, that, either the straight line *CK* is equal to the straight line *ND*, or its ratio to it is not greater than the ratio of *CD* to *DG*; and that the ratio of *GD* to *DB* is greater than the ratio of *ND* to *DE*. But the straight line *DE* is equal to the straight line *KI*; so if the straight line *CK* is equal to the straight line *ND*, then the ratio of *ND* to *DE* is equal to the ratio of *CK* to *KI*. But the ratio of *GD* to *DB* is greater than the ratio of *ND* to *DE*; so the ratio of *GD* to *DB* is greater than the ratio of *CK* to *KI*. But *CD* is greater than *DG* because the angle *DGC* is a right angle; so the ratio of *CD* to *DB* is greater than the ratio of *GD* to *DB* and the ratio of *CD* to *DB* is much greater than the ratio of *CK* to *KI*.

If the ratio of *CK* to *ND* is not greater than the ratio of *CD* to *DG*, then the ratio of *CD* to *DG* will be either equal to the ratio of *CK* to *ND* or greater than the ratio of *CK* to *ND*. But the ratio of *GD* to *DB* is greater than the ratio of *ND* to *DE*, so the ratio of *CD* to *DB* is greater than the ratio of *CK* to *DE* and the ratio of *CD* to *DB* is, in all cases, greater than the ratio of *CK* to *DE*. If we permute, the ratio of *DC* to *CK* is greater than the ratio of *BD* to *DE*; but the arc *DC* is not smaller than the arc similar to the arc *DB*, as has been proved earlier.[54] So the ratio of the arc *DC* to the arc

[54] See p. 302.

CK is greater than the ratio of the arc *BD* to the arc *DE*. If we permute, the ratio of the arc *CD* to the arc *DB* is greater than the ratio of the arc *CK* to the arc *DE*, that is, the arc *KI*; and the ratio of the remainder to the remainder will be greater than the ratio of the arc *CD* to the arc *DB*. So the ratio of the arc *IE* to the arc *EB* is greater than the ratio of the arc *CD* to the arc *DB*. But the ratio of the arc *CD* to the arc *DB* is greater than the ratio of the arc *CK* to the arc *KI*. That is what we wanted to prove.

<15> Similarly, let us return to the figure and from the pole of the sphere let us draw a great circle that cuts the arcs *DK*, *EI*; let the circle be *HLUO*.[55]

I say that if the separation between the circle *ABC* and the circle *LCD* is not greater than a semicircle, then the ratio of the arc *UI* to the arc *UO* is greater than the ratio of the arc *CL* to the arc *LO* and the ratio of the arc *CL* to the arc *LO* is greater than the ratio of the arc *CK* to the arc *KI*.

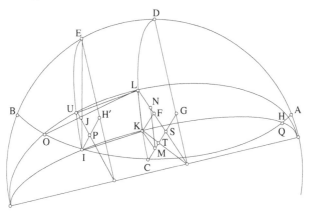

Fig. I.15.1

Proof: We draw the intersections; let the intersection of the circle *HLO* and the circle *ABC* be the straight line *HO*; let the intersection of the circle *QKI* and the circle *ABC* be the straight line *QI*; let the intersection of the circle *HLO* and the circle *DLC* be the straight line *LS*; let the intersection of the circle *QKI* and the circle *DLC* be the straight line *KT* and let the intersection of the circle *HLO* and the circle *EI* be the straight line *UP*. Since the pole of the circle *DLC* is the pole of the equator, the centre of *DLC* will lie on the axis of the World and the two circles *HL* and *QK* cut the circle *DLC* along diameters; so the straight lines *LS* and *KT* are two diameters of the circle *DLC* and they meet one another at its centre. But the

[55] Thus the letter *L* does no longer represents the same point as in the previous part.

straight line *DG* is also a diameter of the circle *DLC*, so it passes through its centre.

If the part of the circle *DLC* that is above the circle *ABC* is smaller than a semicircle, then the centre of the circle *DLC* will be below the circle *ABC* and the figure will be the first figure. The diameter *DG* makes a right angle with the straight line *GC*, so with the straight line *GC* the diameter *LS* makes an acute angle on the side towards the point *C* and the angle *LSC* is thus acute; similarly the angle *KTC* is acute. Now if the diameter *KT* reaches as far as the centre, then it forms a triangle that surrounds the triangle formed by the diameter *LS*;[56] so the angle *LSC* is an exterior angle of the triangle enclosed by the diameters *LS* and *KT* and with its vertex at the centre. So the angle *LSC* is greater than the angle of this triangle that is at the point *T*; but the angle of the triangle that is at the point *T* is equal to the angle *KTC*, so the angle *LSC* is greater than the angle *KTC*. From the point *K* we draw a straight line parallel to the straight line *LS*; it meets the straight line *GC* and it meets the straight line *GC* in a point that does not lie on the straight line *ST*; let this straight line parallel to the straight line *LS* be the straight line *KM*. We draw the straight lines *CL*, *CK*, *KL*, *LU*, *JO*, *UI* and from the point *M* we draw a straight line parallel to the straight line *KL*; then it cuts the straight line *SL*. Since the straight line *GD* is the sagitta of the arc *CLD*, it is consequently the greatest perpendicular falling from [a point on] the arc *DC* onto the straight line *GC*; now, of the perpendiculars, those which are closer to it are greater than those further away; so the perpendicular drawn from the point *K* to the straight line *GC* is smaller than the perpendicular drawn from the point *L* to the straight line *GC*. Similarly, the straight line *KM* is smaller than the straight line *LS*, because it is parallel to it, so the straight line drawn from the point *M* parallel to the straight line *KL* cuts the straight line *LS*; let the parallel line be the straight line *MN*. Then *NL* will be equal to the straight line *MK*. In the same way, since the arc *KL* is similar to the arc *IU*, the ratio of the straight line *IU* to the straight line *KL* is equal to the ratio of the diameter of the circle *EI* to the diameter of the circle *DC*. But *MN* is equal to *KL*; so the ratio of *IU* to *MN* is equal to the ratio of the diameter of the circle *EI* to the diameter of the circle *DC*. But, given that the planes of the circles *DC*, *EI* are parallel, and that the plane of the circle *HLO* cuts them, the straight lines *LS* and *UP* are parallel and the straight lines *SC*, *PI* are parallel; so the angle *LSC* is equal to the angle *UPI*. We draw the straight line *KF* parallel to the straight line *CG*, then the angle *LKF* is equal to the angle *NMS*, because the straight line *NM* is parallel to the straight line *LK* and the straight line *KF* is parallel

---

[56] See the Mathematical commentary (plane figures 2.63.1 and 2.63.2).

to the straight line *GC*; and if we extend the straight line *KF*, it meets the straight line *DG* and forms a right angle with it, because the angle *CGD* is a right angle and the straight line *KF* is parallel to the straight line *CG*; this straight line[57] will then be the sine of the arc *KD*. But the arc *KD* is similar to the arc *IE*, the straight line *IH′* is the sine of the arc *IE*, and the arc *KL* is similar to the arc *IU*; so the angle *LKD* is equal to the angle *UIE*. Now it has been proved that the angle *LKF* is equal to the angle *NMS*, so the angle *NMS* is equal to the angle *UIP*; and it has been proved that the angle *NSM* is equal to the angle *UPI*, so the triangle *NMS* is similar to the triangle *UIP*. So the ratio of *IU* to *MN* is equal to the ratio of *UP* to *NS*; but the ratio of *IU* to *MN* is equal to the ratio of the diameter of the circle *EI* to the diameter of the circle *DC*; so the ratio of *UP* to *NS* is equal to the ratio of the diameter of the circle *EI* to the diameter of the circle *DC*. If the circle *EI* is smaller than the circle *DC*, then the straight line *UP* is smaller than the straight line *NS*. We then prove, as has been proved earlier for the straight line *DG*,[58] that the ratio of *SL* to *LO* is greater than the ratio of *NL* to *LU*.

Similarly, if the circle *EI* is equal to the circle *DC*, we prove that the ratio of *SL* to *LO* is greater than the ratio of *NL* to *LU*; and if the circle *EI* is greater than the circle *DC*, then, if the ratio of *UP* to *PJ* is not smaller than the ratio of the diameter of the circle *EI* to the diameter of the circle *DC*, the ratio of *SL* to *LO* is greater than the ratio of *NL* to *LU*, as has been proved earlier. But the angle *KCG* is acute, because the part of the circle *DLC* that is above the circle *ABC* is smaller than a semicircle; and the angle *KTC* is acute, so the perpendicular dropped from the point *K* onto the straight line *CG* has its foot between the two points *T* and *C*; but the angle *KMC* is also acute and it is greater than the angle *KTC*; so the straight line *KM* is between the straight line *KT* and the perpendicular dropped from the point *K* onto the straight line *TC*; so the point *M* lies between the two points *T* and *C*. The straight line *CL* cuts the straight line *KM*; let it cut it at the point *X*. From the point *X* we draw a straight line parallel to the straight line *KC*, let it be *XO′*; the angle *XO′C* is thus obtuse, because the angle *XO′M* is equal to the angle *KCM*, which is acute, and the straight line *CX* is greater than the straight line *XO′*; so the ratio of *CX* to *XM* is greater than the ratio of *O′X* to *XM*; but the ratio of *CX* to *XM* is equal to the ratio of *CL* to *LS* and the ratio of *O′X* to *XM* is equal to the ratio of *CK* to *KM*; so the ratio of *CL* to *LS* is greater than the ratio of *CK* to *KM*; but the ratio of *SL* to *LO* is greater than the ratio of *NL* to *LU*, that is, the ratio of *MK* to *KI*, because *KM* is equal to the straight line *LN* and *KI* is equal to the straight

[57] See the last lines of fol. 369^r; with *KM* ∥ *DG*, we obtain $\frac{CD}{DG} > \frac{CK}{KM}$.

[58] That is, *KF* extended to meet the diameter *DG*.

line *LU*; so the ratio of *CL* to *LO* is greater than the ratio of *CK* to *KI*. Thus we have proved, as before, that the ratio of the arc *CL* to the arc *LO* is greater than the ratio of the arc *CK* to the arc *KI*.

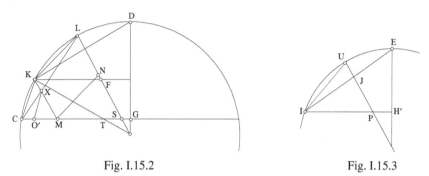

Fig. I.15.2   Fig. I.15.3

If the part of the circle *DLC* that is above the circle *ABC* is a semicircle, then the figure will be the second figure, because the straight line *CG* will be a diameter of the circle *DLC* and the straight line *GD* will also be one of its diameters because the point *G* is the centre of the circle *DLC*. But since the point *G* is the centre of the circle *DLC*, the point *G* will lie on the axis of the sphere and the point *G* will thus lie in the planes of all the circles that pass through the two poles; so the pole of the sphere will be elevated above the plane of the circle *ABC*, because a part of the axis will be above the circle *ABC* since the sphere is inclined towards *B*. But the point *G* lies on the axis, so the point *G* lies in the planes of the circles *HLO* and *QKI*; so the straight lines *QI*, *HO* cut one another at the point *G*. But the point *G* also lies on the straight line *AB*, because the part of the circle *DLC* above the circle *ABC* is a semicircle, so the centre of the circle *DLC* lies in the plane of the circle *ABC*; but it is also in the plane of the circle *ADB*, because the latter is the meridian circle; so the point *G* lies on the [line of] intersection of the circle *ABC* and the circle *ADB*, which is the straight line *AB*; so the point *G* lies on the straight line *AB*; the [line of] intersection of the circle *QKI* and the circle *DLC* is the diameter *KG*; the [line of] intersection of the circle *HLO* and the circle *DLC* is the diameter *LG*, and the [line of] intersection of the circle *HLO* and the circle *EI* is the straight line *UP*; so the angle *KCG* is acute, because the straight line *GC* is a diameter of the circle *DLC* and the straight line *CK* is a chord in it.

But if we extend the straight line *CK* in the direction of *K*, it meets the straight line *GL*, because it meets the straight line *GD*; the straight line *KM* which is parallel to the straight line *LG* will thus lie within the arc *CD* and the point *M* lies between the two points *C*, *G*. But the straight line *KM* is smaller than the straight line *LG*, as we proved earlier, because the perpendicular drawn from the point *K* to the straight line *CG* is smaller than the perpendicular drawn from the point *L* to the straight line *CG*; so the straight line *MN* parallel to the straight line *KL* cuts the straight line *LG* and the straight line *CL* cuts the straight line *KM*; let it cut it at the point *X*. We prove, as was proved in the first figure, that the ratio of *CL* to *LG* is greater than the ratio of *CK* to *KM*, because the angle *XO′C* is obtuse, since the angle *XO′M* is acute. So if the circle *EI* is not greater than the circle *DLC*, then the straight line *UP* is not greater than the straight line *NG*, because the triangle *GNM* is similar to the triangle *PIU* and the straight line *MN* is equal to the straight line *KL*. We have thus proved, as was proved in the first figure, that the ratio of *GL* to *LO* is greater than the ratio of *NL* to *LU*.

So if the circle *EI* is greater than the circle *DLC* and if the ratio of *UP* to *PJ* is not smaller than the ratio of the diameter of the circle *EI* to the diameter of the circle *DLC*, then the ratio of *GL* to *LO* is also greater than the ratio of *NL* to *LU*, as has been also proved in the first figure; so the ratio of *GL* to *LO* is greater than the ratio of *MK* to *KI* and the ratio of *CL* to *LG* is greater than the ratio of *CK* to *KM*. But the ratio of *GL* to *LO* is greater than the ratio of *MK* to *KI*, so the ratio of *CL* to *LO* is greater than the ratio of *CK* to *KI*. We have thus proved, as was proved earlier, that the ratio of the arc *CL* to the arc *LO* is greater than the ratio of the arc *CK* to the arc *KI*. We have proved in both cases that the ratio of the arc *CL* to the arc *LO* is greater than the ratio of the arc *CK* to the arc *KI*, and that the ratio of the remainder to the remainder, which is the ratio of the arc *IU* to the arc *UO*, is greater than the ratio of the arc *CL* to the arc *LO*, which is greater than the ratio of the arc *CK* to the arc *KI*. So the ratio of the arc *IU* to the arc *UO* is greater than the ratio of the arc *CL* to the arc *LO*, and the ratio of the arc *CL* to the arc *LO* is greater than the ratio of the arc *CK* to the arc *KI*. That is what we wanted to prove.

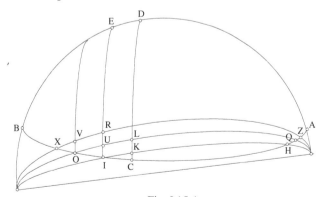

Fig. I.15.4

Similarly, from the pole we draw a circle that cuts the arc *IE* between the two points *U, E* in both cases of the figure, and that cuts the arc *OB*; let it be *XR*. We cause to pass through the point *O* an arc of a circle parallel to the circle *EI*; let it be the arc *OV*. Let the part of the circle *LDC* above the circle *ABC* be no greater than a semicircle and let the circle *LDC* be greater than the circle *IE*. It necessarily follows that the circle *IE* is greater than the circle *OV*. We show, as has been shown earlier from the proof of this proposition, that the ratio of the arc *OV* to the arc *VX* is greater than the ratio of the arc *IR* to the arc *XR* and that the ratio of the arc *IR* to [the arc] *XR* is greater than the ratio of the arc *IU* to the arc *UO*; so the ratio of the arc *OV* to the arc *VX* is greater than the ratio of the arc *IU* to the arc *UO*. But the ratio of the arc *IU* to the arc *UO* is greater than the ratio of the arc *CK* to the arc *IK*.

Similarly, if we draw circles, any number of them, that cut the arc *AB*, and if from their points of intersection we draw arcs parallel to the arc *IE*, then all these arcs we have drawn will be in the ratios we have indicated.

<Second part>

After these introductory lemmas, let us now make a start by proving what we stated concerning what takes place for the seven wandering stars.

<16> Let us begin with the moon.

Let us establish the configuration of the motions of the moon as a whole, then let us state what necessarily follows. Ptolemy proved in his book the *Almagest* that the moon has several motions which are different, having different diameters and different centres. Nevertheless, despite the difference in the motions, the position of the centre of the moon does not depart from the plane of a single circle called the inclined orb. This orb is one of the great circles that lie on the sphere whose centre is the centre of the Universe and it cuts each of the great circles that lie on the sphere of the Universe into two equal parts. So it cuts the circle of the ecliptic, which is the orb of the sun, in two points diametrically opposite one another, which are the two points called the nodes (*jawzahar*),[59] and also cuts the circle of the equator in two points diametrically opposite one another. The moon moves with all its motions in the plane of that inclined orb, that is to say that the centre of the moon does not depart from the plane of this inclined orb. The motion called 'motion of the moon' is the motion of the centre of

---

[59] Ascending node or point of passage to the north; descending node or point of passage to the south.

the moon. Similarly, the motions of all the [wandering] stars are the motions of their centres. The apparent motion of the moon, which is what results from all its motions, is always in the sense of the signs of the Zodiac and in the plane of the inclined circle, and in equal times, it is of unequal magnitude since this apparent motion is the result of [adding] different motions about different centres.[60] Ptolemy established this in detail and gave proofs. However, despite the inequality of its magnitude at various times, this apparent motion of the moon takes place only in the order of the signs of the Zodiac[61] and in the plane of the inclined orb. Since this is so, the apparent motion[62] of the moon is always from west to east and in the plane of the inclined orb. But that orb inclined moves as a whole with a uniform motion about the two poles of the ecliptic and in the sense contrary to the order of the signs of the Zodiac;[63] the entire plane of these circles, that is to say the inclined orb, thus turns about the two poles of the ecliptic, in a sense contrary to the order of the signs of the Zodiac. Ptolemy proved this in his book the *Almagest* and this motion is called the motion of the node. If the entire plane of these circles turns about the two poles of the ecliptic, then all the points we imagine as being on the circumferences of these circles move along parallel circles whose poles are the two poles of the circle of the ecliptic. The two points that are the two nodes move on the circle of the ecliptic itself and do not depart from it, because the motion takes place about the two poles of the circle of the ecliptic; and, of the points that remain, every point we can imagine as being on the circumference of the inclined orb, moves along a circle parallel to the circle of the ecliptic. If this is so, then the inclination of the inclined orb of the moon with respect to the circle of the ecliptic does not change its magnitude by this motion, but it remains constantly in the same state.

The inclination of the inclined orb with respect to the circle of the equator changes its magnitude during this motion; it increases and decreases, because if the inclined orb turns about the two poles of the circle of the ecliptic, then the pole of the inclined orb turns about the two poles of the circle of the ecliptic. But the distance from the pole of the circle of the ecliptic to the pole of the circle of the equator does not change and the position of the one with respect to the other does not change: they are each fixed in the same position, and the great circle that passes through the two

---

[60] We are considering motions on the two circles, the eccentric (deferent) and the epicycle.

[61] That is 'motion in consequence'.

[62] The Arabic term translated here as 'apparent motion' means 'the motion that is visible'.

[63] That is, 'motion in precedence'.

poles is one and the same circle, which is called the circle of the poles. If the pole of the inclined orb turns about the pole of the circle of the ecliptic, and if the pole of the circle of the ecliptic and the pole of the circle of the equator are each fixed in the same position, then the distances from the pole of the inclined orb to the pole of the equator vary and the distance between this pole and the pole of the equator is the magnitude of the inclination of the inclined orb with respect to the circle of the equator. And if the pole of the inclined orb turns about the pole of the circle of the ecliptic, then, in the course of a single rotation, it comes onto the circle of the poles twice; the distance between the pole of the inclined orb and the pole of the circle of the ecliptic does not change in magnitude, because its motion carries it round the pole of the circle of the ecliptic and the distance between these two poles is the magnitude of the inclination of the inclined orb with respect to the circle of the ecliptic; this inclination is much smaller than the inclination of the circle of the ecliptic with respect to the circle of the equator, from what was proved by Ptolemy. But the pole of the inclined orb comes onto the circle of the poles twice in the course of each rotation; for one of them it will be between the pole of the circle of the ecliptic and the pole of the circle of the equator, and for the other one, the pole of the circle of the ecliptic will lie between the pole of the inclined orb and the pole of the equator. If the pole of the inclined orb is between the pole of the circle of the ecliptic and the pole of the equator, it will, in that case, be as close as it can be to the pole of the equator; that distance is the magnitude of the inclination of the inclined orb with respect to the circle of the equator and it is in this case that the inclination of the inclined orb with respect to the circle of the equator will be as small as it can be. If then, after that, the pole of the inclined orb moves away from the circle of the poles, then its distance from the pole of the equator increases and the inclination of the inclined orb with respect to the circle of the equator increases, then the distance from the pole of the inclined orb to the pole of the equator contin-ues to increase until it reaches the circle of the poles for a second time. If it comes onto the circle of the poles in this second case, the pole of the circle of the ecliptic will be intermediate between it and the pole of the equator, and in this case it will be as far as it can be from the pole of the equator and the inclination of the inclined orb with respect to the circle of the equator will be at its greatest. If then, after that situation [has occurred], the pole of the inclined orb moves away from the circle of the poles, then its distance from the pole of the equator diminishes and the inclination of the inclined orb with respect to the circle of the equator diminishes; then the distance from the pole of the inclined orb to the pole of the equator and the inclination of the inclined orb with respect to the equator continually diminishes until the pole of the inclined orb returns to the circle of the poles. The inclination of the inclined orb with respect to the equator varies and is not fixed in a constant state. Half of this inclined orb is always north of the circle of the equator and half is to the south; the point that is the midpoint of the half north of the inclined orb is the one that defines the limit of the inclination of the moon in the northerly direction and the point that is the midpoint of the south half is the one that defines the limit of the inclination of the moon in the southerly direction; however these two points are not [always] the same two points, but change, because every point of the circumference of the inclined orb moves on a circle parallel to the circle of the ecliptic and there is thus no point on the circumference of the inclined orb that moves on the circumference of the circle of the equator. Since this is so, the two points of intersection of the inclined orb with the circle of the equator change. But if these two points change, then the two points at the northern and southern limits of the inclined orb change with respect to the equator and the extreme value of the inclination of the moon in the northerly direction and the extreme value of its inclination in the southerly direction are not always constant at the same magnitude, but vary on account of the variation in the magnitude of the inclination of the inclined orb with respect to the circle of the equator. Nevertheless in all cases of the figure the centre of the moon moves in the plane of its inclined orb from north to south until it reaches the midpoint of the southern half of the inclined orb, that is to say the half of its inclined orb that is cut off by the circle of the equator – by the midpoint of the southern half I mean the point that divides the southern half into two equal parts at the instant when the moon reaches that point – the motion of the moon on the inclined orb will then be from south to north until it comes to the midpoint of the northern half of its orb, that is to say to the point that divides the northern half of its orb into two equal parts at the instant when the moon reaches that point. It [the moon] then moves from north to south, and so on continually.

Since this is so, then the moon moves on its inclined orb from north to south and from south to north. So the motion of the moon in the plane of its inclined orb is from north to south and from south to north, if it is considered in regard to its motion towards the two poles of the equator. If it is considered in relation to the circle of the ecliptic, this same motion will be in the order of the succession of the signs of the Zodiac, and if it is in the order of the succession of the signs of the Zodiac, it will be from west to east. But every point of the inclined orb moves on a circle whose two poles are the two poles of the circle of the ecliptic and in a sense contrary to that of the succession of the signs of the Zodiac. But if it is in the sense contrary to the signs of the Zodiac, it is thus from east to west. The apparent motion of the moon is from west to east, in the order of succession of the signs of the Zodiac and in addition it is inclined towards north or south with respect to the circle of the equator; any point on the circumference of the inclined orb moves from east to west in the sense contrary to the succession of the signs of the Zodiac <and its motion has the same> magnitude as the motion of the ascending node.

This having been proved, let the circle *ABC* be a horizon and let the circle *AHC* be the meridian circle; the arc *ABC* is the eastern half of the circle of the horizon; let the inclined orb of the moon be the circle *BED*; let the arc *BED* of this [circle] be below the horizon, let the position of the moon be the point *B*, let the motion of the moon on its inclined orb be from the point *B* towards the point *E*, let its motion in this case be from north to south, let the pole of the equator be the point *H*. Taking the pole *H* as centre we draw an arc of an hour circle that passes through the point *B*; let it be the arc *BIO*; let the point *I* lie on the meridian circle. If the sphere moves with the rapid motion,[64] then the point *B* moves with the rapid motion. But if the point *B* moves with the rapid motion, during a certain time, then with the motion that is proper to it the moon moves along the arc *BE* and leaves the point *B* of its orb; in addition, with the rapid motion, the moon reaches the meridian circle. Let the position of the moon [when it is] on the meridian circle be the point *N*. If the moon reaches the meridian circle, it will have traversed an arc of its inclined orb and, in this case, it will be moving towards the south and in addition from west to east in the direction of the succession of the signs of the Zodiac. If the centre of the moon arrives at the point *N*, the arc that the moon has traversed on its inclined orb will be to the west of the meridian circle, because the motion of the moon on its inclined orb is from west to east. Thus, the arc it has traversed on its inclined orb will be to the west of the position which it occupies, and the arc traversed by the moon on its inclined orb in the time in which it travelled from the point *B* to the point *N* is to the west of the meridian circle. But that arc is south of the hour circle, because in this case the motion of the moon is from north to south. Let the arc of the inclined orb along which the moon has moved at the moment when it is on the meridian circle be the arc *ON*. But we have seen earlier that any point of the inclined orb always moves on a circle whose two poles are the two

---

[64] Rapid motion, that is, the diurnal motion.

poles of the circle of the ecliptic; the point *B* of the inclined orb[65] is not fixed on the hour circle *BI*, but moves on a circle whose two poles are the two poles of the circle of the ecliptic. Through the point *B* let us draw an arc of a circle whose two poles are the two poles of the circle of the ecliptic; let the arc be *BQ*. So the point *B* of the inclined orb moves on the arc *BQ* with the motion called the motion of the ascending node. Since this is so, then the arc traversed by the moon on the inclined orb in the time during which it has travelled from the point *B* to the point *N* is not, in most cases, the arc *ON* because the point *B* of the inclined orb is not moving on the arc *BO*, but on the arc *BQ*; now the two poles of the arc *BQ* are the two poles of the circle of the ecliptic.

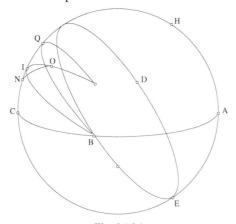

Fig. I.16.1

But if the two poles of the arc *BQ* are the two poles of the circle of the ecliptic and are not the two poles of the circle *BI*, then, either the circle *BQ* touches the circle *BI*, or it cuts it; in fact, either the great circle drawn from the pole of the equator, which is the pole of the circle *BI*, as far as the point *B* passes through the pole of the circle of the ecliptic, or it does not pass through it. If it passes through the pole of the circle of the ecliptic, then the circle drawn through the point *B*, whose pole is the pole of the circle of the ecliptic, will touch the circle *BI*. If the pole of the circle of the ecliptic is, in addition, closer to the point *B* than the pole <of the circle> of the equator, then the circle drawn through the point *B*, whose pole is the pole of the

---

[65] The point *B* is the initial position of the moon. So the letter *B* indicates both a point on the celestial sphere and a point on the inclined orb. The first of these is subject to the diurnal motion and moves along the circle *BI* parallel to the equator while, on account of the motion of the node, the second moves along the circle *BQ* parallel to the ecliptic.

circle of the ecliptic, will lie completely to the north of the circle *BI*. And if the pole of the circle of the ecliptic is further away from the point *B* than the pole of the circle of the equator, then the circle, whose pole is the pole of the circle of the ecliptic, drawn through the point *B*, will lie completely to the south of the circle *BI*. If the great circle drawn through the pole of the equator at the point *B* does not pass through the pole of the circle of the ecliptic, then the circle whose pole is the pole of the circle of the ecliptic [and is] drawn through the point *B*, cuts the circle *BI*, because two circles meet one another in a point, either they touch one another or they cut one another. If the great circle does not pass through the two poles of these two circles, then these two circles do not touch one another; and if they do not touch one another, they cut one another. If the circle drawn through the pole of the equator to the point *B* does not pass through the pole of the circle of the ecliptic, then the pole of the circle of the ecliptic is either above this circle or below it. In saying above and below I mean with respect to the point *I*; thus, if the pole of the circle of the ecliptic is above the great circle drawn from the pole of the equator to the point *B*, then the great circle drawn from the pole of the circle of the ecliptic as far as the point *B* makes an acute angle with the arc *BI* on the side towards the point *I*, that is to say that it will be inclined towards the circle *BI* on the side towards *I*, because the circle drawn through the pole of the equator to the point *B* makes a right angle with the arc *BI* and is perpendicular to it. If the circle drawn from the pole of the circle of the ecliptic to the point *B* makes an acute angle with the arc *BI*, then the circle whose pole is the pole of the circle of the ecliptic, which passes through the point *B* and which cuts the circle *BI*, is such that its upper arc[66] is south of the circle *BI* and its lower arc is north of the circle *BI*. If the pole of the circle of the ecliptic is below the great circle drawn from the pole of the equator to the point *B*, then the great circle drawn from the pole of the circle of the ecliptic to the point *B* makes an obtuse angle with the arc *BI* on the side towards the point *I*. If that angle is obtuse, then the circle whose pole is the pole of the circle of the ecliptic that passes through the point *B* is such that its upper arc is to the north of the circle *BI* and its lower arc is to the south of the circle *BI*.

It is clear, from what we have proved, that the upper arc of the circle *BQ* can lie to the north of the arc *BI* and can lie to the south of it; it can touch it and it can cut it. If the circle *BQ* cuts the circle *BI*, then the [point of] intersection can be on different parts [of it]. The same holds as for the preceding circles, if the two circles are not both great circles. The upper arc of the circle *BQ* can be a semicircle or greater than a semicircle or smaller

---

[66] Upper arc: the arc that cuts the meridian above the horizon.

than a semicircle. If the upper arc of the circle BQ can be smaller than a semicircle without being a specified part,[67] then this arc can be extremely small; so the upper arc of the circle BQ whose endpoints are on the circle BI can be of a certain magnitude such that, by its motion, the node traverses this arc in the time during which the moon travels from the point B to the point N. So if the upper arc of the circle BQ is of this magnitude, then the point B of the inclined orb moves away from the circle BI, moves on the circle BQ and returns to the circle BI at the moment when the moon reaches the point N. If this is so, then, in this configuration, the point O of the inclined orb is the point B of the inclined orb[68] and the arc ON is the arc traversed by the moon in the time it took to travel from the point B to the point N.

If the upper arc of the circle BQ, [the arc] whose endpoints are on the circle BI, is greater than the distance traversed through the motion of the node in the time in the course of which the moon travels from the point B to the point N, then the point B moves away from the circle BI and moves along the circle BQ. When the moon reaches the point N, the point B of the inclined orb has not returned to the circle BI, but will be on the upper arc of the circle BQ not on the circle BI. If the upper arc of the circle BQ is smaller than the distance traversed through the motion of the node in the time in the course of which the moon travels from the point B to the point N, then the point B moves away from the circle BI and moves along the circle BQ; it returns to the circle BI before the moon reaches the point N; next it moves away from the circle BI and moves along the large arc that is the lower arc of the circle BQ. So if the moon reaches the point N, then the point B of the inclined orb will be on the circle BQ and will not lie on the circle BI, but will be below it.

If the moon travels from the point B to the point N, then the point B of the inclined orb can be on the circle BI at the moment when the moon is at the point N and it can be somewhere else. It is on the circle BI in a single case of the figure, when the upper arc of the circle BQ is equal to the distance traversed by the node in its motion in the time in which the moon travels from the point B to the point N. It does not lie on the circle BI in any of the remaining cases for the figure.

If the point B lies on the circle BI at the moment when the moon is at the point N, then the point B of the inclined orb is the point O; if the point B does not lie on the circle BI, then it is on the circle BQ, but it is on the

[67] See Mathematical commentary.
[68] That is, the position reached by the point B.

inclined orb; so it is the point of intersection of the circle *BQ* and the inclined orb.[69]

So it is clear from what we have proved that the arc *ON* of the inclined orb is not, in most cases, the arc traversed by the moon in the time in the course of which it travels from the point *B* to the point *N*; and that the point *O* is not, in most cases, the point *B*. If the point *O* is not the point *B* of the inclined orb, if, at the moment when the moon is at the point *N*, the point *B* of the inclined orb lies outside the circle *BI* and if the point *B* is the point of intersection of the circle *BQ* and the inclined orb, let the point *B*, at the moment when the moon is at the point *N*, then be a point such as *M*. The point *M* can thus lie north of the circle *BI* or south of it, because each upper arc of the circle *BQ* and each lower arc can be north of the circle *BI* or south of it; so the point *M* can be north of the circle *BI* and it can be south of it.

The point *B* of the circle *BQ* always moves along the circle *BI* because the position of the circle *BQ* does not change in regard to the circle *BI*, since their two poles – which are the pole <of the circle> of the equator and the pole of the circle of the ecliptic – have positions that do not change with respect to one another; so the point *B* of the circle *BQ* always moves along the circle *BI*. Let the arc *BS* be the time in the course of which the moon travels from the point *B* to the point *N*. So the point *S* is the point *B* of the circle *BQ*; but the point *B* of the inclined orb is the point *M*, so the points *S* and *M* lie on the circle *BQ*.[70] Let the arc *SM* be an arc of the circle *BQ*. The arc *SM* is thus the arc traversed by the point *B* of the inclined orb through the motion of the node in the time in the course of which the moon travels from the point *B* to the point *N*, because the point *S* is the point *B* which was the [point of] intersection of the inclined orb and the circle *BQ*. But the point *M* of the circle *BQ* is the point reached by the point *B* which lies on the inclined orb; so the arc *SM* is the arc traversed by the point *B* on the circle *BQ* through the motion of the node. But the arc *SM* is to the west of the point *S*, because we have proved that any point on the inclined orb moves from east to west along a circle whose two poles are the poles of the circle of the ecliptic, [that is] in a sense contrary to the succession of the signs of the Zodiac. So the arc *SM* is to the west of the point *S* and the arc *MN* is the arc traversed by the moon on its inclined orb through its proper motion in the course of the time in which it travels from the point *B* to the

---

[69] The position reached by the point *B* is the point of intersection of the circle *BQ* and the inclined orb in the position they have when the moon reaches the point *N* on the meridian circle.

[70] Throughout this paragraph, 'the circle *BQ*' means the position reached by this circle when the moon crosses the meridian at the point *N*.

point *N*, because the point *M* is the point *B* of the inclined orb. But the point *M* can be to the north of the circle *BI*, as it can be to the south of it.

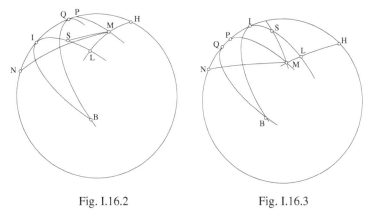

Fig. I.16.2   Fig. I.16.3

We place it in the figure in the two positions and we cause to pass through the point *M* (in each of the two positions) an arc of a great circle that passes through the point *H*. This circle cuts the arc *BIS*; let it cut it at the point *L*. We also cause to pass through the point *M* (in both positions) an arc of an hour circle; let the arc be *MP*. The arc *NP* is thus the inclination of the arc *MN* traversed by the moon in the time in the course of which it travels from the point *B* to the point *N*. The arc *PI* is the inclination of the arc *SM* traversed by the point *B* through the motion of the node in the course of the time we referred to, because the arc *PI* is equal to the arc *ML* and the point *S* is to the west of the meridian circle. Since, when the arc *BI* is moving with the rapid motion, the point *B* of the orb ends up at the point *I*, the moon will be to the east of the meridian circle, because it is moving, with the motion proper to it, from west to east; so it will be to the east of the point *B*, which will have arrived at the point *I*; it [the moon] will be to the east of the meridian circle; and it thus ends up on the meridian circle in the course of a time that is greater than the time *BI*; and, in the course of this increased time, the point *B* will have moved away from the point *I* in the westerly direction, so it will be to the west of the meridian circle. But the point *S* is the point *B*, so the point *S* is to the west of the meridian circle and the arc *BS* is the time in the course of which the moon has travelled from the point *B* to the point *N*, and the arc *BI* will be the time cut off by the meridian circle from the time of the motion of the moon. All this refers to the first case of the figure, in which the motion of the moon on its inclined orb is from north to south.

If the motion of the moon is from south to north, then the arc *BED* on the inclined orb will be to the north with respect to the circle *BI* and will also be below the horizon, because the motion of the moon is from west to east and the motion of the moon will be from the point *B* to the point *E*. If the moon reaches the meridian circle, the inclined orb will be, like the arc *ON*, inclined towards the north with respect to the circle *BI* in the second case of the figure. In the second case of the figure, all the remaining arcs correspond to those found in the first case of the figure. Thus we have proved that the configuration of the motion of the moon and that the configuration of the arcs that the moon traverses in all its motions and the positions the arcs with respect to one another take the two forms shown in the figures that we have drawn.

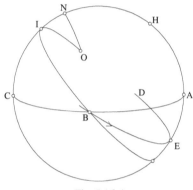

Fig. I.16.4

If the motion of the moon on its inclined orb is from north to south, it is then in accordance with the first case of figure. But if its motion on its inclined orb is from south to north, that is in accordance with the second case of figure. If its motion is from north to south and then becomes from south to north or if it is from south to north and then becomes from north to south, it [the moon] will be at the end of its motion, in all cases for the figure, either to the south of the circle *BI* or to the north of it; or the moon will be at the end of its motion on the circle *BI* itself, that is to say that the moon will have left this circle, to come back to it later on.

If at the end of its motion it [the moon] is to the south of the circle *BI*, then this case is the third case of figure. In this third case of figure, the point *M* is to the south of the circle *BI* and it can be to the north of that circle. If, at the end of its motion, the moon is to the north of the circle *BI*, we again have the third case of figure if it is only the point *N* that is to the north of the point *I* and the point *M* is either to the north of the circle *BI* or to the south of it. If, at the end of its motion, the moon is on the circle *BI*, then the point *N* will be the point *I* and the moon will have no inclination to the circle *BI*; this case is the fourth case of the figure and the point *M* is either to the south of the circle *BI* or to the north of it. The point *M* can lie on the circle *BI* itself at certain moments and this in fact happens if it leaves the circle *BI* to then return to it in the course of the time during which the moon moves from the point *B* to the point *I*.

From all that we have explained it follows necessarily [that we have two possibilities:] that the point *B* is at a height above the horizon or that it is below the horizon, because we have not used the horizon anywhere in what we have explained.

In all cases let us call the arc *BI the required time*, let us call the arc *NI the inclination of the motion of the moon* and let us call the arc *QI the inclination of the motion of the node*. All that we shall need to use in what we shall prove about the properties of the motion of the moon is no more than the required time, the inclination of the motion of the moon and the inclination of the motion of the node; we shall not be concerned with the remaining arcs, or their positions and the variation in their positions. We shall show later how to find the magnitudes of these arcs over known periods of time.

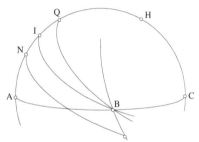

Fig. I.16.5

Let the circle *ABC* also be a horizon, the circle *AHC* a meridian circle and let the pole of the equator be the point *H*. Let the arc *ADC* be the west-ern half of the horizon[71] and let the moon be at the point *N* of the meridian circle. We cause an hour circle to pass through the point *N*; let it be *BND*. Let the arc *NE* lie on the inclined orb of the moon; let us suppose it is to the north of the circle *BND* and to the south of it, that is to say in each of the two positions. Let the arc *NI* lie on the circle whose two poles are the two poles of the circle of the ecliptic; let us suppose that it also is in the two positions, that is to the north and to the south of the circle *BND*. With its proper motion, the moon moves along the arc *NE*; the point *N* of its

---

[71] Lit.: the middle of the day west of the horizon.

inclined orb moves with the motion of the node along the arc *NI*, the point
N of the circle *NI* moves with the rapid motion along the circle *ND*.

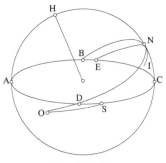

Fig. I.16.6

If the moon reaches the western horizon, the arc traversed by the moon
on its inclined orb will be below the horizon because it will be to the west
of its [the moon's] position. If the motion of the moon is from north to
south, the arc of its inclined orb will be equal to the southern arc *OS*. If its
motion is from south to north, it will be equal to the northern arc *OS*.
Similarly, if its motion is from north to south, then from south to north, or
from south to north then from north to south, and if, at the end of its
motion, it is not on the circle *ND*, then the position of the arc on the
inclined orb is one of the two given positions, that is, the arc *OS* south or
north. If at the end of its motion it is on the circle *ND*, then it will be at the
point *D* and will have no inclination to the circle *ND*. If the moon is to the
north or south of the circle *ND*, at a point such as *S*, then at most instants
the point *N* of its inclined orb will be like the point *M* and at some instants
like the point *O*, and the arc *NI* will be equal to the arc *KM*. So the arc *NK*
is the time in the course of which the moon travels from the point *N* to the
point *S*, at most instants the arc *MS* will be the arc traversed by the moon
on its inclined orb and the arc *KM* will be the arc traversed by the point *N*
which was the position of the moon on the inclined orb with respect to the
arc *NI*; at some instants the arc *OS* will be the arc traversed by the moon on
its orb in the time in the course of which the moon travelled from the point
N to the point *S*. We cause an arc of a great circle to pass through the points
H and *S*; let it cut the hour circle at the point *G*. We cause an arc of an hour
circle to pass through the point *M*; let it cut the arc *HS* at the point *Q*. So
the arc *ND* is the required time, the arc *SG* is the inclination of the motion
of the moon and the arc *QG* is equal to the inclination of the motion of the
node.

All these results necessarily hold if the point *S* is above the horizon.

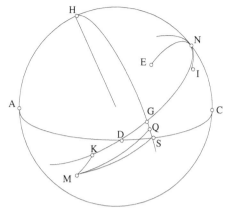

Fig. I.16.7

But if it is below the horizon, that is to say if the moon moves from one position to another with the rapid motion, then, in all cases for the figure, it acquires a required time by this motion; at most instants it has an inclination with respect to the hour circle which it was on and its position[72] has an inclination with respect to the hour circle, namely the inclination that we called the inclination of the motion of the node.

It is with this configuration, and subject to these distinctions, that the motion of the moon at its rising and its setting takes place, as also its motion above the horizon and its motion below the horizon.

<17> As for the sun, its proper motion is a single motion in the order of succession of the signs of the Zodiac from west to east, since the succession of the signs of the Zodiac is from west to east and the centre of the sun is always in the plane of the circle of the ecliptic and is never outside this plane. However the circle of the ecliptic cuts the circle of the equator in two points opposite one another which are the two points of the equinoxes; thus one half of the circle of the ecliptic is always to the north of the circle of the equator and the other half is to the south. These two halves are inclined to the circle of the equator and this inclination is fixed at a constant magnitude that does not change. The two halves on either side of the equator are two halves that, of themselves, do not change because the

---

[72] We are dealing with the position *M* reached by the point *B* of the sphere of the fixed stars, in the course of the displacement that results from the diurnal motion and from the motion of the node.

two points of intersection that are the two points of the equinoxes do not change and do not alter; they are these same two points. Similarly, the two points that divide each of the halves of the circle of the ecliptic on either side of the equator into two halves are [always] the same two points, which do not change. These two points are called *the solstices*, the one to the north is called the *summer solstice* and the one to the south the *winter solstice*. The point of the summer solstice is thus <the point> of maximum inclination of the circle of the ecliptic towards the north with respect to the circle of the equator and the point of the winter solstice is <the point> of maximum inclination of the circle of the ecliptic towards the south with respect to the circle of the equator. If the sun moves round the circle of the ecliptic, which cuts the circle of the equator, and half of which is north of the circle of the equator and half south [of it], then the sun, in its proper motion on the circle of the ecliptic, has an inclination to the circle of the equator, to the north and to the south; the point of the summer solstice is the one that defines the maximum in the declination of the sun in the northerly direction and the point of the winter solstice is the one that defines the maximum in the declination of the sun in the southerly direction. As the sun moves with its proper motion from the summer solstice towards the winter solstice, then it moves from north to south, if we refer its motion to the two poles of the equator; and if it moves from the winter solstice towards the summer solstice, then it moves from south to north with respect to the two poles of the equator. If this same motion – that is, the motion of the sun on the circle of the ecliptic – is referred to the circle of the ecliptic, then the sun moves from west to east because this motion is in the order of succession of the signs of the Zodiac and the succession of the signs of the Zodiac is from west to east. The sun moves with a single motion in the plane of the circle of the ecliptic; this motion is from west to east and in addition it is inclined to the north or to the south.

This having been stated: let the circle *ABCD* be a horizon, let the circle *AH* be the meridian circle, let the pole of the equator be the point *H*. Let the circle of the ecliptic be the circle *BKDL*, let the arc *BKD* be below the horizon and the arc *DLB* above the horizon; let the order of the signs of the Zodiac be taken from the point *B* to the point *K* and to the points following it. Let the position of the sun on the circle of the ecliptic be the point *B*. We cause an arc of an hour circle to pass through the point *B*; let it be *BEM*. Let this arc cut the meridian circle at the point *E*.

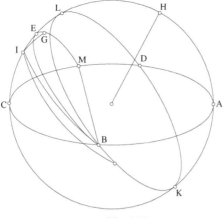

Fig. I.17

If the sphere moves with the rapid motion, then the point B of the circle of the ecliptic moves along the circle BEM and does not depart from it, because the position of the circle of the ecliptic does not change with respect to the equator, nor with respect to any of the circles parallel to the equator; and this is so because the distance between the pole of the circle of the ecliptic and the pole of the circle of the equator does not change and the position of the one with respect to the other does not change; so the point B of the circle of the ecliptic always moves along the circle BEM. In all cases, through the rapid motion of the point B, the sun reaches the meridian circle. But the sun always moves with the motion proper to it in the plane of the circle of the ecliptic and round the circle of the ecliptic. The sun travels from the point B to the meridian circle in a certain time; and in that interval of time the sun, with the motion proper to it, traverses a certain arc on the circle of the ecliptic. If the sun moves round the circle of the ecliptic in the order of the signs of the Zodiac, then it moves along the arc BK from the point B in the direction of the point K; but the arc BK cuts the arc BE, so the sun, in its proper motion, leaves the circle BE and is inclined southward if the arc BK lies south of the circle BE. If the sun reaches the meridian circle, it will be south of the circle BE. Let the position of the sun on the meridian circle be the point I, then the arc that the sun has traversed on the circle of the ecliptic will be to the west of the point I, because the motion of the sun along the circle of the ecliptic is from west to east. But the point B of the circle of the ecliptic does not move off the circle BEM, the arc of the circle of the ecliptic that the sun has traversed in the time in the course of which the sun travelled from the point B to the point I has the same position as the arc GI: so the point G is the point B, and that is so because, while the point *B* travels to the point *E*, the arc *BK* is to the east of the meridian circle. But the centre of the sun is on the circle *BK* and not on the circle *BE* and in this case the position of the centre of the sun will be east of the meridian circle. After that moment the centre of the sun then travels to the meridian circle, so the centre of the sun reaches the meridian circle a certain time after the point *B* has reached the point *E*. That is why, if the centre of the sun reaches the meridian circle, then the point *B* will have moved along the arc *BM* and will be west of the meridian circle; so its position will be at a point such as *G*. If the sun is at the point *B* and if its motion is from north to south, then if it reaches the meridian circle, it will be to the south of the hour circle that passes through the point *B*; the same will be true whether the point *B* is above the horizon or below the horizon. Since the arc *HI* starts from the pole of the equator, the arc *IE* will be the inclination of the arc *GI* of the circle of the ecliptic with respect to the hour circle *BEM* and the arc *BG* is the time in which the sun has travelled from the point *B* to the point *I*. Let us call the arc *BE* the required time and let us call the arc *IE* the inclination of the motion of the sun.

Similarly, we suppose an arc *BLD* north of the circle of the ecliptic, above the horizon, and the arc *BKD* above the horizon and we suppose the position of the sun is the point *B*. The proper motion of the sun is along the arc *BL* starting from the point *B* in the direction towards *L*; if the sun reaches the meridian circle, then the arc traversed by the sun on the circle of the ecliptic will be to the west of the meridian circle and to the north of the circle *BEM*, and will be the inclination of the arc *GS*; so the arc *BE* will be the required time and the arc *SE* will be the inclination of the motion of the sun. The same necessarily holds if the sun starts from the meridian circle and moves to the western horizon, that is to say that there will be a required time and an inclination with respect to the hour circle, which <is the arc> that passes through the point *I* or through the point *S*;[73] it turns out that we can prove this as we proved it for the configuration of the motion of the moon. The same necessarily holds also if the motion of the sun away from the meridian circle is from a point above the western horizon to a point below the western horizon.

It is in this way that we obtain the configuration of the motion of the sun when it rises and when it sets, of its motion above the horizon and of its motion below the horizon. If the motion of the sun in the time in the course of which it travels from the point *B* to the meridian circle is from north to south then from south to north or from south to north and then from north to south, at the end of its motion either it will be outside the

---

[73] The inclination with respect to the hour circle *BEM* is the arc *IE* or the arc *SE*; it is defined by the point *I* or the point *S*.

circle *BEG* or it will be on this same circle *BEG*. If it is outside the circle *BEG*, it is either to the south or to the north of it; if this is so, then its position will be the point *I* or the point *S*, the arc *BE* will be the required time and the arc *IE* or the arc *SE* will be its inclination, at that moment, with respect to the circle *BEG*; if at the end of its motion the sun is on the circle *BEG*, then its position is the point *E*, the required time is the arc *BE* and it has no inclination with respect to the circle *BEG*.

<18> As for the five planets, each of them has an inclined orb like the inclined orb of the moon and each of these orbs cuts the circle of the equator. However, for some of these orbs the inclination does not vary to a perceptible degree with respect to the circle of the ecliptic: these are the orbs of the superior planets, that is Saturn, Jupiter and Mars; and for some the inclination varies with respect to the circle of the ecliptic: these are the orbs of Venus and Mercury. In fact, each of the orbs of these two planets moves as a whole and is inclined to the circle of the ecliptic and becomes coincident with it, then it is inclined towards the other side and attains a limiting inclination, to then return to motion towards the circle of the ecliptic and becomes coincident with it and then becomes inclined in its original direction, and so on continually, as was stated by Ptolemy in his book the *Almagest*. This situation does not, however, prevent each of these two orbs from cutting the circle of the equator or from having an inclination with respect to it; by the motion of each of these orbs towards the circle of the ecliptic and its becoming coincident with it, its inclination to the other side with respect to it and its return to it, the orb will not coincide with the circle of the equator, but by that motion it is only its inclination with respect to the equator that varies, because the inclination of each of these two orbs with respect to the circle of the ecliptic is a small inclination, whereas the inclination of the circle of the ecliptic[74] is a large inclination; thus the inclination of each of these two orbs with respect to the circle of the equator is the inclination with respect to the circle of the ecliptic multiplied many times;[75] thus – because it becomes coincident with the circle of the ecliptic and its inclination is in both directions – neither orb will coincide with the circle of the equator, but [what will happen will

---

[74] With respect to the circle of the equator. The maximum value of the inclination of the inclined orb with respect to the ecliptic is 7° for Mercury and 3°24′ for Venus. For the superior planets, the inclination is more or less constant: 1°51′ for Mars, 1°19′ for Jupiter, 2°30′ for Saturn.

[75] The maximum value of the inclination of Mercury with respect to the ecliptic is 7°, and the maximum value of the inclination ecliptic with respect to the equator is 23°27′.

be that] only the magnitude of its inclination to the circle of the equator
varies. Thus the inclination of the configuration of the orbs of all five
planets with respect to the circle of the equator is like the inclination of the
configuration of the inclined orb of the moon with respect to the circle of
the equator; each of these orbs cuts the circle of the ecliptic in two points
that are opposite [one another]; these two points for each of the orbs of the
five planets are called the nodes. The difference between these nodes and
the two nodes of the moon is that the two nodes of the moon move rapidly,
which becomes perceptible in a single day, whereas the nodes of the
planets move with a slow motion that does not become perceptible either in
a single day, or in several days. In fact, each of the inclined orbs of the five
planets moves as a whole about the two poles of the circle of the ecliptic in
the same configuration as the motion of the inclined orb of the moon;
however, the motion of the orbs of the planets about the two poles of the
circle of the ecliptic is very slow, as Ptolemy and others have proved, and it
is equal to the motion of the fixed stars; in a single day or in several days
this motion does not become perceptible. If each of the five planets moves
around its inclined orb and if we refer its motion around its inclined orb to
the circle of the ecliptic, then its motion will be in the direction of the order
of the signs of the Zodiac if <its motion is> direct. If each of these planets
moves with respect to the circle of the ecliptic in accordance with the order
of the signs of the Zodiac, then it moves from west to east. But if the
inclined orb of each of the five planets cuts the circle of the equator and if
the planet moves around its inclined orb, then each of these planets consequently moves from north to south and from south to north with respect to
the two poles of the equator, in the same way that it happens for the sun
and the moon. The configuration of the motions of each of the five planets,
in its motion from west to east and in its motion from north to south and
from south to north, is like the configuration of the motions of the moon, in
its motion from west to east and in its motion from north to south and from
south to north, except that the configuration of the motions of these five
planets differs from the configuration of the motions of the moon in a
single respect: the epicycle of each of these five planets is inclined to the
plane of the inclined orb towards the north or towards the south; so the
planet departs from the plane of the inclined orb by this inclination,
because the centre of the planet is always on the circumference of the
epicycle; but this is not the case for the moon because the epicycle of the
moon does not depart from the plane of the orb, so the centre of the moon
does not depart from the plane of the inclined orb. Thus the inclination of
these planets with respect to the circle of the equator exceeds the

inclination of the moon with respect to the circle of the equator only by the inclination of the epicycles.

If these planets are in retrograde motion, then the configuration of their motions, when they are in retrograde motion, is no different from the configuration of their motions when they are in the direct sense, except that their motion, which was from west to east with respect to the circle of the ecliptic is [now] from east to west with respect to the circle of the ecliptic; and that difference does not change the form of their inclinations with respect to the circle of the equator or with respect to the hour circles that pass through their centres.

Similarly, if these planets are stationary between retrograde and direct motion, it can happen, for these planets and in particular for two of the superior planets,[76] that there is a halt lasting a perceptible time between the retrograde and the direct motion, that is to say that we find by sight and observation a certain length of time [passes] without the two planets appearing [to have] a motion from west to east or from east to west; however, during this interval of time their inclination can increase or decrease with respect to the circle of the equator. Thus, at the time during which the planet is at station without appearing to have a motion in longitude, it may have a visible motion from north to south or from south to north on account of the inclination of its epicycle; and it will in addition be moving with the rapid motion, while its position on its inclined orb stays the same, as far as is perceptible.

These results having been put in place, let the circle *ABC* be a horizon and let one of the five planets be at the point *B*; let it be in direct motion. We cause one of the hour circles to pass through the point *B*; let <the circle> be *BED*. If the sphere moves, then in all cases for the figure the planet travels towards the meridian circle; but with its proper motion the planet moves about its inclined orb, so it departs from the circle *BED* and has an inclination to the north or to the south. If the planet reaches the meridian circle, then the arc that it has traversed on its orb will be to the west of its position on its inclined orb. If its epicycle is inclined with respect to its inclined orb and if the planet has an inclination on account of its inclination, then the planet will lie to the north of its inclined orb or to the south of it. If this is so, then the position of its inclined orb will be the position of the arc *GH* that we drew in the first case for the figure, whether north of the circle *BED* or south of it, and the position of the epicycle will be like the position of the arc *HI* either north of the arc *GH* or south of it. For these planets the inclination of the motion of the node is not perceptible over this interval of time, so the point *G* is the point *B*. But the arc *BG* is

_____
[76] Lit.: the two superior planets (see p. 337: allusion to Jupiter and Saturn).

the time in the course of which the planet travelled from the point *B* to the point *I*, the arc *BE* is the required time and the arc *IE* is the inclination of the motion of the planet; all this applies in the first case for the figure.

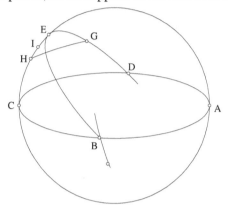

Fig. I.18.1

If the planet is in retrograde motion, then the arc traversed by the planet on its orb is to the east of its position on its orb, and it is to the north of the circle *BE* or to the south of it; the epicycle will be to the north of the inclined orb or to the south of it. The configuration refers to the second case for the figure. If the motion of the planet is from north to south then from south to north, or from south to north then from north to south, then its position will be like the point *I* or the point *S* and the required time will be the arc *BE* in all cases for the figure, as is the case for the sun and the moon.

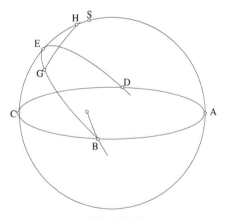

Fig. I.18.2

But if the planet is at station between retrograde and direct motion and if its motion on its inclined orb is imperceptible, then its position on its inclined orb is moving along the circle *BED*; the planet has an inclination with respect to the circle *BED* whose magnitude is only that of the inclination of the epicycle with respect to the inclined orb and this inclination is either to the north or to the south. If this motion is also very slow, that is, the motion of the inclination of the epicycle with respect to the inclined orb, and the magnitude of that inclination is imperceptible, because its magnitude is so small – this can happen for the two planets Saturn and Jupiter –, then the planet does not depart from the circle *BED*. If it reaches the meridian circle, then it will be at the point *E* and the arc *BE* will be the required time, which is the time in the course of which the planet moved from the point *B* to the point *E*. In this case the motion of the planet has no inclination.

So it is clear from all that we have proved for the configuration of the motions of the seven wandering stars that if each of the seven wandering stars moves with the rapid motion in a certain interval of time, then it will acquire, in that interval of time, a required time, which is the arc cut off, on the hour circle the star was on at the start of the time of its motion, by the circle itself starting from the pole of the equator [and going] to the position of the star at the end of the time of its motion, between the circle drawn from the pole of the equator and the position of the star at the beginning of the time of its motion; and that in this interval of time, in most cases, the star will have an inclination with respect to the hour circle which it was on at the beginning of the time of its motion, which is the arc of the circle drawn from the pole of the equator to the position of the star at the end of the time of its motion, the arc between the position of the star at the end of the time of its motion and the hour circle the star was on at the beginning of the time of its motion.

This having been proved, we say: if each of the seven wandering stars is moving at a known moment in a known interval of time, then the arc that is its required time will be known and the arc that is the inclination of its motion will be known.

<19> Let the position of one of the seven wandering stars at a known instant be the point *A*; let this star move in the course of a known time and let its position at the end of the known time be the point *B*; let the north pole of the equator be the point *C*. We cause an arc of a great circle to pass through the points *C* and *A*; let the arc be *CA*. We cause an arc of a great circle to pass through the points *C* and *B*; let the arc be *CB*. We cause an arc of an hour circle to pass through the point *A*; let the arc be *AD*. So the

arc *AD* is the required time and the arc *DB* is the inclination of the motion
of the star.

I say that the arc *AD* is known and that the arc *DB* is known.

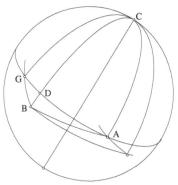

Fig. I.19

Proof: The point *A* is the known position of a star at a known moment.
If that star is the sun, then the point *A* is on the circle of the ecliptic and it
[the point] is known, because it is the position of the sun at a known
moment. Similarly, the point *B* is a known point of the circle of the ecliptic
because the instant when the sun reaches the point *B* is a known instant
since the time between the instant when the sun is located at the point *A*,
which is a known instant, and the instant when it reaches the point *B*, is a
known time, by hypothesis. But since the point *A* is a known point on the
circle of the ecliptic, the arc *CA* is known because the inclination of the
point *A* with respect to the circle of the equator is known; similarly, the arc
CB is known. But since the time between these two instants is known, the
arc traversed by the sun on the circle of the ecliptic is known and it is to the
west of the point *B*. But the point *A* of the circle of the ecliptic moves on
the circle *AD* without departing from it; the arc traversed by the sun on the
circle of the ecliptic, in the time in the course of which it travelled from the
point *A* to the point *B*, is thus between the point *B* and the circle *AD* and its
inclination is the arc *BD* as seen in the first case for the figure. But the arc
GA is known because it is the time in the course of which the sun has
moved from the point *A* to the point *B*, because the point *G* of the circle of
the ecliptic was at the point *A* and has travelled from the point *A* to the
position reached by the sun in the course of the time of its movement from
the point *A* to the point *B*; so the arc *AG* is the known time in the course of
which the sun has moved from the point *A* to the point *B*, and the arc *AD* is

the right ascension of the arc AB;[77] now the arc GD is known; there remains the known arc AD which is the required time. Since the arc CA is known and the arc CB is known, accordingly the difference between them is known; it is the arc BD, which is the inclination of the motion of the sun; so the arc BD is known. If at a given instant the sun is moving, during a known interval of time, then its required time is known and the inclination of its motion in this known time is known.

<**20**> If the wandering star which is at the point A is the moon or one of the five planets, then its position in relation to the circle of the ecliptic[78] is known, because by hypothesis the instant is known; the wandering star's distance from the equator is known and the point where it crosses the circle of the ecliptic is known – the point where it crosses the circle of the ecliptic is the point of intersection of the circle of the ecliptic and the circle that passes through the pole of the equator and through the position of the centre of the wandering star, which in this case is the circle CA; thus the point at which it crosses [the ecliptic] lies on the circle CA.

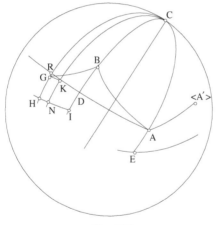

Fig. I.20

Let the point at which it crosses [the ecliptic] be the point E. If the distance from the wandering star that is at the point A to the circle of the equator is known, then its distance from the pole of the equator is known; so the arc CA is known and the point E of the circle of the ecliptic is

---

[77] The right ascension of an arc means the difference δ between the right ascensions of the endpoints of the arc; so we have $\delta(A, B) = \delta(A, D)$ and $\delta(A, D) =$ meas.($\widehat{AD}$) (Proposition 7).

[78] See Supplementary note [6].

known. But since the wandering star was at the point A at a known instant and it then moves, in the course of a known time, so that it travels to point B, the instant at which the wandering star reaches the point B is a known instant; the position of the wandering star in relation to the circle of the ecliptic at the instant when the wandering star arrives at the point B is thus known, its distance from the circle of the equator at that instant is also known and the point at which it crosses [the ecliptic] is also known; but in this case the point at which it crosses lies on the circle CB; let the point at which it crosses be the point I. If the distance from the wandering star to the circle of the equator is known, then its distance from the pole of the equator is known and the arc CB is known, so the point I of the circle of the ecliptic is known. But since the wandering star moves from the point A to the point B in a certain time, accordingly in that interval of time it will have traversed a certain arc on its own orb which is the inclined orb; now that arc is to the west of the point B; for the moon, that is [true] at all times; for the five planets, if they are moving [uniformly], then the point the wandering star occupied on its own orb has <travelled> to the west of the point B. If the wandering star is the moon, then, in most cases, the point of its inclined orb that was at the point A leaves the hour circle AD and has an inclination to north or south with respect to it, [the inclination being] of the magnitude of the inclination of the motion of the node in the known time in the course of which the moon travelled from the point A to the point B. The position of the arc of the inclined orb that the moon has traversed in the course of the time in which it travelled from point A to point B is similar to the arc BG in the second case of figure. So the point G is to the south of the circle AD or to the north of it and the endpoint of the known time in the course of which the moon travelled from the point A to the point B is to the east of the point G, as we have proved for the configuration of the motions of the moon. Let the known time be the arc AK. We cause an arc of a great circle to pass through the two points C and G; let the arc be CG <which cuts the hour circle AD at the point R>.[79] Since the point R of the arc AR was at the point A and it has moved with the motion along the hour circle AD, the arc CR will be the arc CA and the crossing point, which is E, has been displaced by the movement of R to reach the arc CG; let the point for crossing the arc CG be the point H. So the point H on the circle of the ecliptic is known, the point I on the circle of the ecliptic is known and the point H has reached the arc CG at the instant when the point I of the circle of the ecliptic has reached the arc CB; so the points H and I are the endpoints of a known arc of the circle of the ecliptic. We cause to pass

---

[79] See Supplementary note [7].

through the points H and I the arc of the circle of the ecliptic whose end-points they are; let the arc be HI. We cause an arc of a great circle to pass through the points C and K; let <the arc> be CK. Let this circle cut the arc HI at the point N. But the arc GK is parallel to the circle of the ecliptic and through the point G there passes a circle [that is] drawn from the pole of the circle of the ecliptic and[80] which passes through the position of the wandering star when it was at the point A; this position is a known point of the circle of the ecliptic. The circle drawn from the pole of the circle of the ecliptic, and that passes through the point K, [also] passes through a known point of the circle of the ecliptic, because the arc GK is known; in fact, it has the magnitude of the motion of the node in the known time, which is the time AK, and the latitude of the point K with respect to the circle of the ecliptic is known because it is equal to the latitude of the point G, which is known; so the point K is a point whose position is known with respect to the circle of the ecliptic, and the crossing point for the point K is known. But the arc CK is known because it is equal to the arc AC and the crossing point for the point K is the point N; so the point N of the circle of the ecliptic is known, and the arc NI of the circle of the ecliptic is known. But the arc KD is the right ascension of the arc NI, so the arc KD is known and the arc AK is known; so the arc AD is known and it is the required time. But the arc CD is equal to the arc CA, which is known, and the arc CB is known; so the arc BD is known and it is the inclination of the motion of the moon in the known time.

All that we have explained and proved follows necessarily if the point G is to the south of the circle ADK or if it is to the north of it. If the point B which is the position of the moon at the second instant is on the circle ADK, that is to say the position of the moon with respect to the hour circle is the point D, then the method used to prove the magnitude of the required time is the same method we have [already] described without any difference from it for any of the results we have described, except that the motion of the moon in this known time has no inclination with respect to the hour circle which it has reached, which is the circle ADK. So if the moon moves in the course of a known interval of time, beginning at a known instant, then its required time is known and the inclination of its motion is known.

---

[80] Ibn al-Haytham means that through the point G there passes a circle drawn from the pole of the ecliptic, like the circle that passes through the position of the wandering star when it was at the point A.

<21> If the planet that is at the point *A* is Mercury or Venus, then the figure for its required time is similar to the figure for finding the required time of the moon. In fact, the inclined orb of these two stars moves towards the circle of the ecliptic until it is coincident with it and then leaves it and becomes inclined towards the other side; then it returns to it, later leaving it and becoming inclined towards the side it was inclined to at first and so on repeatedly. Thus, on account of this situation, the inclination of the inclined orb with respect to the circle of the equator is variable. But if the inclination of the inclined orb with respect to the circle of the equator varies, then the distance from the point of its inclined orb occupied by the star to the pole of the equator varies; the point of the orbs of Mercury or of Venus, which is the point *G*, will thus in most cases not lie on the circle *AD*, but will be to the north or the south of it; however, the pole of this motion is the point of the node. In the known time, every point of the inclined orb moves on an arc of a circle that cuts the hour circles because, for each of these two stars, the node is not situated at the pole of the equator. So in the course of the known time the point *G* moves on an arc of a circle that cuts the circle *AD*. This motion is sometimes in the order of succession of the signs of the Zodiac and sometimes in the reverse of the order of the signs of the Zodiac. We shall prove later at what moments the motion of the point *G* and its homologues is in the order of succession of the signs of the Zodiac and when it is in the reverse to the direction of the signs of the Zodiac. We shall show the magnitude of this motion in the known time.

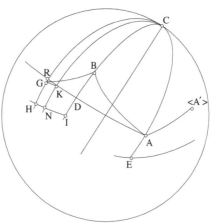

Fig. I.21

The arc along which the point *A* moves in the course of the known time, [an arc] that belongs to the orbs of Mercury or of Venus, is known and its direction is known. This being so, this case reduces to a figure similar to that for the motion of the moon: the arc of a circle that cuts the hour circle which is between the point G and the point reached by the point A of the circle AD, which is the arc GK corresponding to the arc GK in the motion of the moon, is thus known. The proof to show the magnitude of the required time for each of these two stars is completed in the same way as the proof for the required time for the moon. So the required time for the planets Mercury or Venus is known and the case of figure for these two planets is the third one.

The inclination of the motion of each of these two stars is the difference of the two arcs CA and CB, which is the arc BD. This inclination is made up of the inclination of the inclined orb with respect to the circle of the ecliptic and the inclination of the epicycle with respect to the inclined orb; however, resulting from these two inclinations, at any known instant the star will have a known latitude with respect to the circle of the ecliptic. If the latitude of the star is known and if its position with respect to the circle of the ecliptic is known, then the distance from the star to the equator or to the pole of the equator is known. The two arcs CA and CB, which are the distances from the star to the pole of the equator at two known instants, are known. Their difference is known; now this difference is the arc BD, which is the inclination of the motion of the star with respect to the circle AD. If the second arc, which is CB, is greater than the first arc, which is CA, then the inclination is to the south with respect to the circle AD; but if the first arc, which is CA, is greater than the second arc, which is CB, then the inclination is to the north. So if each of the two planets, Mercury and Venus, moves in the course of a known time, starting at a known instant, then its required time is known. But the required time, for each of the seven stars, is always to the east of the inclination of the motion of the star, because the time in the course of which the star moves is always greater than the right ascension of the arc the star traverses on its inclined orb.[81]

And if the star that is at the point A is Saturn, Jupiter or Mars, and if the time in the course of which the star moves from the point A to the point B is a single circuit of an hour circle or a part of a circuit, then the point G is on the circle AD since the nodes for these stars do not move in a time that is a single revolution or a part of a revolution of a perceptible magnitude on the circles parallel to the circle of the ecliptic that pass through the point G, and that are homologues of the arc KG in the second case of figure. The point G, which was at the point A in the course of the time that is a single revolution or a part of a revolution, does not perceptibly depart from the

[81] This conclusion is valid when the motion of the planet on its orb is direct. The conclusion when the motion is retrograde is given later for the five planets (p. 344).

circle AD;[82] at the instant when the star is at the point B, so the point G will be on the hour circle AD and the point G will be to the west of the point B if the star is moving with direct motion. So the arc AG is the known time[83] in the course of which the star has moved from the point A to the point B and the two crossing points which are the points H and I are known, as has been proved earlier. So the arc HI that belongs to the circle of the ecliptic is known; but the arc DG is the right ascension of the known arc HI, so the arc GD is known and the arc AG is known; there remains the known arc AD which is the required time. So the required time for each <of the planets> Saturn, Jupiter and Mars is known.

The inclination of the motion of each of these three planets is in the same state as that of the inclination of Mercury and of Venus. In fact, the inclination of these stars is also made up of the inclination of their inclined orb <with respect to the ecliptic> and of the inclination of their epicycle; however, their latitude with respect to the circle of the ecliptic is known at any known instant and the point at which they cross [it] is known; so the arcs CA and CB are known and their difference is known; but the difference between these two arcs is the inclination of the motion of these stars, and the direction of their inclination is like the direction of that of the two planets Venus and Mercury. So the inclination of the motion of each of these three stars is known and the direction of their inclination is known. If each <of the planets> Saturn, Jupiter and Mars moves in the course of a known time starting at a known instant, then its required time is known and the inclination of its motion with respect to the hour circle which it was on at the first instant is known.

If the star that is at the point A is one of the five planets and if it is in retrograde motion, then the point G will be to the east of the point B and the arc AG which is the known time will be smaller than the arc AD; and the proof as a whole will be like the preceding one.

If the star is at station between retrograde motion and direct motion, and if it appears that it has no motion in longitude, then its second position with respect to the circle of the ecliptic is [the same as] its first position and its required time is the known time in the course of which it has moved from the first position to the second position.

---

[82] The motion of the node is extremely slow, so the great circle CG cuts the hour circle AD in a point that is identified with D. The arc RG obtained in the case of the moon is here effectively zero, $G \cong R$, so G is identified with a point of the circle AD. Consequently, G and K are also identified. The known time, which was AK in the case of the moon, here becomes AG.

[83] See preceding note.

It is clear from what we have proved that if each of the seven wandering stars moves in the course of a known time beginning at a known instant, then its required time is known and the inclination of its motion with respect to the hour circle it was on at the first instant is known. That is what we wanted to prove.

The first case of figure is that for the sun, the second is that for the moon, the third is that for Venus and Mercury and the fourth is that for Saturn, Jupiter and Mars.

<22> I also say that the maximum inclination[84] of the inclined orb for each of the seven wandering stars with respect to the circle of the equator is known for any known instant and the position in which this inclination in relation to the circle of the ecliptic is a maximum is known.

The inclined orb of the sun is in fact the circle of the ecliptic and the maximum inclination of the circle of the ecliptic with respect to the circle of the equator is known; it has the magnitude indicated by Ptolemy. The magnitude of that inclination is fixed, always the same, and does not vary; the position of the northern limit is the summer solstice which is the first point of Cancer and the position of the southern limit is the winter solstice which is the first point of Capricorn.

For the moon, its inclined orb cuts the circle of the ecliptic and cuts the circle of the equator, because any great circle on a sphere, cuts any great circle of the sphere and cuts it into two equal parts. If the inclined orb cuts the circle of the ecliptic and cuts the circle of the equator, then it is inclined with respect to the circle of the ecliptic and with respect to the circle of the equator. As for the relation of the inclined orb to the circle of the ecliptic, Ptolemy found its magnitude [of the inclination to the ecliptic] and proved that this magnitude is unchanging and fixed in the same state. However, the position of the maximum inclination of this orb, that is, the orb of the moon, with respect to the circle of the ecliptic, is variable; in fact the entire plane of this inclined orb moves as a whole about the poles of the circle of the ecliptic, and each point of the circumference of this inclined orb changes its position with respect to the circle of the ecliptic; the point that <corresponds to> the northern limit with respect to the circle of the ecliptic accordingly moves and its position with respect to the circle of the ecliptic varies; similarly the point that <corresponds to> the southern limit, and similarly the two points of the nodes; and this motion is in the sense opposite to that of the succession of the signs of the Zodiac, as Ptolemy has

---

[84] That is, the inclination of the north or south limits of the orb with respect to the equator.

proved; it is called the motion of the nodes. The magnitude of the inclination of this orb with respect to the circle of the ecliptic is not changed by
this motion, because this motion is about the poles of the circle of the
ecliptic; so the distance between the pole of this circle and the pole of the
circle of the ecliptic does not vary and this distance is the maximum value
of the inclination of the orb of the moon with respect to the circle of the
ecliptic. If the plane of this orb moves about the poles of the circle of the
ecliptic, then the inclination of this orb with respect to the circle of the
equator varies. In fact, if this entire circle moves and its position changes
with respect to the circle of the ecliptic, then the two poles of this circle
move and rotate about the poles of the circle of the ecliptic and they coincide once with <points on> the circumference of the circle that passes
through the two poles of the circle of the ecliptic and through the two poles
of the circle of the equator, and that is called the circle of the poles; they
move away from them once. The coincidence of the two poles of the inclined orb <with points> of the circle of the poles happens twice for each
revolution. We have mentioned this result earlier and if we have repeated it
here, that is in order to point out the magnitude of the inclination. The
magnitude of the inclination of this orb with respect to the circle of the
ecliptic is smaller than the magnitude of the inclination of the circle of the
ecliptic with respect to the circle of the equator. But the magnitude of the
inclination, one to another, of two great circles that intersect on a sphere is
equal to the magnitude of the arc between their poles, which forms part of
the great circle that passes through their four poles. The magnitude of the
arc between the two poles of the circle of the ecliptic and the pole of the
inclined orb of the moon is smaller than the magnitude of the arc between
the pole of the circle of the ecliptic and the pole of the circle of the equator.
If the pole of the inclined orb moves about the two poles of the circle of the
ecliptic, and if it coincides twice with <a point> of the circle of the poles,
then it will, once, be further from the pole of the equator <than from the
pole of the ecliptic> and, once, closer to the pole of the equator <than to the
pole of the ecliptic>. In fact, if the pole of the inclined orb coincides with
<a point> of the circle of the poles, it will, once, be between the pole of the
circle of the ecliptic and the pole of the circle of the equator, and the pole
of the circle of the ecliptic will, once, be between the pole of the circle of
the equator and the pole of the inclined orb. If the pole of the inclined orb
coincides <with a point> of the circle of the poles and if the three poles
come to lie on a circle, then the maximum inclination of the inclined orb
with respect to the circle of the ecliptic is an arc of this circle and the point
where we find the maximum inclination of the inclined orb with respect to
the circle of the ecliptic is the point where we find the maximum

inclination of the inclined orb with respect to the circle of the equator. But we have seen earlier that the magnitude of the maximum inclination between the two circles is the arc between the two poles. So if the pole of the inclined orb coincides <with a point> of the circle of the poles and if it is between the pole of the circle of the ecliptic and the pole of the circle of the equator, then the magnitude of the maximum inclination of the inclined orb with respect to the circle of the equator is the magnitude of the inclination of the circle of the ecliptic with respect to the circle of the equator, minus the magnitude of the inclination of the inclined orb with respect to the circle of the ecliptic. If the pole of the inclined orb coincides <with a point> of the circle of the poles and if the pole of the circle of the ecliptic is between the pole of the circle of the equator and the pole of the inclined orb, then the magnitude of the maximum inclination of the inclined orb with respect to the circle of the equator is the magnitude of the inclination of the circle of the ecliptic with respect to the circle of the equator, plus the magnitude of the inclination of the inclined orb with respect to the circle of the ecliptic. If the pole of the inclined orb coincides with <a point of the> circle of the poles, the position of the maximum inclination of the inclined orb with respect to the circle of the equator, in regard to the circle of the ecliptic, is the point of a solstice, because the circle that passes through the limiting point for the inclination of the inclined orb, in this case, passes through the pole of the circle of the ecliptic; so it is the one that gives the limit for the position of the maximum inclination in regard to the circle of the ecliptic. But the positions on the circle of the ecliptic through which the circle of the poles passes at the instant when the pole of the inclined orb coincides <with a point> of the circle of the poles, are the two points of the solstices. But since the positions of the two nodes which are the head and tail[85] of the orb of the moon on the circle of the ecliptic are known, the position of the northern limit of the inclined orb in regard to the circle of the ecliptic is known and the position of the southern limit of the orb in regard to the circle of the ecliptic is known. By the northern limit and the southern limit I mean, here, the point of the inclined orb that is furthest away from the circle of the ecliptic. But if the positions of the northern and southern limits of the inclined orb in regard to the circle of the ecliptic are the two points of the solstices, the magnitude of the inclination of the inclined orb with respect to the circle of the equator will be known; in fact, if the northern limit of the inclined orb in regard to the circle of the ecliptic is at the first point of Cancer, the pole of the inclined orb will be further away from the pole of the circle of the

---

[85] Head: ascending node; tail: descending node. See Mathematical commentary, p. 192.

equator than from the pole of the circle of the ecliptic; in this case the three poles all lie on a circle, which is the circle of the poles. The pole of the circle of the ecliptic will thus lie between the pole of the circle of the equator and the pole of the inclined orb. The magnitude of the inclination of the inclined orb with respect to the circle of the equator is the magnitude of the inclination of the circle of the ecliptic with respect to the circle of the equator, plus the inclination of the inclined orb with respect to the circle of the ecliptic. If the southern limit in regard to the circle of the ecliptic is at the first point of Cancer, the pole of the circle of the ecliptic is further away from the pole of the equator than from the pole of the inclined orb; so the pole of the inclined orb will lie between the pole of the circle of the equator and the pole of the circle of the ecliptic. The magnitude of the inclination of the inclined orb with respect to the circle of the equator is the same as the magnitude of the inclination of the circle of the ecliptic with respect to the circle of the equator, from which we subtract the inclination of the inclined orb with respect to the circle of the ecliptic.

If the northern limit of the inclined orb with respect to the circle of the ecliptic is at the first point of Cancer, the ascending node will be at the first point of Aries because the first point of Aries is the pole of the circle of the poles. So if the point of <maximum> inclination of the inclined orb with respect to the circle of the ecliptic, which is the northern limit, lies on the circle of the poles, then the point of the ascending node is the first point of Aries. If the southern limit of the inclined orb with respect to the circle of the ecliptic is at the first point of Cancer, the descending node is at the first point of Aries. So it is clear from this that if the ascending node is at the first point of Aries, then the magnitude of the inclination of the inclined orb of the moon with respect to the circle of the equator is the same as the magnitude of the inclination of the circle of the ecliptic with respect to the circle of the equator, plus the inclination of the inclined orb with respect to the circle of the ecliptic. But if the descending node is at the first point of Aries, then the inclination of the inclined orb of the moon with respect to the circle of the equator is the same as the magnitude of the inclination of the circle of the ecliptic with respect to the circle of the equator, from which we subtract the inclination of the inclined orb with respect to the circle of the ecliptic. But if the descending node is at the first point of Aries, then the ascending node will be at the first point of Libra.

It is clear from what we have proved that the magnitude of the inclination of the inclined orb of the moon with respect to the circle of the equator, at the two moments at which the ascending node is at the two points of the equinoxes, is known and that the two positions of the northern limit and of the southern limit of the inclined orb with respect to the circle of the equator at these two moments are known.

It remains to prove that the magnitude of this inclination is known and that the position of the limit of this inclination is known, if the ascending node is not at one of the two equinoctial points.

Let *ABC* be the circle of the ecliptic, *ADC* the inclined orb and let each of the arcs *CB* and *CD* be a quarter of a circle. We cause a great circle to pass through the points *B* and *D*; let the circle be *KBDE*. The arc *BD* will then be the limit of the inclination of the inclined orb with respect to the circle of the ecliptic, because the magnitude of this inclination does not change and the pole of the circle of the ecliptic and the pole of the inclined orb are on the circle *KBE*; let the point *N* be the pole of the circle of the ecliptic and the point *E* the pole of the inclined orb. The point *D* is the most northerly or southerly position with respect to the circle of the ecliptic referred to the circle of the ecliptic, and the two points *A* and *C* are the two nodes. So the point *B* of the circle of the ecliptic is known, because its distance from the point of the node is a quarter of a circle, and the position of the node is known. But if the positions of the two nodes are not the two points of the equinoxes, then the two points *A* and *C* are not the two points of the equinoxes, while still being known. Let the two points of the equinoxes be the two points *L* and *M*.

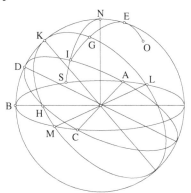

Fig. I.22.1

<a> Let the point *M* lie on the arc *CB*. We construct the circle of the equator through *L* and *M*; let it be the circle *LKMH*; let the point *K* lie on the circle *DB* and the point *H* lie on the circumference of the inclined orb. Thus the point *H* will be somewhere other than at the points *C* or *A*, because each of the arcs *CM* and *MA* is smaller than a semicircle. But since the point *B* of the circle of the ecliptic is not the point of a solstice and the point *N* is the pole of the circle of the ecliptic, the circle *KBNE* is not the

circle of the poles; so the pole of the circle of the equator does not lie on the circle *KBNE*. Let the pole of the equator be the point *O*. We cause a great circle to pass through the points *E* and *O*; let it cut the inclined orb at the point *I* and let it cut the circle of the equator at the point *G*. So the arc *IG* is the limit of the inclination of the inclined orb with respect to the circle of the equator, because the circle *EG* passes through the pole of the inclined orb and through the pole of the circle of the equator and is such that the arc *GI* is equal to the arc *EO* and the arc *EI* is a quarter of a circle: the point *E* is in fact the pole of the inclined orb. So each of the arcs *HG* and *HI* is a quarter of a circle. Similarly, since the point *B* of the circle of the ecliptic is known, because it is the position of the northern or southern limit of the inclined orb, <referred to the ecliptic>,[86] and the point *M* is the point of the equinox, then the arc *MB* of the circle of the ecliptic is known. The arc *MC* is known because the arc *CB* is a quarter of a circle. But since the circle *KBE* passes through the pole of the circle of the ecliptic, it will be perpendicular to the circle of the ecliptic; but since the circle *EBK* is perpendicular to the circle *ABC* and the arc *MB* is known, then if we insert the arc *MB* into the table of right ascensions, and if we take the appropriate part of the circle of the ecliptic,[87] that will be equal to the arc *MK*; so the arc *MK* is known; if we insert the known arc *MK* into the table of the inclination and we take the appropriate part of the inclination, that will be equal to the arc *KB*; so the arc *KB* is known, each of the arcs *MK* and *KB* is known and the arc *BD* is known because it is the limit of the inclination[88] of the inclined orb with respect to the circle of the ecliptic; so the arc *KD* is known. But since, between the two arcs *KD* and *CD*, two arcs *HK* and *CB* cut one another at the point *M*, accordingly the ratio of the sine of the arc *KD* to the sine of the arc *DB*, a known ratio, is compounded of the ratio of the sine of the arc *KH* to the sine of the arc *HM*, a known ratio,[89] and the ratio of the sine of the arc *CM* to the sine of the arc *CB*. But the arc *KM* is known, so the arc *HM* is known; the complete arc *KH* is known and the arc *HG* is a quarter of a circle; so the arc *KG* is known. Similarly, we have proved that the arc *KD* is known and that the arc *DE* is a quarter of a circle. But since, between the two arcs *EG* and *HG*, two arcs *EK* and *HI* cut one

[86] See Supplementary note [6].

[87] Since *N* is taken as the pole, the arc *MB* of the ecliptic is the right ascension of the arc *MK* of the equator and the arc *KB* is the inclination of the arc *MK* with respect to the ecliptic (see Propositions 7 and 5).

[88] Here, the inclination of one of the extremities of the orb; so it is the arc of a great circle that is the measure of the dihedral angle between the plane of the orb and the plane of the ecliptic, that is, the ecliptic latitude of the limit concerned.

[89] See Mathematical commentary for an argument in support of this.

another at the point *D*, the ratio of the sine of the arc *EI* to the sine of the arc *IG* is compounded of the ratio of the sine of the arc *ED* to the sine of the arc *DK*, a known ratio, and the ratio of the sine of the arc *KH* to the sine of the arc *HG*, a known ratio; thus the ratio of the sine of the arc *EI* to the sine of the arc *IG* is known. But the arc *EI* is a quarter of a circle, so the arc *IG* is known and it is the limit of the inclination of the inclined orb with respect to the circle of the equator. The point *I* is the northern or southern limit of the inclined orb with respect to the equator: if the point *O* is the north pole, the point *I* will be the northern limit and if the point *O* is the south pole, then the point *I* will be the southern limit.

Similarly, we cause a great circle to pass through the point *N*, which is the pole of the circle of the ecliptic, and through the point *I*, which is the northern or southern limit of the inclined orb with respect to the circle of the equator; let the circle be *NIS*. Let this circle cut the circle of the ecliptic at the point *S*. Since the ratio of the sine of the arc *KB* to the sine of the arc *BD*, a known ratio, is compounded of the ratio of the sine of the arc *KM* to the sine of the arc *MH*, a known ratio, and the ratio of the sine of the arc *HC* to the sine of the arc *CD*; the ratio of the sine of the arc *HC* to the sine of the arc *CD* is thus known. But the arc *CD* is a quarter of a circle, so the arc *HC* is known. But since the arc *CD* is a quarter of a circle and the arc *HI* is a quarter of a circle, the arc *DI* will be equal to the arc *CH*; so the arc *DI* is known. But since the arc *DI* is equal to the arc *HC*, the arc *DI* will be smaller than a quarter circle; so the point *I* lies between the two points *A* and *D* and the arc *AI* is smaller than a quarter of a circle. But since the circle *NI* is a great circle, and it cuts the circle *AIC* at the point *I*, it accordingly cuts it in another point, between the two points *A* and *C*, opposite the point *I*. If the circle *NI* cuts the circle *AIC* in a point opposite the point *I*, then it cuts the arc *AB* in a point between the two points *A* and *B*; so the point *S* lies between the two points *A* and *B* and the arc *CS* is smaller than a semicircle. Since, between the two arcs *NS* and *CS*, the two arcs *NB* and *CI* cut one another, the ratio of the sine of the arc *CS* to the sine of the arc *SB* is compounded of the ratio of the sine of the arc *CI* to the sine of the arc *ID*, a known ratio, and the ratio of the sine of the arc *DN* to the sine of the arc *NB*, a known ratio; so the ratio of the sine of the arc *CS* to the sine of the arc *SB* is known. But the arc *CB* is a quarter of a circle, so the arc *SB* is known and the point *B* on the circle of the ecliptic is known; so the point *S* on the circle of the ecliptic is known and it is the position[90] of the point *I*, which is the northern limit or the southern limit of the inclined orb. All this proof is concerned with the first case of figure.

---

[90] Referred to the ecliptic.

<b> If the point *M* that is the point of the equinox lies on the arc *AB*, then the proof is the same as the proof we have just given, except that the two arcs *EG* and *NI* are on the side towards the point *C*.

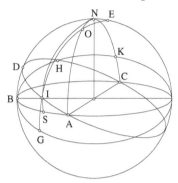

Fig. I.22.2

<c> If the point of the equinox is the point *B*, then we construct the circle of the equator to pass through the point *B* as in the second case of figure; let it be *BH*. We cause a great circle to pass through *E* and *O*, let it be *EOIG*; let there be the point *I* on the inclined orb and the point *G* on the circle of the equator, as in the first case of figure; thus the arc *IG* will be the limit of the inclination of the inclined orb with respect to the circle of the equator and the point *I* will be the northern or southern limit of the inclined orb with respect to the circle of the equator. We cause a great circle to pass through the two points *N* and *I*; let it cut the circle of the ecliptic at the point *S*, as in the first case of figure. The point *S* will then be the position of the northern or southern limit in regard to the circle of the ecliptic. If the point *B* is the point of the equinox, then the point *C* will be the point of the solstice. We cause a great circle to pass through the two points *N* and *O*; this circle is thus the circle of the poles and accordingly passes through the point *C*; let it cut the circle of the equator at the point *K*. So the arc *CK* is the inclination of the circle of the ecliptic with respect to the circle of the equator, so it is known. The arc *NC* is a quarter of a circle, there remains the known arc *KN* and the ratio of the sine of the arc *BD* to the sine of the arc *DN*, a known ratio, is compounded of the ratio of the sine of the arc *BH* to the sine of the arc *HK* and the ratio of the sine of the arc *KC* to the sine of the arc *CN*, a known ratio; so the ratio of the sine of the arc *BH* to the sine of the arc *HK* is known. But the arc *BK* is a quarter of a circle; so the arc *HK* is known. But the arc *HB* is known and the arc *HG* is a quarter of a circle, because the point *H* is the pole of the circle *GOE*; so the arc *BG* is equal to the arc *HK*, which is known. Similarly, the ratio of the sine of the

arc *EI* to the sine of the arc *IG* is compounded of the ratio of the sine of the arc *ED* to the sine of the arc *DB*, a known ratio, and the ratio of the sine of the arc *BH* to the sine of the arc *HG*, a known ratio. So the ratio of the sine of the arc *EI* to the sine of the arc *IG* is known; but the arc *EI* is known, so the arc *IG* is known and it is the limit of the inclination of the inclined orb with respect to the circle of the equator. Similarly, it being given that the two arcs *BN* and *ND* are known and that the two arcs *BK* and *KH* are known, the two arcs *CH* and *HD* are known from the compounded ratio. So the arc *DI* is known because it is equal to the arc *HC*.[91] Similarly, since the arcs *CI* and *ID* are known and the two arcs *DN* and *NB* are known, the ratio of the sine of the arc *CS* to the sine of the arc *SB* is known. But the arc *CB* is a quarter of a circle; so the arc *BS* is known and the point *B* is known because it is the point of the equinox; so the point *S* on the circle of the ecliptic is known and it is the position, referred to the circle of the ecliptic, of the point *I* which is the northern or southern limit.

From all that we have explained in this section, it is clear that the position of the northern or southern limit of the inclined orb of the moon with respect to the circle of the equator, referred to the circle of the ecliptic, is known for any known instant and that the magnitude of the inclination of this orb with respect to the equator, for any known instant, is known.

By means of the same proof, we find the position of the northern or southern limit of the inclined orb of each of the three superior planets with respect to the circle of the equator, referred to the circle of the ecliptic, and the magnitude of the inclination of each of these orbs with respect to the circle of the equator; that is what we wanted to prove in this section.

<23> The inclination of the orbs of Venus and Mercury varies with respect to the circle of the equator as a consequence of the inclination of these two orbs with respect to the circle of the ecliptic. However the magnitude of the inclination of each of these two orbs with respect to the circle of the equator, at a known instant, is known and the positions of the northern limit and the southern limit in relation to the circle of the equator, referred to the circle of the ecliptic, are known. In fact, even though the inclination of the inclined orb of each of these two planets with respect to the circle of the ecliptic varies, its variation is nevertheless known; thus its magnitude at any known instant is known and the position of the limit in relation to the circle of the ecliptic is known. As for the position of the limit in relation to the circle of the ecliptic, its distance from the position of the ascending node is always a quarter of the circle of the ecliptic. But the position of the

---

[91] See Mathematical commentary.

ascending node on the circle of the ecliptic at each known instant is known; the positions of the northern and southern limits, referred to the circle of the ecliptic, for each of the planets Venus and Mercury, at any known moment, are known. As for the magnitude of the inclination of the inclined orb with respect to the circle of the ecliptic, it is when the centre of the epicycle of each of these two planets is at its apogee or at its perigee on the eccentric that the inclination of the inclined orb with respect to the circle of the ecliptic is at its maximum. Now its maximum inclination with respect to the circle of the ecliptic is of known magnitude; Ptolemy in fact demonstrated it and found its magnitude.

If, on the other hand, the centre of the epicycle is at one of the two points of intersection, whatever the position of these two points may be, then, in this case, the inclined orb has no inclination with respect to the circle of the ecliptic since, in this case, the inclined orb will be coincident with the circle of the ecliptic, as Ptolemy explained. And if the centre of the epicycle is between the apogee and the point of intersection, then the inclination of the inclined orb with respect to the circle of the ecliptic will be smaller than the maximum inclination and its ratio to the maximum inclination will be equal to the ratio of the arc of the inclined orb, which lies between the centre of the epicycle and the point of intersection, to a quarter of a circle; and that, indeed, because the inclined orb moves from its maximum inclination until it coincides with the circle of the ecliptic in the time during which the centre of the epicycle, by the motion in longitude, traverses a quarter of a circle.[92] The position of the planet in longitude, which is the mean position, is known at any known instant, its distance from the position of the apogee, on the eccentric, is known and that distance is referred to the centre of the Universe. The distance of the centre of the epicycle, which is the mean position[93] of the star starting from the point of intersection which is the point of the ascending node – by ascending node I mean the ascending node which shares in the motion in longitude[94] – is in this case of known magnitude at any known instant. So the ratio of this distance to a quarter of the circle will be a known ratio; this ratio is the ratio of the magnitude of the inclination of the inclined orb with respect to the circle of the ecliptic at this known moment to its maximum inclination with respect to the circle of the ecliptic, which is a known magnitude. Thus the magnitude of the inclination of the inclined orb for the planets Venus and Mercury with respect to the circle of the ecliptic at any known moment, is known. But if the magnitude of the inclination of the

[92] See Supplementary note [8].

[93] That allows us to find the mean position.

[94] See Mathematical commentary, pp. 201ff.

inclined orb with respect to the circle of the ecliptic is known and if the position of the maximum inclination in relation to the circle of the ecliptic, referred to the circle of the ecliptic, is known, then the magnitude of the inclination of the inclined orb with respect to the circle of the equator is accordingly known and the position of the northern limit or the southern limit in relation to the circle of the equator and referred to the circle of the ecliptic will be determinate by the method we indicated earlier for the orb of the moon.

If, on the other hand, the two nodes[95] are at the two points of the equinoxes, then the positions[96] of the northern limit and the southern limit of the inclined orb in relation to the circle of the equator are the two points of the solstices, as we have proved for the orb of the moon.

As for the magnitude of the inclination of the inclined orb with respect to the circle of the equator: if the position of the northern limit is the first point of Cancer, then the magnitude of the inclination is the magnitude of the inclination of the circle of the ecliptic with respect to the equator, to which we add the inclination of the inclined orb with respect to the circle of the ecliptic at the known instant; and if the position of the southern limit is the first point of Cancer, then the magnitude of the inclination is the magnitude of the inclination of the circle of the ecliptic with respect to the circle of the equator, from which we subtract the magnitude of the inclination of the inclined orb with respect to the circle of the ecliptic at that known instant.

If the positions of the two nodes are not the two points of the equinoxes, then the magnitude of the inclination as well as the two positions of the northern and southern limits are found by the same method as that we described, for the orb of the moon.

As for the two points of intersection of the inclined orb with the circle of the equator, they move about the points of the two nodes, that is why the two points of intersection vary. Similarly, any point on the circumference of the inclined orb moves about the points of the two nodes. In fact, if the inclined orb moves until it is coincident with the circle of the ecliptic, to [then] separate itself from the circle of the ecliptic and to return to it, and if the two points of intersection do not move with this motion, then this motion takes place about the two points of intersection, which are the nodes, and these two points are the poles of this motion. And if this motion is about these two poles, then any point of the inclined orb moves with this motion along a circle whose two poles are the points of the two nodes. So if

---

[95] The ascending and the descending nodes.

[96] Understood: referred to the ecliptic.

the planet Venus or the planet Mercury move for a known time with a
<diurnal> temporal motion, then any point on the circumference of the
inclined orb moves along a circle whose poles are the points of the two
nodes. But any point of the quarter of the inclined orb that is between the
point of the head and the apogee on the eccentric – by point of the head I
mean the point after which the centre of the epicycle moves upwards
towards the apogee – and any point of the quarter opposite this quarter,
apart from the points of the limits – which are the two points of the nodes
and the two points of the northern and southern limits – moves according to
the succession of the signs of the Zodiac, if the motion of the apogee on the
eccentric is, for the planet Venus, from the north towards the circle of the
ecliptic and, for the planet Mercury, from the south towards the circle of
the ecliptic. If the inclined orb then moves and leaves the circle of the
ecliptic by going in the other direction, then for the planet Venus the
apogee moves from the circle of the ecliptic towards the south; and for the
planet Mercury it moves from the circle of the ecliptic towards the north;
and in this case, any point of the two quarters we mentioned before moves
in the sense opposite to the succession of the signs of the Zodiac. As for the
two remaining quarters, any point on each of them has a motion contrary to
the motion on the first two quarters. If, on the other hand, the motion of the
apogee for the planet Venus takes place from the north towards the circle of
the ecliptic and for the planet Mercury from the south towards the circle of
the ecliptic, then any point on these last two quarters has a motion opposite
to the succession of the signs of the Zodiac. If the inclined orb later leaves
the plane of the circle of the ecliptic by moving in the other direction, then
any point on these two quarters moves according to the succession of the
signs of the Zodiac. If the inclined orb later moves so as to return to the
circle of the ecliptic and from the circle of the ecliptic to the first limit, then
the motions of the points are in the opposite sense; those of them that move
according to the succession of the signs of the Zodiac thus in this case
move contrary to the succession of the signs of the Zodiac; and those of
them that were moving contrary to the succession of the signs of the Zodiac
in this case move in accordance with the succession of the signs of the
Zodiac.

Let us show all this in a proof. Let the circle of the ecliptic be the circle
ABCD and the inclined orb *AECG*. Let the point *E* be the northern limit of
the planet Venus or the southern limit of the planet Mercury. Let the point
G be the southern limit of the planet Venus or the northern limit of the
planet Mercury. Let the succession of the signs of the Zodiac be from the
point *A* towards the point *B* and beyond *B*; so the point *A* will be the point

of the head and the point C the point of the tail,[97] because the apogee of the
planet Venus is always at the northern limit and that of the planet Mercury
is always at the southern limit. We cause a great circle to pass through the
points E and G and [also] through the pole of the circle of the ecliptic; let
the point H be the pole of the circle of the ecliptic. We take any point on
the arc AE; let the point be I. Similarly, we take any point on the arc CG;
let it be the point I'.[98] We cause a great circle to pass through the points I
and H; let the circle be HLI; let this circle cut the circle of the ecliptic at the
point L. This circle is accordingly at right angles to the circle of the
ecliptic; the angle ALI is a right angle and the angle $CL'I'$ is also a right
angle, so the arc AI is greater than the arc AL and the arc CI', on the
opposite side, is also greater than the arc CL'. We take the point A as pole
and with distance AI we draw a circle. It cuts the arc AB in a point between
the points L and B; let it cut it at the point K. It also cuts the arc HL in a
point between the points H and L, because the arc AL is perpendicular to
the arc HI. But the arc AL is smaller than a quarter of a circle; let the circle
IK cut the arc HL at the point R. Similarly, if we take the point C as pole
and with distance CI' we draw a circle, it cuts the arc CD in a point
between the points C and D – let it cut it at the point K' – and it cuts the arc
HL' in a point that lies beyond the point L' – let it cut it at the point R'. We
cause a great circle to pass through the points H and K; let it be HK.

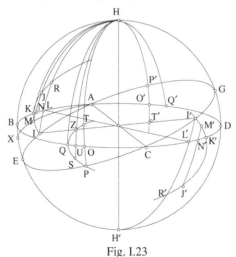

Fig. I.23

Since the point H is the pole of the circle of the ecliptic, each of the
arcs HB and HK is a quarter of a circle; and since the point H is the pole of

---

[97] That is, the ascending node and the descending node.
[98] See note 23 in the Mathematical commentary, p. 198.

the circle *BK* and the point *H* is on each of the circles *HB* and *HK*, the poles of the circles *HB* and *HK* lie on the circle *BK*. But the point *A*, which is the pole of the circle *HB*, is the pole of the circle *IKR*; so the circle *BK* passes through the pole of the circle *IK*. But since the circles *HK* and *IKR* meet one another at the point *K*, a great circle, which is the circle *BKA*, passes through the point *K* and the poles of the circles *HK*, *IKR* lie on the circle *BK*, [accordingly] the circle *HK* touches the circle *IKR* at the point *K*. But if the circle *HK* touches the circle *IKR* at the point *K*, then any circle drawn from the point *H* to a point of the arc *IK* between the points *I* and *K*, or to a point of the arc *KR* between the points *K* and *R*, cuts the arc *KL*. On the arc *IK* let us mark an arbitrary point; let the point be *M*. We cause a great circle to pass through this point and through the point *H*; let the circle be *HM*; this circle cuts the arc *KL* and cuts the arc *KR*; let it cut the arc *KL* at the point *N* and let it cut the arc *KR* at the point *J*. The point *N*, which lies on the circle of the ecliptic, is the position of the point *M* <referred to the ecliptic> and the point *L* is the position of the point *I*. So if the inclined orb is at its maximum inclination, then the position of the point *I* is the position which it occupies[99] and the position of the point *I*, referred to the circle of the ecliptic, is the point *L*. If the inclined orb next moves towards the circle of the ecliptic, the point *I* moves along the circle *IKR*, because this motion takes place about the pole *A*. If the point *I* reaches the point *M*, the position of the point *I*, referred to the circle of the ecliptic, will be the point *N*. But the point *N* is further away from the point *A* than the point *L* and the succession of the signs of the Zodiac is from the point *A* towards the point *B*, and beyond *B*. So, in this case, the motion of the point *I* is in accordance with the succession of the signs of the Zodiac. Similarly, if the point *I* travels from the point *M* to the point *K*, its position[100] will be the point *K* after having been the point *N*. The motion of the position of the point *I* until it reaches the point *K* is in accordance with the succession of the signs of the Zodiac. If the inclined orb next moves in the other direction <with respect > to the circle of the ecliptic, the point *I* moves from the point *K* in the direction of the point *R*; then the point *I* reaches the point *J*; if it reaches the point *J*, its position will be the point *N*. But given that, when it was at the point *K*, its position was the point *K*, accordingly it will have returned from the point *K* to the point *N* and thus its motion will be contrary to the succession of the signs of the Zodiac. Similarly, if it reaches the point *R*, its position will be the point *L*, and if it reaches the point *R*, it will have

---

[99] The initial position of the point *I* is a point of the arc *AE* of the inclined orb when the orb has its maximum inclination.

[100] The term position must often be taken to mean 'position referred to the ecliptic'.

attained its maximum inclination, because the arc *RL* is equal to the arc *LI*. In fact, the arc of the great circle drawn from the point *A* to the point *R* is equal to the arc *AI* since the point *A* is the pole of the circle *IKR* and the arc *AI* makes right angles with the arc *RI*; so the arc *RL* is equal to the arc *LI*. If the inclined orb later moves so as to return to the circle of the ecliptic, the point *I* moves from the point *R* towards the point *K*; thus its position moves from the point *L* to the point *N* and later to the point *K*. So its position moves in accordance with the succession of the signs of the Zodiac. If the inclined orb later moves so as to return in the direction of the point *E*, then the point *I* moves from the point *K* to the point *I*; so its position moves from the point *K* to the point *L*. The motion of its position is in the sense contrary to the succession of the signs of the Zodiac.

Similarly, we prove that, for the arc *I'K'R'* cut at the point *J'*, the circle *H'I'R'* touches the circle *I'K'R'*; thus, if the point *I'* moves along the arc *I'K'* from the point *I'* to the point *K'*, its position moves from the point *L'* to the point *K'*; so its motion will be in accordance with the succession of the signs of the Zodiac. If the point *I'* moves along the arc *K'R'*, its position moves from the point *K'* to the point *L'*, so its motion will be in the sense contrary to the succession of the signs of the Zodiac. If the inclined orb moves so as to return to the circle of the ecliptic, the point *I'* moves from the point *R'* to the point *K'*; the motion of its position will be in accordance with the succession of the signs of the Zodiac. If the point *I'* moves from the point *K'* to the point *I'*, the motion of its position will be from the point *K'* to the point *L'*, so its motion will be in the sense contrary to the succession of the signs of the Zodiac.

Similarly, we take an arbitrary point on each of the arcs *EC* and *GA*; let it be the point *P*.[101] We cause a great circle to pass through this point and through the point *H*, let it be *HP*; let this circle cut the circle of the ecliptic at the point *O*. So this circle will be set at right angles to the circle of the ecliptic; so the angle *COP* is a right angle and the arc *CP* is greater than the arc *CO*. Similarly, the angle *AO'P'*, on the opposite side, is a right angle and the arc *AP'* is greater than the arc *AO'*. We take the point *C* as pole and with distance *CP* we draw a circle. This circle cuts the arc *CB* in a point between the points *O* and *B*; let it cut it at the point *Q*. But this circle cuts the circle *HO* in a point between the points *H* and *O*; let it cut it at the point *T*. Similarly, we take the point *A* as pole and with distance *AP'* we draw a circle. This circle cuts the arc *O'D* in a point between the points *O'* and *D* and cuts the circle *HO'*; let it cut the arc *O'D* at the point *Q'* and the arc *HO'* at the point *T'*. We cause a great circle to pass through the points *H*

---

[101] We take *P* on *EC* and *P'* on *GA*.

and Q; let the circle be HQ. We prove, as we proved earlier, that the circle HQ touches the circle PQT. Any circle drawn from the point H to a point of the arc PQ or to a point of the arc QT cuts the arc QO of the circle of the ecliptic. We take an arbitrary point on the arc PQ; let it be the point S. We cause a great circle to pass through it and through the point H; let the circle be HS. This circle cuts the arc QO, it cuts the arc QP and the arc QT; let it cut the arc QO at the point U and the arc QP at the point S and let it cut the arc QT at the point Z. If the inclined orb moves towards the circle of the ecliptic, the point P moves along the arc PQ from the point P towards the point Q, so if the point P reaches the point S, the position of the point P, referred to the circle of the ecliptic, will be the point U. But the position of the point P when it was in its <initial> position, which is close to its maximum inclination,[102] is the point O of the circle of the ecliptic. The position of the point P was thus the point O and later travelled to the point U. But the point U is closer to the point A than the point O; so the motion of the position of the point P is, in this case, in the sense contrary to the succession of the signs of the Zodiac; and its condition will be the same until it reaches the point Q, its position on the circle of the ecliptic will then be the point Q. If the inclined orb then moves and leaves the circle of the ecliptic, in the other direction, the point P moves along the arc QT; it travels from the point Q towards the point Z; its position will then be the point U, after having been the point Q; in this case, the motion of its position takes place according to the succession of the signs of the Zodiac and the same will be true until it arrives at the point T, in this case its position will then be the point O. If the inclined orb next moves in the direction of the circle of the ecliptic, the point P moves along the arc TQ, from the point T to the point Q; if it arrives at the point Z, its position will be the point U, so its position moves from the point O to the point U, so its motion is in the sense contrary to the succession of the signs of the Zodiac and the same will be true until it arrives at the point Q; and if it arrives at the point Q, the inclined orb will then coincide with the circle of the ecliptic. If the inclined orb later leaves the circle of the ecliptic in the direction of the point E, the point P moves from the point Q towards the point P. If it reaches the point S, its position arrives at the point U; now its position was the point Q, so in this case the motion of its position takes place in the sense of the succession of the signs of the Zodiac, and the same will be true until it returns to the point P. It is in this way that the motion of the point P' takes place in the opposite direction, that is <when it is> on the arc AG.

---

[102] The initial position of the point P is a point of the inclined orb when this orb has its maximum inclination.

I also say that the motion of any point on the circumference of the inclined orb about the two points of the nodes will be of known magnitude in a known time on parallel circles, whose two poles will be the two points of the nodes; and the motion of its position on the circle of the ecliptic in the known time will be known.

Proof: The motion of the inclined orb from its maximum inclination until it coincides with the circle of the ecliptic takes place in a known time because it occurs in the time during which the centre of the epicycle traverses a quarter of a circle on the inclined orb.[103] If the centre of the epicycle moves from the apogee of the eccentric in the direction of the descending node, the inclined orb moves in the direction of the circle of the ecliptic. If the centre of the epicycle traverses a part of the eccentric, the point E will have traversed a part of the arc EB whose ratio to the arc EB is equal to the ratio of the part traversed by the centre of the epicycle on the eccentric to the arc of the eccentric intercepted by a right angle which has its vertex at the centre of the Universe. But this arc is known. It follows necessarily that if the position of the centre of the epicycle on the circumference of the inclined orb[104] is known, the position of any point of the circumference of the inclined orb on the circle on which this point is moving, a circle homologous to the circle IKR, is known. If this is so, for any known time with known starting time, any point of the circumference of the inclined orb will have traversed a known arc on its circle during this time. So if in a known time the point I traverses the arc IM, the arc IM will be known. If the point I later moves during a known time, once it arrives at the point M, the part it has traversed on the circle IKR will then be a known arc.

I also say that, during a known time whose starting time is known, the position of the point I traverses an arc of the circle of the ecliptic of known magnitude.

Proof: We cause an arc of a great circle to pass through the points A and M; let it be AMX. If the point I of the arc AE is at the point I of the circle IK at a given moment that is the starting time of the known time, the arc EB, which is the inclination of the inclined orb to the circle of the ecliptic, is known.

---

[103] The centre of the epicycle traverses an arc on the eccentric that corresponds to a quarter of a circle on the inclined orb. So what we are concerned with here is the apparent position on the inclined orb. Ibn al-Haytham sometimes speaks of the true position and sometimes of the apparent position.

[104] See previous note.

If, at that known instant, the centre of the epicycle is at the apogee of the eccentric, the arc *EB* is then the maximum inclination; so it is known. But if it is not at the apogee, it is at a point whose distance from the apogee is known; so its distance from the point of intersection is a known arc, because it is the remainder of the arc intercepted by a right angle whose vertex is the centre of the Universe. The ratio of this remainder to the arc of the eccentric intercepted by a right angle whose vertex is the centre of the Universe is a known ratio; so the ratio of the arc *EB* to the maximum inclination is known and the arc *EB* is known. If the point *I* later moves during a known time so as to traverse the arc *IM*, then the point *E* will traverse the arc *EX*; so the arc *EX* will be known and what remains is the known arc *XB*. But the arc *XE* is similar to the arc *MI*, because the two circles with the same pole, which is *A*, are parallel; so the arc *IM* is known and the arc *MK* is known; but the arc *AM* is equal to the arc *AI*, so the arc *AM* is known because the arc *AI* is known. If the point *I* of the inclined orb is known and the arc *EB* is known, then the point *L* of the circle of the ecliptic will be known. In fact, the ratio of the sine of the arc *HB* to the sine of the arc *BE* is compounded of the ratio of the sine of the arc *HL* to the sine of the arc *LI* and the ratio of the sine of the arc *IA* to the sine of the arc *AE*, a known ratio; so the ratio of the sine of the arc *HL* to the sine of the arc *LI* is known. But the arc *HL* is a quarter of a circle, so the arc *LI* is known. Similarly, the ratio of the sine of the arc *HE* to the sine of the arc *EB*, a known ratio, is compounded of the ratio of the sine of the arc *HI* to the sine of the arc *IL*, a known ratio, and the ratio of the sine of the arc *LA* to the sine of the arc *AB*; so the ratio of the sine of the arc *LA* to the sine of the arc *AB* is known. But the arc *AB* is a quarter of a circle, so the arc *AL* is known; but the point *A* is known, because it is the position of the ascending node, which is the starting point; so the point *L* is known and it is the position of the point *I* on the circle of the ecliptic at a known instant. Let the point *I* later move for a known time and let it traverse the arc *IM*; so the arc *IM* will be known and the arc *EX* will be known, as has been proved earlier; there remains the known arc *XB*. But the arc *AM* is known because it is equal to the arc *AI*; so the ratio of the sine of the arc *HB* to the sine of the arc *BX*, a known ratio, is compounded of the ratio of the sine of the arc *HN* to the sine of the arc *NM* and the ratio of the sine of the arc *MA* to the sine of the arc *AX*, a known ratio. So the ratio of the sine of the arc *HN* to the sine of the arc *NM* is known, and the arc *HN* is a quarter of a circle, so the arc *NM* is known. Similarly, the ratio of the sine of the arc *HX* to the sine of the arc *XB*, a known ratio, is compounded of the ratio of the sine of the arc *HM* to the sine of the arc *MN*, a known ratio, and the ratio of the sine of the arc *NA* to the sine of the arc *AB*; so the ratio of the sine of the arc *NA* to the sine of the arc *AB* is known; but the arc *AB* is a quarter of a circle, so the arc *AN* is known; then the point *A* of the circle of the ecliptic is known; so the point *N* is known and it is the position of the point *M*. So if it arrives at the point *M* in a known time, the position of the point *I* is known, and the arc *LN* traversed by the position of the point *I* in the known time is known, because each of the points *L* and *N* of the circle of the ecliptic is known.

It is clear from what we have proved that, in a known time whose starting time is a known instant, any point of the inclined orb traverses a known arc on the circle whose pole is the point of the ascending node, and that its position <referred to the ecliptic> traverses a [distance of] known magnitude on the circle of the ecliptic in a known time.

We have proved that the motions of the points on the circumference of the inclined orb can take place according to the succession of the signs of the Zodiac and can take place in the sense contrary to the succession of the signs of the Zodiac. We have proved when their motion takes place according to the succession of the signs of the Zodiac and when their motion takes place in the sense contrary to the succession of the signs of the Zodiac. That is what we wished to prove in this section, and these are the results that we promised to prove in Proposition 20.

<24> Similarly, we say that if each of the seven wandering stars moves in a known time such that its motion starts from the side of the northern limit [of its inclined orb], with respect to the circle of the equator, towards the side of the southern limit [of its inclined orb], and if its mean motion is from the apogee on the eccentric in the direction towards the perigee, and if its anomalistic motion is accelerated, that is, if its motion during a second interval of time is faster than in the first interval, then in this case there exists for this star a known ratio that is greater than any ratio of any required time for its motion between the two extremities of the interval during which it moves, to the part of the inclination of its motion proper to this required time throughout the time during which it moves.

The time during which each of the seven wandering stars moves in the way that we defined is a known time. For the sun, this occurs if it moves from the first point of Cancer, where it has its maximum northern inclination with respect to the circle of the equator, towards the perigee of its eccentric: if the sun moves on that arc, then it moves from north to south and in addition it moves from the apogee of its eccentric towards its perigee; so in equal times it traverses unequal arcs on the circle of the ecliptic, so that the second of any two arcs is always greater than the first. We prove this result, that is the inequality of these arcs, from Proposition 8 of this treatise. The configuration of the motion of the sun from the first point of Cancer to the perigee of its eccentric is the configuration that we have defined.

The moon, for its part, completes approximately one circuit on its inclined orb in the course of each month; in fact in the course of each month it moves from the northern limit of its inclined orb to the southern limit; I mean by the two limits the limit of the inclination of its inclined orb with respect to the equator, to the north and to the south. The apogee of its eccentric also completes a circuit each month and its motion is in the sense contrary to the succession of the signs of the Zodiac. It also moves, in the course of each month, from the southern limit that we mentioned earlier, to the northern limit. If the mean motion of the moon is from the apogee of the eccentric towards the perigee, then its motion on its inclined orb is always accelerated, that is on its inclined orb the moon traverses unequal arcs in equal times and the second [arc] is always greater than the first. In fact we have proved this result in Proposition 8.

If the correction made by the epicycle is additive, then the motion of the moon is accelerated and if the correction made by the epicycle is a subtraction, while at the same time being smaller than the increase brought about by the eccentric, then the motion of the moon is likewise accelerated. So if the mean motion of the moon is from the apogee of the eccentric towards the perigee, if the motion of the apogee is from the southern limit towards the northern limit with respect to the circle of the equator and if its motion is accelerated, then in this case the configuration of its motion is the configuration that we have defined. But we have proved that the positions of the northern limit and the southern limit of the inclined orb of the moon, referred to the equator, are known for every known instant; so, the position of the apogee, which moves, is also known at any known instant.

For the five planets, we have proved that the northern and southern limits of their inclined orbs are known in position, as measured from the circle of the ecliptic, at known instants, and the positions of the apogees of these planets, that is to say the apogees of their eccentrics, are known at any known instant. We have proved that the motions of these apogees are slow and are insensible in short intervals and that for the planets these apogees can, after a long time, become displaced from their positions, and that from north to south. So if the centre of the epicycle for each of the five planets is in motion on the eccentric, away from the apogee in the direction of the perigee – whose positions we have proved are known – if in addition the centre of the epicycle moves from north to south and if its motion is accelerated in the manner we have described for the motion of the moon, then in this case the configuration of the motion of the planet is the configuration that we have defined.

Proof of what we have stated: We take the inclined orb for each of the seven wandering stars to be the circle *ABC*, the circle of the equator the circle *GEC* and the north pole of the equator the point *D*. Let the point *C* be the point of intersection of the inclined orb and the circle of the equator. Let *A* be the point of the northern limit or a point of the arc between the northern limit and the point of intersection. Let the motion of the star be from the apogee of its eccentric towards the perigee; let it move from the point *A* towards the point *B* in a known time; let the time be *IA*; let it be the arc *IA* of the hour circle whose pole is the point *D*. Let its motion be accelerated, that is to say that what we obtain from its corrections is greater than the mean motion from the start of its motion in the known time until its end. In the known time *IA* let it traverse the arc *AB* on its inclined orb; let it traverse the arc *HB*, part of the arc *AB*, in the time *KH*. We cause great circles to pass through the point *D*, which is the north pole of the circle of the equator, and [similarly] through each of the points *A*, *H*, *B*; let the circles be *DAG*, *DHL* and *DBE*. Let the circle *DBE* cut the arcs *IA* and *KH* at the points *N* and *M*; so the arc *NE* is the inclination of the arc *CA* with respect to the circle of the equator, because it is equal to the arc *AG*; the arc *ME* is the inclination of the arc *CH* with respect to the circle of the equator, because it is equal to the arc *HL*, and the arc *BE* is the inclination of the arc *CB* with respect to the circle of the equator. So the arc *NB* is the excess in inclination of the arc *AB*[105] and the arc *MB* is the excess in inclination of the arc *HB*; so the arc *NB* is the one that we have called the inclination of the motion of the star in the time *IA*, and the arc *MB* is the inclination of the motion of the star in the time *KH*; and the arc *KM* is the required time for the motion of the star from the point *H* to the point *B*;[106] so it is one of the required times for the star that are between the two extremities of the time interval *IA*.

I say that the ratio of the known time *IA* to the arc *NB*, which is the inclination of the motion of the star in the known time – we have proved earlier that it is known – is greater than the ratio of the arc *KM*, which is the required time, to the arc *MB*, part of the arc *NB*, proper to the arc *KM*.

---

[105] That is to say the difference between the inclinations of *A* and of *B*.

[106] See note 25 on the required time *KM*, p. 204.

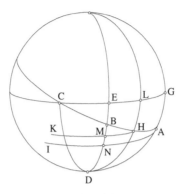

Fig. I.24

Proof: The motion of the star is from the apogee of the eccentric towards the perigee; so if it traverses equal parts in equal times on the eccentric, on the inclined orb it traverses unequal parts, the smallest of which is on the side towards the point *A* and the greatest on the side towards the point *B*, as we have proved in Proposition 8. The ratio of the arc traversed by the star in its mean motion on its eccentric in the time *IA*, which is the arc of the eccentric cut off by the two straight lines drawn from the centre of the inclined orb to the points *A* and *B*, to the arc traversed by the star in its mean motion on its eccentric in the time *KH*, which is the arc of the eccentric cut off by the two straight lines drawn from the centre of the inclined orb to the points *H* and *B*, is greater than the ratio of the arc *AB* to the arc *BH*, as has been proved in Proposition 9 of this treatise. But the ratio of any arc that the star traverses in its mean motion on its eccentric in a certain time to any arc that the star traverses in its mean motion on its eccentric in a different time is equal to the ratio of the time to the time, because the mean motion on the eccentric is uniform and self similar, taking place along a circumference of a circle, whose parts are similar. And for any moving object, in uniform and self similar motion across an interval whose parts are similar, the ratio of the distance it traverses in a certain time to the distance it traverses in a different time is equal to the ratio of the time to the time. The ratio of the arc of the eccentric traversed with the mean motion of the star, which is the motion of the centre of the epicycle on the circumference of the eccentric in the time *IA*, to that traversed with the mean motion in the time *KH*, is equal to the ratio of the time *IA* to the time *KH*. But the ratio of the arc that the mean motion traverses on the eccentric in a time *IA* to the arc traversed by the mean motion on the eccentric in a time *KH*, as we have proved,[107] is greater

---

[107] In Proposition 9.

than the ratio of the arc *AB* to the arc *BH*; so the ratio of the time *IA* to the time *KH* is greater than the ratio of the arc *AB* to the arc *BH*.

Similarly, we have proved in Proposition 5 that the excesses[108] in the inclinations of the equal parts for two circles that intersect are unequal; and that <the excess for the arc> further from the point of intersection is smaller than <the excess for the arc> closer to the point of intersection. We shall prove in the following proposition that for two arcs of the inclined circle, the ratio of the one further from the point of intersection to [the one] closer to the point of intersection is greater than the ratio of the two excesses in their inclinations one to another. So the ratio of the arc *AH* to the arc *HB* is greater than the ratio of the arc *NM* to the arc *MB*. By composition, the ratio of the arc *AB* to the arc *BH* is greater than the ratio of the arc *NB* to the arc *BM*. But the ratio of the time *IA* to the time *KH* is greater than the ratio of the arc *AB* to the arc *BH*; and the ratio of the arc *AB* to the arc *BH* is greater than the ratio of the arc *NB* to the arc *BM*; so the ratio of the time *IA* to the time *KH* is much greater than the ratio of the arc *NB* to the arc *BM*. But the ratio of the arc of the great circle similar to the arc *IA* to the arc of a great circle, that is to say the circle of the equator, similar to the arc *KH*, is equal to the ratio of the time *IA* to the time *KH*; the ratio of the arc of the great circle similar to the arc *IA* to the arc of the great circle similar to the arc *KH* is thus greater than the ratio of the arc *NB* to the arc *BM*. If we permute, the ratio of the arc of a great circle similar to the arc *IA* to the arc *NB* is greater than the ratio of the arc of a great circle similar to the arc *KH* to the arc *MB*. So the ratio of the time *IA* to the arc *NB* is greater than the ratio of the time *KH* to the arc *MB*. And the ratio of the time *KH* to the arc *MB* is greater than the ratio of the time *KM* to the arc *MB*; so the ratio of the time *IA* to the arc *NB* is much greater than the ratio of the time *KM* to the arc *MB*.

Similarly, we prove that, for any required time for the star between the two points *A* and *B*, the ratio of *IA* to *NB* is greater than the ratio of the required time to the inclination proper to this required time. But the ratio of the time *IA* to the arc *NB* is known because each of them is known.

<25> Similarly, let the inclined orb be *ABC*, the circle of the equator *AMG* and the north pole of the equator the point *D*. Let *A* be the point of intersection of the inclined orb and the circle of the equator. Let the point *C* be the southern limit of the inclined orb with respect to the circle of the equator. Let the motion of the star be from the apogee towards the perigee and let it move from the point *B* towards the point *H* in a known time; let the time be *IB*. Let the arc *IB* lie on the hour circle whose pole is the point

---

[108] See above, p. 365, note 105.

D; let the motion of the star be accelerated; in the known time *IB* let it traverse the arc *BH* of the inclined orb and let it traverse the arc *EH*, which is a part of the arc *BH*, in the course of the time *KE*. We cause great circles to pass through the point *D*, which is the north pole of the circle of the equator, and through each of the points *B*, *E*, and *H*, *C*; let the great circles be *DMB*, *DNE*, *DOH* and *DGC*. Let the circle *DOH* cut the two arcs *IB* and *KE* at the two points *P* and *Q*.

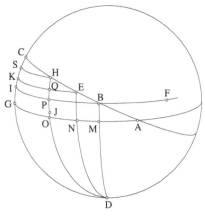

Fig. I.25

So the arc *PO* is the inclination of the arc *AB* with respect to the circle of the equator, because it is equal to the arc *BM*; and the arc *QO* is the inclination of the arc *AE* because it is equal to the arc *EN*. The arc *HO* is the inclination of the arc *AH*; so the arc *HP* is the inclination of the motion of the star in the time *IB* and the arc *HQ* is the inclination of the motion of the star in the time *KE*. The arc *KQ* is the required time for the star in its motion from the point *E* to the point *H*; so it is one of the required times for the star which lie between the two endpoints of the time *IB*. The arc *BP* is the right ascension of the arc *BH* and the arc *EQ* is the right ascension of the arc *EH*.[109] We cause an arc of an hour circle to pass through the point *H*; let it cut the arc *DC* at the point *S*. The arc *HS* is thus known, and the arc *SC* is known because the arc *HC* is known. In fact, the point *B* is known, because it is the position of the star at the beginning of the known time during which the star moved, and the point *H* is known, because it is the position of the star at the end of the known time. So the arc *HC* is known, so its ascension is known. The excess of its inclination[110] is known, because

---

[109] The right ascension of an arc must be understood as the difference between the right ascensions of its two endpoints.

[110] See above, p. 365, note 105.

any great circle cuts the circle of the equator in such a way that its maximum inclination with respect to it is known; the right ascensions of its known parts are thus known, and the excesses of the inclinations of its known parts are known. We in fact proved in Proposition 22 that the inclination of the inclined orb of each of the seven wandering stars with respect to the circle of the equator is known at any known instant. So the arc *HS* is known, because it is the right ascension of the arc *HC* at the known instant. But the arc *SC* is known, because it is the excess of its inclination. But the arc *BP* is known, because it is the ascension of the known arc *BH*; but the arc *PH* is known, because it is the excess of the inclination of the known arc *BH*. We put the ratio of the known arc *BP* to the arc *PJ* equal to the ratio of <the arc> *HS* to <the arc> *SC*, a known ratio; so the arc *PJ* is known, and the ratio of <the arc> *HP* to <the arc> *PJ* is known. Let us put the ratio of the arc *FI* to the arc *IB*[111] equal to the ratio of <the arc> *HP* to <the arc> *PJ*, a known ratio. But the arc *IB* is known, so the arc *FI* is known.

I say that the ratio of the known arc *FI* to the known arc *PH* is greater than the ratio of the time *KQ* to the arc *QH*.

Proof: The ratio of the time *IB* to the time *KE* is greater than the ratio of the arc *BH* to the arc *HE*, as has been proved in the previous section; but the ratio of the arc *BH* to the arc *HE* is greater than the ratio of the arc *BP*, which is the ascension of the arc *BH*, to the arc *EQ*, which is the ascension of the arc *EH*, from what we proved in Proposition 7. So the ratio of the arc *IB* to the arc *KE* is much greater than the ratio of the arc *BP* to the arc *EQ*. But the ratio of the arc *HS* to the arc *SC* is greater than the ratio of the arc *EQ* to the arc *QH*, from Proposition 6,[112] because the circle *HS* is smaller than the circle *IB*; the circle *DGC* is perpendicular to the circle *ABC*, because it passes through its pole. But the ratio of *HS* to *SC* is equal to the ratio of *BP* to *PJ*; so the ratio of *BP* to *PJ* is greater than the ratio of *EQ* to *QH*; but the ratio of *IB* to *KE* is greater than the ratio of *BP* to *EQ*. If we permute, the ratio of *IB* to *BP* is greater than the ratio of *KE* to *EQ*. But the ratio of *BP* to *PJ* is greater than the ratio of *EQ* to *QH*; so the ratio of *BI* to *PJ* is much greater than the ratio of *KE* to *QH*. But the ratio of *FI* to *IB* is equal to the ratio of *HP* to *PJ*, so the ratio of *FI* to *PH* is equal to the ratio of *BI* to *PJ*. But the ratio of *BI* to *PJ* is greater than the ratio of *KE* to *QH*, so the ratio of *FI* to *PH* is greater than the ratio of *KE* to *QH*; but the ratio of *KE* to *QH* is greater than the ratio of *KQ* to *QH*, so the ratio of *FI* to *PH*

---

[111] By hypothesis *IB* is the arc that represents the time for motion along the arc *BH*, so *FI* also represents a time.

[112] See note 30 in the Mathematical commentary, p. 213.

is much greater than the ratio of KQ to QH. Now the ratio of FI to PH is a known ratio, because each of the two arcs FI and PH is known.

Similarly, we prove for any required time, obtained <for a point that lies> between the two points H and B, that the known ratio is greater than the ratio of this required time to the part of the inclination of the motion of the wandering star proper to this required time.

If the wandering star moves on the arc of its inclined orb from the northern limit towards the point of intersection of the inclined orb and the circle of the equator, then we prove, as has been proved by the demonstration set out for the first case, that for the star there can exist a known ratio greater than the ratio of any required time that it can have between the two endpoints of the time of its motion, to the part of the inclination of its motion proper to this required time. And if its motion is from the point of intersection towards the southern limit, but without actually reaching the southern limit, we show, by the proof that we set out in the second case that for the wandering star there can exist a known ratio greater than the ratio of any required time that it can have between the two endpoints of the time of its motion, to the part of the inclination of its motion proper to that required time.

This result necessarily holds for the wandering star in its motion from the northern limit towards the southern limit, without reaching the southern limit.

This same result is also necessarily true if the motion of the wandering star on its inclined orb is from the southern limit towards the northern limit, without its reaching that same northern limit, but only [progresses] until there remains a certain distance between the two, even if this latter [distance] is extremely small; if its motion is on the eccentric, from the apogee towards the perigee; and if its motion is accelerated, that is to say if, in this half also, in each part of the time in the course of which it moves on this half of its inclined orb, it has a known ratio greater than any ratio of any required time that it can have between the two endpoints of that part of time to the part of the inclination of the motion of the wandering star proper to the required time, because the proof we mentioned is necessary, itself, for the other half of the inclined orb. It follows from this that this result necessarily holds for the wandering star, in its motion on its inclined orb, if its motion is accelerated, except for the two parts close to the two endpoints, even if these two parts are extremely small.

<26> Similarly, let the inclined orb be ABC, the circle of the equator DEC and the north pole of the equator the point H. Let the point A be the southern limit of the inclined orb with respect to the circle of the equator,

let the point *C* be the point of intersection of the inclined orb and the circle of the equator, let the motion of the wandering star be from the perigee of the eccentric towards the apogee and let its motion on its epicycle be accelerated, that is to say let the correction imposed by its epicycle be added to its position on its inclined orb.[113] Let it move from the point *A* towards the point *B* in a known time; let the time be *IA*. Let the arc *QB* belong to the hour circle. In the time *IA* let it [the wandering star] traverse the arc *AB* on the inclined orb; let it traverse the arc *MB* which is a part of the arc *AB* in the time *KM*. We cause great circles to pass through the point *H*, which is the pole of the circle of the equator, and through each of the points *A, M, B*, let the circles be *HDA, HGM, HEB*. Let the circle *HEB* cut the arcs *IA* and *KM* at the points *N* and *L*. So the arc *NB* is the inclination of the motion of the wandering star for the known time in the course of which the wandering star moved from the point *A* towards the point *B*; so the arc *NB* is known. The arc *LB* is the inclination of the motion of the wandering star in the time in the course of which it [the wandering star] moved from the point *M* towards the point *B*; the arc *LK* is the required time for the motion of the wandering star from the point *M* to the point *B*.

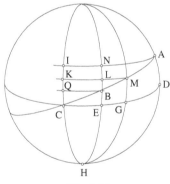

Fig. I.26

I say that there can exist a known ratio greater than the ratio of the time *KL* to the arc *LB*.

Proof: We have proved in Proposition 9 that the ratio of the arc of the eccentric cut off by the two straight lines drawn from the centre of the inclined orb to the points *B* and *M* to the arc of the eccentric cut off by the two straight lines drawn from the centre of the inclined orb to the points *M* and *A*, is greater than the ratio of the arc *BM* to the arc *MA*. Conversely, the ratio of the arc *AM* to the arc *MB* is greater than the ratio of the arc of the

---

[113] We add this correction to the arc traversed by the wandering star on its orb.

eccentric cut off together with the arc AM to the arc of the eccentric cut off together with the arc MB. If on the circle ABC we take two arcs similar to the two arcs cut off together with the two arcs AM, MB, then the ratio of the one to the other will be equal to the ratio of the two arcs of the eccentric cut off together with the arcs AM, MB.

<a> If each of the arcs AM, MB is greater than the arcs cut off together with them on the eccentric, it then becomes the ratio of the difference between the arc AM and the arc cut off together with it to the difference between the arc MB and the arcs cut off together with it, [is] greater than the ratio of the two arcs cut off together with them on the eccentric, the one to the other. So if we cut off from the excess of the arc AM an arc whose ratio to the excess of the arc MB is equal to the ratio of the arc of the eccentric cut off together with the arc AM to the arc cut off together with the arc MB, from the arc AM we are left with an excess in such a way that the ratio of what is left of the arc AM after this excess, to the arc MB, is equal to the ratio of the arc of the eccentric cut off together with the arc AM, to the arc cut off together with the arc MB. And the difference between any arc whatsoever cut off on the eccentric and the arc cut off together with it on the inclined orb is the correction. For each of the orbs of the seven wandering stars, the correction imposed by the eccentric is always smaller than the arc of the eccentric that imposed that correction and is smaller than the arc corrected by that correction.[114] That is why the ratio of the arc of the eccentric cut off together with the arc AM to the arc cut off together with the arc MB is greater than the ratio of the excess that is left from the arc AM to the arc MB, because that excess is smaller than the correction for the arc AM. It necessarily follows that the ratio of twice the arc of the eccentric cut off together with the arc AM to the arc of the eccentric cut off together with the arc MB is greater than the ratio of the arc AM to the arc MB. The ratio of twice the arc cut off on the eccentric together with the arc AM, plus the arc cut off together with the arc MB, to the arc cut off together with the arc MB, is greater than the ratio of the arc AB to the arc BM; the ratio of twice the arc cut off together with the arc AB to the arc cut off together with the arc MB is accordingly much greater than the ratio of the arc AB to the arc BM. But we have proved in the preceding proposition that the ratio of each arc of the eccentric to each arc of the eccentric is equal to the ratio of the time in the course of which the wandering star moves with a mean motion

---

[114] If α and β are the arcs cut off together with $\overset{\frown}{AM}$ and $\overset{\frown}{MB}$, we know that $|\overset{\frown}{AM} - \alpha| < \alpha$ and $|\overset{\frown}{AM} - \alpha| < \overset{\frown}{AM}$. If $\overset{\frown}{AM} > \alpha$, we then have $\alpha < \overset{\frown}{AM} < 2\alpha$; which is the case treated in this paragraph. If $\overset{\frown}{AM} < \alpha$, we then have $\overset{\frown}{AM} < \alpha < 2\overset{\frown}{AM}$.

along one of the two arcs to the time in the course of which the wandering star moves with a mean motion along the other arc. So the ratio of twice the time *IA* to the time *KM* is greater than the ratio of the arc *AB* to the arc *BM*.

<b> But if the arcs *AM*, *MB* are smaller than the two arcs cut off on the eccentric together with them, the ratio of what is lacking for the arc *AM* to what is lacking for the arc *MB* is smaller than the ratio of the arc of the eccentric cut off together with the arc *AM* to the arc cut off together with the arc *MB*. The ratio of what is lacking for the arc *AM* to a certain part of what is lacking for the arc *MB* is equal to the ratio of the arc *AM* to the arc *MB*, and what remains will be smaller than all that is lacking from the arc *MB*[115] with respect to the arc cut off together with it on the eccentric. But all that is lacking from *MB* is smaller than *MB*; so that difference is much smaller than what remains of the arc *MB*. So if we add to the arc cut off together with the arc *AM* an arc whose ratio to that difference is equal to the ratio of the arc *AM* to the arc *MB*, this last arc is smaller than the arc *AM*; so this arc is much smaller than the arc cut off together with the arc *AM* on the eccentric; the ratio of twice the arc cut off together with the arc *AM* to the arc cut off together with the arc *MB* is greater than the ratio of *AM* to *MB*;[116] the ratio of twice the arc cut off together with the arc *AB* to the arc cut off together with the arc *MB* is accordingly much greater than the ratio of *AB* to *MB* and the ratio of twice the time *IA* to the time *KM* is thus greater than the ratio of *AB* to *BM*.

<c> If the difference for *AM* is additive and the difference for *MB* is subtractive, which can occur close to the mean distance, we make <the arc> *JM* similar to the arc cut off together with the arc *AM*; we make <the arc> *MQ* similar to the arc cut off together with the arc *MB* and we make the ratio of *SM* to *MB* equal to the ratio of *JM* to *MQ*; so it remains that the ratio of *JS* to *BQ* is equal to the ratio of *JM* to *MQ*. But *AJ* is smaller than *JM* because it is the correction to the arc *MJ*, imposed by the eccentric; so the whole [arc] *AM* is smaller than twice *JM*. So the ratio of twice *JM* to *MB* is greater than the ratio of *AM* to *MB*. But the arc *BQ* is smaller than the arc *BM*, since *BQ* is the correction of the arc *MQ*. But twice *JM* is greater than *AM*, so the ratio of twice *JM* to *BQ* is greater than the ratio of *AM* to *MB*. So the ratio of four times *JM* to *MQ* is greater than the ratio of *AM* to *MB*. By composition, the ratio of four times *JM* plus *MQ* to *MQ* is greater than the ratio of *AB* to *BM*, the ratio of four times *JQ* to *QM* is thus much greater than the ratio of *AB* to *BM*. But the arc *JQ* is similar to the arc

---

[115] See Mathematical commentary, pp. 212–13.
[116] See note 30 of the Mathematical commentary, p. 213.

of the eccentric cut off together with the arc AB and the arc MQ is similar to the arc of the eccentric cut off together with the arc MB. So the ratio of four times the arc of the eccentric cut off together with the arc AB to the arc cut off together with the arc MB is greater than the ratio of AB to BM. So the ratio of four times the time IA, [which is] known, to the time KM is greater than the ratio of AB to BM.

In the various situations for making the correction, there can exist a known time whose ratio to the time KM is greater than the ratio of the arc AB to the arc BM. But the ratio of the arc AB to the arc BM is greater than the ratio of the arc NB to the arc BL, arcs that are the differences of the two inclinations; the ratio of the known time, whose magnitude has been established, to the time KM is thus much greater than the ratio of the arc NB to the arc BL. If we permute, the ratio of the known time to the arc NB, a known ratio, is greater than the ratio of the time KM to the arc LB; so the ratio of the known time to the arc NB, which is a known ratio, is greater than the ratio of the time KM to the arc LB.

If the circle ABC is the orb of the sun, then we have proved what we wanted, and if the circle ABC is the inclined orb of the moon or the inclined orb of one of the five planets, then we have proved what was required in regard to the correction imposed by the eccentric. But there remains the correction imposed by the epicycle. Now we have stipulated that the correction imposed by the epicycle shall be added. If the corrections for the arcs AM and MB are added, then, either the ratio of the correction for the arc AM to the correction for the arc MB is equal to the ratio of the two corrected arcs; or it is greater than the ratio of the two corrected arcs; or it is smaller than the ratio of the two corrected arcs.

If the ratio of the correction for the arc AM to the correction for the arc MB is equal to the ratio of the two corrected arcs, then this correction has no effect on the first ratio established for the two corrected arcs; so the first ratio established from the correction for the eccentric does not change if the ratio of the two corrections imposed by the epicycle, the one to the other, is equal to the ratio of the two corrected arcs.

If the ratio of the correction for AM to the correction for MB, the corrections imposed by the epicycle, is smaller than the ratio of the two corrected arcs, then this ratio increases the first ratio, that is to say that the ratio of the known time to the time KM is much greater than the ratio of AB to BM, if the ratio of the two corrections imposed by the epicycle is smaller than the ratio of the two corrected arcs, because this ratio has the effect that the ratio of the arc AM to the arc MB is smaller than the ratio of the two corrected arcs.

If the ratio of the correction for AM to the correction for MB is greater than the ratio of the two corrected arcs – let the correction of the arc AM be the arc AS and the correction of the arc MB be the arc BO – we then have two arcs SM and MO corrected for the two arcs of the inclined orb which have been corrected by the correction imposed on the eccentric, since the two corrected arcs on the inclined orb have their beginning at the point A;[117] we have taken only the two arcs SM and MO, which are equal to them, so that the two corrections shall be separated and in two different directions, and thus it will be clearer and easier to discuss them. But since the two arcs SM and MO are equal to the two corrected arcs, accordingly the ratio of the known time, whose magnitude has been established, to the time KM, is greater than the ratio of SO to OM. But the ratio of AS to OB is greater than the ratio of SM to MO. We make AU equal to OB; so the arc US is known, because it is the correction of the whole known arc SO, and the ratio of US to SO is known. We put the ratio of some arbitrary time to the known time – whose ratio to the time KM is greater than the ratio of SO to OM – equal to the known ratio of US to SO; so this [arbitrary] time will be known, and the ratio of the sum of the two times to the first time is equal to the ratio of UO to OS. But the ratio of the first known time to the time KM is greater than the ratio of SO to OM; the ratio of the sum of the two known times to the time KM is accordingly greater than the ratio of UO to OM. But the ratio of UO to OM is greater than the ratio of AO to OM and the ratio of AO to OM is greater than the ratio of AB to BM; the ratio of the known time composed of the two known times to the time KM is accordingly much greater than the ratio of AB to BM. But the ratio of AB to BM is greater than the ratio of NB to BL; so the ratio of the known time composed of the two known times to the time KM is much greater than the ratio of NB to BL. If we permute, the ratio of the known time composed [of the two known times] to the arc NB is greater than the ratio of the time KM to the arc LB; the ratio of the known time composed [of the two known times] to the arc NB is thus much greater than the ratio of the time KL to the arc LB.

Similarly, we show that for any required time for the wandering star between the two points A and B whose inclination is NB, the ratio of the known time, whose magnitude has been definitely established, to the arc NB is greater than the ratio of this required time to the <part of that> arc NB proper to this required time.

[117] A is the initial position of the planet which will traverse the two arcs AM and MB.

<27> Similarly, let *ABC* be the inclined orb, *ADG* the circle of the equator, the point *E* the north pole of the circle of the equator. Let *A* be the point of intersection of the inclined orb and the circle of the equator, let the point *C* be the northern limit of the inclined orb with respect to the circle of the equator. Let the motion of the wandering star on the inclined orb be from the point *A* towards the point *C*; let its motion on the eccentric be from the perigee towards the apogee, let its motion on the epicycle be accelerated and let it move from the point *B* to the point *I* in a known time, let the time be *SB*; in the time *SB* let it traverse the arc *BI* and let it traverse the arc *HI*, which is a part of the arc *BI*, in the time *HO*. Let the two arcs *SB* and *OH* belong to hour circles, which are parallel to the equator. We cause great circles to pass through the point *E* and through each of the points *B*, *H*, *I*, *C*; let the circles be *EBD*, *EHK*, *EIL*, *ECG*. Let the circle *EIL* cut the two arcs *BS* and *HO* at the two points *N* and *U*. So the arc *NI* is the inclination of the motion of the wandering star in the known time in the course of which the wandering star has travelled from the point *B* to the point *I*; so the arc *NI* is known. The arc *UI* is the inclination of the motion of the wandering star in the time in the course of which it has travelled from the point *H* to the point *I*; the arc *O U* is the required time for the motion of the wandering star from the point *H* to the point *I*.

I say that we can find a known ratio greater than the ratio of the time *OU* to the arc *UI*.

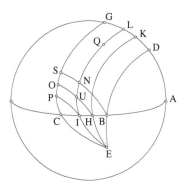

Fig. I.27

Proof: We prove, as was proved in the preceding proposition, that there can exist a known time whose ratio to the time *OH* is greater than the ratio of the arc *BI* to the arc *IH*. But the arc *BN* is the right ascension of the arc *BI* and the arc *HU* is the right ascension of the arc *HI*; so the ratio of the arc *BI* to the arc *IH* is greater than the ratio of the arc *BN* to the arc *HU*, as we proved in Proposition 7. We cause an arc of an hour circle to pass through the point I; let the arc be IP. So the arc IP is known, and the arc PC is known, because the arc IC is known and the arc IP is the ascension of the known arc IC. But the arc PC is the inclination of the arc IC, so the ratio of the arc IP to the arc PC is known; but the ratio of the arc IP to the arc PC is greater than the ratio of the arc HU to the arc UI, as we proved in Proposition 14. We put the ratio of the known arc BN to the arc NQ equal to the ratio of IP to PC, a known ratio; so the arc NQ is known, and the ratio of BN to NQ is greater than the ratio of HU to UI. We put the ratio of an arbitrary time to the known time – whose magnitude has been demonstrated in the preceding proposition and whose ratio to the time OH is greater than the ratio of BI to IH – equal to the ratio of the known arc IN to the known arc NQ; so this time will be known. Since the ratio of the first known time – whose magnitude has been demonstrated in the preceding proposition – to the time OH is greater than the ratio of BI to IH and since the ratio of BI to IH is greater than the ratio of BN to HU, we have that the ratio of the first known time to OH is much greater than the ratio of BN to HU. If we permute, we have that the ratio of the first known time to the arc BN is greater than the ratio of the time OH to the arc HU. But the ratio of the arc BN to the arc NQ is greater than the ratio of the arc HU to the arc UI; so the ratio of the first known time to the arc NQ is greater than the ratio of the time OH to the arc UI. But the ratio of the second known time to the first known time is equal to the ratio of the arc IN to the arc NQ. If we permute, we have that the ratio of the second known time to the arc NI is equal to the ratio of the first known time to the arc NQ; but the ratio of the first known time to the arc NQ is greater than the ratio of the time OH to the arc UI; so the ratio of the second known time to the arc NI is greater than the ratio of the time OH to the arc UI. But the second time is known and the arc NI is known; so the ratio of the one to the other is known; but the ratio of OH to UI is greater than the ratio of OU to UI; the ratio of the second known time to the arc NI, which is a known ratio, is accordingly greater than the ratio of OU, which is the required time, to the arc UI, which is the inclination proper to the time OU.

Similarly, we prove that the ratio of the known time to the arc NI, a known ratio, is greater than the ratio of any required time between the two points B and I, to the part of the arc NI that is proper to that required time.

We have proved, in this proposition and in the proposition before it, that if the wandering star is moving on its inclined orb from the southern limit towards the northern limit, without actually reaching that northern limit, if its motion on its eccentric is from the perigee towards the apogee and if its motion on its epicycle is accelerated, then for any part of the time in the course of which the wandering star moves on that half of its inclined orb, there is a known ratio greater than any ratio, for any required time that there can be between the two endpoints of that time interval to the part of the inclination of the motion of the wandering star proper to this required time. This same result is necessary if the motion of the wandering star on its inclined orb is from the northern limit towards the southern limit, if its motion on its eccentric is from the perigee towards the apogee and if its motion on its epicycle is accelerated. Thus this result is necessarily true for the wandering star in its motion on its complete inclined orb from which we cut off two parts close to the two limits, even if they are extremely small.

We have thus proved, in the four propositions that we have established, that if each of the seven wandering stars moves, for a known time, whatever the place of its motion on its inclined orb, except for a part close to the southern limit and a part close to the northern limit, though they are extremely small, and whatever the place of its motion on its eccentric – for the sun there is no additional condition; for the moon and for each of the five planets, the correction imposed by its epicycle is added to its motion – then the wandering star has a known ratio that is greater than any ratio for any required time that there can be between the two endpoints of this known time in the course of which it has moved, to the part of the inclination proper to that required time, whether the motion of the wandering star on its eccentric is from the apogee towards the perigee or its motion is from the perigee towards the apogee. This ratio is the ratio of a greater magnitude to a lesser magnitude. This result also necessarily holds for the moon and <each> of the planets, even if the correction imposed by its epicycle is subtracted from its motion, whatever the place of its motion on its inclined orb – if the correction imposed by the eccentric is additive, and if the excess it acquires is that imposed by the eccentric; this result again necessarily holds if there are numerous corrections, sometimes an addition and sometimes a subtraction, and if the subtraction is smaller than the addition imposed by the eccentric.

Similarly if for each of the quarters of the inclined orb we follow the method we have adopted in the quarter that is next to it, that is to say that if in the first quarter, which is between the northern limit and the point of intersection,[118] we adopt the method of proof that we adopted in the last quarter, which is between the point of intersection and the northern limit, if we adopt in the last quarter the method that we adopted in the first quarter, if we adopt in the second quarter, which is between the point of intersection and the southern limit, the method that we adopted in the third quarter, which is between the southern limit and the point of intersection, and if we

---

[118] That is, the point of intersection of the orb with the equator.

adopt in the third quarter the method that we adopted in the second quarter, in each of these quarters we shall obtain a known ratio greater than any ratio of any required time for the planet between the two endpoints of the time in the course of which it moves, to the part of the inclination of the motion of the planet proper to this required time. These inclinations proper to these required times are on the side towards the beginning of the motion and the inclinations proper to the required times mentioned before are on the side towards the parts of the motion.[119]

If we have proved all that, it necessarily follows that for each of the seven wandering stars, if it moves in the manner that we have described in the course of a certain time, whatever time that is, whether we know it or do not know it, it in fact has a ratio greater than any ratio of any required time for the wandering star between the endpoints of this time in the course of which it moved to the inclination proper to that required time, whether this ratio is known to us or is not known to us, whether we determine this ratio or whether we do not determine it. And for any ratio that we have determined and for which we have proved that it is greater than any ratio of any required time to the inclination proper to that required time, there can exist numerous ratios greater than it is, because for any ratio there can exist numerous ratios each of which is greater than this ratio.

If this is the case, then for each star among the seven wandering stars, if it moves in the manner that we have described, in the course of a certain time, whatever this time is, then there are numerous ratios whose number is infinite, [and] each of them is greater than any ratio of any required time for that wandering star, between the two endpoints of this time in the course of which it moved, to the inclination proper to this required time.

It is this result that we wanted to prove for the four propositions that we have established.

<28> That result having been proved, we say: if each of the seven wandering stars moves from the eastern horizon towards the meridian circle for any horizon where the sphere is inclined towards the south, or is right; and if the position of the wandering star on the meridian circle has an inclination to the south with respect to the pole of the horizon and if its motion on its inclined orb is from the northern limit towards the southern limit, with respect to the circle of the equator, without its actually reaching the southern limit – for the sun, that requires no additional condition; for the moon, it is when its anomalistic motion is accelerated, that is to say that

---

[119] Ibn al-Haytham has proved that, in all cases, there exists a minimum value of the mean motion of the right ascension. On this conclusion see Mathematical commentary, p. 221.

its motion in equal times is of unequal magnitudes so that its magnitude in
the second time is larger than that in the first time or only when the motion
of its epicycle is accelerated, that is to say that the correction imposed by
its epicycle is additive for its motion; for each of the five planets, it is when
its anomalistic motion is accelerated or when the motion of only its
epicycle is accelerated and in addition the motion of the inclination of its
epicycle or the departure of its epicycle [from the plane of the eccentric]
which causes an increase in latitude is inclined in the southerly direction –
then the wandering star has, in the east, equal heights, equal two by two, it
has, in the east, unequal heights such that the second is smaller than the
first and it [the planet] has a height before its meridian passage[120] equal to
the height at its meridian passage, and its equal heights are greater than the
height at its meridian passage.

Let the circle *ABC* be one of the horizons mentioned earlier.

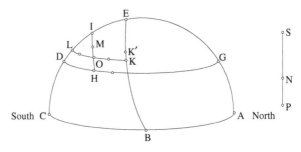

Fig. I.28.1

Let the arc *ABC* be the western half of the horizon and let the centre of
the sun or the centre of [any] one of the seven wandering stars rise from the
point *B*; let it reach the meridian circle *ACD* and let its centre arrive at the
point *D* on the meridian circle. Let the point *D* have an inclination to the
south with respect to the pole of the horizon. We cause to pass through the
point *B* an arc of an hour circle that passes through the point *E*, let the arc
be *BE*. We draw from the point *D* a circle parallel to the horizon; let the
circle be *DHG*. So the circle *DHG* will be one of the *muqanṭarāt* of height,
the arc *BE* will be the required time for the wandering star in its motion
from the point *B* towards the point *D* and the arc *ED* is the inclination of
the motion of the wandering star. Now we have proved that if the
wandering star moves, for a certain time, whatever this time is, we have for
it ratios greater than any ratio, for any required time that it has between the
two endpoints of this time in the course of which it moved, to the

---

[120] In this volume, *niṣf nahārihi* or *intiṣāf nahārihi* is translated as the meridian
passage or simply meridian.

inclination proper to that required time, a part of the inclination of the motion of the wandering star. Let one of these ratios that are greater than any ratio for any required time to the inclination that is proper to that required time be the ratio of *SN* to *NP*. Since the circle *ADC* passes through the pole of the circle *DHG*, it cuts it orthogonally into two equal parts; so the arc *ED* is perpendicular to the diameter of the circle *DHG*.[121] Let us draw the arc *HI* parallel to the arc *BE* so that the ratio of the chord of the arc *HI* to the chord of the arc *ID* is greater than the ratio of *SN* to *NP*, as we have proved in Proposition 10.[122] So if the point *I* lies between the points *E* and *D* – if it does not we draw any arc parallel to the arc *HI* between the points *E* and *D* to make the ratio of the chord of this second arc to the chord of what it cuts off from the arc *ED* greater than the ratio of the chord of the first arc to the chord of what it cuts off from the arc *ED*, as was proved in Proposition 11 and Proposition 12 –, then the ratio of these chords, the one to the other, is greater than the ratio of *SN* to *NP* and the ratio of the two arcs subtended by these two chords is greater than the ratio of the two chords. So the ratio of the two arcs, the one to the other, is greater than the ratio of *SN* to *NP*. Let the ratio of *HI* to *ID* be greater than the ratio of *SN* to *NP*, which is greater than any ratio of any required time for <the motion> that takes place between the two points *E* and *D* to the part of the arc *ED* proper to this required time. So the ratio of *HI* to *ID* is greater than any ratio of any required time for the wandering star between the points *B* and *D*, to the part of the arc *ED* proper to this required time. The required time whose inclination is the arc *ID* is smaller than the arc *HI* and the required time, will always lie to the east with respect to the inclination of the motion of the wandering star. Let this required time be the arc *MI*. Since the wandering star moves from the point *B* towards the point *D*, it will cross any hour circle lying between the two points *E* and *D*; consequently the wandering star will cross the circle *HI*. But if the wandering star reaches the circle *HI*, the arc of the circle *HI* which lies between the position of the wandering star and the arc *ED* is the required time whose inclination is the arc *ID*, in the same way that the arc *BE* is the required time whose inclination is the arc *ED*. But the required time, whose inclination is the arc *ID*, is the arc *MI*; consequently if the wandering star reaches the circle *HI*, then it reaches the point *M*. So the wandering star, which was at the point *B*, later reaches the point *M*; but the point *B* is below the *muqanṭara DHG* and the point *M* is above the *muqanṭara DHG*. Thus

---

[121] He means that the arc *ED* is in the plane *ADC* which is perpendicular to the plane of the circle *DHG*.

[122] $\frac{SN}{NP}$ is a given ratio, we must take $\frac{SN}{NP} > 1$ (see Proposition 10).

on its motion from the point *B* to the point *M* the wandering star will cross the *muqanṭara DHG* in a point between the two circles *BE* and *HI*, since the planet moves from being in a northerly direction to being in a southerly one. The wandering star then in fact travels to the point *D* of the meridian circle; but the point *D* is on the *muqanṭara DHG*, thus in its motion from the point *B* to the point *D* the wandering star will, on two occasions, come onto the *muqanṭara DHG*; its two heights at these two instants are equal and these two heights are equal to the arc *CD* of the meridian circle; however it has reached it in its motion from the eastern horizon towards the south and from north to south on its inclined orb; in this case its height is to the east [of the meridian]. So if the wandering star moves from the eastern horizon towards the meridian circle, it will have a height in the east equal to the height at its meridian passage. Similarly, we mark an arbitrary point on the arc *MH*; let the point be *O*. We draw through the point *O* a *muqanṭara* parallel to the *muqanṭara DHG*; let the *muqanṭara* be *LOK*. Since the point *M* is higher than the *muqanṭara LOK*, and the point *B* is lower than the *muqanṭara LOK* and the wandering star has moved from the point *B* towards the point *M*, the wandering star crosses the *muqanṭara LOK* before reaching the point *M*, and it crosses it between the two arcs *BE* and *HI*. But since the point *M* is higher than the *muqanṭara LOK*, and the point *D* is lower than the *muqanṭara LOK* and the wandering star has moved from the point *M* towards the point *D*, the wandering star crosses the *muqanṭara LOK* before reaching the point *D*. But in its motion from the point *M* towards the point *D* it does not cross the *muqanṭara LOK* at the same point at which it crossed it in its motion from the point *B* to the point *M*, because the hour circle which it will be on at this second instant will be closer to the point *D* than the circle *HI*, since the wandering star is moving from the north towards the south and the hour circles are parallel. Thus the point of the circle *LOK* to which the wandering star came in its motion from the point *M* towards the point *D* is different from the point of the circle *LOK* to which the wandering star came in its motion from the point *B* towards the point *M* and it is not true that the point of the circle *LOK* to which the wandering star came in its motion from the point *M* towards the point *D* is to the west of the meridian circle, because, if it were to the west of the meridian circle, then the wandering star would have crossed the meridian circle before reaching this point and then later coming to the meridian circle when it reaches the point *D*; it would thus have crossed the meridian circle above the horizon[123] twice, in a time shorter than the time for one complete revolution;[124] which is impossible. Indeed for any arc traversed

---

[123] Lit.: above the earth.
[124] *zamān nahārihi* (lit.: the time of its day).

by each of the seven wandering stars on its inclined orb, in a certain time, its right ascensions are much smaller than the time in the course of which the wandering star traversed the arc of its orb. So if the wandering star crosses the meridian circle above the horizon,[125] it does not return to it above the horizon[126] except in the second revolution. The two points of the circle *LOK* at which the wandering star encounters it in its motion from the point *B* towards the point *D* are both to the east of the meridian circle. In consequence the wandering star has two equal heights that are equal to the height of the *muqanṭara LOK* which is the arc *LC*, and these two heights are greater than the height at the meridian, which is the arc *DC*.

Similarly, any *muqanṭara* cuts the arc *MH* between the two points *M* and *H*, which arc the wandering star has encountered twice in its motion from the point *B* towards the point *D*; so it [the wandering star] will have two equal heights that are also equal to the height of this *muqanṭara*. If the wandering star reaches a *muqanṭara* between the two points *M* and *O*, then its height will be greater than its height if it is on the *muqanṭara LOK*; and if the wandering star reaches a *muqanṭara* between the two points *M* and *O*, it reaches it before coming to the *muqanṭara LOK* a second time. Thus if the wandering star is on a *muqanṭara* between the two points *M* and *O* and later reaches the *muqanṭara OLK*, then its second height will be smaller than its first height; the two heights are to the east and all its heights that are above the *muqanṭara* at its meridian passage are greater than the height on the meridian.

So we have proved, from what we have established, that each of the seven wandering stars, if it moves from the eastern horizon towards the meridian circle, [and] if its motion on its inclined orb is from the northern limit towards the southern limit and if its motion takes place in the manner we have defined, it then has equal heights to the east, that is equal two by two, and it has a height to the east equal to the height on the meridian and it has unequal heights to the east the second of which is smaller than the first and, of all its equal heights to the east, each is greater than the height on the meridian. That is what we wanted to prove.

We say that if the motion of the wandering star is in the manner we have described and if its heights to the east are as we have explained, then, if it moves from the meridian circle towards the western horizon, none of what we have referred to takes place for the wandering star, instead <all>

---

[125] Lit.: above the earth.
[126] Lit.: above the earth.

its heights will be unequal, the second of them always being smaller than the first.

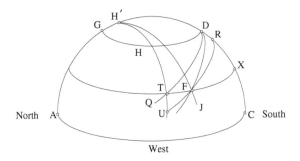

Fig. I.28.2

Proof: We cause to pass through the point *D* an arc of an hour circle that passes through the point *Q*; let the arc be *DQ*. Thus the circle *DQ* will be tangent to the circle *DHG*, because their two poles lie on the meridian circle, which is the circle *ADC*. So if the sphere moves in its rapid daily motion, then the point *D* which is the position of the wandering star on the higher orb moves on the circle *DQ*. But if the point *D* moves on the circle *DQ*, then the wandering star moves with its proper motion; so it departs from the point *D*, its motion giving it an inclination to the south. If the wandering star moves, its motion gives it an inclination towards the south, then it leaves the circle *DQ* and comes to a circle with a greater inclination to the south than the circle *DQ*. But any hour circle among the circles that lie between the point *D* and the western horizon, [and] with a greater inclination to the south than the circle *DQ*, cuts the arc *DC* in a point between the two points *D* and *C*. The greatest height for points on this hour circle is the arc this circle cuts off on the arc *DC*, and any point on this hour circle, except the point that is on the arc *DC*, has a height less than the height of the point that is on the arc *DC*, so that the wandering star arrives at the point *F* after having left the point *D*. We cause an hour circle to pass through the point *F*; let it be *RFU*. The greatest height of the points that are on the arc *RU* is then the arc *RC*; so the arc *RC* is smaller than the arc *DC* and the height of the point *R* is smaller than the height of the point *D*; so the height of the point *F* is much smaller than the height of the point *D*. But since the wandering star always has an inclination in the southerly direction with respect to the circle *DQ*, and the circle *DQ* touches the circle *DHG*, the wandering star accordingly does not return to the circle *DHG*. We cause to pass through the point *F* a *muqanṭara* parallel to the plane *DHG*; let the *muqanṭara* be *TFX*. Let the arc *DQ* cut this *muqanṭara* at the point *T*. Let the pole of the equator be the point H'. We cause a great circle to pass through the points H' and F – let the circle be $H'FJ$ – and through the points H' and T a great circle – let it be $H'T$. Since the arcs TD and FR are to the south with respect to the pole of the horizon, the arc TD will be greater than the arc similar to the arc FR, as has been proved in Proposition 13. So the circle $H'FJ$ cuts the arc TD; but if it cuts the arc TD, then it cuts the *muqanṭara TFX* at the point F and the great circle $H'T$ cuts the *muqanṭara TFX* in a point north of the point F. If this is so, the arc FR lies to the east of the circle $H'FJ$. But since the wandering star moves with the diurnal motion from the point F towards the western horizon, accordingly its proper motion does not bring it back onto the circle $H'FJ$, because the circle $H'FJ$ is one of the meridian circles. Now we have proved earlier that if the wandering star leaves a meridian circle, then it does not return to this half of the circle <which is above the horizon> except in a second revolution. But if the wandering star does not return to the circle $H'FJ$, then it does not arrive on the arc FT of the *muqanṭara*. But since the wandering star has, by its motion, an inclination to the south, accordingly it does not return to the circle FU; but if it does not return to the circle FU, then it does not arrive on the arc FX of the *muqanṭara*; so the wandering star does not arrive on the *muqanṭara TFX* except at the point F. Similarly, any *muqanṭara* through which the wandering star passes will be lower than the *muqanṭara DHG*. So the wandering star passes through it only once. So the wandering star does not have two equal heights in the westerly direction if it is moving in the manner we described earlier; but all its heights to the west are unequal, the second being smaller than the first. That is what we wanted to prove.

<29> We say similarly that each of the seven wandering stars, if it moves from the meridian circle towards the western horizon, for any horizon for which the sphere is inclined towards the south or is right, if the position of the wandering star with respect to the meridian circle has an inclination to the south with respect to the pole of the horizon, and if its motion on its inclined orb is from the southern limit towards the northern limit with respect to the circle of the equator without actually reaching the northern limit – for the sun that is so without any additional condition; for the moon, it is so when its anomalistic motion is accelerated or when its motion on its epicycle is accelerated; for each of the five planets, it is so according to the same conditions [for the motion] as for the motion of the moon; and if, in addition, the motion of the inclination of its epicycle or the departure of its epicycle [from the plane of the eccentric], which causes an increase in latitude, is inclined in the northerly direction – then it has equal

heights to the west, equal two by two, and to the west it has unequal heights, the second of them being larger than the first; it has a height after its meridian passage equal to the height on the meridian, and its equal heights to the west are larger than the height on the meridian.

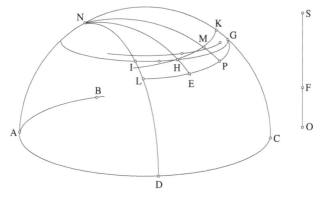

Fig. I.29.1

Let the circle *ABCD* be one of the horizons and let the circle *AGC* be the meridian circle; let the arc *ABC* be the eastern half of the horizon and the arc *ADC* the western half of the horizon. Let the centre of the sun or the centre of one of the seven wandering stars rises at the point *B* and reaches the meridian circle, let the centre *E* be at the point *G* of the meridian circle and let the point *G* have an inclination towards the south with respect to the pole of the horizon. Let the centre of that wandering star set at the point *D* and let the pole of the equator be the point *N*. We cause a great circle to pass through the points *N* and *D*, let it be *ND*. We cause an hour circle to pass through the point *G*; let it be the circle *GL*. Let this circle cut the circle *ND* at the point *L*; the arc *GL* is the required time for the motion of the wandering star and the arc *DL* is the inclination of the motion of the wandering star. So the wandering star has numerous ratios [associated with it], each of which is greater than any ratio of any required time for the wandering star between the two endpoints of its motion from the point *G* to the point *D*, to the part of the inclination of the motion of the wandering star proper to this required time, which is *LD*, on the side towards the beginning of the motion.[127] Let one of these ratios be the ratio of *SF* to *FO*. We draw the arc *HK* parallel to the arc *GL*[128] in such a way that the ratio of *HK* to *KG* is greater than the ratio of *SF* to *FO*. We cause a great circle to

---

[127] The origin for each arc traversed is on the side towards *G*.

[128] We take *H* on the circle *GHI*, which is the *muqanṭara* of *G*.

pass through the points N and H, let it be NHE; let this circle cut the arc GL at the point E, then the arc EG will be similar to the arc HK and the arc EH will be equal to the arc KG. So the ratio of GE to EH is greater than the ratio of SF to FO; so the ratio of GE to EH is greater than any ratio of any required time in the course of which the wandering star moved between the two endpoints of its motion from the point G to the point D, to the part of the inclination of the motion of the wandering star proper to any required time on the side towards the beginning of the motion; so if the wandering star moves in for the time GE, the inclination of its motion is greater than the arc EH; the time in the course of which the wandering star moves in such a way that the inclination of its motion in the course of this time is an arc equal to the arc EH is thus smaller than the time GE. Let the time in the course of which the wandering star acquires an inclination of an arc equal to the arc EH be the time GP. We cause a great circle to pass through the two points N and P; let it be NMP. Let this circle cut the arc HK at the point M. So if the wandering star moves for the time GP, the inclination of its motion will be the arc PM; so in the course of the time GP the wandering star has arrived at the point M; so the wandering star travels from the point G to the point M. But the point M is higher than the *muqanṭara GHI*, and this wandering star sets at the point D; so this wandering star travels from the point M to the point D and it crosses the *muqanṭara GHI*; it crosses it in a point between the two points H and I, because this wandering star moves so as to acquire an inclination towards the north, and it thus does not return to the circle KH between the points K and H; so it crosses the *muqanṭara* in a point between the two points H and I. But this wandering star was at the point G which is on this *muqanṭara*, so this wandering star arrives on the *muqanṭara GHI* twice; so it will have a height equal to the height on the meridian.

We prove, as was proved in the preceding proposition, that it arrives twice on any *muqanṭara* situated between the two points M and H; it thus has equal heights to the west, equal two by two; these heights are greater than the height on the meridian and it [the wandering star] has unequal heights such that the second of them is greater than the first, and all of them are to the west. That is what we wanted to prove.

I say that if the wandering star has a motion in the way we have just described, then, among its heights in the east, there are no two that are equal, but they are <all> unequal, the second of them being always greater than the first.

Proof: We cause an arc of an hour circle to pass through the point B; let it be BQ. So the point Q has a greater inclination towards the south than the point *G*, because by its motion the wandering star acquires an inclination in the northerly direction; any circle among the hour circles that the wandering star has crossed from the instant it left the point *B* until it reached the point *G* is thus to the south of the circle *GL*. But the circle *GL* touches the circle *HI*; so the circle *GHI* does not meet any of the hour circles which have been crossed by the wandering star in its motion from the point *B* to the point *G*. So the wandering star does not have a height in the east equal to the height of the *muqanṭara GHI*. So the height *GC* is the greatest of its heights in the east.

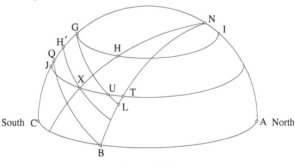

Fig. I.29.2

Similarly, we draw a *muqanṭara* closer to the horizon than the *muqanṭara GHI*; let the *muqanṭara* be *JXT*. In its motion from the point *B* to the point *G* the wandering star crosses any *muqanṭara* closer to the horizon than the *muqanṭara GHI*; so it crosses the *muqanṭara JXT*; let it cross it at the point *X*. We cause an arc of an hour circle to pass through the point *X*; let it be *XH′*. Let this arc cut the meridian circle at the point *H′*. We extend the arc *LG* until it cuts the *muqanṭara JXT*; let it cut it at the point *U*. Since the two arcs *UG*, *XH′* are two arcs of hour circles, which have a greater inclination towards the south with respect to the pole of the horizon, the arc *GU* will be greater than the arc similar to the arc *H′X*, as has been proved in Proposition 13. We cause a great circle to pass through the points *N* and *X*; let the circle be *NX*. This circle cuts the arc *GU*, because the arc *GU* is greater than the arc similar to the arc *H′X*; so the arc *XU* lies to the east of the circle *NX*, and in its motion from the point *X* to the point *G*, the wandering star does not arrive at any <point> of the arc *XU*. But the arc *UT* lies to the east of the meridian circle; if the wandering star comes to the point *G*, it will not return to any <point> of the arc *UT*, but it does not arrive at any <point> of the arc *UT* before reaching the point *G*, because the arc *UT* lies to the north of the circle *UG* and the motion of the wandering star is from south to north. But the arc *XJ* is to the south of the arc *H′X*; the wandering star thus does not return to it, because by its motion the wandering star acquires an inclination in the northerly direction. The wandering star arrives on the arc *JXT* only at the point *X* alone. So in the east it has only one single height equal to the height *JC*.

Similarly, we prove that, in any *muqanṭara* through which the wandering star passes on the eastern side [of the meridian], the wandering star arrives on it only once; so this wandering star does not have two equal heights in the east; indeed its heights in the east are unequal, the second of them being greater than the first. That is what we wanted to prove.

<30> Let us return to the figure of the heights in the east.

We say that the greatest of the heights of the wandering star in the east, if it has equal heights in the east, is a single height, and that the wandering star has no other height that is equal to it, and that it [the height] is greater than the height at its meridian passage.

Proof: It has been proved that the wandering star passes across many *muqanṭarāt* which are higher than the *muqanṭara* of the meridian; then, after having reached these *muqanṭarāt*, the wandering star returns to the point *D*, which is on the *muqanṭara* of the meridian. Thus, either the wandering star reaches a maximum height, then comes down again from it to the point *D*; or [the wandering star] reaches a maximum height only to go beyond it to <another> higher than it is; so if it only reaches a maximum height after having gone past <another> higher than it is, it will never return to the point *D*, since the point *D* is the lowest <point> among the positions to which the wandering star has risen. But the wandering star returned to the point *D*; so it necessarily follows that, in height, the wandering star reaches a limit from which it comes down again to the point *D*. So this limit does not lie on the meridian circle, nor to the west of the meridian circle, because if the wandering star had crossed the meridian circle at a point higher than the point *D*, it would not have returned to the point *D* – we have in fact proved that the wandering star does not encounter any of the meridian circles twice in the time of its day. The limit of its height is not to the west of the meridian circle because, if it were to the west, the wandering star would have crossed the meridian circle before coming west of the meridian circle, and thus the wandering star would have crossed the meridian circle before reaching the point *D*, and then later returning to the point *D*; which is impossible. So the limit of the height of the wandering star does not lie on the meridian circle, nor to the west of the meridian circle. Consequently, it lies to the east of the meridian circle.

Let the limit of the height of the wandering star lie on the *muqanṭara* *OKI*; and let the pole of the equator be the point *N*. We draw through the points *N* and *S* a great circle that touches the circle *OKI*; let the circle be *NS*; let the point of contact be the point *S*.

I say that the wandering star never passes through the arc *SI*.

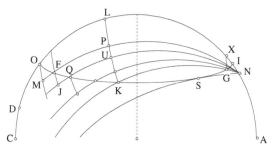

Fig. I.30

If that were possible, let it pass through the point *G* of the arc *SI*. We cause a great circle to pass through the points *N* and *G*; it cuts the *muqanṭara OKI* in two points; one is the point *G*; let the other point be the point *K*. We cause to pass through the two points *G* and *K* two hour arcs; let the arcs be *GX* and *KL*. If the wandering star passes through the point *G*, then the arc *GX* is the required time whose inclination is the arc *XD*, because the wandering star travels from the point *G* to the point *D*. But the arc *GX* is similar to the arc *KL*; the required time whose inclination is *LD* is thus a part of the time *KL*. Let the required time whose inclination is *LD* be the time *UL*; in its motion from the point *G* to [the point] *D* the wandering star crosses the hour circle *LK*; if it reaches the circle *LK*, then its distance from the meridian circle is the required time whose inclination is *LD*; consequently, the wandering star passes through the point *U*, and the point *U* is higher than the *muqanṭara OKI*. Consequently, the wandering star moves so as to be higher than the *muqanṭara OKI*; which is impossible, because the *muqanṭara OKI* is by hypothesis the highest *muqanṭara* that the wandering star reaches. So the wandering star does not pass through the point *G*, nor through any other point of the arc *SI*. So the position of the wandering star on the *muqanṭara OKI* is either the point of contact <*S*> or <a point> to the south of the point of contact. Let the wandering star pass through the point *K*, and let the point *K* be a point such that the wandering star has not passed through any other point of the *muqanṭara OKI* before *K*.

I say that the wandering star does not arrive on the *muqanṭara OKI* at any point other than the point *K*.

Indeed, we cause an arc of an hour circle to pass through the point *K*; let the arc be *KL*. Let this arc cut the arc of the meridian circle at the point *L*. Since the wandering star travels from the point *K* to the point *D*, the arc

KL is the required time whose inclination is the arc *LD*. We cause an arc of an hour circle to pass through the point *O*; let the arc be *OM*. Let the arc *MO* be the required time whose inclination is the arc *OD*. We cause a great circle to pass through the point *M* and through the pole of the equator, which is the point *N*; let the circle be *NM*. Let this circle cut the arc *KL* at the point *P* and let it cut the arc *KO* at the point *F*; so the arc *PL* is similar to the arc *MO*; so the time *MO* is equal to the time *PL* and the arc *OD* is the inclination of the time *PL*. But the arc *LD* is the inclination of the time *KL*, accordingly there remains the arc *LO*, the inclination of the time *PK*. But the arc *PM* is equal to the arc *LO*, so the arc *PM* is the inclination of the time *PK*.

Similarly, we cause an arc of an hour circle to pass through the point *F*; let the arc be *FJ*. Let the arc *FJ* be the required time whose inclination is the arc *FM*. We cause a great circle to pass through the points *N* and *J*; let this circle cut the arc *PK* at the point *U* and cut the arc *KF* at the point *Q*. So the arc *FM* is the inclination of the time *PU*, and there remains the arc *PF*, the inclination of the time *UK*. But *PF* is equal to *UJ*, so the arc *UJ* is the inclination of the time *UK*.

Similarly, if we cause to pass through the point *Q* an arc of an hour circle equal to the time whose inclination is the arc *QJ*, and if we cause a great circle to pass through its endpoint and through the point *N*, [then] on the side towards the point *K*, it cuts off from the arc *UK* an arc whose inclination is the arc cut off on the great circle between the arc *UK* and the arc of an hour circle drawn from the point *Q*.

If we continue to proceed in this way, we prove from this that for any arc cut off from the arc *KL* on the side towards the point *K*, its own inclination is greater than the arc cut off on the great circle drawn from the pole of the equator between the arc *KL* and the *muqanṭara OKI*, corresponding to the two arcs *UQ* and *PF*. We prove from this that the ratio of the arcs cut off on the arc *KL* to the arcs cut off on the great circles between the two arcs *KL* and *KO* is greater than the ratios of the arcs cut off on the arc *KL* to their own inclinations; and if the triangles corresponding to the triangle *KUQ* become smaller and smaller, then there will be no difference between them and the triangles with rectilinear sides, neither in regard to magnitude nor in regard to ratios. Now if these triangles have rectilinear sides, then they are similar, because the angles that are at the points *P*, *U* and those that correspond to them are right angles.[129] The ratios of the arcs cut off on the arc *KL*, to the arcs cut off on the great circles between the two arcs *KU* and *KQ*, are equal to the ratio of the arc *KU* to the arc *UQ*. But if the arcs

---

[129] See Mathematical commentary.

of hour circles are small and if they are contiguous and consecutive, then their ratios to their inclinations are not unequal, because they are so small and are close to one another. So the ratios of the arcs cut off on the arc *UK* to their inclinations are equal to the ratio of *KU* to *UQ*.[130] But the ratio of *KU* to *UQ* is greater than the ratio of *KU* to *UJ*; the ratios of the arcs cut off on the arc *KU* to the arcs of great circles cut off between the two arcs *KU* and *KO* are thus greater than the ratios of these same arcs, cut off on the arc *KU*, to their own inclinations. Now, it has been proved that the ratios of the arcs *KP* and *KU* and their homologues, to the arcs *PF* and *UQ* and their homologues, are greater than the ratios of the arcs *KP*, *KU* and their homologues to the arcs *PM*, *UJ* and their homologues, which are the inclinations of the arcs *KP*, *KU* and their homologues. So the ratio of each of the parts of the arc *KL* to the arc of a great circle cut off between it and the arc *KO* is greater than the ratio of that same part of the arc *KL* to its proper inclination. If this is so, then, if the wandering star moves after having reached the point *K*, it will always be below the *muqanṭara OKI*, because, for each part, among the part of the time during which the wandering star moves, during these parts of the time it will be on the endpoint of the inclination proper to this time. But it has been proved that the inclinations of the parts of the time *KL* lie below the *muqanṭara OKI*. So if the wandering star moves from the point *K*, then, whatever the amount of time in the course of which it moves, it will be below the *muqanṭara OKI*.

So the wandering star does not arrive on the arc *KI* of the *muqanṭara*,[131] because its motion gives the wandering star an inclination in the southerly direction and the wandering star does not arrive on the *muqanṭara OKI* at any point other than the point *K*. If this is so, the wandering star does not arrive on the *muqanṭara OKI* at any points other than the point *K*.

Similarly, we prove that, if we suppose the position of the wandering star is the point *S*, then the wandering star will not have another height

---

[130] See Mathematical commentary.

[131] It has been proved in paragraph a) (p. 228) that the wandering star does not reach the arc *SI*. On the other hand, it does not reach the arc *KS*; in fact, by hypothesis it passes through the point *K* and if it arrived at a point *W* of the arc *KS*, we could prove, by an argument like that used for the point *K*, that the wandering star would not pass through any point of *WO* and thus would not pass through the point *K*; which is contrary to our hypothesis. So the wandering star does not pass through any point of the arc *KI*. The only point of the horizontal circle *OKI* through which the wandering star passes is the point *K*.

If the point *S*, the point of contact of the circle *OI*, of height h_m, with a great circle passing through the pole *N*, is the point where the wandering star reaches the circle *OI*, then this point *S* is the only point of height h_m the wandering star reaches.

equal to the greatest of its heights in the east. That is what we wanted to prove.

Having proved this, we say that, for any horizon for which it always rises in the east and sets in the west, [and] if it has equal heights in the east, equal two by two, the wandering star then does not have a third height equal to the two equal heights; that if two of the wandering star's heights in the east are equal, then they are greater than the height on the meridian; and that for every height it has in the east, smaller than that on the meridian, there is no more than one of them.

<31> Let us return to the figure for Proposition 28, in which we explained the heights in the east. Since the wandering star has travelled from the point *B*, at which it rose, towards the point *M* on the arc *HI*, whose ratio to the arc *ID* is greater than any ratio for any required time for this wandering star between the two endpoints of its motion, to the part of the inclination of the motion of the wandering star proper to this required time, so the wandering star has crossed the *muqanṭara DHG* in a point between the two hour circles *BE* and *HI*. Let the point where the wandering star crossed the *muqanṭara DHG* be the point *K*. So the wandering star arrived on the *muqanṭara DHG* at the points *K* and *D*.

I say that it does not arrive on the *muqanṭara DHG* in any point other than these two points.

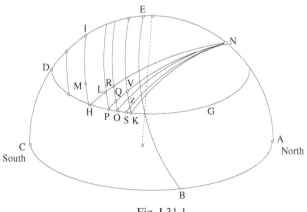

Fig. I.31.1

Proof: If, from any point of the arc *HD*, we draw a straight line to the arc *ID* in a circle parallel to the circle *HI*, its ratio to the chord of the arc that the line cuts off from the arc *ID* is greater than the ratio of the chord of the arc *HI* to the chord of the arc *ID*, as has been proved in Propositions 11 and 12; and the ratio of the chord of the arc *HI* to the chord of the arc *ID* is greater than any ratio for any required time for the wandering star between the two endpoints of its motion to the part of the inclination of the motion of the wandering star proper to that required time. The ratio of any straight line drawn from the arc *HD* to the arc *ID* in a circle parallel to the circle *HI* to the chord of the part that straight line cuts off from the arc *ID* is accordingly greater than any ratio of any required time for the wandering star between the endpoints of its motion to the part of the inclination of the motion of the wandering star proper to this required time; and the ratio of the arc that is subtended by this straight line to the arc that the line cuts off from the arc *ID* is much greater than the ratio of any required time for the wandering star between the endpoints of its motion to the part of the inclination of the motion of the wandering star proper to that required time. So between the arc *HD* and the arc *ID* there is no hour arc such that it is one of the required times for the wandering star. Consequently, the wandering star does not arrive on the arc *HD* of the *muqanṭara DHG*. Nor does the wandering star arrive on the arc *KG* of this *muqanṭara*, because its motion always gives the wandering star an inclination in the southerly direction and the arc *KG* is to the north with respect to the point *K*; so what remains from the *muqanṭara* is the arc *KH*.

We cause a great circle to pass through the point *H* and through the pole of the equator – let the point be *N*; let the circle be *NH*. Since the wandering star travelled from the point *K* to the point *M*, accordingly it crossed the circle *NH*, and it does not cross it at the point *H*, because it does not arrive on the arc *HI* twice, since its motion always gives it an inclination in the southerly direction; so it does not arrive more than once on any hour circle, and it does not cross the circle *NH* at the point *H*. But on the other hand it does not cross it in a point to the south other than the point *H*, because if it travels to the south of the circle *HI*, it does not return to it [the circle] in the course of this motion; so it does not return from the south to the point *M*. If the wandering star travels from the point *K* to the point *M*, then it crosses the circle *NH* and it does not cross it either at the point *H* or at a point to the south other than the point *H*; so it crosses it only in a point of the arc *NH*; let it cross it at the point *L*.

We cause an arc of an hour circle to pass through the point *L*; then this arc cuts the *muqanṭara DHG* in a point between the two points *K* and *H*, because the point *L* is to the south of the point *K* and to the north of the point *M*. Let the hour arc cut the *muqanṭara DHG* at the point *P*, which lies between the two points *K* and *H*. Since the wandering star has travelled from the point *K* to the point *L*, accordingly it does not pass through the point *P*, because it does not arrive on the arc *LP* twice; but it does not, either, pass through the arc *PH*, because the arc *PH* is to the south of the arc *LP*; so it does not return from the arc *PH* to the point *L* and the wandering star thus does not arrive at any point of the arc *PH* in its motion from the point *K* towards the point *L*. But it does not, either, arrive on the arc *PH* in its motion from the point *L* towards the point *M*, because after having reached the circle *NH*, it cannot move to the east of it. So in its motion from the point *K* towards the point *M* the wandering star does not arrive at any point of the arc *PH*.

We cause a great circle to pass through the points *N* and *P*, let the circle be *NP*. Since the wandering star travelled from the point *K* to the point *L*, it crosses the circle *NP* and it does not cross it either at the point *P* or in a point to the south of the point *P*, for the reason given for the circle *NH*; so it crosses the circle *NP* in a point of the arc *NP*, let the point be *Q*.

We cause an arc of an hour circle to pass through the point *Q*, then it cuts the arc *KP* between the points *K* and *P*, for a reason analogous with that given for the point *P*; let the arc be *QO*; let this arc cut the arc *NL* at the point *R*; so the arc *QR* is the required time whose inclination is the arc *RL*.

We cause an arc of an hour circle to pass through the point *K*; let it cut the arc *NQ* at the point *V*, let the arc be *KV*; so the arc *KV* is the required time whose inclination is the arc *VQ*. Since the wandering star travelled from the point *K* to the point *Q*, accordingly it does not pass either through the point *O* or through the arc *OP*, for a reason analogous to that given for the arc *PH*. So, in its motion from the point *K* to the point *Q*, the wandering star does not arrive at any point of the arc *OP*.

Similarly, if we draw the arc *NO* of a great circle, we [can] prove that the wandering star arrives on the circle *NO* at a point between *N* and *O*. So if we draw an hour arc through the point at which it arrives on the arc *NO*, this arc cuts the arc *KO* and we prove that the wandering star does not arrive at any point of the arc of the *muqanṭara* cut off by the hour arc on the side towards the point *O*.

The same is true for all the arcs homologous to the arcs *OP*, *PH*, cut off by the triangles homologous to the triangles *OQP*, *PLH*; many triangles homologous to the triangles *OQP*, *PLH* are generated in this way and, as these triangles approach the point *K*, they become smaller; and the arc that remains between them and the point *K* becomes smaller; and we prove that the wandering star does not pass through the parts of the arc *KH* contained in these triangles. And if these triangles become smaller, there will be no difference between them and the rectilinear triangles, and thus there will be no difference between the ratios of the arcs that enclose these triangles and the ratios of the straight lines, because these triangles are extremely small;

the arc *ED* is in fact the inclination of the motion of the wandering star in the time in the course of which it travelled from the point *B* to the point *D* and this time is a part of the day;[132] in the majority of cases, it is a quarter of a revolution or approximately that, but it never reaches half a revolution. The inclination of the motion of the sun and of the five planets in a single day is several minutes; the size of the inclination of the motion of the sun and the five planets for a quarter of a revolution and for less than half a revolution is several minutes.[133] So for the sun and for the five planets the size of the arc *ED* is some of these minutes. As for the moon, the size of the inclination of its inclined orb with respect to the circle of the equator is a maximum of twenty-nine degrees, which is the maximum inclination of the sun, to which we have added the latitude of the moon in relation to the circle of the ecliptic. This inclination occurs only on rare occasions, but apart from this instant of time, its inclination is smaller than this magnitude and the inclination proper to the motion of the moon in a single day never reaches more than four degrees from the twenty-nine degrees, or something close to it, either more or less, if we start from the twenty-nine degrees.[134] The inclination of the motion of the moon, referred to the circle of the equator, for a quarter of a day or something close to it, does not reach more than a single degree with a small addition. So for the moon, the arc *ED* is at most approximately a degree, and this on rare occasions and for the places where the inclination is in the east; for other instants and for the horizons of habitable latitudes, the inclination does not reach a single degree. The inclinations, which are *HL*, *PQ* and their homologues, are the inclinations proper to the required times between the two points *K* and *L*[135] and each of them [the inclinations] is a small part of the arc *ED*; the small triangles generated between the points *K* and *H* in the way we have described, and [which are] homologous with the triangle *LPH*, are extremely small, and the same holds for their inclinations and for the required times; because if

---

[132] See notes 133 and 134 below.

[133] The inclination of the Sun with respect to the equator goes from 0 to $23°27'$ in approximately 90 days, that is it changes by $15'$ to $16'$ per day. Between sunrise and meridian passage, the interval is 6 hours at an equinox; we then have $\overset{\frown}{ED} \cong 4'$.

[134] The inclination of the orb of the moon with respect to the equator has been investigated in Propositions 16 and 22. The maximum inclination is close to $29°$ and is reached very rarely (periodicity 18 years 8 months). The moon makes a complete circuit of its orb in a Lunar month, of about 29 days, so the inclination of the moon goes from $0°$ to $29°$ in a quarter of that period, so in the course of a day it varies by about $4°$; and in 6 hours, on an equinox, it varies by about $1°$.

[135] They are the inclinations *LR* and *VQ*, which are the inclinations proper to the required times *QR* and *KV*, which correspond, respectively, to the passage of the wandering star from the point *Q* to the point *L* and from the point *K* to the point *Q*.

the inclinations are extremely small, then their required times are equally small, since there is no difference, in regard to their magnitudes and their ratios, between the perimeters of these triangles and straight lines. If these arcs are extremely small, then there is no difference between the ratios of the arcs that they form,[136] [arcs] which are the required times, to the inclinations which belong to them and the ratio of the time *KV* to the inclination *VQ*, because the two arcs *KV* and *VQ* are among these small arcs and the arcs that these latter include, which are the sides of the small triangles; so they are smaller than them. But if the required times are small and are contiguous, then there is no difference between the ratio of each of them to the inclination proper to it and the ratio of the required time that follows it to the inclination proper to it. So there is no difference between the ratio of the time *KV* to the arc *VQ* and the ratios of the required times that are the sides of the small triangles to the inclinations proper to them.

This being proved, we say: if the wandering star moves from the point *K* to the point *Q*, it will not pass through any point of the arc *KO*.

Now, we take any point on the arc *KO*; let the point be *S*. We cause a great circle to pass through the point *S* and through the pole *N*; let the circle be *NS*. This circle cuts the arc *KV*; let it cut it at the point *Z*. We form the triangle *KZS*; this triangle is right-angled because the circle *NZS* is orthogonal to the circle *KV*. Similarly, the triangle *KVP* is right-angled because the circle *NVP* is orthogonal to the circle *KV*. So the angle *KVP* is a right angle, the angle *KZS* is a right angle and there is no difference between the arcs of the two triangles *KVP* and *KZS* and straight lines. But the triangle *KZS* is <constructed> on a side of the triangle *KVP* and the two angles *KZS* and *KVP* are right angles; so the two triangles are similar and the ratio of *KV* to *VP* is equal to the ratio of *KZ* to *ZS*. Consequently, in most cases, the ratio of *KV* to *VP* is greater than the ratio of *KZ* to *ZS*; and, in some cases, the ratio of *KV* to *VP* is equal to the ratio of *KZ* to *ZS*. So in all cases the ratio of *KV* to *VP* is not smaller than the ratio of *KZ* to *ZS*. But the ratio of *KV* to *VQ* is greater than the ratio of *KV* to *VP*, so the ratio of *KV* to *VQ* is greater than the ratio of *KZ* to *ZS* in all circumstances. But the ratio of the time *KZ*, which is the required time, to the inclination of the motion of the wandering star proper to this time *KZ*, is equal to the ratio of the time *KV* to the arc *VQ*, as has been established earlier; but, among these arcs, *KZ* is smaller than *VQ*; so the ratio of the time *KZ* to the inclination proper to the time *KZ* is greater than the ratio of the time *KZ* to the arc *ZS*, and the inclination proper to the time *KZ* is smaller than the arc *ZS*. So, in its motion from the point *K* to the point *Q*, the wandering star crosses the arc *ZS* and does not pass through the point *S*. If it does not pass through the

---

[136] See Mathematical commentary, pp. 232–8.

point *S*, then it does not pass through the arc *SO*, as has been proved for the arc *OP* and the arc *PH*.

Similarly, we have proved, for any point we take on the arc *KO*, that the wandering star does not pass through it, as has been proved for the point *S*; so the wandering star does not pass through any point of the arc *KO*. But we have proved that it does not pass through the arcs *OH*, *HD*, *KG*; so the wandering star does not pass through any point of the *muqanṭara DHG* except the two points *K* and *D*; this if the circle *NH* is higher than the point *K*.

But if the circle *NH* passes through the point *K*, the arc *KH* will lie to the east of the circle *NH* and the wandering star reaches it at the point *K*, so the wandering star does not pass through any point of the arc *KH*.

If the circle *NH* is below the point *K*, then we cause a great circle to pass through the two points *N* and *K*; it cuts the hour arc *HM*; so the arc *KH* will lie to the east in relation to the great circle that passes through the point *K* and the wandering star reaches it at the point *K*, so the wandering star does not pass through any point of the arc *KH*.

In all cases of the figure, the wandering star does not arrive on the *muqanṭara DHG* except in only two points.

I say that it does not arrive more than once on any of the *muqanṭarāt* closer to the horizon than the *muqanṭara DHG*.

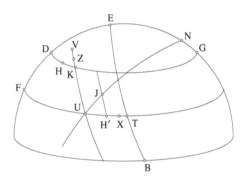

Fig. I.31.2

Let the *muqanṭara FXT* be closer to the horizon than the *muqanṭara DHG*. This *muqanṭara* cuts the circle *BE* and cuts the circle *VZK*. Let us extend the arc *VZK* until it cuts this *muqanṭara*; let it cut it at the point *U*. In its motion from the point *B* to the point *K*, the wandering star crosses the *muqanṭara FXT* and crosses it in a point to the south of the arc *BE* and to the north of the arc *VU*; let it cross it at the point *X*. Let us cause a great circle to pass through the points *U* and *N*; let the circle be *NU*. So if the wandering star moves from the point *X* to the point *K*, then it crosses the circle *NU* without crossing it at the point *U*, or in a point to the south of the circle *KU*, as was proved earlier; so it crosses it in a point of the arc *NU*; let it cross it at the point *J*. The wandering star does not arrive on the arc *UF* in its motion from the point *X* to the point *K*, because the arc *UF* is to the south of the point *K*. So if the wandering star has arrived on the arc *FU*, it would not have come to the point *K*. The wandering star does not arrive on the arc *XT* of the *muqanṭara* because it is to the north of the point *X*; so all that remains of the *muqanṭara* is the arc *XU*.

But, in its motion from the point *X* to the point *K*, the wandering star crosses the circle *NU* at the point *J*; so the point *J* is to the south of the point *X* and is to the north of the point *F*. We cause an arc of an hour circle to pass through the point *J*; let it cut the *muqanṭara* *FXT* at the point *H'*; so this gives rise to the triangle *H'JU* and we prove that the wandering star does not arrive on the arc *H'U* belonging to the triangle *H'JU*, as has been proved for the arc *OP* of the triangle *OQP*. We prove that it does not arrive on the arc *XH'*, as has been proved for the arc *KO*. If the great circle *NU* passed through the point *X* or under the point *X*, the arc *XU* would lie to the east of the great circle that passes through the point *X*; then the wandering star does not pass through the latter, and the wandering star thus does not arrive on the part of the *muqanṭara* *FXT* east of the meridian except at the point *X*. Similarly, for any *muqanṭara* between the horizon and the *muqanṭara* *DHG*, the wandering star arrives on it only in a single point.

<32> Let us return to the figure; let the point *K* be the highest point attained by the wandering star. Let the *muqanṭara* *FXT* lie below the point *K* and above the *muqanṭara* *DGH*. We cause an arc of an hour circle to pass through the point *K*; let it be *KL*. This arc cuts the *muqanṭara* *FXT*, because if the circle *BE* cuts the eastern horizon, then the circle *LK* cuts the eastern horizon; and if this circle *LK* cuts the eastern horizon, then it cuts the *muqanṭara* *FXT*; let the arc *LK* cut the *muqanṭara* *FXT* at the point *I*. Since the wandering star has travelled from the point *B* to the point *K*, accordingly it crosses the *muqanṭara* *FXT*; let it cross it at the point *X*. The point *X* is to the north of the point *K* and to the south of the point *T*; so it lies between the two circles *BE* and *LI*. The wandering star, after having reached the point *K*, then returns to the point *D*; thus the wandering star has crossed the *muqanṭara* *FXT* in a point other than the point *X*, because the point *X* is to the north of the point *K* and the other point is to the south of the point *K*; let the other point be the point *M*. So the wandering star arrives on the *muqanṭara* *FXT* at the two points *X* and *M*.

Then I say that it does not arrive on this *muqanṭara* at a third point.

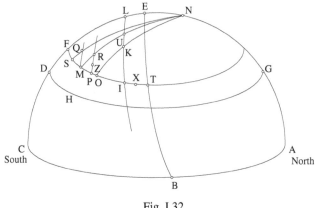

Fig. I.32

It is clear that it does not arrive on the arc *XT*, because the arc *XT* is to the north of the point *X* and its motion always gives the wandering star an inclination in the southerly direction; accordingly it does not arrive on the arc *XT*.

That it does not arrive on the arc *XI*: this is proved as we proved in the preceding proposition that it does not arrive on the arc *KH* of the *muqanṭara DHG*, the [arc concerned] in the preceding proposition.

That it does not arrive on the arc *IM*: this is proved as we shall describe. We cause a great circle to pass through the points *N* and *K*; so it cuts the arc *FI*; let it cut it at the point *O*; let the circle be *NKO*. The wandering star does not arrive on the arc *IO*, because it will not travel to the east of the circle *NKO*; and it will not travel to the point *O*, because it will not arrive twice on the circle *NK*; so the wandering star does not arrive on the arc *IO*.

We cause a great circle to pass through the points *N* and *M*; let it be *NM*; let it cut the arc *LK* at the point *U*. So the arc *KU* is the required time and the arc *UM* is the inclination proper to this time *KU*. We take any point on the arc *OM*; let the point be *P*. We cause an hour arc to pass through the point *P*; then it cuts the arc *UM*; let the arc be *PR*. The ratio of the arc *PR* to the arc *RM* is equal to the ratio of the arc *IU* to the arc *UM*, because these arcs are extremely small; so there is no difference between them and the straight lines which are their chords. But the two triangles *IUM* and *PRM* are right-angled, because each of the two angles *IUM* and *PRM* is a right angle; so the ratio of the arc *PR* to the arc *RM* is equal to the ratio of the arc *IU* to the arc *UM*. But the ratio of the arc *IU* to the arc *UM* is greater than the ratio of *KU* to *UM*, so the ratio of *PR* to *RM* is greater than the ratio of *KU* to *UM*; but the ratio of *KU* to *UM* is equal to the ratio of the time *PR* to the inclination proper to *PR*, because these times are close and extremely small; accordingly there is no difference between their ratios to their inclinations, so the ratio of *PR* to *RM* is greater than the ratio of *PR* to the inclination proper to *PR*; so the inclination proper to *PR* is greater than the arc *RM* and the time whose inclination is the arc *RM* is smaller than the time *PR*. Let this time be the time *ZR*, the arc *ZR* is the required time whose inclination is the arc *RM*. But the wandering star, in its motion from the point *K* to the point *M*, passed through the circle *PR*, to arrive later at the point *M*. But when the wandering star arrived on the circle *PR*, then the arc which lies between it and the circle *NRM* is the required time whose inclination is *RM*; but the required time whose inclination is *RM* is the arc *ZR*; consequently, when the wandering star arrives on the circle *PR*, it arrives at the point *Z*. But if the wandering star arrives at the point *Z*, then it does not arrive on the circle *PR* at another point, so the wandering star does not pass through the point *P*.

Similarly, for any point of the arc *OM*, we prove that the wandering star does not pass through this point, as we have proved for the point *P*; so the wandering star does not arrive on the arc *OM*.

We also mark a point *S* on the arc *MF* and we cause a great circle to pass through the points *N* and *S*; let it be *NS*. We cause an hour arc to pass through the point *M*; let it cut the circle *NS* at the point *Q*. So we have that the ratio of *MQ* to *QS* is equal to the ratio of *PR* to *RM* because of these triangles being small; thus there is no difference between them and straight lines. But the ratio of *PR* to *RM* is greater than the ratio of *ZR* to *RM*, so the ratio of *MQ* to *QS* is greater than the ratio of *ZR* to *RM*. But *ZR* is a required time on account of the inclination of the arc *RM*; so the ratio of the time *MQ* to the inclination proper to it is the ratio of *ZR* to *RM*; the ratio of the time *MQ* to the arc *QS* is greater than the ratio of the time *MQ* to the inclination proper to the time *MQ*; so the inclination proper to the time *MQ* is greater than the arc *QS* and the limit in that inclination is below the *muqanṭara TX*. So if the wandering star moves from the point *M* for the time *MQ*, its centre arrives on the circle *NQS* at a point of it below the point *S*. And if the wandering star arrives on the circle *NQS* at a point other than the point *S*, then it does not pass through the point *S*, but passes below the point *S*.

Similarly, we prove, for any point of the arc *MF*, as has been proved for the point *S*, that the wandering star does not pass through it but below it. So the wandering star does not arrive at any point of the arc *MF*. But if it moves from the point M towards the point D, it will always be below the *muqanṭara FXT*.

Thus we have proved that the wandering star does not arrive on the arc *IMF* except at the point M and does not arrive on the arc *IT* except at the point X. So the wandering star does not arrive on the *muqanṭara FXT* except at only two points.

Similarly, for any *muqanṭara* above the *muqanṭara DGH* and below the point K, the wandering star does not arrive on it except at only two points. Among its equal heights in the east, the wandering star has no more than two equal heights in the east.

So, if it has equal heights in the east, each of the seven wandering stars will have not more than two equal heights; all these heights will be greater than the height on the meridian, it [the wandering star] will have another height that is only equal to the height on the meridian and any height that it has in the east <before it crosses the meridian> will be smaller than its height on the meridian. So it has no other height in the east that is equal to it; that is what we wanted to prove in the last two propositions.

<33> Let us return to the figure for heights in the west.

I say that the greatest of the heights of the wandering star in the west is a single height.

That the wandering star has a height that is the greatest of its heights in the west: this is proved as it was proved in the proposition about heights in the east; that in the west the wandering star does not have a height equal to the greatest of its heights: this is proved as we shall describe.

Let the highest *muqanṭara* reached by the wandering star in its motion from the meridian to the western horizon be the *muqanṭara OKE*. We cause an hour arc to pass through the point O; let it be OP. We draw from the pole of the equator, which is the point N, a great circle that touches the *muqanṭara OKE*; let the circle be NF; let this circle touch the *muqanṭara OKE* at the point F. We prove, as has been proved in the proposition on the heights in the east, that the wandering star does not arrive at any point of the arc FE, because if we draw through any point of the arc FE a great circle from the point N, then it cuts the *muqanṭara OKE* in another point and it cuts the circle OP in such a way that the arc of a great circle that lies between the point of the arc FE and the arc OP is the inclination of the time cut off on the arc OP; the two hour arcs drawn from the two points of intersection of the *muqanṭara* and the great circle as far as the meridian are equal. From this there follows the impossibility demonstrated in the proposition on the heights in the east, which is Proposition 30; so the wandering star does not arrive at any point of the arc FE. The position of the wandering star on the *muqanṭara OKE* is either the point *F* or lies to the south of the point *F*. Let the position of the wandering star be the point *K* and let the point *K* be the point after which the wandering star does not arrive at any point of the *muqanṭara OKE*.

I say that the wandering star does not arrive on the *muqanṭara OKE* at any point other than the point *K*.

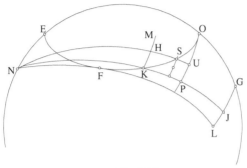

Fig. I.33

Proof: We cause a great circle to pass through the two points *N* and *K*; let it cut the arc *OP* at the point *P* and let this circle cut the arc *GL* at the point *J*.[137] Since the wandering star moves from the point *G* to the point *K*, the arc *GJ* is the required time whose inclination is the arc *KJ*; but the arc *OP* is similar to the arc *GJ* and the arc *KP* is a part of the arc *KJ*; so the required time whose inclination is the arc *KP* is smaller than the time *OP*. Let this required time be the arc *UP*; so if the wandering star moves from the point *G* to the point *K*, then it crosses the circle *OP*; but if it crosses the circle *OP*, it crosses it at the point *U*. We cause an hour arc to pass through the point *K*; let the arc be *KM*. And we cause a great circle to pass through the points *N* and *U*; let the circle be *NU*. This circle cuts the arcs *KM* and *KO*; let it cut the arc *KM* at the point *H* and let it cut the arc *KO* at the point *S*. So the arc *HS* is smaller than the arc *HU* and the arc *HU* is the inclination of the time *HK*, which is similar[138] to *UP*; so the ratio of *KH* to *HS* is greater than the ratio of *KH* to *HU*. If we cause an hour arc to pass through the point *S*, it cuts the arc *KP*; this hour arc will be similar to the arc *UP* and the arc cut off from the arc *KP* is equal to the arc *HS*; so the required time whose inclination is the arc cut off from the arc *KP* is a part of the hour arc drawn from the point *S* as far as the circle *KP*. So a part of this

---

[137] The arc *GL* is an arc of the hour circle of *G*, *G* is the point where the wandering star crosses the meridian.

[138] Lit.: equal.

hour arc is the required time whose inclination is the arc cut off from the
arc KP.

We prove, by this procedure, as has been proved in the proposition on
the heights in the east, that, in its motion from the point U to the point K,
the wandering star will always be below the *muqanṭara* OKE; so the
wandering star does not arrive at any point of the arc KO; but it does not,
either, arrive at any point of the arc KE, because the point K is the point
after which the wandering star does not arrive at any point of the
muqanṭara OKE. So the wandering star does not arrive on the *muqanṭara*
OKE in any point other than K.

Similarly, if we suppose that the position of the wandering star is the
point F, we prove by an analogous demonstration that the wandering star
does not arrive at any point of the *muqanṭara* OKE.

So, in all cases for the figure, the wandering star does not arrive on the
muqanṭara OKE in any point other than K, and the wandering star thus has
no other height in the west equal to the greatest of its heights in the west.
That is what we wanted to prove.

This being proved, we say that, for any horizon for which the
wandering star always rises in the east and sets in the west, if it has equal
heights in the west, equal two by two, then it has not got a third height
equal to the two equal heights; and that for any height it has in the west
which is smaller than that on the meridian, it will have no more than a
single one.

<34> Let us return to the proposition on the heights in the west. Let the
ratio of the chord of the arc HK to the chord of the arc KG be greater than
any ratio for any required time that occurs for the wandering star in its
motion from the point G to the point D, to the inclination proper to that
required time on the side towards the beginning of the motion; so the ratio
of the arc HK to the arc KG is much greater than any ratio of any required
time to the inclination proper to that required time; now the ratio of the arc
KH to the arc EH is the ratio of the arc HK to the arc KG. Let the required
time whose inclination is the arc KG, be the time KM; so the wandering
star travels from the point G to the point M. Let the second point of the
muqanṭara GHI through which the wandering star passes be the point F.

I say that the wandering star does not arrive on the *muqanṭara* GHI
except at the two points G and F.

Proof: If from any point of the arc HG we draw a straight line to the arc
KG in a circle parallel to the circle KH, then its ratio to the chord of the part
it cuts off from the arc KG is greater than the ratio of the chord of the arc

HK to the chord of the arc *KG*; the ratio of any arc drawn between the two arcs *HG* and *KG*, parallel to the arc *HK*, to the arc cut off from the arc *KG* is thus greater than any ratio for any required time that occurs for the wandering star to the inclination proper to this required time; so the wandering star does not pass through any point of the arc *GH*, but in its motion from the point *G* to the point *M* it will always be above the *muqanṭara HI*. We cause an hour circle to pass through the point *F*; let the circle be *FS*. But the wandering star has travelled from the point *M* to the point *F*, so it crosses the circle *NH* and it does not cross it at the point *H*, because it does not arrive twice on the arc *KH*; it does not cross it in any point to the south other than the point *H*, because it does not go below the arc *HG*; it does not cross it at the point *S*, because it does not pass though the point *S* of the arc *FS* and it does not cross it in a point of the arc *NS* because it would be to the north of the arc *FS* and it would not return to the point *F*. But since it arrives at the point *F*, it does not arrive on the circle *NH* except in a point between the points *S* and *H*; let the point of the arc *SH* through which the wandering star passes be the point *P*. If we draw an hour arc from the point *P*, it cuts the arc *FH*; a triangle is formed on the side towards the point *H*. We prove that the wandering star does not arrive at any point of the arc contained in this triangle on the side towards the point *H*, because the arc – a part of the arc *FH* contained in this triangle – lies to the south of the hour arc. We prove, by the procedure we employed in the proposition about heights in the east, that the wandering star does not arrive at any point of the arc *FH*.

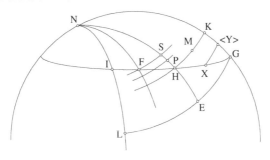

Fig. I.34.1

Similarly, we cause a great circle to pass through the two points *N* and *F*; let the circle be *NF*. If this circle touches the *muqanṭara GHI* at the point *F*, then the wandering star, after moving from the point *F*, does not arrive at any points of the arc *FI*, because the arc *FI* is to the east of the circle *FN*. If the circle *NF* cuts the *muqanṭara GHI*, then it cuts it in two points. If the point *F* is the northern point of these two points of intersection, then the wandering star does not arrive at any point of the arc *FI*, because the arc *FI* will lie to the east of the circle *NF*. But if the point *F* is the southern point of the two points of intersection, then we draw through the point *N* a circle that touches the *muqanṭara IHG*; let the circle be *NXU*; let the point *X* be the point of contact. We extend the hour arc *SF* so that it cuts the circle *NXU* and cuts it at the point *U*. We cause an hour arc to pass through the point *X*; let it cut the circle *NF* at the point *R*. So we have that the ratio of the arc *FS* to the arc *SH* is equal to the ratio of the arc *XR* to the arc *RF*, as has been proved in Proposition 31. But the ratio of *FS* to *SP*, which is the ratio of the required time to the inclination that is proper to it, is greater than the ratio of *FS* to *SH*; so the ratio of the time *FS* to the inclination *PS* is greater than the ratio of the arc *XR* to the arc *RF*. But the ratio of the time *XR* to the inclination proper to the time *XR* is the ratio of *FS* to *SP*, because these times are small and close to one another; so their ratios to their inclinations are not unequal and the ratio of the time *XR* to the inclination that is proper to it is greater than the ratio of *XR* to *RF*. So the inclination proper to the time *XR* is smaller than the arc *RF*; but the time *XR* is the time *FU* and the arc *XU* is equal to the arc *RF*; so the ratio of *FS* to *SP* is greater than the ratio of the arc *UF* to the arc *UX*; so if the wandering star moves from the point *F* for the time *FU*, then it arrives on the arc *UX* between the two points *U* and *X*. The same is true for any point of the arc *XF*; if through such a point we draw a great circle from the point *N*, then it cuts the arc *FU*. If we cause an hour arc to pass through <each of> these points, then it cuts the arc *RF*. Thus we prove, as was proved for the arc *RF*, that the arc cut off by the hour arc on the arc *RF* is greater than the inclination of that hour arc. It necessarily follows from this that the endpoints of the inclinations that are generated between the point *F* and the circle *NU* are all below the arc *FX*. From this it is clear that the wandering star does not arrive at any point of the arc *XF*. If the wandering star arrives on the circle *NU*, it does not arrive at any point of the arc *XI*, because the arc *XI* lies to the east of the circle *NU*, so the wandering star does not arrive at any point of the arc *FI*. But we have proved that it does not arrive at any point of the arc *FG*, so the wandering star does not arrive at any point of the *muqanṭara GHI*, which is the *muqanṭara* on the meridian, except in the two points, *G* and *F*.

Similarly, we draw one of the *muqanṭarāt* below the *muqanṭara FHI*, let it be the *muqanṭara OUT*; let the wandering star arrive on this *muqanṭara* at the point *H'*. It is clear that the point *H'* is to the north of the arc *SF*, because its motion gives the wandering star an inclination in the northerly direction.

I say that the wandering star does not arrive on the *muqanṭara OUT* at a point other than the point *H'*.

Proof: We extend the hour arc *SF* so as to make it cut the *muqanṭara OUT*; let it cut it at the point *U*. We cause a great circle to pass through the two points *N* and *H'*; let the circle be *NH'*. Then the circle *NH'* either touches the *muqanṭara OUT* at the point *H'*, or cuts it in two points one of which is the point *H'*.

If it touches the *muqanṭara*, then the wandering star does not arrive on the arc *H'T* of the *muqanṭara*, because the arc *H'T* lies to the east of the circle *NH'*. It does not arrive on the arc *H'O* of the *muqanṭara* because the arc *H'O* lies to the south of the point *H'*, which is the position of the wandering star; so the wandering star does not arrive on the *muqanṭara OUT* except at the point *H'*.

Similarly, if the circle *NH'* cuts the *muqanṭara OUT* in a point *H'* and in a point to the south of the point *H'*, the arc *H'T* will lie to the east of the circle *NH'* and the arc *H'O* will lie to the south of the wandering star.

If the circle *NH'* cuts the *muqanṭara OUT* at the point *H'* and in another point to the north of the point *H'*, as shown in the figure, then we cause a great circle to pass through the two points *N* and *U*; let the circle be *NU*. This circle also cuts the *muqanṭara OUT* in two points, one of which is the point *U* and the other lies to the north of the point *U*, from what is shown in the figure. We cause an hour arc to pass through the point *H'*; let it be *H'J*. The wandering star travels from the point *F* to the point *H'*; so it crosses the circle *NU* and it crosses it only in a point to the south of the arc *H'J* and to the north of the arc *FU*; so it crosses it in a point of the arc *JU* between the two points *J* and *U*. So if we cause an hour arc to pass through this point, then it cuts the arc *H'U*. We prove, as has been proved for the arc *FX*,[139] by considering the small triangles that are formed inside the triangle *H'JU*, that the wandering star does not pass through any point of the arc *H'U*; and we prove that it does not arrive at any point of the arc *H'T*, as has been proved for the arc *FI*. But the wandering star does not arrive at any point of the arc *UO*, because the arc *UO* lies to the south of the arc *FU*. So the wandering star does not arrive at the *muqanṭara OUT* except only at the point *H'*.

---

[139] See Fig. I.34.1.

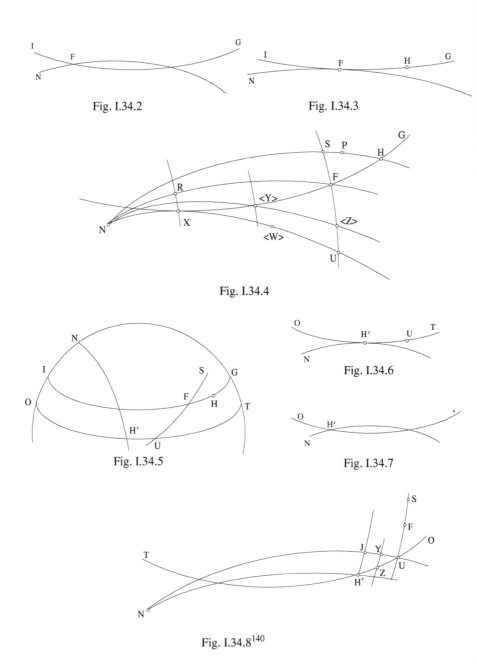

Fig. I.34.2

Fig. I.34.3

Fig. I.34.4

Fig. I.34.5

Fig. I.34.6

Fig. I.34.7

Fig. I.34.8[140]

[140] Figures I.34.2 to I.34.8 are added by the editor.

Similarly, we prove, for any *muqanṭara* between the *muqanṭara GHI* and the horizon, that the star does not arrive on it except in a single point. Accordingly, for all heights of the star in the west that are less than that on the meridian, there will be only a single such height.

<35> Let us return to the figure; let the point *U* be the highest point reached by the wandering star. Let the *muqanṭara OST* be higher than the *muqanṭara GHI* and lower than the point *U*.

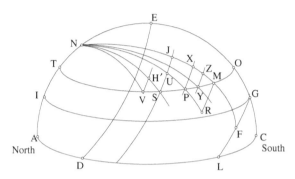

Fig. I.35

We have proved earlier, for any *muqanṭara* higher than the *muqanṭara* for its meridian passage, that if the wandering star goes above it, then the wandering star arrives on it at two points. So the wandering star arrives on the *muqanṭara OST* in two points. We cause an hour arc to pass through the point *D*; then it cuts the meridian circle; let it cut it at the point *E*; so the circle *DE* lies to the north of the point *U*, because the point *D* is to the north of the point *U* and the circle *GL* lies to the south of the point *U*; so the point *U* lies between the two circles *ED* and *GL*. But these two circles cut the horizon; so the hour circle that passes through the point *U* cuts the horizon. But the point *U* is higher than the *muqanṭara OST*; so the hour circle that passes through the point *U* cuts the *muqanṭara OST*. We cause an arc of the hour circle to pass through the point *U*; let it cut the *muqanṭara OST* at the point *S*. We cause a great circle to pass through the two points *N* and *S* – let the circle be *NS*; and a great circle through the points *N* and *U* – let the circle be *NU*. Let this circle cut the *muqanṭara OST* at the point *P*. Since the wandering star travelled from the point *G*, which is below the *muqanṭara OST*, to the point *U*, which is above the *muqanṭara OST*, accordingly it crosses the *muqanṭara OST*, and it does not cross it in a point of the arc *SP*; indeed if it passed through a point of the arc *SP*, it would not arrive at the point U, because the point U lies to the east of any point of the arc SP; it does not pass through the point P either, because it does not arrive on the circle NU at two points and the wandering star does not pass through the point S either, because it does not arrive on the arc US at two points. The wandering star does not arrive at any point of the arc SP; but it crosses the *muqanṭara* OST before reaching the point U; so it crosses the *muqanṭara* on the arc OP; let it cross it at the point M. We cause a great circle to pass through the points M and N; let the circle be NM. So the arc GF is the required time whose inclination is the arc FM; let the circle NF cut the arc SU in a point J. So the wandering star has travelled from the point G to the point M; let the point M be the first point at which the wandering star arrives on the *muqanṭara* OST. So the wandering star does not arrive at any point of the arc MO, because the arc MO lies to the east of the circle NM. We cause an hour arc to pass through the point M; let it cut the circle NUR at the point R. So the arc MR will be the required time whose inclination is the arc RU. So the ratio of UJ to JM is the ratio of the required time to the inclination that is proper to it. We cause an hour arc to pass through the point P; let it cut the circle NM at the point X. So the arc PX is similar to the arc UJ and the arc JM is greater than the arc XM; so we have that the ratio of PX to XM is greater than the ratio of UJ to JM. If from any point of the arc MP we draw an hour arc that cuts the arc XM, we prove that, for the small triangles, as was proved in the earlier proposition, their ratios to the parts they cut off from the arc XM[141] are equal to the ratio of PX to XM. We prove from this that, for any hour arc drawn from a point of the arc PM to the arc XM, its ratio to the part it cuts off from the arc XM is greater than its ratio to the inclination proper to it. If we cause to pass through the point N, and through a point of the arc PM, a great circle that cuts off from the arc UJ an arc similar to the arc that is inside the arc PM and such that the arc of the great circle situated between the two arcs PM and MR is equal to the arc cut off from the arc XM, then we prove from this that, for any part among the parts of the time MR, its inclination is greater than the arc situated between this part[142] and the arc MR. So the wandering star does not pass through any point of the arc MP, but it will pass above it. It is clear that it does not pass through the arc SM.

Let the second point of the *muqanṭara* OST through which the wandering star passes be the point V. So the point V lies between the two circles SU and DE. We cause an hour arc to pass through the point V, let it

---

[141] For each of the triangles obtained this way, we consider the ratio of the sides homologous to the sides PX and XM.

[142] Ibn al-Haytham means the hour arc drawn from a point of the arc PM to the arc XM.

be *VH'*. In its motion from the point *U* to the point *V*, the wandering star crosses the arc *H'S* between the two points *H'* and *S*. We prove, as was proved in the preceding proposition, that the wandering star does not arrive at any point of the arc *VS* except for the point *V*. We prove, as was proved in the preceding proposition, that the wandering star does not arrive at any point of the arc *VT*, because the great circle that passes through the two points *N* and *V* either touches the *muqanṭara* at the point *V*, or cuts it in this point. We prove, by the same method we followed in the preceding proposition, that the wandering star does not arrive at any point of the arc *TV*; so the wandering star does not arrive at any point of the western part of the *muqanṭara OST* except at the points *M* and *V*.

Similarly, we prove, for any *muqanṭara* that is higher than the *muqanṭara GHI* and lower than the point *U*, that the wandering star does not arrive on it at more than two points.

Thus for each of the seven wandering stars, if it has equal heights in the west, then it will not have more than two equal heights; and for any height it has in the west and which is smaller than its height on the meridian, then it will not have any other height in the west that is equal to it, it will not have any height in the west equal to its greatest height and all its equal heights are greater than the height on the meridian. That is what we wanted to prove.

It is clear from all that we have established that each of the seven wandering stars can have, in a single day, equal heights in the east, equal two by two, and that it can have, in a single day, equal heights in the west, equal two by two; and that this happens if its meridian passage is to the south of the pole of the horizon.

But if the position of its meridian passage is to the north of the pole of the horizon, then it will be able to have equal heights, in the east and in the west; it is possible for it not to have any, depending on the horizon [concerned].

<36> For horizons for which the sphere is right, it is possible that each of the seven wandering stars has equal heights in the east or equal heights in the west. In fact, the position of the hour arcs starting from the *muqanṭara* after the meridian passage of the wandering star, in the direction north of the pole of the horizon or in the direction south of the pole of the horizon, is the same position, because the hour circles are orthogonal to the plane of the *muqanṭara*. So for any hour arc that is above the *muqanṭara* on the meridian and to the south of the pole of the horizon, there is, in the direction north of the pole of the horizon, an hour arc above

the *muqanṭara* on the meridian, which is equal to it and which, on the side toward the point of the meridian, cuts off from the meridian circle an arc equal to the arc of the meridian circle cut off by the hour arc to the south on the side towards the meridian. It follows that there exist, in the northerly direction in relation to the pole of the horizon, hour arcs whose ratio to the part they cut off from the meridian circle on the side towards the point of the wandering star's meridian passage is greater than any ratio for any required time for the wandering star to the inclination of the motion of the wandering star.

If this is so, then in the northerly direction in relation to the pole of the horizon, and if [the point of] the wandering star's meridian passage lies to the north of the pole of the horizon, what occurs is analogous to what occurs in the southerly direction. So the wandering star has equal heights in the east and equal heights in the west for the horizons for which the sphere is right, whether [the point of] the wandering star's meridian passage lies to the south of the pole of the horizon or to the north of the pole of the horizon.

For horizons for which the sphere is inclined towards the south, then the wandering star, if its [point of] meridian passage is to the north in relation to the pole of the horizon, can have equal heights in the east or in the west, but they will be few and close to one another and this occurs in places close to the equator for which the sphere is slightly inclined towards the south. For strongly inclined horizons, that is to say those for which the sphere is inclined with a large inclination, then this result does not occur. The cause of this is that, for horizons for which the sphere is inclined towards the south, with a large inclination, the hour arcs to the south of the pole of the horizon are inclined towards the south. Thus the arcs of the meridian circle cut off by the hour arcs are small; so the ratios of the hour arcs to these latter will be ratios of a considerable size.[143] It is possible that there are among them ratios that are greater than any ratio for any required time that occurs for the wandering star to the inclination of the motion of the wandering star. But the hour arcs that lie to the north of the pole of the horizon, for horizons for which the sphere is inclined with a large inclination towards the south, are also inclined with a large inclination towards the south. The arcs cut off by these hour arcs on the meridian circle on the side towards the northern point of the wandering star's meridian passage are much greater than the arcs of the meridian circle cut off by the hour arcs in the south. In most cases, the ratios of the hour arcs in the north to the parts they cut off from the meridian circle are not greater

---

[143] Lit.: of a great magnitude.

than any ratio for any required time for the wandering star to the inclination of the motion of the wandering star. That is why it rarely happens that the heights in the east are equal and that the heights in the west are equal for horizons that are strongly inclined towards the south, if the wandering star's [point of] meridian passage is to the north of the pole of the horizon. Here I mean by 'for horizons that are strongly inclined' the horizons whose inclination, despite its size, remains smaller than the greatest inclination of the wandering star with respect to the circle of the equator. In these locations the meridian passage of the wandering star is sometimes to the north of the pole of the horizon and sometimes to the south of the pole of the horizon. But the locations for which the meridian passage of the wandering star is sometimes to the north of the pole of the horizon and sometimes to the south of the pole of the horizon are the locations for which the height of the pole above their horizon is smaller than the inclination of the inclined orb of the wandering star with respect to the circle of the equator. Indeed, in these locations a part of the inclined orb of the wandering star turns in accordance with hour circles to the north of the pole of the horizon, and a part of the inclined orb turns in accordance with hour circles to the south of the pole of the horizon. For locations for which the heights of the pole are greater than the inclination of the inclined orb of the wandering star with respect to the circle of the equator, the inclined orb of the wandering star, as a whole, turns in accordance with hour circles [that are] all to the south of the pole of the horizon. In these locations the meridian passage of the wandering star is always to the south of the pole of the horizon. For horizons where the height of the pole is greater than the inclination of the inclined orb of the wandering star, the wandering star always has some equal heights in the east and some equal heights in the west in the times that we have determined. For horizons where the height of the pole is smaller than the inclination of the inclined orb of the wandering star, for them the wandering star has some equal heights in the east and some equal heights in the west, if its meridian passage is to the south of the pole of the horizon. But if its meridian passage is to the north of the pole of the horizon, then it is possible for this to happen to it, if it is in the same ratio that occurs when its meridian passage is to the south, and this [happens] for horizons with a slight inclination. But it is possible that this result does not occur for it, and this for horizons with a strong inclination.

But if the meridian passage of the wandering star is at the pole of the horizon itself, then, in this day, the wandering star will have neither two equal heights in the east nor two equal heights in the west, whether the sphere is right or whether it is inclined. We shall establish this by a proof.

Let the horizon be the circle *ABC*, let the meridian circle be *ADC*, and the pole of the horizon the point *D*. Let the wandering star reach the meridian at the point *D*.

I say that in this day the wandering star has neither two equal heights in the east nor two equal heights in the west, but that any height it has in the east is always unique and that any height it has in the west is always unique.

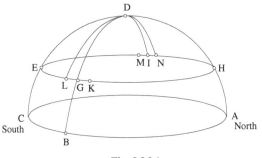

Fig. I.36.1

Proof: We draw one of the *muqanṭarāt* of height; let it be *EGHI*. We cause an hour arc to pass through the point *D*; it cuts the *muqanṭara EGHI* in two points, one to the east and the other to the west, whether the sphere is right or inclined. If the wandering star rises in the east and sets in the west, let it cross the *muqanṭara* at the two points *G* and *I*; let the point *G* be to the east and the point *I* be to the west. So in its motion from the eastern horizon towards the point *D*, the wandering star crosses the *muqanṭara EGHI*. If its motion is from the north towards the south, then it arrives on the *muqanṭara EGHI* in a point to the north of the circle *DG*; let this point be the point *K*. And if the motion of the wandering star is from the south towards the north, it arrives on the *muqanṭara* in a point to the south of the circle *DG*; let that point be the point *L*. If the wandering star then descends towards setting, and if its motion is from the north towards the south, then it arrives on the *muqanṭara* in a point to the south of the circle *DI*; let this point be the point *M*. And if its motion is from the south towards the north, then it arrives on the *muqanṭara* in a point to the north of the circle *DI*; let this point be the point *N*. So if the position of the wandering star is the point *K*, we prove, as was proved at the end of Proposition 31, that the wandering star arrives on the eastern part of the *muqanṭara* only in a single point; if the position of the wandering star is the point *L*, then we prove, as was proved at the end of Proposition 29, that the wandering star arrives on the eastern part of the *muqanṭara* on it in a single point; for the position of the wandering star in the west is the point *M*, we prove, as was proved at the end of Proposition 28, that the wandering star arrives on the western part of the *muqanṭara* only in a single point; and if the position of the wandering star is the point *N*, we prove, as was proved at the end of Proposition 34, that the wandering star arrives on the western part of the *muqanṭara* only in a single point.

The same holds for each of the *muqanṭarāt* of height. The day when the meridian passage of the wandering star takes place at the pole of the horizon, then the wandering star has neither two equal heights in the east nor two equal heights in the west. That is what we wanted to prove.

We also say that each of the seven wandering stars, if it moves from the northern limit of its inclined orb towards the point of intersection of its inclined orb and the circle of the equator, and if its motion is accelerated, then, in the course of each of the days that pass between the limits of its motion, in certain places on the Earth it sets on the eastern horizon and later rises from the eastern horizon after having set on that horizon. Let us establish this by proof.

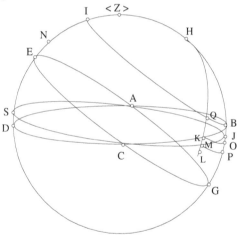

Fig. I.36.2

Let the circle *ABCD* be one of the horizons for which the height of the pole is equal to the complement of the inclination of the orb of the wandering star with respect to the circle of the equator, that is to say the magnitude by which a quarter of a circle exceeds the inclination of the orb of the wandering star. Let the circle *BED* be the meridian circle. Let the arc *BCD* be the eastern half of the horizon and the arc *DAB* the western half of the horizon. Let the arc *DE* be the maximum inclination of the orb of the wandering star with respect to the circle of the equator; so the circle of the equator passes through the point E if the height of the pole is equal to the complement of the inclination. Let the circle of the equator be the circle $AECG$, and let the point G be the intersection of the circle of the equator and the meridian circle, that is to say the northern intersection. So the arc BG is equal to the arc DE. Let the pole of the equator be the point H. We draw an hour circle centred on the pole H and with distance HB; let the circle be BQI; so the arc IE is equal to the arc BG which is equal to the inclination of the orb of the wandering star. The circle BQI is the path of the limit of the inclination of the orb of the wandering star, which for the orb of the sun is the tropic of Cancer; and for each of the remaining wandering stars, it will be the circle corresponding to the tropic of Cancer. So the horizon $ABCD$ touches the circle BQI. Since the circle BQI is one of the paths on which the wandering star performs its revolution, thus, on the day when it is at the northern limit of its orb, the wandering star arrives at a point of the circle BQI. Let the point at which the wandering star arrives on the circle BQI be the point B. But the circle that passes through the point B and through the pole of the equator is the circle BED, the circle BED is the meridian circle for several horizons, and the great circle that touches the circle BQI at the point B is one of the horizons whose meridian circle is the circle BED. If the wandering star reaches its maximum inclination with respect to the circle of the equator, it is then on the circumference of one of the horizons whose meridian is the circle that passes through the centre of the wandering star and the pole of the equator; let the circle $ABCD$ be this horizon. We draw the arc KO parallel to the circle of the equator,[144] so that the ratio of the arc KO to the arc OB is greater than any ratio of any required time for the wandering star in its motion from the point B to the point I to the inclination of the motion of the wandering star. Since the ratio of KO to OB is greater than any ratio of any required time for the wandering star in its motion from the point B to the point I to the inclination of the motion of the wandering star, the ratio of KO to OB is greater than the ratio of the time KO to the inclination proper to the time KO; so the inclination proper to the time KO is greater than the arc BO; let the inclination proper to the time KO be the arc BP. We cause an hour arc to pass through the point P; let it be PL. And we cause a great circle to pass through the points H and K; let this circle cut the circle PL at the point L and let it cut the circle BQI at the point Q. So the arc BQ is equal to the time OK and the arc QL is equal to the arc BP; so the arc QL is the inclination <proper> to the time BQ. If the wandering star passes through the point B, then moves for a time equal to the arc BQ, then it travels to the

---

[144] K is on the horizon and O on the meridian.

point *L*. We make the arc *BJ* smaller than the arc *OP* and smaller than the arc *BO* and we cause a great circle to pass through the points *A*, *C*, *J*; let the circle be *AJCS*. This circle cuts the arc *ED*; let it cut it at the point *S*. Thus the arc *SD* is equal to the arc *BJ*; but this circle, that is to say the circle *AJCS*, cuts the arc *KO* and cuts the arc *KL*; let it cut the arc *KL* at the point *M*. So the point *M* lies between the two points *K* and *L*. In fact, the arc *KM* is the inclination of the arc *KC* with respect to the circle *AJC* and the arc *BJ* is the maximum inclination of the circle *ABC* with respect to the circle *AJC*; so the arc *KM* is smaller than the arc *BJ* and the arc *BJ* is smaller than the arc *OP*; so the arc *KM* is much smaller than the arc *OP*. But the arc *OP* is equal to the arc *KL*, so the arc *KM* is much smaller than the arc *KL*. So the point *M* lies between the points *K* and *L*, so the point *L* is below the circle *AJCS* and the point *B* is above the circle *CJ*. For the place whose horizon is *AJCS*, the point *B* will be above its horizon and the point *L* below its horizon. If the wandering star moves from the point *B* for the time *BQ*, it travels to the point *L*; if the wandering star passes through the point *B* that is above the horizon *AJCS*, then moves for the time *BQ*, it goes below the horizon *AJCS*. But the motion of the wandering star is from the direction from *B* towards *Q*; consequently, the wandering star sets at a point on the arc *JM*. But the arc *JCS* is the eastern half of the horizon; consequently, the wandering star sets on the eastern horizon.

Similarly, the arc *IE* is the inclination of the orb of the wandering star with respect to the circle of the equator; so it is much greater than the inclination proper to the time in which the wandering star completes half a revolution. If the wandering star moves for a time equal to the arc *BQI*, which is half a revolution, then the inclination of its motion is much smaller than the arc *IE*; let the inclination of the motion of the wandering star in the course of the time *BQI* be the arc *IN*. So if the wandering star moves for the time *BQI*, it arrives at the point *N*. Since the wandering star travels from the point *L* towards the point *N*, and the point *L* is below the horizon and the point *N* above the horizon, consequently the wandering star rises from [a point on] the arc *MC*, because in the course of this day, it does not reach the circle of the equator. So if the wandering star reaches the northern limit of its inclined orb, then it sets on one of the horizons, on the eastern horizon, then rises from the eastern horizon after having set on it.

Similarly, if it passes through a point of the arc *BG* close to the point *B*, then its state will be the same state [that has just been described]. In fact, if we cause to pass through this point a horizon which we proceed to deal with as we dealt with the horizon *ABC*, we obtain a figure like the figure that we have explained. Similarly, the behaviour of the wandering star in the second day, as it descends from the northern limit, passes through a point of the meridian circle between the points B and G. If we cause an hour circle to pass through this point, it will correspond to the circle BQI. If we cause to pass through the point of the meridian circle through which the wandering star passes, and the hour circle passes, a great circle that touches the hour circle, it will correspond to the circle $ABCD$. If we cause to pass through <a point of> this great circle an hour arc corresponding to the arc KO, whose ratio to the arc corresponding to the arc OB is greater than any ratio for any required time for the wandering star to the inclination of the motion of the wandering star, and if we suppose there is a point corresponding to the point J through which we cause a great circle to pass, as also through the two points A and C, we prove for this figure by similar reasoning that the wandering star sets on the eastern half of this circle, then rises on the same eastern half, [and we shall also prove] that this circle that passes through the point corresponding to the point J is in all cases a horizon of a place on the Earth, and it is one of the northern horizons since the north pole is less than a quarter circle above it. So, at this place on the Earth, the wandering star sets on the eastern horizon and rises from the eastern horizon. Similarly, for any point of the circle parallel to the plane of the equator that passes through this position on the Earth, the wandering star sets to the east of the horizon and rise from it, if the point [on the orb] for the path of the wandering star to be at its smallest is its point of contact with the horizon.

This is how the motion of each of the seven wandering stars takes place on each of the days in the course of which the wandering star moves from the northern limit towards the circle of the equator until its path comes close to the equator; between it and the equator there is less than the inclination of half a revolution. It is in the course of this day that the meridian passage of the wandering star will be at a point south of the circle of the equator. So this point will lie on the circle BED and [will be] to the south of the point E. So if the whole arc that corresponds to the arc ES is greater than the inclination proper to half a revolution, then the wandering star rises on the eastern horizon because its point of meridian passage will be above the horizon, and it will rise either from [a point on] the arc corresponding to the arc MC, or from [a point on] the arc corresponding to the arc CS.

Thus we have proved from what we have established that, for each of the seven wandering stars, if its motion is from the northern limit towards the circle of the equator, and if its motion is accelerated, then for any day between the two endpoints of this motion, in certain places in the north of the Earth, it sets on the eastern horizon and then it rises from the eastern horizon. That is what we wanted to prove.

Similarly, we say, for each of the seven wandering stars, that, if it is moving from the southern limit towards the point of intersection of its inclined orb and the circle of the equator, and if its motion is accelerated, then for any day between the two endpoints of its motion, in certain places on the Earth, it rises from [a pint on] the western horizon and, after its rising, sets on the western horizon. Let us establish that by means of a proof.

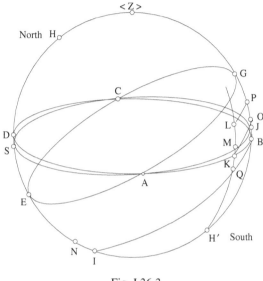

Fig. I.36.3

Let the circle *ABCD* be one of the horizons for which the height of the pole above it is equal to the complement of the inclination of the orb of the wandering star with respect to the circle of the equator. Let the circle *DEBG* be a meridian circle; let the arc *DEB* be the half below the horizon *ABCD*; let the arc *BAD* be the western half of the horizon; let the south pole of the equator be the point *H′*; then the arc *H′B* is the complement of the inclination of the orb of the wandering star and the arc *H′B* is the depression of the pole[145] with respect to the horizon. Let the arc *BG* be the inclination of the orb of the wandering star, then the point *G* will lie on the circle of the equator. Through the point *G* we draw the circle of the equator, let the circle be *AGCE*. We draw the arc *KO*[146] such that the ratio of the arc *KO* to the arc *OB* is greater than any ratio for any required time

---

[145] That is, the negative height of the pole.

[146] The arc *KO* is an arc of an hour circle, *K* is on the horizon, *O* is on the meridian.

for the wandering star in its motion from the point B to the complement of the semicircle to the inclination of the motion of the wandering star on the side towards the beginning of its motion. We take the point H' as a pole and, with distance $H'B$, we draw an hour circle; let the circle be BQI. This circle is the path of the southern limit which, for the orb of the sun, is the tropic of Capricorn and for the orbs of the remaining wandering stars is the circle corresponding to the tropic of Capricorn. We cause a great circle to pass through the two points H' and K; it cuts the circle BQI; let it cut it at the point Q. So the arc BQ is similar to the arc OK and the inclination proper to the time BQ is greater than the arc BO; let the inclination proper to the time BQ be the arc BP. We cause an hour arc to pass through the point P; let it be PL; let it cut the circle $H'K$ at the point L. We take the arc BJ, smaller than the arc OP and smaller than the arc BO. We cause a great circle to pass through the two points A and C and through the point J; let the circle be $AJCS$; let it cut the circle $H'KL$ at the point M. We prove that the point M lies between the two points K and L, as has been proved in the preceding proposition. Thus the point L will be above the circle AJC, the point B will be above the circle AJC, and the arc QL is the inclination of the time BQ. So if the wandering star passes through the point B, then moves in the course of the time BQ, it arrives at the point L; but the point K is above the circle AJC and the point L is above the circle AJC. So if the wandering star moves in the course of the time BQ, then it crosses the arc JM; but the arc JAS is the western half. Thus, for the place whose horizon is the circle $AJCS$, if the wandering star arrives at the southern limit of its orb, it rises, on it, from the western horizon, from [a point of] the arc JMS. Similarly, the arc IH' is the inclination of the orb of the wandering star because it is equal to the arc BG; so the arc IH' is much greater than the inclination proper to a single half revolution. Let the inclination proper to a single half revolution be the arc IN. So if the wandering star moves in the course of the time BQI, it arrives at the point N; so the wandering star travels from the point L to the point N; but the point L is above the horizon and the point N is below the horizon; so the wandering star crosses the horizon and goes below the horizon, and the wandering star sets at [a point of] the arc MA, because in the course of this day it cannot reach the circle of the equator. Thus the wandering star, on the day it arrives at the southern limit, rises on the west of this horizon and sets in the west of this horizon.

Similarly, it rises on the west of each of the horizons of the points of the circle parallel to the equator, [a circle] that is on the surface of the Earth and that passes through the first place of the Earth,[147] if the point of the path that is the smallest of the paths of the wandering star is the point of contact with this horizon. These horizons are the same as the horizons that have been used in the preceding proposition.

We prove, as we proved in the preceding proposition, that for any day between the two endpoints of its motion, from the southern limit towards the circle of the equator, the wandering star rises on the western horizons of northern places and sets in the west of these horizons. That is what we wanted to prove.

We have demonstrated from what we have established in this book, the configuration of the motions of each of the seven wandering stars and we have proved that each of the seven wandering stars can, at certain times, in the same day and [when] in the easterly direction, have equal heights in all places on the Earth in which the day for the wandering star has been divided into two halves; that at certain times it can have, in the same day, in the westerly direction, equal heights in all places in the Earth in which the day for the wandering star has been divided into two halves; that in certain places on the Earth, at certain times it sets on the eastern horizon and on the same day rises on the eastern horizon and that, at certain times in the same place on the Earth, it rises on the western horizon and in the same day sets on the western horizon. That is what we wanted to prove.

Completed.

---

[147] This is a place on the Earth chosen initially for the purpose of observation, a place mentioned earlier (p. 419); but the period of observation is different. The argument is identical with that for the previous case, concerning rising and setting in the east.

CHAPTER III

THE VARIETY OF HEIGHTS: A PROPÆDEUTIC TO THE *CONFIGURATION OF THE MOTIONS OF THE SEVEN WANDERING STARS*

3.1. INTRODUCTION

In *The Configuration of the Motions of Each of the Seven Wandering Stars*, Ibn al-Haytham devotes a substantial part of his work to investigating the variation in height of the wandering star between its rising and setting. With Ibn al-Haytham, the height of a body in the course of its observed movement became one of the principal subjects for astronomical research. It is thus important to retrace the development of his research on this topic, at least in part – the more so since he himself states in the introduction to *The Configuration of the Motions* that he had already begun to investigate this problem in some writings that were now out of date;[1] as he puts it, he wrote on the height and related questions in 'the way of mathematicians' and according to 'known principles'. By this he means that in the past he considered heights using the method traditional in astronomy and according to its accepted standards, and that in *The Configuration of the Motions* he is returning to the matter of these (now superseded) writings, which he treats using a new method and new principles.

This puts the historian in a privileged position, and one rarely encountered in the history of the mathematical sciences in Arabic. The historian would be able to actually see how far Ibn al-Haytham had travelled between the obsolete texts and *The Configuration of the Motions*, and could thus come to a better understanding of what is new about the latter treatise. In addition it would be possible to put Ibn al-Haytham's different versions into order. However, Ibn al-Haytham does not give any exact title for these earlier writings, but merely speaks of 'our books'. We know, however, from the list

---

[1] See above, p. 260.

of his works supplied by old biobibliographers,[2] that he had devoted at least two treatises to height. For the first one, whose title is *On the Ratios of the Hour Arcs to their Heights* (*Fī nisab al-qusīy al-zamāniyya ilā irtifaʿātihā*), no copy is known to exist. The second is called *On the Variety that Appears in the Heights of the Wandering Stars* (hereafter *On the Variety of Heights*) and has come down to us in a single manuscript, which is difficult to read. This document is the more valuable for being the solitary piece of evidence that allows us to see something of the development of Ibn al-Haytham's thought on the subject of height.

According to what we are told in the lists of the old biobibliographers, Ibn al-Haytham composed these two treatises before 1038. That would confirm (if confirmation were needed) that, at the same time that he was writing his books criticizing Ptolemaic astronomy, he was pursuing an innovative programme of research on celestial motions – the winding motion, the motion of the Moon – and on the height of observed motions.

Let us now turn to the book *On the Variety of Heights*. It does not include any preliminary material in which Ibn al-Haytham might have explained his intentions and purpose, instead it begins immediately with definitions. These are followed by seven propositions in plane geometry – lemmas – all of which are used later on. After these lemmas we have nine propositions concerning height. They are interconnected and are proved with the help of the lemmas. The logical implications provide a coherent structure, and confirm that nothing extraneous has slipped into this initial part of the text. Nor has anything been omitted in this preliminary section, except perhaps some kind of introduction or (which is Ibn al-Haytham's usual custom) an explanation of the purpose of the work. But there is no positive reason to support any claim that such preliminary material has been lost. All the definitions given at the beginning of the book relate to a given place on the terrestrial sphere, taking it as known that the radius of this sphere is negligibly small compared to that of the celestial sphere. We assume, also, that the given place is the centre of the celestial sphere, and for every place we have an associated horizon and a point Z on the celestial sphere called the zenith.

---

[2] *Ibn al-Haytham and Analytical Mathematics*. A History of Arabic Sciences and Mathematics, vol. 2, Culture and Civilization in the Middle East, London, 2013, pp. 414–15 and 401–2.

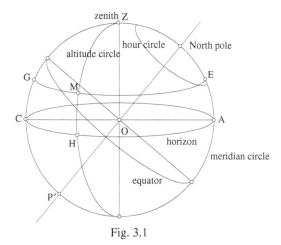

Fig. 3.1

Throughout the investigation in this book, the planes of reference are the horizon AC and the local meridian plane. In the research he undertakes here, Ibn al-Haytham does not need to refer to a system of coordinates. It is a different matter in *The Configuration of the Motions*, where he has recourse to several systems of coordinates and in particular to the system of equatorial coordinates. This is an important difference, which for the moment we shall merely bear in mind.

Several of the definitions given at the beginning of this treatise relate to the notion of height (or altitude) of a wandering star. Ibn al-Haytham considers a general point M on the celestial sphere. With this point, we associate a great circle ZM of the celestial sphere; this is the *altitude circle*. The circle meets the horizon at the point H; the arc HM is called the *arc of altitude*, or simply the *altitude* or *height* of the point M. The horizontal plane through M cuts the meridian circle in two points G and H, for which we have $\widehat{HM} = \widehat{CG} = \widehat{AE}$ (see Proposition 8). The height of the point M is given by one of the arcs $\widehat{CG}$ or $\widehat{AE}$. The *height of an arc* is the difference between the heights of its two endpoints.

The definitions are followed by seven propositions in plane geometry. The propositions are concerned with geometrical properties of the circle that will allow Ibn al-Haytham to proceed from an equality of areas to an inequality – or an equality – of arcs. So it is clear that Ibn al-Haytham has proved these propositions because he intends to compare heights. We shall comment on them in more detail below, but by way of example let us take a look at the fifth one. Ibn al-Haytham proves that if, in a circle, a chord BD is divided at two points K and E in such a way that $KB \cdot KD = \frac{1}{4} BD \cdot DE$, and if we draw parallel straight lines KM and EC so that the angle BEC is less than a right angle (we take M and C on the smaller arc BD), then $\widehat{BM} < \widehat{MC}$.

Once he has proved these lemmas, Ibn al-Haytham turns to the real subject of his book with nine propositions. He considers two positions of a point that moves on an hour circle. The size of the arc separating the two positions defines the time the moving point takes to traverse this arc, since the motion is circular and uniform. From Proposition 9 onwards, Ibn al-Haytham uses the term 'time' to designate the arc, and the expression 'height of the time' to designate the height of the arc. As we have noted, he also measured time as an arc in *The Configuration of the Motions*, an identification which allowed him to apply the theory of proportions.

Ibn al-Haytham considers a point that moves on an hour circle, describing an arc whose midpoint is given. He compares the height associated with the first half of the trajectory to the height associated with the second half. He carries out this investigation for a variety of places. He begins with a place where the celestial sphere is a right sphere, then one where it is an inclined sphere. He distinguishes between the various possible positions of the hour circle: an hour circle that cuts the horizon, an hour circle completely above the horizon, and the special case where the hour circle passes through the zenith.

From the tenth proposition of the treatise onwards, the statements of the propositions are in kinematic terms: the wandering star is treated simply as a point moving on the celestial sphere. Ibn al-Haytham then investigates increments in height for equal increments of time. In other words, what he is examining is the fact that the height is a concave function of the time. But he does not deal with this variation in continuous terms. Here, Ibn al-Haytham considers only three points: the origin, the endpoint that lies on the meridian and the midpoint of the corresponding arc. In *The Configuration of the Motions*, he was to go beyond this point by point investigation and deliberately set about engaging in a continuous investigation of the variation.

This is not the only difference between *The Variety of Heights* and *The Configuration of the Motions*. In addition to the ones we have already pointed out, we may note another one, which is equally important. We have mentioned that in the first treatise Ibn al-Haytham investigates the uniform motion of a point that is constrained to describe an hour circle. In contrast, in *The Configuration of the Motions* he considers the apparent motion of a planet, a motion that is composed from three uniform circular motions. This compound apparent motion does not take place along an hour circle.

Investigating the variation of this apparent motion in continuous terms requires mathematical methods that go far beyond those of plane geometry, employed in *The Variety of Heights*. For example we find an investigation of the variation of trigonometrical expressions such as $\frac{\sin x}{x}, \frac{\tan x}{x}, \frac{\sin kx}{x}$.

A comparison between the two treatises reveals unequivocal evidence of the lines of development of Ibn al-Haytham's work on heights and also, indirectly, his work in astronomy. The motion with which he is concerned is no longer that of a point along an hour circle, but the apparent motion of a wandering star; the investigation of variations in height is no longer carried out point by point but continuously; the mathematics to which he appeals is no longer simple plane geometry but geometry involving infinitesimals.

3.2. MATHEMATICAL COMMENTARY

In this commentary we refer the reader to the actual propositions and to the figures provided. Here, we shall merely endeavour to indicate what ideas are to be found in the propositions and what difficulties, if any, are to be encountered. That is, the commentary is not intended to be read separately from the propositions presented and translated in this volume. To avoid repetition, we have decided – because the propositions are straightforward – not to write them out in detail.

Proposition 1. — The idea behind the proof is as follows: the similarity that transforms the figure *ABC* into the figure *DEG* transforms *H* to *I*, since

$$\frac{AH}{HC} = \frac{DI}{IG}.$$

$A\hat{H}B > D\hat{I}E$; so the similarity transforms *B* to *E*. It follows that

$$\frac{\widehat{AB}}{\widehat{BC}} = \frac{\widehat{DE}}{\widehat{EG}}.$$

Proposition 2. — In this proposition, which plays an important part in the treatise, we assume that

$$BG \cdot GD = \frac{1}{4} BD \cdot DE;$$

and we want to deduce from this that $\widehat{BI} = \widehat{HC}$. We have

$$AH^2 = 4GH^2 = 4BG \cdot GD = BD \cdot BE = BC^2,$$

hence

$$AH = BC \quad \text{and} \quad \widehat{BH} = \widehat{AB} = \widehat{HC}.$$

Conversely, if we have these equalities, we deduce from them that $\widehat{AH} = \widehat{BC}$, hence $AH = BC$ and $BD \cdot BE = AH^2 = 4GH^2 = 4BG \cdot GD$.

So it is a matter of transforming an equality of arcs into an equivalent equality of areas; these areas are products of segments cut off on the diameter by the ends of the arcs in question.

Proposition 3. — The assumptions for this proposition are the same as in the previous proposition, apart from the fact that BD is now a chord smaller than a diameter; the conclusion is thus that $\widehat{AP} > \widehat{PB}$, whereas in the previous case we had an equality.

In the course of the proof, this proposition is reduced to the previous one by constructing the semicircle with diameter BD. So we have $\widehat{KI} = \widehat{IB}$ and the arcs $\widehat{AP}$, $\widehat{PB}$ are found from the arcs $\widehat{KI}$ and $\widehat{IB}$ by projection of $\widehat{BIK}$ onto $\widehat{BPA}$ (the projection being orthogonal to BD).

Using another auxiliary construction, we introduce the arc BLK similar to the arc APB and we now need to prove that $\widehat{KL} > \widehat{LB}$, that is, that L lies between Q and B. We first establish that O, on the arc KLB, lies between Q and B, then that L lies between O and B (see Fig. III.3, p. 444).

To do this it is sufficient to note that angle BHI is obtuse, since angle BHL is a right angle. Thus we can see that the arc BLK is the transformed version of the arc BIK, the transformation concerned being a combination of the previous transformation and a similarity transformation with centre B. We reduce proving the inequality $\widehat{AP} > \widehat{PB}$ to proving $\widehat{KL} > \widehat{LB}$.

In essence, Ibn al-Haytham's argument in this proof consists of using this transformation to get two arcs of a circle with the same chord BK.

Proposition 4. — This proposition is analogous to the previous proposition, with angles HGD and AED being acute instead of being right angles.

The proof is different. Here we make use of the diameter BK through the point B, which is perpendicular to the straight lines HM and AL.

We prove that $KB \cdot BL = 4HG \cdot GI$ by appealing to the hypothesis and to the fact that we have a circle – using the power of the point G with respect to the circle. But we know that $HG \cdot GI < HM \cdot MI$, so $HG \cdot GI < BM \cdot MK$ (power of the point M). We then construct the point N on BM such that

$$KN \cdot NB = \frac{1}{4}KB \cdot BL = HG \cdot GI.$$

In Proposition 2, we have $\overset{\frown}{BP} = \overset{\frown}{PA}$, and since H lies between P and A, we have $\overset{\frown}{BH} > \overset{\frown}{HA}$.

Note: In the statement of this proposition Ibn al-Haytham imposes the following condition: the chord AC cuts off an arc no greater than a semicircle. This condition is not repeated in the ensuing formal setting-out of the proposition (*ekthesis*). Let us look at this latter in detail.

The chord BD cuts AC in E, and the point G on BE is taken in such a way that

$$DG \cdot GB = \frac{1}{4}DB \cdot BE.$$

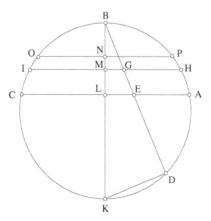

Fig. 3.2

We project B, G, E into B, M, L on the diameter BK, and we have

$$KM \cdot MB = \frac{1}{4}KB \cdot BL.$$

If NP is parallel to AC, we know that $\widehat{BP} = \widehat{PA}$ when N lies between B and L (Proposition 2). We draw MH parallel to AC; if we know that N lies between B and M, we can conclude that $\widehat{BH} > \widehat{HA}$.

Let $d = BK$ the diameter of the circle and let $a = BL$; the abscissa $x = BN$ of the point N on the axis BK is determined by the equation

(1) $x\,(d-x) = \frac{1}{4}ad.$

This equation has two roots between 0 and d.

Since $a(d-a) - \frac{1}{4}ad = a\,(\frac{3}{4}d - a)$, we can see that a is between the two roots if $a \leq \frac{3}{4}d$, but lies outside the interval between the roots if $a > \frac{3}{4}d$. Thus, when $a \leq \frac{3}{4}d$, only the smaller root gives us N between B and L, but when $a > \frac{3}{4}d$, both roots do so.

Let $y = BM$ be the abscissa of M; the assumption that $KM \cdot MB > \frac{1}{4}KB \cdot BL$ can be written $y\,(d-y) - \frac{1}{4}ad > 0$. Thus y is always between the roots of equation (1), so we can always choose a point N (determined by the smaller root of (1)) lying between B and M, and can thus complete the argument; and we can see that the condition in the statement of the proposition is superfluous.

We may also observe that, where $a \leq \frac{3}{4}d$, that is where AC cuts off an arc greater than two-thirds of a circle, the condition 'N lies between B and L' implies 'N lies between B and M'.

We may also wonder why Ibn al-Haytham gave this condition in the statement of the proposition and did not repeat it in the formal setting-out of the result prior to giving his proof. Did he need it in some of the propositions that were to follow?

Proposition 5. — We have already mentioned this proposition (p. 426). All we need do here is to point out that the data and the hypotheses are analogous to those in the previous proposition, the only difference being that the lines of projection CE, MK are no longer perpendicular to the diameter through the point B.

The proof again consists in reducing this case to the one in which BD is a diameter of a circle; but this time the circle is not the one we were given and BD does not change.

Proposition 6. — The proposition establishes that the condition $AG < AD$ implies $\widehat{CE} < \widehat{BE}$, that $AG = AD$ implies $\widehat{CE} = \widehat{BE}$ and that $AG > AD$ implies that $\widehat{CE} > \widehat{BE}$.

So we are concerned with a proposition on the variation of the ratio $\dfrac{\widehat{CE}}{\widehat{BE}}$ as a function of the ratio $\dfrac{AG}{AD}$, where AC is a diameter of the circle, the point B lies on the upper semicircle, with $AB > BC$, the point G lies on AB and GE is parallel to AC.

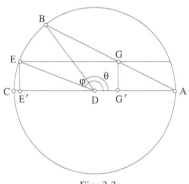

Fig. 3.3

We can give an analytic proof of this proposition: let $A\hat{D}B = \theta \in \left[\dfrac{\pi}{2}, \pi\right]$ and $A\hat{D}E = \varphi \in [\theta, \pi]$. Let us put $r = AD$, the radius of the circle, and $h = EE' = GG'$, the distance from EG to AC. We have $D\hat{A}B = \dfrac{\pi}{2} - \dfrac{\theta}{2}$, $E\hat{D}C = \pi - \varphi$ and $B\hat{D}E = \varphi - \theta$; thus $h = AG\cos\dfrac{\theta}{2} = r\sin\varphi$.

The inequality $\widehat{CE} < \widehat{EB}$ is equivalent to $\pi - \varphi < \varphi - \theta$, that is $\pi + \theta < 2\varphi$. Since $\pi + \theta$ and 2φ are in the interval $[\pi, 2\pi]$ in which the cosine is an increasing function, the inequality $\widehat{CE} < \widehat{EB}$ is equivalent to $-\cos\theta < \cos 2\varphi$, or $0 < \cos\theta + \cos 2\varphi = 2\left(\cos^2\dfrac{\theta}{2} - \sin^2\varphi\right) = \dfrac{2h^2}{r^2 AG^2}\left(r^2 - AG^2\right)$ where again $AG < r$. In the same way $\widehat{CE} = \widehat{EB}$ is equivalent to $AG = r$ and $\widehat{CE} > \widehat{EB}$ is equivalent to $AG > r$.

We may also note that, if $\lambda = \dfrac{\widehat{CE}}{\widehat{EB}} = \dfrac{\pi - \varphi}{\varphi - \theta}$, then $\varphi = \dfrac{\lambda\theta + \pi}{\lambda + 1}$, which is a decreasing function of λ because $\theta < \pi$; since $\sin\varphi$ is a decreasing function of φ in the interval $\left[\dfrac{\pi}{2}, \pi\right]$, we can see that $\dfrac{AG}{r} = \dfrac{\sin\varphi}{\cos\dfrac{\theta}{2}}$ is an increasing function of λ (θ being constant). Thus $\dfrac{\widehat{CE}}{\widehat{EB}} = \lambda$ and $\dfrac{AG}{AD} = \dfrac{AG}{r}$ change in the same direction. For $\lambda = 1$, $\varphi = \dfrac{\theta}{2} + \dfrac{\pi}{2}$ and $\dfrac{AG}{r} = 1$, so $\lambda < 1 \Leftrightarrow AG < r$ and $\lambda > 1 \Leftrightarrow AG > r$.

Proposition 7. — In this proposition, the arc ABC is no longer greater than a semicircle and the chord GE parallel to AC cuts AB and BC in their midpoints D and K respectively. We then have $\widehat{BE} > \widehat{EA}$ and $\widehat{BG} > \widehat{GC}$; so we can draw conclusions about inequalities of arcs from equalities of line segments.

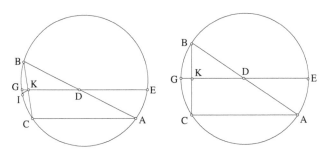

Fig. 3.4

Note: In the statement of this proposition as well as in the formal setting-out that follows, Ibn al-Haytham gives the condition $\widehat{ABC} \leq \frac{1}{2}$circle, which implies that $C\hat{A}B$ and $A\hat{C}B$ are acute.

If $\widehat{ABC} > \frac{1}{2}$circle, it is possible for one of the angles CAB or ACB to be obtuse or a right angle. If $A\hat{C}B$ is obtuse, then KI cuts the arc GC and we have $BG < GC$. If $A\hat{C}B$ is a right angle, then D is the centre of the circle, the points G and I coincide and $CG = GB$.

If $A\hat{C}B$ and $A\hat{B}C$ are acute, we have, as in the case treated in the proposition, $\widehat{CG} < \widehat{GB}$.

So if $\widehat{ABC} > \frac{1}{2}$circle, we have *three* possibilities for the arcs BG and GC.

This marks the end of the first group of propositions that are lemmas in the investigation of height. The second group of propositions to be employed in the investigation of height is made up of the nine propositions that follow. These propositions deal with the investigation of the height of a point that moves on an arc.

Proposition 8. — Ibn al-Haytham first establishes that the height of a point G on the celestial sphere can be measured along the local meridian, as the arc between the horizon and the circle through G parallel to the horizon. This follows from the fact that the local zenith, E, is the pole both of the horizon and of the parallel circle concerned.

Proposition 9. — Here, Ibn al-Haytham measures the height of an arc of an hour circle along the local meridian, which is between the circles through the ends of the arc parallel to the horizon. This follows from the previous proposition if we consider the difference between two arcs.

Proposition 10. — The statement of this proposition asserts that heights and times are the same if we have a right sphere. In fact, in this case, the altitude circle is a meridian circle. The proof considers only the case in which the point E lies midway between B and D.

We may note here that the statement is in kinematic terms, a wandering star being treated as a point moving along the equator on the celestial sphere.

Proposition 11. — In this proposition, Ibn al-Haytham again investigates the motion of a wandering star, which is treated as a point moving on the celestial sphere, and supposes the point describes an hour circle *BIH* parallel to the equator but not identical with it.

If *I* is the midpoint of $\widehat{BH}$, then the height of $\widehat{IH}$ is greater than the height of $\widehat{BI}$. Thus, in equal increments of time the heights decrease from *H* to *B*.

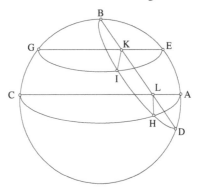

Fig. 3.5

We may note that, here too, Ibn al-Haytham considers only the midpoint *I* of the arc *BH* and, as in Proposition 10, he does not prove that the height is a concave function of the time, which he could have proved had his investigation of the variation been carried out in continuous terms, as he was to do later in *The Configuration of the Motions of the Seven Wandering Stars*. The present investigation considers only discrete points in the motion.

Here too, the proof depends on an inequality being implied by an equality. The equality of the time intervals is transformed into an equality of areas by the converse of Proposition 2, applied to the circle *BHD*; then that equality of areas implies an inequality of arcs, by Proposition 3, applied to the circle *BCD*.

Proposition 12. — In this proposition, Ibn al-Haytham considers the motion of a point moving on an hour circle *BIH* that is inclined with respect to the horizon – so the celestial sphere is no longer right; but he supposes that the point *B* is at the zenith.

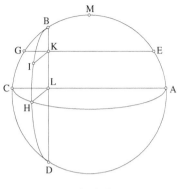

Fig. 3.6

If I is the midpoint of $\widehat{BH}$, the height of $\widehat{IH}$ is less than the height of $\widehat{BI}$. Thus, for equal times of travel, the heights increase. As in the previous proposition, the investigation is carried out in discrete terms rather than continuous ones. The latter, as in the other cases, will appear only in *The Configuration of the Motions*.

The proof is analogous to that of the previous proposition: the equality of times is expressed as an equality of areas that, by Proposition 4, implies an inequality of arcs.

We may note that in the proof, but not in the statement of the proposition, Ibn al-Haytham calls upon the superfluous hypothesis found in Proposition 4 (AC cuts off an arc greater than a semicircle).

Proposition 13. — The statement is analogous to that of the previous proposition except that the point B is no longer at the zenith. Here we suppose B is situated between the equator and the zenith. This condition is not made explicit in the statement, but it is necessary to ensure that $\widehat{AB} > \widehat{BC}$. We also suppose that $\widehat{ABC}$ is not greater than a semicircle, so that we can use Proposition 5. This result allows us to deduce that $\widehat{CG} > \widehat{GB}$, working from the equality of areas that, as in the previous propositions, expresses the equality of the times $\widehat{HI}$ and $\widehat{IB}$. Proposition 6 is not used.

Proposition 14. — The motion considered is that of a point starting from a point H on the horizon and arriving at the point B on the meridian by moving along an hour circle HIB. The point I is the midpoint of $\widehat{HB}$, so the times $\widehat{HI}$

and $\widehat{IB}$ are equal, but the corresponding heights $\widehat{CG} = \widehat{EA}$ and $\widehat{GB}$ or $\widehat{EB}$, depending on the figure, may be equal or unequal, with either being the greater. This follows from Proposition 6.

Proposition 15. — The set-up is analogous to that in the previous proposition, except that this time the hour circle of the motion touches the horizon at the starting point A. We suppose that $\widehat{AB} = \widehat{DB}$ and, thanks to Proposition 7, we can draw the conclusion that the height $\widehat{CH} = \widehat{AG}$ of $\widehat{AD}$ is less than the height of $\widehat{DB}$ (which may be $\widehat{HB}$ or $\widehat{GB}$).

Proposition 16. — In this final proposition, Ibn al-Haytham wishes to compare the height of the time $\widehat{EI}$ with that of time $\widehat{IB}$.

In the first case he investigates ($\widehat{AB} < \widehat{BC}$), we have $\widehat{AG}$ as the height of the point I and $\widehat{EG}$ the height of the time $\widehat{EI}$; now $\widehat{EG} < \widehat{AG}$.

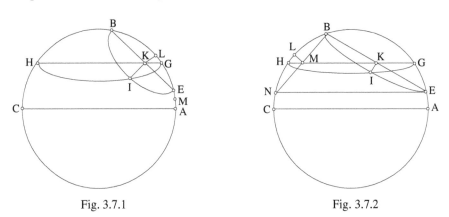

Fig. 3.7.1   Fig. 3.7.2

Ibn al-Haytham distinguishes three cases:

1. $\widehat{ME} = \widehat{GL}$, so $\widehat{AG} = \widehat{GB}$ and $\widehat{EG} < \widehat{GB}$; and $\widehat{GB}$ is the height of the time $\widehat{IB}$. So the height of the time $\widehat{EI}$ is less than that of the time $\widehat{IB}$; it cannot be equal to it.

2. $\widehat{ME} < \widehat{GL}$, so $\widehat{AG} < \widehat{GB}$ and $\widehat{EG} < \widehat{GB}$. The height of the time $\widehat{EI}$ is less than that of the time $\widehat{IB}$.

3. $\widehat{ME} > \widehat{GL}$, so $\widehat{AG} > \widehat{GB}$; we cannot draw any conclusion, because a comparison between $\widehat{EG}$ and $\widehat{GB}$ depends on the position of the point E, that is, on the particular position of the hour circle concerned.

In the second version of the figure ($\widehat{AB} > \widehat{BC}$), the problem is the same. We have $\widehat{CH}$ as the height of the point I, $\widehat{NH}$ the height of the arc $\widehat{EI}$ and $\widehat{HB}$ the height of the arc $\widehat{IB}$. We again have three cases:

1. $\widehat{CN} = 2\widehat{LH}$; here again the height of the time $\widehat{EI}$ is less than that of the time $\widehat{BH}$.

2. $\widehat{CN} < 2\widehat{LH}$; we prove that the height of the time $\widehat{EI}$ is less than that of the time $\widehat{BH}$.

3. $\widehat{CN} > 2\widehat{LH}$; in this case we have $\widehat{CH} > \widehat{BH}$. We cannot draw any conclusion because a comparison between $\widehat{NH}$ and $\widehat{BH}$ depends on the position of the point E, that is, on the position of the chosen hour circle.

Thus, in both the first and second versions of the figure, the height of the time $\widehat{EI}$ is less than the height of the time $\widehat{IB}$ and in the third case we have three possibilities, depending on the position of E.

Ibn al-Haytham may have written up this proposition a bit too quickly, which would explain his accidentally confusing the height of the point I with that of the arc $\widehat{EI}$.

3.3. HISTORY OF THE TEXT

In the three old lists that give titles of works Ibn al-Haytham had written before 1038, lists that are transmitted by al-Qifṭī, Ibn Abī Uṣaybiʻa and the anonymous author of the Lahore manuscript, we find the title *Fī al-ikhtilāf fī irtifāʻāt al-kawākib*, 'On the variety of heights of the <wandering> stars'. This book has come down to us under a fuller title: *Fī mā yaʻriḍ min al-ikhtilāf fī irtifāʻāt al-kawākib*, 'On the variety that appears in the heights of the <wandering> stars'.[3] This last title is most probably the one that Ibn al-Haytham wanted to give his book, a title that was then slightly shortened by

---

[3] R. Rashed, *Ibn al-Haytham and Analytical Mathematics*, p. 400.

the old biobibliographers – a procedure which is by no means unusual. In any case, the text exists in only one manuscript. It forms part of the collection Fātiḥ no. 3439 (fols 151ʳ–155ʳ) of the Süleymaniye Library in Istanbul. This collection contains other texts by Ibn al-Haytham, such as his *Exhaustive Treatise on the Figures of Lunes*. The manuscript was copied in 806/1403–1404. The text is difficult to read because the ink is pale, which, with the passage of time, has made certain parts difficult to decipher. The copyist's writing is in rather untidy *naskhī*, and the text has about twenty places where a single word has been omitted and five omissions of a group of more than two words. We have also noticed numerous errors of transcription, notably of letters for geometrical points, together with a certain number of errors in Arabic – in particular grammatical ones – which, on examination, seem to be due to the copyist. Nevertheless, once a text has been established, these accidents and mistakes do not in any way obscure the meaning.

TRANSLATED TEXT

Al-Ḥasan ibn al-Haytham

On the Variety that Appears in the Heights of the Wandering Stars

In the name of God, the Compassionate the Merciful
Lord, make our task easy and do not raise up difficulties

TREATISE BY AL-ḤASAN IBN AL-ḤASAN
\<IBN AL-HAYTHAM\>

On the Variety that Appears in the Heights
of the Wandering Stars

The horizon is a great circle that divides the sphere of the Universe into two equal parts. The meridian circle is the great circle that passes through the zenith and through the two poles around which the sphere moves; and it is perpendicular to the horizon. The circles of altitude are great circles that pass through the zenith and are perpendicular to the horizon.

The two poles of the sphere are two endpoints of one of its diameters, and the two poles remain fixed while the sphere moves. The pole of a circle is a point such that all the straight lines that are drawn from it to the circumference of the circle are equal; and any circle on a sphere has two poles.

The hour circles are circles of unequal sizes generated by the motion of the sphere, and among them there is a single great circle. They are all parallel and their two poles, for all of them, are the two poles of the sphere, around which it moves; those of them that are the closest to the pole are the smallest. The hour \<arcs\> are parts of these circles.

The circle of the equator is the greatest circle that the sphere describes by its motion.

The arc of altitude of a circle of altitude is \<the arc\> between the point that has a height with respect to the horizon, and the horizon.

A right sphere is one whose two poles lie on the circumference of its[1] horizon. An inclined sphere is one for which one of the two poles is visible above the ground,[2] and the other hidden beneath it. Any point on the sphere has a height with respect to the horizon and with respect to any plane

---

[1] That is the horizon of the place considered.

[2] He means the horizon of the place considered.

parallel to the horizon in a given time and at a given altitude;[3] I call the altitude *the height for this time*. And if its height is more than that altitude, then I call the excess of the second altitude over the first altitude *the height of the second time*.

<Proposition 1>: If we divide the bases of two similar segments of circles in the same ratio and if we draw from the points of division two straight lines at equal angles, then they [these lines] divide the two arcs in the same ratio.

Example: Let ABC, DEG be two similar segments; we put the ratio of AH to HC equal to the ratio of DI to IG and we draw HB and IE at two equal angles. I say that the ratio of the arc AB to the arc BC is equal to the ratio of the arc DE to the arc EG.

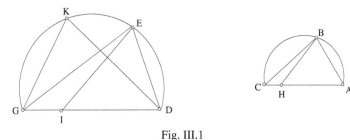

Fig. III.1

Proof: It cannot be otherwise. If that were possible, then let the ratio of the arc AB to the arc BC be equal to the ratio of the arc DK to the arc KG. We join AB, BC, DK, KG. Since the arcs ABC, DEG, which are similar, have been divided in the same ratio, the arcs BC and KG are similar. So the angles BAC, KDG are equal. But the angles ABC and GKD are equal because the two points B and K are corresponding points. So the two triangles ABC, DKG are similar. So the ratio of BA to AC is equal to the ratio of DK to DG. But the ratio of AC to AH is equal to the ratio of GD to DI, by our hypothesis. So, using the ratio of equality, we have that the ratio of AB to AH is equal to the ratio of KD to DI. But the two angles BAH and KDI are equal; so the triangles BAH and KDI are similar and their angles are equal. So the angle AHB is equal to the angle DIK. Which is not possible.[4]

So the ratio of the arc DK to the arc KG is not equal to the ratio of the arc AB to the arc BC; so the ratio of the arc AB to the arc BC is equal to the ratio of the arc DE to the arc GE. That is what we wanted to prove.

---

[3] The translation is literal. The meaning becomes clearer in what follows.

[4] If $K \neq E$, because by hypothesis the angles AHB and DIE are equal.

<**Proposition 2**>: If in a circle we draw one of the diameters; if we take any two points such that the product of the two parts cut off by the first point, one part [multiplied] by the other, is equal to a quarter of the area enclosed by the whole diameter and the part cut off by the second point and if we draw from these two points two perpendiculars to the diameter that end on the circumference, then on the side towards the end of the diameter they cut off two equal arcs.

Example: In the circle *ABCD* we draw the diameter *BD* and on it we take two points *E* and *G* such that the product of *BG* and *GD* is equal to a quarter of the area enclosed by the whole diameter and *BE*; if we draw *GH* and *EC* at right angles [to *BE*], I say that the arc *BH* is equal to the arc *HC*.

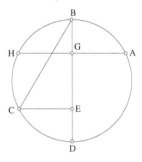

Fig. III.2

Proof: We extend *HG* to *A* and we join *BC*. Since *BD* is a diameter of the circle and *GH* is a perpendicular to the diameter, the straight line *AG* is equal to the straight line *GH*; so the square of *AH* is four times the square of *GH*. But the product of *DG* and *GB* is equal to the square of *GH*; so the square of *AH* is four times the product of *DG* and *GB*. But the product of *DB* and *BE* is four times the product of *DG* and *GB*; so the product of *DB* and *BE* is equal to the square of *AH*. But, since *DB* is a diameter and *CE* is a perpendicular, the product of *BD* and *BE* is equal to the square of *BC*. So the square of *AH* is equal to the square of *BC* and *AH* is equal to *BC*. So the arc *AH* is equal to the arc *BC*. If we take away the common arc *BH*, there remains the arc *AB* equal to the arc *CH*. But the arc *BH* is half of the arc *AH*, because *AH* is perpendicular to the diameter; so the arc *BH* is equal to the arc *HC*.

Using this method, we prove that, if we cut off from the circumference two equal arcs and if we draw from them two perpendiculars to the diameter, then they [the perpendiculars] divide the diameter in this ratio.[5] That is what we wanted to prove.

<Proposition 3>: If in a circle we draw a chord that cuts off from it an arc smaller than a semicircle; if we take on the chord two points such that the product of the two parts into which it is cut by the first point, [the product of] the one and the other, is equal to a quarter of the area enclosed by the whole straight line and the part cut off by the other point; if we draw from the two points two perpendiculars that end on the arc, then on the side of the end of the chord they [the perpendiculars] cut off two unequal arcs such that the smaller is on the side towards the endpoint *<B>* of the chord.

Example: In the circle *ABCD*, there is the chord *BD* which cuts off from it an arc *BAD* that is smaller than a semicircle. We take two points *G* and *E* on the chord in such a way that the product of *DG* and *GB* is equal to a quarter of the area enclosed by *DB* and *BE*. We draw from the points *E* and *G* the two perpendiculars *AE* and *GP*; I say that the arc *AP* is greater than the arc *PB*.

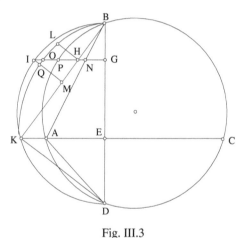

Fig. III.3

Proof: We construct on the straight line *BD* a semicircle *BKD* and we draw *EA* as far as *K* and *GP* as far as *I*; we join *BHK* and *BNA*, *AD* and *DK*.

Since *BD* is a diameter of the circle *BKD* and we have put the product of *DG* and *GB* equal to a quarter of the area enclosed by *DB* and *BE*, and *KE* and *IG* are two perpendiculars, we have that the arc *IB* is equal to the arc *IK*, from what has been shown in the preceding proposition. But since

---

[5] That is, the division has the property described in the statement of the proposition.

the angle *KDE* is greater than the angle *ADE*, the arc *KB* is greater than the arc similar to the arc *AB*.

We construct on the straight line *BK* an arc *BLK* similar to the arc *AB*. We draw *IM* perpendicular to *BK*; it cuts the straight line *KH* because the angle *IHK* is acute – it is in fact equal to the angle *BKE*. Since the arc *KI* is equal to the arc *IB*, and *IM* is a perpendicular, so *KM* is equal to *MB*. But *MQ* is a perpendicular; so the arc *KQ* is equal to the arc *QB*. So the arc *KO* is greater than the arc *OB*; but the angle *BHI* is greater than the angle *BGH* because it is exterior to the triangle. We cut off from it an angle *BHL* equal to *BGH*. So the straight line *HL* divides the arc *OB*. But we have proved that the arc *KO* is greater than the arc *OB*; so the arc *KL* is much greater than the arc *LB*. Since the straight line *HG* is parallel to the straight line *AK*, the ratio of *KH* to *HB* is equal to the ratio of *AN* to *NB*. But the angle *BHL* is equal to the angle *BGH* and the arc *KLB* is similar to the arc *APB*. So the ratio of the arc *KL* to the arc *LB* is equal to the ratio of the arc *AP* to the arc *PB*. But the arc *KL* is greater than the arc *LB*. So the arc *AP* is greater than the arc *PB*. That is what we wanted to prove.

<**Proposition 4**>: If in a circle we draw a chord that cuts off from it an arc that is not greater than a semicircle;[6] if we divide the arc into two equal parts and if we draw from the point of division a straight line that meets the chord at an acute angle and ends on the circumference of the circle; if on the straight line we take a point [that lies] between the arc and the first chord, in such a way that we make the product of the two parts of the whole straight line one and the other <equal to> a quarter of the area enclosed by the whole straight line and the part that ends at the chord; and if we draw from this point a straight line parallel to the initial chord, then it divides the two arcs that are on either side of it into two unequal parts such that the greater of them is on the side towards the endpoint <*B*> of the segment.

Example: In a circle *ABCD* we draw the chord *AC*. We divide the arc *ABC* into two equal parts at the point *B*. We draw the straight line *BED*, at an acute angle;[7] we take a point *G* on this line and we make the product of *DG* and *GB* equal to a quarter of the area enclosed by the straight lines *DB*, *BE*. We draw the straight line *GHI* parallel to *CA*. I say that the arc *BH* is greater than the arc *HA*.

---

[6] See Mathematical commentary.

[7] He means that we draw the straight line *BED* to be such that the angle *BEC* is acute.

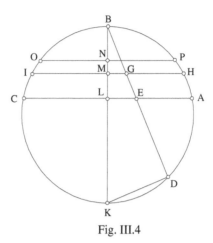

Fig. III.4

Proof: We draw the straight line *BML* perpendicular to *AC* and we extend it to *K*; we join *DK*. Since the arc *AB* is equal to the arc *BC* and *BL* is a perpendicular, the straight line *BK* is a diameter of the circle and the angle *BDK* is a right angle. But since *BL* is a perpendicular and the angle *BDK* is a right angle, the two triangles *BDK*, *BLE* are similar. So the ratio of *KB* to *BD* is equal to the ratio of *EB* to *BL*, and the product of *DB* and *BE* is equal to the product of *KB* and *BL*. But the product of *DB* and *BE* is four times the product of *DG* and *GB*; so it is four times the product of *HG* and *GI*. But since *HI* is parallel to *CA*, *IM* is perpendicular to *BK*. But *BK* is a diameter. So the straight line *HM* is equal to the straight line *MI*. So the product of *IM* and *MH* is greater than the product of *IG* and *GH*. So the product of *IM* and *MH* is greater than a quarter of the product of *KB* and *BL*. But the product of *IM* and *MH* is equal to the product of *KM* and *MB*. So the product of *KM* and *MB* is greater than a quarter of the product of *KB* and *BL*. Accordingly, we put the product of *KN* and *NB* equal to a quarter of the product of *KB* and *BL* and we draw the straight line *NPO* parallel to *CA*. Since the straight line *BK* is a diameter of the circle, the product of *KB* and *BL* is four times the product of *KN* and *NB* and the straight lines *CLA* and *ONP* are perpendicular to the diameter, [accordingly] the arc *CO* is equal to the arc *PA* and the arc *PH* is equal to the arc *OI*. So the arc *BH* is greater than the arc *HA* and the arc *BI* is greater than the arc *IC*. That is what we wanted to prove.

<Proposition 5>: If in a circle we draw a chord that cuts off from it an arc that is not greater than a semicircle, if we take on this a point that divides the arc into two unequal parts and if from this point we draw a straight line that meets the chord at an acute angle on the side of the smaller part; if, on that side, this straight line cuts off from the circle a segment that is not greater than a semicircle, if we then take on that [segment] a point that lies between the two arcs and the chord, in such a way that we put the product of the two parts of the whole straight line, the one and the other, [equal to] a quarter of the area enclosed by the whole straight line and the part that ends on the chord, and if we draw from this point a straight line parallel to the chord, then it [the straight line] divides the minor arc into two unequal parts such that the smaller part is on the side towards the endpoint <*B*> of the arc.

Example: In a circle *ABCD*, there is the chord *AC*, the arc *AC* being no greater than a semicircle. On it we take a point *B* in such a way that the arc *AB* is greater than the arc *BC*; we draw the straight line *BED* such that the angle *BEC* is acute and the arc *BCD* is not greater than a semicircle; we put the product of *DK* and *KB* [equal to] a quarter of the area enclosed by *DB* and *BE* and we draw *KM* parallel to *AC*. I say that the arc *CM* is greater than the arc *MB*.

Proof: We draw the straight lines *EH*, *KI* at a right angle.

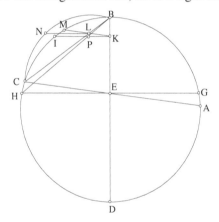

Fig. III.5.1

First of all, let the arc *BCD* be a semicircle; we extend *HE* to *G* and we join *BLC* and *BPH*. Since the product of *DK* and *KB* is a quarter of the area enclosed by *DB* and *BE* and the two straight lines *HE* and *KI* are perpendicular to *BD*, the arc *BI* is equal to the arc *IH*. But the arc *BH* is greater than the arc *BC*. So on the straight line *BC* we construct an arc similar to the arc *BH*. Let the arc be *BNC*. Since *KI* is parallel to *EH*, the ratio of *BP* to *PH* is equal to the ratio of *BK* to *KE*. In the same way the ratio of *BL* to *LC* is equal to the ratio of *BK* to *KE*. So the ratio of *BP* to *PH* is equal to the ratio of *BL* to *LC*. But the arc *BG* is smaller than the arc *AB*;

so the angle *BHE* is smaller than the angle *BCE*; so the angle *IPH* is smaller than the angle *MLC*. So we cut off the angle *CLN* equal to the angle *HPI*. Since the arc *BH* is similar to the arc *BNC*, the ratio of *BP* to *PH* is equal to the ratio of *BL* to *LC*. But the angle *HPI* is equal to the angle *CLN*. So the ratio of the arc *BN* to the arc *NC* is equal to the ratio of the arc *BI* to the arc *IH*. But the arc *BI* is equal to the arc *IH*; so the arc *BN* is equal to the arc *NC*. But the angle *CLN* is acute, and the perpendicular drawn from the point *N* to the straight line *BC* divides the arc *BC*[8] into two equal parts; it follows that the arc *BM* is smaller than the arc *MC*. That is what we wanted to prove.

Alternatively, let the arc *BCD* be smaller than a semicircle. I say that the arc *BM* is smaller than the arc *MC*.

Proof: On the straight line *BD* we construct a circle with diameter *BD*. It falls outside the arc *BCD* because the latter is smaller than a semicircle; and it lies inside the arc *BAD* because this latter is greater than a semicircle. We draw *EH*, *KI* perpendicular to *BD*; we extend *EH* to *G* and we join *DH*, *DC*, *DG*, *DA*, *BG*, *HB*, *CH*, *OH*.[9] So the angle *HDB* is greater than the angle *CDB*; so the arc *BIH* is greater than <the arc> similar to the arc *BMC*.

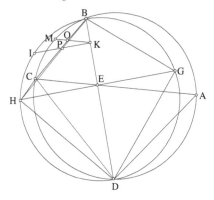

Fig. III.5.2

But the angle *GDB* is smaller than the angle *ADB*; so the angle *BHE* is smaller than the angle *BCE*, because they are equal to the first two angles. So the angle *HPI* is smaller than the angle *COM* and the ratio of *BP* to *PH* is equal to the ratio of *BO* to *CO*. But the arc *BIH* is greater than <the arc> similar to the arc *BMC*, the ratio of *BP* to *PH* is equal to the ratio of *BO* to

---

[8] This is the arc *BC* of the given circle.
[9] *O* is the point of intersection of *BC* and *KM*; one may note that the segment *OH* plays no part in the argument.

OC and the angle *HPI* is smaller than the angle *COM*. So the arc *BI* is equal to the arc *IH* and the arc *BM* is thus smaller than the arc *CM*, as was proved in the preceding proposition. That is what we wanted to prove.

<**Proposition 6**>: If in a circle we draw one of its diameters, if we take any point on its circumference, if we join the ends of the greater of the two arcs with a straight line from which we cut off a segment equal to the semidiameter of the circle and if from the point of division we draw a straight line parallel to the diameter, then it divides the other arc into two equal parts.

Example: In the circle *ABC* we draw the diameter *AC*; its centre is *D*. On it we take a point *B* and we join *AB*. We cut off *AG* equal to the semidiameter. We draw the straight line *EG* parallel to *CA*. I say that the arc *BE* is equal to the arc *EC*.

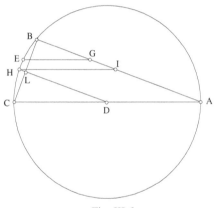

Fig. III.6

Proof: It cannot be otherwise. If it were possible, let them be unequal.

We divide the arc *BC* into two equal parts at the point *H*. We draw the straight line *HI* parallel to *CA* and we join *BC*. Since the arc *BH* is equal to the arc *CH* and *D* is the centre of the circle, accordingly *DL* is perpendicular to *BC*. So the angle *DLC* is a right angle. But the angle *ABC* is a right angle because it is in a semicircle. So the straight line *DLH* is parallel to *BA*. But the straight line *HI* is parallel to *CA*. So the straight line *AI* is equal to the straight line *DH*. But *DH* is a semidiameter; so the straight line *AI* is a semidiameter. But *AG* is a semidiameter. So the arc *BE* is equal to the arc *EC*.

By this proof, we have shown that if *AG* is smaller than a semidiameter, then the arc *CE* is smaller than the arc *EB*; and if it is greater, then the arc is greater. That is what we wanted to prove.

<Proposition 7>: If in a circle we draw a chord that cuts off from it an arc not greater than a semicircle; if on the arc we take a point from which we draw to one end of the chord a straight line that we divide into two equal parts and we draw from it[10] a straight line parallel to the chord, then it divides each of the two arcs into two unequal parts such that the greater part is on the side of the vertex of the arc.

Example: In the circle *ABC* we draw a chord *AC* <which cuts off from it an arc that is not greater than a semicircle>. On the arc we take a point *B* from which we draw the straight line *AB* that we divide into two equal parts at the point *D*, from which we draw a straight line parallel to *CA*. I say that the arc *BE* is greater than the arc *EA*, and that the arc *BG* is greater than the arc *GC*.

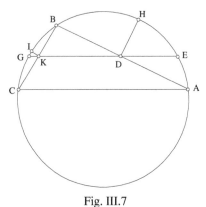

Fig. III.7

Proof: We join *BKC* with a straight line and we draw from the points *D* and *K* two perpendiculars *DH* and *KI*. They then cut the two angles *EDB* and *GKB*,[11] because the two angles *EDA* and *GKC* are acute. They are in fact equal to the angles *BAC* and *BCA*, which are acute, since each of the arcs *AB* and *CB* is smaller than a semicircle. So the straight lines *DH* and *KI* cut the arcs *BE* and *BG*. But the straight line *AD* is equal to *DB* and *DH* is a perpendicular. So the arc *BH* is equal to the arc *HA* and the arc *BE* is greater than the arc *EA*. In the same way the arc *BI* is equal to the arc *IC* and the arc *BG* is thus greater than the arc *GC*. That is what we wanted to prove.

---

[10] That is starting from the point we have obtained, the mid point of the chord.

[11] The straight line *HD* cuts the angle *EDB*, because $E\hat{D}B = \pi - A\hat{D}E$ and *KI* cuts the angle *GKB* because $G\hat{K}B = \pi - G\hat{K}C$.

<Proposition 8>: If the circle *ABC* is one of the meridian circles and the arc *AHD* is the semicircle of the horizon; if the point *G* lies on the surface of the sphere and if we draw through it a plane parallel to the horizon that cuts the meridian circle at the point *C*, I say that the arc *CD* is the height of the point *G*.

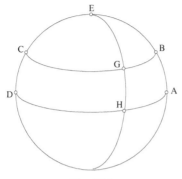

Fig. III.8

Proof: We draw from the zenith, let it be *E*, an arc of a great circle to the point *G*; let the arc be *EG*. Let it meet the horizon at the point *H*. Since the circle *ABC* is a meridian circle and since the circle *AHD* is the circle of the horizon, the circle *ABCD* is perpendicular to the circle *AHD* and the point *E* is the mid point of the arc *BEC*. So the point *E* is the pole of the circle *AHD* and the arcs drawn from the point *E* to the arc *AHD* are equal. But since the plane drawn through the point *G* is parallel to the horizon, it gives us a circle parallel to the circle of the horizon; the point *E* is its pole and the arcs drawn from the point *E* to the circumference of the circle *BGC* are equal. So the two arcs *EH* and *ED* are equal, because they are drawn from the pole to the circumference of the circle of the horizon; and the two arcs *EG* and *EC* are equal, because they are drawn from the pole of the circle *BGC* to its circumference. It remains that, the arc *GH* is equal to the arc *CD*, and it is the height of the point *G* because it is an arc of the altitude circle between the point *G* and the horizon. So the arc *CD* is the height of the point *G* with respect to the horizon. If the point lies on the circumference of the meridian circle, as for the point *C*, we show that the arc *CD* is the height of the point *C* because the meridian circle is one of the altitude circles – it is in fact drawn from the zenith and ends on the horizon, and if it passes through the point *C*, the arc *CD* is the height of the point *C*. That is what we wanted to prove.

<Proposition 9>: If the circle *ABC* is one of the meridian circles and *ADC* is the horizon of this circle; if the arc *BD* is one of the hour circles and from it we cut off the arc *LE* and if through *L* and *E* we draw two planes parallel to the horizon that cut the meridian circle at the points *K* and *H*, I say that the arc *KH* is the height for the time *LE*.

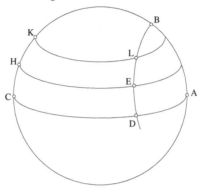

Fig. III.9

Proof: The arc *HC* is the height of the time *ED*, from what has been proved in the preceding proposition. But the arc *KC* is the height of the time *DL*. The amount by which the height *KC* exceeds the height *HC* is the height *KH*. So the height *KH* is the height of the time *EL*. That is what we wanted to prove.

<Proposition 10>: If a point moves on the circle of the equator in the right sphere, then its height increases by equal amounts in equal times.

Let there be a meridian circle *ABC* in the right sphere, let the circle *ADC* be a horizon, or alternatively parallel to the horizon; let the arc *BD* be an arc of the circle of the equator and let the time *DE* be equal to the time *EB*. I say that the point that moves on the arc *DB* rises in the course of the time *DE* by the same amount as it rises in the course of the time *EB*.

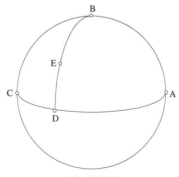

Fig. III.10

Proof: The circle *ABC* is a meridian circle in the right sphere and the arc *DB* forms part of the circle of the equator. So the point *B* is the zenith, in the right sphere. But we have drawn through this point the arc *BC*; so it [the arc] forms part of a great circle. But the arc *BD* forms part of one of the altitude circles; thus the height of the point *E* is the arc *ED* and the height of the point *B* is the arc *BD*. So the height of the time *DE* is the arc *DE* and the height of the time *EB* is the arc *EB*. But they are equal. So the equal times in the circle of the equator have equal heights. That is what we wanted to prove.

<**Proposition 11**>: If in the right sphere a point moves along a circle parallel to the circle of the equator, and ends on the meridian circle so that its hour arc is divided into two equal parts, then the height of the first time is greater than the height of the second time.

Let the circle *ABCD* be a meridian circle in the right sphere. Let there be an arc *BH* which is part of one of the hour circles parallel to the equator. Let the point *I* move on it [the hour circle] so that it passes through the time [*i.e.* the arc] *HB*. Let the time *BI* be equal to the time *IH*. I say that the height of the time *HI* is greater than the height of the time *BI*.

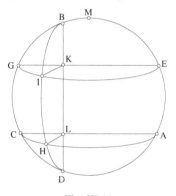

Fig. III.11

Proof: We consider two planes passing through the points *I* and *H* and parallel to the horizon. Let them cut the meridian circle at the points *E, G* and *A, C*. Let their lines of intersection with the meridian circle be *EG* and *AC*. Let the line of intersection of the circle *DHB* and the meridian circle be the straight line *BKLD* and the lines of intersection of this circle and the circles *AHC* and *EIG* be the straight lines *HL* and *IK*. Let the point *M* be the zenith. Since the circle *ABCD* is the meridian and since the circle *BHD* is parallel to the equator, the arc *BHD* is a semicircle and the straight line *BD* is its diameter. Since the circle *ABCD* is perpendicular to the horizon

– because it is the meridian circle – the horizon is perpendicular to this circle. So the planes parallel to the horizon are perpendicular to this circle and the planes *EIG* and *AHC* are perpendicular to the circle *ABCD*. But the circle *BHD* is also perpendicular to the circle *ABCD* and it cuts the planes *EIG* and *AHC*. But if these two planes are perpendicular to a [third] plane and they cut one another, then their line of intersection is perpendicular to this plane. So the straight lines *HL* and *IK* are perpendiculars to the straight line *BD*. But the straight line *BD* is a diameter of the circle *BHD* and the arcs *BI* and *IH* are equal; and we have proved earlier that, if in a circle we have a diameter and if we cut off from the circumference two equal arcs from [points of] which we draw two perpendiculars to the diameter, then they divide the diameter into parts such that the product of the two parts cut off by the first perpendicular, [the product of] the one and the other, is equal to a quarter of the area enclosed by the whole diameter and the straight line cut of by the second perpendicular. So the product of *DB* and *BL* is four times the product of *DK* and *KB*.

In the same way, since the circle *BHD* is one of the circles parallel to the equator in the right sphere, it is perpendicular to the horizon and to all the planes parallel to the horizon. So the circle *BHD* is perpendicular to the plane of the circle *AHC*. In the same way, the circle *ABCD* is perpendicular to the plane *AHC*, so the straight line *BL* is perpendicular to the plane *AHC*. So *BL* is perpendicular to *AC*. In the same way, it is perpendicular to *GKE* which is parallel to *CA*. But, since the circle *ABCD* is the meridian circle and the point *M* is the zenith, the point *M* is the pole of the circle of the horizon and it is the pole of all the circles that are parallel to it. So the arc *MA* is equal to the arc *MC*. So the arc *AB* is greater than the arc *BC*. But we have drawn *BLD* to be perpendicular to *AC*, so the arc *BCD* is smaller than a semicircle. But the product of *DB* and *BL* is four times the product of *DK* and *KB*; and *KG* and *LC* are two perpendiculars, so the arc *GC* is greater than the arc *BG*, as in the third proposition. But, since the planes *EIG* and *AHC* are parallel to the horizon, the arc *GC* is the height of the time *IH* and the arc *BG* is the height of the time *IB*, from what we proved in the ninth proposition. But the arc *GC* is greater than the arc *BG*; so the height of the time *HI* is greater than the height of the time *BI*. That is what we wanted to prove.

<Proposition 12>: If a point moves on one of the hour circles, in the inclined sphere and such that it passes through the zenith; if this point, by its motion, reaches the zenith and its hour arc is divided into two equal parts; then the height of the first time is smaller than the height of the second time.

Let the circle *ABCD* be the meridian circle in the inclined sphere, let the point *B* be the zenith and the circle *BHD* an hour circle on which the point moves to traverse the time [arc] *HB*. Let the time *BI* be equal to the time *IH*. I say that the height of the time *IH* is smaller than the height of the time *BI*.

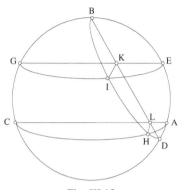

Fig. III.12

Proof: We draw from the points *I* and *H* two planes parallel to the horizon; let them be *EIG* and *AHC*. Let their lines of intersection be *EG* and *AC*. The line of intersection of the circles *ABCD* and *BHD* is the straight line *DLKB*, and the two lines of intersection of the circle *BHD* with the circles *EIG* and *AHC* are the straight lines *IK* and *LH*. Since the circles *BHD*, *EIG* and *AHC* are perpendicular to the plane of the circle *ABCD*, the straight lines *IK* and *HL* are perpendicular to the plane of the circle; so they are perpendiculars to the straight line *BD*. But the straight line *BD* is a diameter of the circle *BHD* and the arc *BI* is equal to the arc *IH*. So the product of *DB* and *BL* is four times the product of *DK* and *KB*. But, since the point *B* is the zenith and since the circle *AHC* is parallel to the horizon, the arc *AB* is equal to the arc *BC*. And, since the sphere is inclined, the circle *BHD* is inclined to the planes parallel to the horizon and the straight line *BL* makes an acute angle with the straight line *ALC*.

Since the arc *ABC* is not greater than a semicircle,[12] since the arc *AB* is equal to the arc *BC*, since the straight line *BL* makes an acute angle with the straight line *LC*, since the product of *DB* and *BL* is equal to four times the product of *DK* and *KB*, since the straight line *KG* is parallel to the straight line *AC*, the arc *GC* is smaller than the arc *BG*, as we have proved in the fourth proposition. Now the arc *GC* is the height of the time *IH* and the arc *BG* is the height of the time *BI*; so the height of the time *IH* is smaller than the height of the time *BI*. That is what we wanted to prove.

<Proposition 13>: If a point moves in the inclined sphere on one of the hour circles, either the circle of the equator or a circle that is parallel to it on the side to which the sphere is inclined; if in the course of its motion it reaches the meridian circle and its hour arc is divided into two equal parts, then the height of the first time is greater than the height of the second time.

Let the circle *ABCD* be the meridian circle in the inclined sphere and let the circle *BHD* be one of the hour circles; let this be either the circle of the equator or a circle that is parallel to it on the side to which the sphere is inclined; let the point move on it so as to traverse the time [arc] *HB* and let the time *BI* be equal to the time *IH*. I say that the height of the time *IH* is greater than the height of the time *BI*.

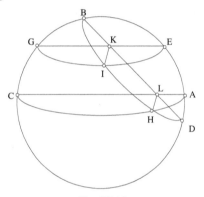

Fig. III.13

Proof: We draw from the points *I* and *H* two planes parallel to the horizon. Since the sphere is inclined, since the circle *BHD* is one of the hour circles and since the circle *AHC* is parallel to the horizon, the circle *BHD* is inclined to the plane *AHC* and the straight line *BL* makes an acute angle with the straight line *ALC*; let the angle be *BLC*. Since the circle *BHD* is either the circle of the equator or is parallel to it on the side to which the sphere is inclined, the arc *BC* is smaller than the arc *AB*; but the arc *BED* is not greater than a semicircle and *DB* is a diameter of the circle *BHD*, the arc *BI* is equal to the arc *IH* and the two planes *AHC* and *EIG* are parallel and are perpendicular to the circle *ABCD*; and the product of *DB* and *BL* is four times the product of *DK* and *KB*, as we proved earlier. But the arc *ABC* is not greater than a semicircle, the arc *AB* is greater than the arc *BC* and the angle *BLC* is acute, the arc *BHD* is not greater than a semicircle, the product of *DB* and *DL* is four times the product of *DK* and *KB* and *KG* is parallel to *CA*; so the arc *GC* is greater than the arc *GB*, as we have proved in the fifth [and sixth] propositions. But the arc *GC* is the height of the time *HI* and the arc *BG* is the height of the time *BI*. So the height of the time *HI* is greater than the height of the time *BI*. That is what we wanted to prove.

<Proposition 14>: If a point moves in the inclined sphere on one of the hour circles that is parallel to the equator on the side where the pole is visible; if in its motion from the horizon it reaches the meridian circle and if its hour arc is divided into two equal parts, then the height of the first time can be equal to the height of the second; it can be smaller than its height and it can be greater.

Let *ABCD* be the meridian circle in the inclined sphere, let the circle *BHD* be parallel to the equator on the side where the pole is visible and let the circle *AHC* be a horizon; and let the time *BI* be equal to the time *HI*. I say that the height of the time *IH* can be equal to the height of the time *IB*; it can be less than it and it can be more than it.

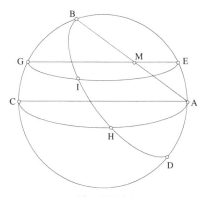

Fig. III.14.1

Proof: We draw the chord of whichever is the greater of the two arcs *AB* and *BC*; in the first case for the figure let it be the straight line *AMB* and in the second case the straight line *BMC*. We consider a plane passing through the point *I* parallel to the horizon; let its line of intersection[13] be the straight line *EMG*.

---

[13] That is, the line of intersection of the plane and the meridian circle.

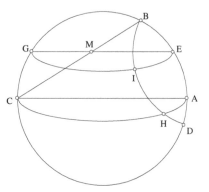

Fig. III.14.2

The straight line *MA*, or *CM*, is either equal to the semidiameter of the circle, or greater than it, or smaller than it. Since the arc *ABC* is a semicircle in which we have the chord *AB*, or *BC*, and since *MG* is parallel to the diameter, accordingly, when the straight line *AM*, or *MC*, is equal to the semidiameter of the circle *ABCD*, the arc *AE*, or *CG*, is equal to the arc *EB* or *GB*; and when the straight line is smaller than a semidiameter, then the arc is smaller than the arc; when the straight line is greater [than a semidiameter], then the arc is greater than the arc, as has been proved in Propositions 6 [and 8]. But the arc *CG*, or *AE*, is the height of the time *HI*; and the arc *GB*, or *EB*, is the height of the time *IB*. So if the straight line *AM*, or *CM*, is equal to the semidiameter, then the height of the first time is equal to the height of the second time; if it is smaller then the height of the first time is smaller; and if it is greater, then the height of the first time is greater. That is what we wanted to prove.

<Proposition 15>: If in the inclined sphere a point moves along one of the hour circles parallel to the equator on the side where the pole is visible and such that the whole circle is visible at the horizon <and raised above it> and is tangent to it; if the motion of the point takes it to the meridian circle so that its hour arc is divided into two equal parts, then the height of the first time is smaller than the height of the second time.

Let the circle *ABC* be the meridian circle in the inclined sphere, let the circle *AIC* be the circle of the horizon and the circle *BDA* the hour circle on which the point *A* moves. Let the arc *AD* be equal to the arc *DB*. I say that the height of the time *AD* is smaller than the height of the time *DB*.

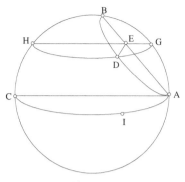

Fig. III.15

Proof: We take a plane passing through the point *D* and parallel to the horizon. Let the plane cut the circle *ABC* along the straight line *GEH*; let the circle *ADB* cut the circle *ACB* along the straight line *AEB* and let *DE* be the line of intersection of the circles *ADB* and *GDH*. Since the circles *ADB*, *GDH* are perpendicular to the circle *ABC*, their line of intersection is a perpendicular to the circle. So the straight line *DE* is a perpendicular to the circle *ABC*; so it is a perpendicular to the straight line *AB*. But since the arc *AD* is equal to the arc *DB* and since *DE* is a perpendicular, the straight line *AE* is equal to the straight line *EB*. But since *AE* is equal to *EB* and since the straight line *GEH* is parallel to the straight line *AC*, the arc *CH* is smaller than the arc *HB* and the arc *AG* is smaller than the arc *GB*, as has been proved in Proposition 7. So if the arc *CB* is not greater than the arc *AB*, the arc *HC* is the height of the time *AD* and the arc *HB* is the height of the time *DB*. And if the arc *AB* is smaller than the arc *BC*, the arc *AG* is the height of the time *AD* and the arc *GB* is the height of the time *DB*. But the arcs *AG* and *HC* are smaller than the arcs *HB* and *GB*; so the height of the time *AD* is smaller than the height of the time *DB*. That is what we wanted to prove.

<Proposition 16>: If a point moves in the inclined sphere along one of the hour circles parallel to the equator on the side where the pole is visible, [an hour circle] such that it is completely visible at the horizon and raised above it; and if in the course of its motion the point reaches the meridian circle and its hour arc is divided into two equal parts, then the height of the first time can be equal to the height of the second time; it can be less than it or more than it.

Let the circle *ABC* be a meridian circle in the inclined sphere, let the circle *ADC* be the horizon, the circle *BIE* one of the hour circles; let it be raised above the horizon by the magnitude of the arc *AE*; let *EI* be equal to

IB. I say that the height of the time *EI* can be equal to the height of the time *IB*; it can be less than it and it can be more than it.

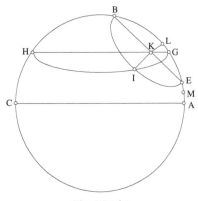

Fig. III.16.1

Proof: We consider a plane passing through the point *I* and parallel to the horizon. Let it cut the circle *ABC* along the straight line *GKH*; let the line of intersection of the hour circle and the meridian circle be the straight line *BKE* and let the line of intersection of the hour circle and the circle parallel to the horizon be the straight line *IK*.

It has been proved in the preceding proposition that the straight line *BK* is equal to the straight line *KE*. First, let the arc *AB* be smaller than the arc *BC*; we draw from the point *K* a perpendicular line *KL*, then the arc *EL* is equal to the arc *LB*. But the arc *AE* is either twice the arc *GL*, or smaller than twice it, or greater than it. We divide the arc *AE* into two equal parts at the <point> *M*. If the arc *AE* is twice the arc *GL*, then the arc *ME* is equal to the arc *GL*. But the arc *GE* is common; so the arc *MG* is equal to the arc *EL*.[14] But *EL* is equal to *LB*; so *MG* is thus equal to *LB*. But *AM* is equal to *GL*; so *AG* is equal to *GB*. But *AG* is the height of the time *EI*[15] and *GB* is the height of the time *IB*; so the height of the time *EI* is equal to the height of the time *IB*.

If the arc *AE* is smaller than twice the arc *GL*, then the arc *ME* is smaller than the arc *GL*. But *EG* is common; so *MG* is smaller than *EL*. But *EL* is equal to *LB*. So *MG* is smaller than *LB*. But *AM* is smaller than *GL*; so *AG* is smaller than *GB*. So the height of the time *EI* is smaller than the height of the time *IB*.

---

[14] That is, if we add *GE* to *ME* and to *GL*, we obtain *MG* = *EL*.

[15] The arc *AG* if the height of the point *I*; the height of the time *EI* is the arc *EG*, so the height of the arc *EI* is smaller than that of the arc *IB*.

In the same way, if *AE* is greater than twice *GL*, the height of the time *EI* is greater than the height of the time *IB*. That is what we wanted to prove.

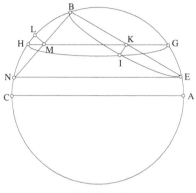

Fig. III.16.2

In the same way, let the arc *BC* be smaller than the arc *AB*. We draw from the point *E* a straight line *EN* parallel to the straight line *AC*, we join *BMN* and we draw the perpendicular *ML*. Since the straight line *EK* is equal to *KB*, the straight line *BM* is equal to the straight line *MN* and the arc *BL* is equal to the arc *LN*. But the arc *CN* is either twice the arc *LH*, or smaller, or greater.

If it is twice it, the arc *CH* is equal to the arc *HB*; if it is smaller than twice it, the arc *CH* is smaller than the arc *HB*; and if it is greater, it is greater, from what has been proved for the first case of the figure. But the arc *CH* is the height of the time *EI*[16] and the arc *HB* is the height of the time *BI*; so it can be equal to the height of the time *IB*, or it can be smaller, or it can be greater. That is what we wanted to prove.

The treatise on the heights of the wandering stars is completed.
Praise be to God, Lord of the worlds

---

[16] The arc *CH* is the height of the point *I*; the height of the time *EI* is the arc *NH*, so the height of the arc *EI* is smaller than that of the arc *IB*.

PART II

INTRODUCTION

INSTRUMENTS AND MATHEMATICS:
HOUR LINES, SUNDIALS AND COMPASSES FOR LARGE CIRCLES

Under the rule of the Caliph al-Ma'mūn (813–833), research in astronomy was pursued with a vigour without precedent since second-century Alexandria. One of the characteristics of this new beginning is that it occurred both in mathematical astronomy and in observational astronomy at the same time. In the early part of this period we find such illustrious figures as al-Fazārī, Yaḥyā ibn Abī Manṣūr, al-Farghānī and Ḥabash, among many others. That is a historical fact, known and described by historians. But it has not been sufficiently emphasized that this intense activity set other activities in train, among them research work on instruments, and in particular on instruments that could be used by astronomers. The testimony provided by the tenth-century biobibliographer al-Nadīm is eloquent on the matter. Reading the entries-in *al-Fihrist* that he devotes to the astronomers from the ninth century onwards, we may remark that the great majority of them did work on astrolabes, or on sundials, or on the armillary sphere, or on more than one of these types of instrument. Al-Nadīm himself points out that there is a close connection between progress in research in astronomy, notably in observational astronomy, and the study and making of instruments. He writes:

> The domain of makers (of these instruments) has become larger under the Abassid State from al-Ma'mūn up to our own time <end of the tenth century>. When al-Ma'mūn wanted astronomical observations to be carried out, he turned to Ibn Khalaf al-Marwarrūdhī, who constructed for him an armillary sphere that is in the keeping of certain scholars in our land; al-Marwarrūdhī also constructed an astrolabe.[1]

There is a long list of astronomer mathematicians who, from al-Fazārī onwards, wrote on this subject. We find it in *al-Fihrist* and in the biobibliographical books old and modern that derive from it.[2] All we need

---

[1] Al-Nadīm, *Kitāb al-fihrist*, ed. R. Tajaddud, Teheran, 1971, p. 342.

[2] C. Brockelmann, *Geschichte der arabischen Literatur*, 2nd ed., Leiden, 1937–1942 and F. Sezgin, *Geschichte des arabischen Schrifttums*, vol. V, Leiden, 1976; vol. VI, 1978.

to say here is that these astronomer mathematicians formed an established tradition. But there was another tradition that ran alongside it, one that can be identified through the names of authors and the titles of their books, a tradition to which al-Nadīm gave its own separate title: 'instruments and their makers'. This is his designation for a complete tradition of artisans who made scientific instruments, and he even went so far as to give the name 'astrolabists' to practitioners of a new craft that had an independent existence. The history of this tradition of making scientific instruments remains to be written.

For the moment, the titles and contents of works help us to identify some of its characteristics. Some books deal with the art of constructing the instrument and using it; others describe how the instrument is used, with a fairly short explanation of the rules governing the way it works; others deal with the theory of the instrument and set out the mathematical proofs required; finally, others take this as an occasion to develop mathematical tools. Thus many books about the astrolabe are of the first two types, whereas *al-Kāmil* by al-Farghānī is of the third type, since the author, a mathematician, begins it with a rigorous study – the earliest we know – on the subject of stereographic projection. As for the last type, it is represented by al-Qūhī and by Ibn Sahl, who started from the astrolabe and went on to develop a serious study of projections, in which stereographic projection is merely one example. Even from this rapid description, it is obvious that a dialogue has been set up between research on instruments and on mathematics, and that it is a dialogue the historian should not neglect.

The study of sundials has the same history as that of astrolabes. In the ninth century, scholars as distinguished as al-Khwārizmī, Ḥabash, Banū al-Ṣabbāḥ, al-Farghānī, Thābit ibn Qurra and, later on, al-Māhānī wrote about sundials and their use. But there is every indication that it was only with Ibrāhīm ibn Sinān (296/909–339/946) that the first substantial text on sundials appeared. Ibn Sinān sets out a theory of sundials that is solidly based in geometry.[3] As for the astrolabe, the mathematical study of sundials also became a discipline in which specialists (*aṣḥāb al-aẓlāl*) were, as Ibn al-Haytham says, identified by a special name, and the specialist artisans were recognized as such.

This literature on sundials was already substantial by the end of the ninth century and the beginning of the tenth, notably in including Ibn Sinān's book.[4] Together with the body of craftsmen who constructed sundials, it constituted the starting point for Ibn al-Haytham. First, let us

---

[3] A detailed account of of Ibn Sinān's contribution has been given elsewhere.

[4] *Fī ālāt al-aẓlāl*, in R. Rashed and H. Bellosta, *Ibrāhīm ibn Sinān. Logique et géométrie au Xᵉ siècle*, Leiden, 2000, Chap. IV.

note that Ibn al-Haytham, like other great mathematicians of the time, was willing to write books designed for teaching and for practical uses. This cultural characteristic is worth emphasizing, because it is a distinctive feature of the scientific activity of the period. For example, we may note that Thābit ibn Qurra wrote an elementary treatise on the measurement of surfaces and volumes, which is manifestly intended for beginners; that his grandson Ibn Sinān wrote a manual without proofs that is addressed to artisans who make sundials; and that Abū al-Wafā' al-Būzjānī wrote two treatises for the same readership.[5] The list is long and substantial. If we confine our attention to Ibn al-Haytham, we find he is the author of a complete treatise on practical geometry addressed to surveyors.[6]

On the other hand, we may observe that Ibn al-Haytham, like Ibn Sinān before him and many others among his predecessors or contemporaries (for example al-Bīrūnī), took a very particular interest in the study of mathematical instruments, and even of the way they were made. It is in this context that a new type of literature arose: the manual addressed to craftsmen written by a mathematician. In such works, the author's concerns were twofold: to study the instrument for the purpose of constructing a mathematical theory of it; and gaining a mathematical understanding that then opened up the possibility of inventing an instrument which might also be useful to society at large. We find Ibn al-Haytham working on both these aspects at the same time.

It is very probable that he modelled himself on al-Qūhī and Ibn Sahl in devoting a complete treatise in two books to the drawing of conic sections. The work, now lost, probably dealt with an instrument designed to draw the conic sections, like the perfect compasses of al-Qūhī.[7] Indeed, in his book

---

[5] See his book *What the Artisan Needs for Geometrical Constructions* (*Fīmā yaḥtāju ilayhi al-ṣāni' min a'māl al-handasa*), ms. Istanbul, Aya Sofia 2753, as well as *What Administrators and Functionaries Need to Know of the Art of Calculation* (*Fīmā yaḥtāju ilayhi al-kuttāb wa-al-'ummāl min 'ilm al-ḥisāb*), ed. A. S. Saidan in *Arabic Arithmetic* (*'Ilm al-ḥisāb al-'arabī*), Amman, 1971; see also Seyyed Alireza Jazbi, *Applied Geometry*, Teheran, 1991.

[6] *Fī uṣūl al-misāḥa*, in R. Rashed, *Les Mathématiques infinitésimales du IXᵉ au XIᵉ siècle*, vol. III: *Ibn al-Haytham. Théorie des coniques, constructions géométriques et géométrie pratique*, London, 2000, Chap. IV; English translation by J. V. Field: *Ibn al-Haytham's Theory of Conics, Geometrical Constructions and Practical Geometry. A History of Arabic Sciences and Mathematics*, vol. 3, Culture and Civilization in the Middle East, London, Centre for Arab Unity Studies, Routledge, 2013.

[7] See R. Rashed, *Geometry and Dioptrics in Classical Islam*, London, 2005, Chap. V.

on parabolic burning mirrors, Ibn al-Haytham makes reference to such an instrument (*āla*).[8]

He also wrote a short treatise on another instrument, compasses for drawing large circles. This text deals with the geometry of the compasses and also describes the procedure for making the instrument.

Ibn al-Haytham has also given us two treatises on the sundial, which exemplify the two forms of interest we referred to earlier. The first treatise – *On Horizontal Sundials* (*Fī al-rukhāmāt al-ufuqiyya*) – is obviously addressed to artisans who make the dials. Contrary to Ibn al-Haytham's usual habits, the book is written in a descriptive style rather than being concerned with proof. It mainly deals with the procedures for making sundials, which are given a rational basis in a preliminary account of some relevant astronomy. In these pages Ibn al-Haytham intends, as he puts it, 'to explain the subject as a whole and the principles on which the construction of sundials is based, and to draw attention to the manner of the construction and the parts where we need these ideas to which reference is not often made except in the books of specialists on shadows (*aṣḥāb al-azlāl*)'.[9] It is at the end of this book that Ibn al-Haytham promises to write a second one, on 'shadow instruments in which we shall give an exhaustive treatment of all the ideas, as well as their purposes and the constructions required in this art'.[10] This is a book on 'shadow instruments', a subject that echoes the title of the book by Ibn Sinān. But this time the purpose is different, because such a book must give proofs. We have every reason to suppose that the book in question is the treatise called *On the Hour Lines* (*Fī khuṭūṭ al-saʿāt*), which has come down to us and is presented in this volume.

---

[8] Ibn al-Haytham, *Majmūʿ al-rasāʾil*, Osmānia Oriental Publications Bureau, Hyderabad, 1938–1939, p. 11: 'We have established in a treatise [...] the finding of all the conic sections by means of the instrument'.

[9] *Fī al-rukhāmāt al-ufuqiyya*, below, p. 581; Arabic text in *Les Mathématiques infinitésimales*, vol. V, p. 849, 1–3.

[10] *Ibid.*, p. 581; Arabic text in *Les Mathématiques infinitésimales*, vol. V, p. 849, 4–5.

CHAPTER I

HOUR LINES

1.1. INTRODUCTION

In his book *On the Hour Lines*, Ibn al-Haytham is carrying forward a project of Ibn Sinān's whose purpose was, essentially, to develop the theory of sundials to the highest possible level. And it is indeed clear that Ibn al-Haytham based this second treatise on the work of Ibn Sinān, but also directed remarks against his predecessor. His scientific methods led him to new results in several fields, among them plane trigonometry. One of the trigonometrical results he obtains here reappears in two other treatises, each of which is of fundamental importance in its own field: in dioptrics his treatise on *The Burning Sphere*, and in astronomy his book on *The Configuration of the Motions of the Seven Wandering Stars*.[1] So we shall begin with the book *On the Hour Lines*, and then return to examine Ibn al-Haytham's first book *On Sundials*.

To understand Ibn al-Haytham's intentions, but also to see how far he distanced himself from Ibn Sinān, we need to look back briefly to the latter's treatise *On Instruments for Shadows*. In this work Ibn Sinān has several purposes: to lay out a geometrical basis for a unified theory of sundials, no longer, like his predecessors, to be content merely to draw the instrument but instead to give proofs of the principles underlying its construction and its use, and, finally, to point out the errors of his predecessors. Ibn Sinān himself lets us know in concrete terms what he conceived his task to be:

> It is said that the ancients, and their successors up to the present day, constructed a special sundial for each plane, determining the lines by a method proper to that plane. I have sought and solidly established a universal method for any plane, using a single proof, which I have demonstrated.[2]

In other words, for a given place *L*, there are several planes that need to be considered – the plane of the horizon, the plane of the meridian, the

---

[1] See above, Part 1, Chapter I.

[2] *Fī ālāt al-aẓlāl*, in R. Rashed and H. Bellosta, *Ibrāhīm ibn Sinān. Logique et géométrie au Xᵉ siècle*, Leiden, 2000, p. 342; Arabic text p. 343, 1–3.

plane of the altitude (the vertical plane) and so on. The idea that allowed Ibn Sinān to construct a unified theory is the following: each of the planes considered for a place L is parallel to the horizontal plane of another place L', whose location is to be found, a place lying somewhere in the northern hemisphere – which is the only hemisphere Ibn Sinān considers.

There is every indication that he was the first to imagine such an undertaking. In any case, that is what he himself claims, in no uncertain terms:

> He judges correctly who ascribes to us all that we have determined about this matter; and for him who suspects that we have relied upon what was done by our predecessors, which is to be found in what they wrote, we have in addition collected up their writings, which had been scattered, and we have supplied proofs for them all. Indeed, I know of none of them who has provided a proof for what they have done, as regards its essentials, but it was said that they gave a description of how to construct sundials, a mere description. And we have moreover added interesting things, matters in which we have no predecessors.[3]

As far as we know, there is no reason to doubt the truth of these statements.

The surviving fragment of the second book of Ibn Sinān's work includes a list of contents. This tells us that this second book is made up of seventeen chapters, the first of which has the title 'On the incorrectness of what preceding mathematicians have used to draw lines for the hours'.[4] In this chapter Ibn Sinān sets out to show that these predecessors were mistaken when they stated that, on a sundial, the points that correspond to a particular hour h for each day of the year lie on a straight line. This criticism had already been expressed by Ibn Sinān's grandfather Thābit ibn Qurra. This latter writes in his treatise *On Sundials*:

> In connection with these sundials (set up in the plane of the horizon), we need to know [the length of] the shadow and the azimuth, for the hours, or for the hours and their subdivisions, whether the hours are seasonal or equinoctial,[5] in either case, you have the possibility of drawing them on the dial, doing this for the first point of Capricorn and the first point of Cancer, then drawing the lines for the hours, as straight lines between these points, or instead to do it in the same way for the other signs; the lines for the hours will

---

[3] *Fī ālāt al-aẓlāl*, p. 340; Arabic text p. 341, 23–28.

[4] *Ibid.*, p. 414; Arabic text p. 415, 8–9.

[5] Equinoctial hours are such that each one represents the length of time in which the celestial sphere turns through 15°. Seasonal hours are unequal and each represents one twelfth of the day and one twelfth of the night. Such hours vary through the year.

then [that is, in the latter case] be exactly correct, they are not straight lines.[6]

So it really seems as if Ibn Sinān wanted to sharpen up the detail in his grandfather's text and prove that the lines are not straight. But we need to ask who are these predecessors Ibn Sinān criticizes. Does he mean al-Māhānī, Abū Saʿīd al-Ḍarīr, or others such as al-Kindī, or even Diodorus?

We have provided this roll call to help us to put Ibn al-Haytham's book into context. Ibn al-Haytham does, in fact, start from what is in Ibn Sinān's book, and goes on to develop a theory regarding a sundial that works correctly at any geographical latitude (what is now called a 'universal' sundial). So his purpose is to go further than Ibn Sinān did, and also to point out an error he made in the proof of one proposition in his second book. Finally, he argues that Ibn Sinān's criticisms of his predecessors were unjust.

This orderly plan of action is another example of a tendency we have already remarked upon in Ibn al-Haytham's scholarly works, whether we turn to his researches in infinitesimal geometry or in the geometry of conics, his studies on analysis and synthesis, or other works. His purpose is to extend the tradition of research to which he belongs as far as possible, to exhaust all its logical possibilities and, if possible, to take it to its limit and complete it. This time the tradition goes back to Ibn Sinān and, beyond him, to the mathematicians of the ninth century. Thus the tradition has already been reshaped by Ibn Sinān.

1.2. MATHEMATICAL COMMENTARY

Let us now turn to Ibn al-Haytham's treatise. It is made up of eleven propositions, which can be divided into two distinct groups. We have a group of geometrical and trigonometrical propositions that make up very nearly half the treatise. These are lemmas required in the proofs of the five propositions of the second group, which constitute Ibn al-Haytham's theory of the sundial. To get a better grasp of the part played by the first group of six lemmas, let us look at what Ibn al-Haytham says:

Before the treatise we have given lemmas that are themselves new results, results that none of those who preceded us has mentioned – as it seems to

---

[6] *Fī ālāt al-sāʿāt allatī tusammā rukhāmāt*, in *Thābit ibn Qurra, Œuvres d'astronomie*, edited and translated by R. Morelon, Paris, 1987, p. 134; Arabic text p. 134, 11–16.

us – and thanks to these lemmas we can go on to derive all the ideas that we have expressed in this treatise.[7]

This text is clearly important, since it tells us that these lemmas are new and original to Ibn al-Haytham, that they constitute a treatise (that is give new results) within the treatise *On the Hour Lines*, and they are fundamental to the theory of sundials. We shall consider these lemmas in detail later. For the time being we merely need to note that – to use terms unknown at the time – some of them deal with the variation of trigonometric functions.

The six lemmas are followed by the five propositions, numbered 7 to 11. The seventh proposition is one that had been stated and proved by Ibn Sinān. He shows, in fact, that the positions of the sun that correspond to a seasonal hour *h* on the circle of the equator and on two parallel circles placed symmetrically with respect to the equator, lie on a great circle. So in the plane of the sundial they correspond to three points on the same straight line Δ.

Ibn al-Haytham's proof is different from the one given by Ibn Sinān. We shall discuss it later, noting its divergences from that given by his predecessor.

The eighth proposition is the one that – according to Ibn al-Haytham (Ibn Sinān's text is lost) – Ibn Sinān 'was not able to establish it for every hour line'.[8] Ibn al-Haytham considers an hour circle – let it be Γ – between the circle of Cancer and the equator, and shows that, if *V* is the point of Γ corresponding to the seasonal hour *h* of the seventh proposition, then, when the sun is at *V*, the shadow of the point *E*, the tip of the gnomon of the sundial, does not lie on the straight line Δ.

So if we consider the 91 circles of parallels associated with the ecliptic longitudes in one-degree steps from 0° to 90°, for a given seasonal hour *h*, we obtain a point on each of these circles. The point we obtain on the equator together with any one of the 90 other points defines a great circle, whose plane cuts the horizontal plane in a straight line Δ. This gives us 90 straight lines for a given hour *h*: $\left(\Delta_{i,h}\right)_{i=1}^{90}$.

In the ninth proposition, Ibn al-Haytham proves that the angle that the straight line $\left(\Delta_{90,h}\right)$ – associated with the solstices – makes with any general straight line $\left(\Delta_{i,h}\right)$, is negligibly small. The statement is by no means merely

---

[7] *Fī khuṭūṭ al-sā'āt*, see below, p. 517 ; Arabic text p. 737, 23–26.

[8] *Fī khuṭūṭ al-sā'āt*, see below, p. 532; Arabic text in *Les Mathématiques infinitésimales du IXᵉ au XIᵉ siècle*. Vol. V: *Ibn al-Haytham: Astronomie, géométrie sphérique et trigonométrie*, London, 2006, p. 771, 16–17.

qualitative, since his calculation is for a particular place, whose latitude is specified by giving the diurnal arc at the summer solstice as 210°, and taking the inclination of the ecliptic to the equator to be 24°, and the radius of the celestial sphere as 60°.

Ibn al-Haytham starts by taking $h = 1$ and looks at the position of parallel straight lines to $(\Delta_{90,1})$ and $(\Delta_{i,1})$, for instance the straight lines EP and EJ that lie in the horizontal plane and pass through the tip of the gnomon, E (see Fig. 1.9.5 and 1.9.6). He first finds the position of EP with respect to the line of the meridian, EN, by calculating PN and EN. He then investigates their ratio $\dfrac{JP}{EP}$ and proves that, for $h = 1$, we have $\dfrac{JP}{EP} < \dfrac{1}{174}$, for any i.

Ibn al-Haytham then repeats the calculation for $h = 5$ and proves that we obtain the same inequality for any i. Finally, he proves that, at the latitude concerned, we have the same inequality for any h. So we conclude that JP is negligibly small with respect to PE.

This method, using calculation, leads to the same conclusion when applied to any latitude. Ibn al-Haytham thus arrives at a general result, and, more importantly, has an inequality of ratios that allows him to note the degree of the approximation.

The tenth proposition concerns the calculation of the length of the longest shadow on a horizontal sundial.

In the last two propositions, Ibn al-Haytham deduces that if the point Q is the shadow of E, the tip of a gnomon, at the end of the first hour of any day, the distance from Q to the line of intersection of the plane of the dial and the great circle that marks the first hour on the equator and on the two circles of Cancer and Capricorn, is less than $\dfrac{1}{30}$ of the height of the gnomon; that is, it is less than 'a distance whose magnitude is not such as the senses can perceive'.[9] Whatever the value of i, a straight line $(\Delta_{i,1})$ drawn on the plane of the sundial cannot be distinguished from the straight line as drawn $(\Delta_{90,1})$. Although they are distinct as mathematical entities, these two straight lines are not distinct as physical ones.

In this same Proposition 11, Ibn al-Haytham employs an identical argument for any non-zero latitude and for any hour. The hour lines are straight lines 'such as the senses can perceive'. For places with latitude zero, the hour lines are all straight lines and parallel to one another.

---

[9] *Fī khuṭūṭ al-sā'āt*, see below, p. 545; Arabic text in *Les Mathématiques infinitésimales du IXᵉ au XIᵉ siècle*, vol. V, p. 799, 22–23.

Accordingly, we can understand the criticism Ibn al-Haytham directs to the work of Ibn Sinān:

> Similarly, he has not shown what is the magnitude of the distance by which the tips of the shadows at the seasonal hours depart from the <straight> line given for that hour. It is possible that the tips of the shadows depart very little from the straight line given for that hour, so that this deviation is insensibly small. And the proof indeed depends on the fact that a mathematical straight line is a length that has no width, whereas the line drawn on the plane surface of the sundial is one that has a noticeable width, which could take in the deviation of the shadows, if this deviation is insensibly small, or is less than it <sc. the width of the line> by some negligibly small amount.[10]

Thus, according to Ibn al-Haytham, Ibn Sinān's error is to have seen the lines of the shadows only as mathematical lines. This criticism is indicative of some of the guiding principles behind what Ibn al-Haytham proposes to do, as a great mathematician who is also a great natural philosopher. The ruling principle of this work is indeed that he proposes 'combining mathematics and natural philosophy' when studying nature. We must, of course, use mathematics in 'physics', but we cannot reduce a shadow to a straight line any more than we can reduce a ray of light to the straight line along which light is propagated. It is precisely this separation that leads us to accept an approximate truth. The facts of the matter are, of course, established with all requisite rigour, but the process includes our permitting some approximation. The question is thus one of knowing how to assess that approximation and correct the earlier mistakes. This is precisely what Ibn al-Haytham turns his attention to doing.

So, to sum it up, in this treatise Ibn al-Haytham develops a general theory of the sundial and the hour lines drawn on it; he proves that the same sundial will work in any location if we allow a negligible error. The principal instruments used by this theory are theorems in plane trigonometry concerning the variation of functions such as $\frac{\sin\varphi}{\varphi}$, and theorems in spherical geometry that lead to certain inequalities of ratios, which are used specifically to put an upper limit on the error made in the approximations. This concern to assess errors seems to be unprecedented. So the theoretical, mathematical and also epistemological implications of the work are very considerable.

We shall now analyse and comment upon each of the propositions in this book.

---

[10] *Fī khuṭūṭ al-sā'āt*, see below, p. 516; Arabic text *Les Mathématiques infinitésimales du IXe au XIe siècle*, vol. V, p. 735, 11–17.

Lemma 1. — Let there be in a circle two parallel chords EG and BD on the same side of the centre, $\widehat{EG} < \widehat{BD} < \pi$. A perpendicular to these chords cuts the arc EG in A, the chord EG in H and the chord BD in I; then

(1) $$\frac{AI}{IH} < \frac{\widehat{AD}}{\widehat{DG}}$$

and

(2) $$\frac{AI}{AH} > \frac{\widehat{AD}}{\widehat{AG}}.$$

We have $BA < BG$ and $BG < BD$. The circle (B, BA) cuts BG in K, BD in L and EG in M, where M lies between A and K.

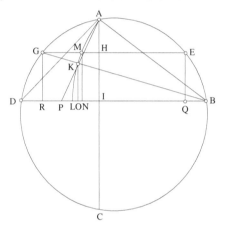

Fig. 1.1

$\widehat{AL}$ is intercepted by the angle at the centre $A\hat{B}L$; that angle is inscribed in the original circle and intercepts $\widehat{AD}$; in the same way $\widehat{KL}$ is intercepted by the angle at the centre $G\hat{B}D$ which is inscribed in the original circle and intercepts $\widehat{GD}$, so

$$\frac{\widehat{AL}}{\widehat{KL}} = \frac{\widehat{AD}}{\widehat{DG}}.$$

AK cuts BD in P. We have

$$\frac{\text{sect.}\,(BAK)}{\text{sect.}\,(BKL)} > \frac{\text{tr.}\,(BAK)}{\text{tr.}\,(BKP)},$$

so

$$\frac{\overwideparen{AK}}{\overline{KL}} > \frac{AK}{KP},$$

hence

$$\frac{\overwideparen{AL}}{\overline{LK}} > \frac{AP}{KP} \quad \text{and} \quad \frac{\overwideparen{AL}}{\overline{LK}} > \frac{AI}{KO} > \frac{AI}{MN}$$

(KO and MN are perpendicular to BD). So

$$\frac{\overwideparen{AL}}{\overline{LK}} > \frac{AI}{IH}$$

and it follows that

(1) $$\frac{\overwideparen{AD}}{\overline{DG}} > \frac{AI}{IH}.$$

From which we deduce

(2) $$\frac{AI}{AH} > \frac{\overwideparen{AD}}{\overline{AG}}.$$

In the same way we prove that

$$\frac{IA}{AH} > \frac{\overwideparen{BA}}{\overline{AE}} \quad \text{and} \quad \frac{\overwideparen{AB}}{\overline{BE}} > \frac{IA}{IH}.$$

Corollary: Let C be the part of AI cut off by the circle

1) If the arc $ABC \le \pi$ and $EQ \perp DB$, then

$$\frac{IB}{BQ} > \frac{\overwideparen{AB}}{\overline{BE}}.$$

2) If the arc $ADC \le \pi$ and $GR \perp DB$, then

$$\frac{ID}{RD} > \frac{\overwideparen{AD}}{\overline{DG}}.$$

To prove this, we merely need to exchange the parts AC and BD play in the figure; then Q (with respect to R) plays the part of H.

Here we wish to prove

$$\frac{\overline{AL}}{\overline{LK}} > \frac{AI}{IH}.$$

Now $AI = AB \sin A\hat{B}L$ and $IH = MN = GB \sin K\hat{B}L$; so the second ratio is equal to

$$\frac{AI}{IH} = \frac{AB \sin A\hat{B}L}{BG \sin K\hat{B}L} < \frac{\sin A\hat{B}L}{\sin K\hat{B}L} < \frac{A\hat{B}L}{K\hat{B}L} = \frac{\overline{AL}}{\overline{KL}}$$

because the function $\varphi \mapsto \dfrac{\sin \varphi}{\varphi}$ decreases for $0 < \varphi \le \dfrac{\pi}{2}$ and from $0 < K\hat{B}L < A\hat{B}L < \dfrac{\pi}{2}$, since $\widehat{EG} < \widehat{BD} < \pi$.

We finish with the observation that

$$\frac{\overline{AD}}{\overline{DG}} = \frac{\overline{AL}}{\overline{KL}}.$$

We can see that this proposition results from the decrease in $\dfrac{\sin \varphi}{\varphi}$.

Lemma 2. — On a circle we are given the arcs $\widehat{BA}$ and $\widehat{AD}$ such that $\widehat{AD} = \dfrac{1}{2}\widehat{AB}$ and $\widehat{AB} \le \dfrac{\pi}{2}$. If E on the arc $\widehat{BA}$ and G on the arc $\widehat{AD}$ are such that $\dfrac{\overline{AB}}{\overline{AE}} = \dfrac{\overline{AD}}{\overline{AG}} \left(AG = \dfrac{AE}{2} \right)$, then

$$\frac{\sin \widehat{AD}}{\sin \widehat{AG}} > \frac{\sin \widehat{AB}}{\sin \widehat{AE}}.$$

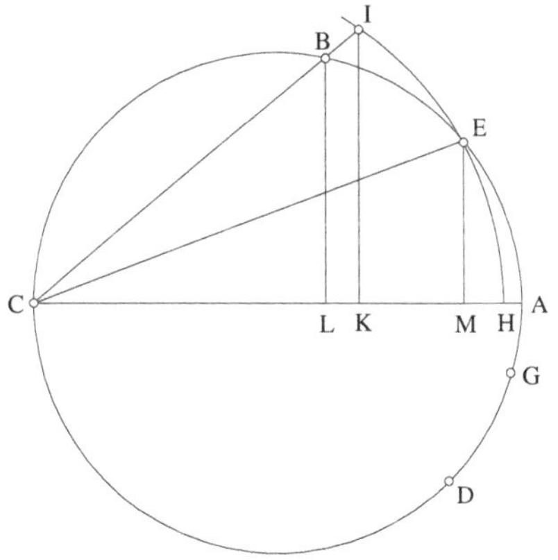

Fig. 1.2

If we put $\widehat{AD} = \alpha_1$ and $\widehat{AG} = \alpha_2$, and if $\alpha_2 < \alpha_1 < \dfrac{\pi}{4}$, then the preceding relation can be written:

$$\frac{\sin \alpha_1}{\sin \alpha_2} > \frac{\sin 2\alpha_1}{\sin 2\alpha_2},$$

and it is accordingly equivalent to $\cos \alpha_2 > \cos \alpha_1$, that is, to $\cos \alpha$ decreasing for $0 \le \alpha \le \dfrac{\pi}{4}$ (which also applies for $\alpha \le \pi$).

Let AC be the diameter from A. The arc $\widehat{AE} < \widehat{AB} \Rightarrow BC < EC < AC$.

The circle (C, CE) cuts AC in H and CB in I. We draw BL, IK and EM perpendicular to AC; we have $IK > BL$ (because $CI > CB$) and $BL > EM$ (because $\widehat{AB} > \widehat{AE}$), hence

$$\frac{IK}{EM} > \frac{BL}{EM}, \quad \frac{IK}{EM} = \frac{\sin \widehat{HEI}}{\sin \widehat{HE}}, \quad \frac{BL}{EM} = \frac{\sin \widehat{AB}}{\sin \widehat{AE}},$$

so

$$\frac{\sin \widehat{IH}}{\sin \widehat{HE}} > \frac{\sin \widehat{AB}}{\sin \widehat{AE}}.$$

But $\widehat{IEH}$ is intercepted by the angle at the centre $I\hat{C}H$, which is inscribed in the given circle and intercepts the arc $\widehat{BA}$ of that circle; so $\widehat{IEH}$ is similar to $\dfrac{1}{2}\widehat{BA}$ which is equal to $\widehat{AD}$; in the same way $\widehat{HE}$ is similar to $\dfrac{1}{2}\widehat{AE}$ which is equal to $\widehat{AG}$, so

$$\frac{\sin \widehat{IH}}{\sin \widehat{HE}} = \frac{\sin \widehat{AD}}{\sin \widehat{AG}},$$

and it follows that

$$\frac{\sin \widehat{DA}}{\sin \widehat{AG}} > \frac{\sin \widehat{BA}}{\sin \widehat{AE}}.$$

Lemma 3. — On a circle let there be the points A, B, C, such that $\dfrac{\pi}{2} \ge \widehat{AB} > \widehat{BC}$; and let there be D on $\widehat{BA}$ and E on $\widehat{BC}$ such that $\dfrac{\overline{BD}}{\overline{BE}} = \dfrac{\overline{BA}}{\overline{BC}}$, then

$$\frac{\sin \widehat{BD}}{\sin \widehat{BE}} > \frac{\sin \widehat{BA}}{\sin \widehat{BC}}.$$

This lemma corresponds to Proposition 2 of the *Treatise on the Configuration of the Motions of Each of the Seven Wandering Stars*.

Let F be the centre of the circle; the straight line FB cuts AC in H, DE in I and the circle in P. The tangents to the circle at A and C cut one another in G, because $\widehat{ABC} < \pi$. There are two cases.

1ˢᵗ case: Let us assume $\widehat{AB} < \dfrac{\pi}{2}$, then $\widehat{AP} > \dfrac{\pi}{2}$.

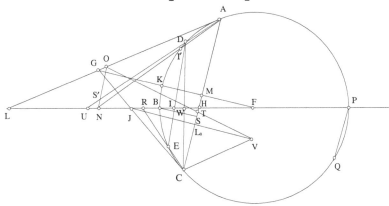

Fig. 1.3.1

Let us draw FG to cut AC in its midpoint M and $\widehat{AC}$ in its midpoint K. We have $\widehat{CP} > \widehat{AP}$, because $\widehat{BC} < \widehat{AB}$. Let Q be such that $\widehat{CQ} = \widehat{AP}$, then $PQ \parallel AC$ and $B\hat{H}C = B\hat{P}Q$. We have $\widehat{QC} = \widehat{AP} > \widehat{AB}$, so $\widehat{QB} > \widehat{AC}$, and it follows that $B\hat{P}Q > G\hat{A}C$ (angles inscribed in a circle); so $B\hat{H}C > G\hat{A}C$, the straight line AG meets the extension of HB in L and HL cuts CG in J. We have

$$\frac{AH}{HC} = \frac{\sin \widehat{AB}}{\sin \widehat{BC}} \quad \text{and} \quad \frac{DI}{IE} = \frac{\sin \widehat{BD}}{\sin \widehat{BE}}.$$

But

$$\frac{AB}{BC} > \frac{\sin \widehat{AB}}{\sin \widehat{BC}}.$$

In fact, if we consider the double arcs $\widehat{BA'} = 2\widehat{AB}$, $\widehat{BC'} = 2\widehat{BC}$ and their chords, we have

$$\widehat{BC'} < \widehat{BA'} < \pi \quad \text{and} \quad \frac{\widehat{BA'}}{\widehat{BC'}} > \frac{BA'}{BC'}.$$

This property was proved by Ptolemy.[11]

Which gives us $\dfrac{\overline{AB}}{\overline{BC}} > \dfrac{AH}{HC}$; similarly, $\dfrac{\overline{BD}}{\overline{BE}} > \dfrac{DI}{IE}$.

Let $BS \perp AC$, then $\dfrac{MC}{CS} > \dfrac{\overline{KC}}{\overline{BC}}$, from Lemma 1 (in which the points that here are called C, B, K, M, S correspond to the points called A, G, D, I, H in Lemma 1), hence

$$\frac{AC}{CS} > \frac{\overline{AC}}{\overline{BC}} \quad \text{and} \quad \frac{AS}{CS} > \frac{\overline{AB}}{\overline{BC}}.$$

So the point T of AC such that $\dfrac{AT}{TC} = \dfrac{\overline{AB}}{\overline{BC}}$ lies between A and S.

Let $JL_a \perp AC$; L_a lies between S and C, hence

$$\frac{AL_a}{L_aC} > \frac{AS}{SC} > \frac{AT}{TC}.$$

Let $CV \parallel AG$, let V be a point on JL_a; we have $A\hat{C}V = C\hat{A}G = A\hat{C}G$, hence $L_aV = L_aJ$ and $CV = CJ$. The straight line VT cuts AG in O; we have

$$\frac{AO}{CV} = \frac{AT}{TC} = \frac{\overline{AB}}{\overline{BC}},$$

hence

$$\frac{AO}{CJ} = \frac{\overline{AB}}{\overline{BC}} = \frac{\overline{BD}}{\overline{BE}} = \frac{\overline{AD}}{\overline{CE}}.$$

The parallel to AC drawn through O meets FL in N, $AT > TC$; hence $AO > CJ$. So the point N lies beyond J. We have $A\hat{N}O = N\hat{A}C$, an acute angle, hence $A\hat{O}N$ is obtuse.

Let I' be the part of AN cut off by the circle. There are three cases in regard to the point D:

a) D between A and I'.

The straight line AD cuts ON in S' and FL in U (see Fig. 1.3.1). We have

[11] *Composition mathématique de Claude Ptolémée*, French transl. N. Halma, 2 vols, Paris, 1813, vol. I, pp. 34–5.

$$AU > AS' > AO \text{ and } \frac{AU}{CJ} > \frac{AO}{CJ},$$

hence

$$\frac{AU}{CJ} > \frac{\widehat{AD}}{\widehat{CE}}.$$

The straight line CE cuts FL in R; $C\hat{B}H$ is acute, hence $C\hat{B}R$ is obtuse, $C\hat{R}J$ is obtuse, $CJ > CR$, and

$$\frac{AU}{CR} > \frac{AU}{CJ} > \frac{\widehat{AD}}{\widehat{CE}} > \frac{AD}{CE} \Rightarrow \frac{AU}{AD} > \frac{CR}{CE} \Rightarrow \frac{UD}{AD} > \frac{RE}{CE} \Rightarrow \frac{DU}{UA} > \frac{ER}{RC}.$$

The straight line DC cuts BH at the point W. Applying Menelaus' theorem to the triangle ADC with transversal UWH gives

$$\frac{HC}{HA} \cdot \frac{UA}{UD} \cdot \frac{WD}{WC} = 1;$$

and we can write

$$\frac{CH}{HA} = \frac{CW}{WD} \cdot \frac{DU}{UA}$$

and

$$\frac{CW}{WD} = \frac{CH}{HA} \cdot \frac{UA}{UD}.$$

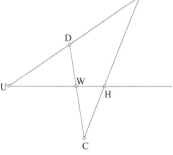

Fig. 1.3.2

Applying the same theorem to the triangle DEC with transversal RIW gives

$$\frac{WD}{WC} \cdot \frac{RC}{RE} \cdot \frac{IE}{ID} = 1,$$

hence

$$\frac{WC}{WD} = \frac{IE}{ID} \cdot \frac{RC}{RE}.$$

So we have

$$\frac{ID}{IE} \cdot \frac{ER}{RC} = \frac{AH}{HC} \cdot \frac{DU}{UA}.$$

But

$$\frac{DU}{UA} > \frac{ER}{RC},$$

hence

$$\frac{DI}{IE} > \frac{AH}{HC};$$

and it follows that

$$\frac{\sin \widehat{DB}}{\sin \widehat{BE}} > \frac{\sin \widehat{BA}}{\sin \widehat{BC}}.$$

b) *D* is at the point *I′*.

In this case, $S′ = N = U$ and $AN > AO$, and the proof is as before.

c) *D* lies between *I′* and *B*.

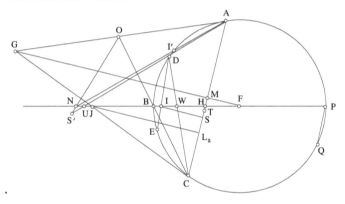

Fig. 1.3.3

In this case, we require to find an integer n such that if $\widehat{BD′} = 2^n \widehat{BD}$, the point $D′$ lies between $I′$ and A; we associate with it the point $E′$ such that $\widehat{BE′} = 2^n \widehat{BE}$. The argument used before in relation to $D′$ and $E′$ gives

$$\frac{\sin \widehat{BD′}}{\sin \widehat{BE′}} > \frac{\sin \widehat{BA}}{\sin \widehat{BC}}.$$

But by applying Lemma 2, we have

$$\frac{\sin \widehat{BD}}{\sin \widehat{BE}} > \frac{\sin 2\widehat{BD}}{\sin 2\widehat{BE}} > ... > \frac{\sin 2^n \widehat{BD}}{\sin 2^n \widehat{BE}},$$

so

$$\frac{\sin \widehat{BD}}{\sin \widehat{BE}} > \frac{\sin \widehat{BD'}}{\sin \widehat{BE'}} > \frac{\sin \widehat{BA}}{\sin \widehat{BC}}.$$

2nd *case*: If $\widehat{AB} = \dfrac{\pi}{2}$, then $\widehat{AP} = \widehat{QC} = \dfrac{\pi}{2}$.

The straight line GA, the tangent at A, is then parallel to FB. Whatever the position of D, the straight line AD meets FB at the point U. The proof is as before.

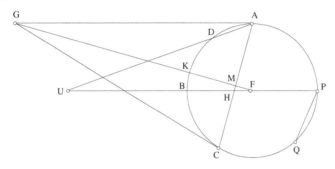

Fig. 1.3.4

This is the proof Ibn al-Haytham gives for Lemma 3. For any D such that $\widehat{BD} < \widehat{BA}$, we can establish that

$$\frac{ID}{IE} \cdot \frac{RE}{RC} = \frac{HA}{HC} \cdot \frac{UD}{UA}.$$

But, in conclusion, we need to prove that $\dfrac{UD}{UA} > \dfrac{RE}{RC}$.

Ibn al-Haytham distinguishes three cases:

$D \in \widehat{AI'}$; in this case, $AU > AO$,
$D = I'$; we again have $AU > AO$;

in these two cases, we may draw the required conclusion.

But if $D \in \widehat{I'B}$, the point S' lies on the extension of ON and U lies between N and B; we still have $AN > AO$, but $AU < AN$; so we can have $AU < AO$, $AU = AO$ or $AU > AO$, and the argument used in the first case does not apply. That is why Ibn al-Haytham tries to get round the difficulty, as we note in part c).

In addition, the matter of the existence of n raises a new difficulty. If we put $\widehat{BA} = \alpha$, then $\widehat{BI'} = \beta$ $\left(\beta < \alpha < \dfrac{\pi}{2}\right)$ and $\widehat{BD} = \gamma$. If $\gamma < \beta$, we need to find an integer n such that $\gamma_n = 2^n \cdot \gamma$ satisfies the double inequality

$$\beta < \gamma_n < \alpha \ (\text{mod } 2\pi).$$

Contrary to what Ibn al-Haytham thought, the problem does not always have a solution. For example let us take $\gamma = 3°$; the series $(\gamma_n)_{n \geq 0} = (3 \cdot 2^n)$ gives: 3, 6, 12, 24, 48, 96, 192, 384... We have $\gamma_7 = 3 \cdot 2^7 = 24$ (mod 360), hence $\gamma_8 = 48$ (mod 360). If we use the name D_n for the point associated with γ_n, we have

$$D_3 = D_7, \ D_4 = D_8, \ ..., \ D_n = D_{n+4}.$$

Consequently, for whatever $\beta \in \ ]48, \ \alpha[, \ \alpha \leq 90°$, for example $\beta = 50°$, it is impossible to find D_n between I' and A.

These difficulties might explain those later encountered by al-Fārisī in making his edition of this proposition. As he puts it: 'But since the copy was badly damaged, I was not able to decipher it: thus I simply recalled the statement. If I am able to decipher it later on, I will add the wording in this place'.[12]

The same al-Fārisī was given pause in his commentary by the condition Ibn al-Haytham formulated in this treatise on *Hour Lines*, but that he had, strangely, left out of his treatise on *The Burning Sphere*, that is

$$\widehat{BC} < \widehat{AB} \leq \frac{\pi}{2}.$$

Now this condition is not necessary. Moreover, Ibn al-Haytham himself uses his Lemma 3 in Propositions 3 and 4 of *The Burning Sphere*, where the arc concerned $\widehat{TJ}$ can be greater than $\dfrac{\pi}{2}$ for certain values of the angle of incidence i, because $\widehat{TJ} > 4d$ (where d is the angle of deviation – something he could not fail to notice).

---

[12] *Fī al-kura al-muḥriqa*, in R. Rashed, *Geometry and Dioptrics in Classical Islam*, London, 2005, p. 258; Arabic text p. 259, 10–11.

Now let us return to the statement of the proposition and put $\widehat{BD} = \beta_1$, $\widehat{BE} = k\beta_1$, $\widehat{BA} = \alpha_1$, $\widehat{BC} = k\alpha_1$ with $k < 1$; the condition set by Ibn al-Haytham may be rewritten as

$$\beta_1 < \alpha_1 < \frac{\pi}{2}.$$

Now it is sufficient to take $\alpha_1 = 120°$, $\beta_1 = 90°$ and $k = \dfrac{1}{2}$ – we obtain $\dfrac{\sin \beta_1}{\sin k\beta_1} = \sqrt{2}$, $\dfrac{\sin \alpha_1}{\sin k\alpha_1} = 1$ – to see that the condition $\alpha_1 < \dfrac{\pi}{2}$ is overly restrictive.

Further, we can show that the proposition remains true for $\beta_1 < \alpha_1 < \pi$. Indeed, let us put

$$f(x) = \frac{\sin x}{\sin kx}, \qquad\qquad \text{with } k < 1;$$

and prove that the function f defined in the interval $]0, \pi[$ decreases in that interval. We have

$$f'(x) = \frac{\cos x \cdot \sin kx - k\cos kx \cdot \sin x}{\sin^2 kx}$$

$$= \left\{ \sin(kx - x) + \frac{1-k}{2}[\sin(x + kx) - \sin(x - kx)] \right\} \frac{1}{\sin^2 kx},$$

$$= \left[\frac{1+k}{2}\sin(kx - x) + \frac{1-k}{2}\sin(x + kx) \right] \frac{1}{\sin^2 kx}$$

$$= \frac{1-k^2}{2\sin^2 kx}\left[\frac{\sin x(1+k)}{1+k} - \frac{\sin x(1-k)}{1-k} \right].$$

Let us put

$$g(x) = \frac{\sin x(1+k)}{1+k} - \frac{\sin x(1-k)}{1-k},$$

we have $g(0) = 0$ and $g'(x) = -2\sin x \cdot \sin kx$.

But $x \in \;]0, \pi[$, and $k < 1$, hence $kx \in \;]0, \pi[$, and consequently $g'(x) < 0$ in the interval $]0, \pi[$; so g decreases if we start from $g(0) = 0$. So we have $g(x) < 0$, hence $f'(x) < 0$, and consequently f decreases in the interval $]0, \pi[$. Accordingly, the inequality

$$\frac{\sin \beta_1}{\sin \beta_2} > \frac{\sin \alpha_1}{\sin \alpha_2}$$

is satisfied for $\beta_1 < \alpha_1 \leq \pi$, if we put $\alpha_2 = k\alpha_1$ and $\beta_2 = k\beta_1$ $(0 < k < 1)$.

Finally, we may note that, in this treatise, as in *The Configuration of the Motions of Each of the Seven Wandering Stars*, Ibn al-Haytham extends his proposition to two similar arcs in two different circles. But whereas Ibn al-Haytham does not repeat this extension in his treatise on *The Burning Sphere*, al-Fārisī recalls it when making his edition of the work.

• Similar arcs in different circles: they are intercepted by equal angles at the centre:

$$\frac{\sin \widehat{BD}}{\sin \widehat{BE}} = \frac{\sin \widehat{B'D'}}{\sin \widehat{B'E'}}.$$

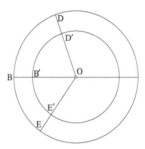

Fig. 1.3.5

• Extension of Lemma 3 to arcs of different circles (arcs smaller than a quarter of the circle):

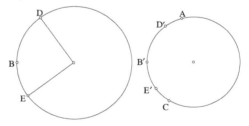

Fig. 1.3.6

$$\widehat{B'A} > \widehat{B'D'}, \ \widehat{B'D'} \text{ similar to } \widehat{BD}$$

$$\widehat{B'C} > \widehat{B'E'}, \ \widehat{B'E'} \text{ similar to } \widehat{BE}$$

$$\frac{\widehat{B'A}}{\widehat{B'C}} = \frac{\widehat{BD}}{\widehat{BE}} \Rightarrow \frac{\widehat{B'A}}{\widehat{B'C}} = \frac{\widehat{B'D'}}{\widehat{B'E'}}.$$

In the second circle, we have

$$\frac{\sin \widehat{B'D'}}{\sin \widehat{B'E'}} > \frac{\sin \widehat{B'A}}{\sin \widehat{B'C}},$$

hence

$$\frac{\sin \widehat{BD}}{\sin \widehat{BE}} > \frac{\sin \widehat{B'A}}{\sin \widehat{B'C}}.$$

In the treatise on *The Configuration of the Motions of Each of the Seven Wandering Stars*, the proof is specifically concerned with similar arcs in different circles, which may suggest that this latter treatise was written later than the treatise on *The Hour Lines*.

Lemma 4. — Let AC and BD be two chords in the same circle which cut one another in a point E; if $\dfrac{\widehat{AB}}{\widehat{ABC}} = \dfrac{\widehat{CD}}{\widehat{CDA}}$, then $\dfrac{A\hat{E}B}{\pi} = \dfrac{\widehat{AB}}{\widehat{ABC}} = \dfrac{\widehat{CD}}{\widehat{CDA}}$.

We have

$$\frac{A\hat{C}B}{C\hat{A}B} = \frac{\widehat{AB}}{\widehat{BC}} \quad \text{and} \quad \frac{C\hat{A}D}{A\hat{C}D} = \frac{\widehat{CD}}{\widehat{DA}};$$

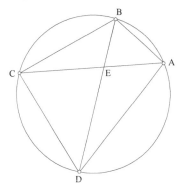

Fig. 1.4

from this hypothesis we deduce

$$\frac{\widehat{AB}}{\widehat{BC}} = \frac{\widehat{CD}}{\widehat{DA}},$$

and on the other hand $C\hat{A}D = C\hat{B}D$ and $A\hat{C}D = A\hat{B}D$, hence

$$\frac{A\hat{C}B}{C\hat{A}B} = \frac{C\hat{B}D}{A\hat{B}D} = \frac{A\hat{C}B + C\hat{B}D}{C\hat{A}B + A\hat{B}D} = \frac{A\hat{E}B}{B\hat{E}C},$$

so

$$\frac{\overline{AB}}{\overline{BC}} = \frac{A\hat{E}B}{B\hat{E}C} \quad \text{and} \quad \frac{\overline{AB}}{\overline{AB}+\overline{BC}} = \frac{A\hat{E}B}{A\hat{E}B+B\hat{E}C},$$

so

$$\frac{\overline{AB}}{\overline{ABC}} = \frac{A\hat{E}B}{\pi}.$$

If α and β are the angles at the centre subtended by the arcs AB and BC, those subtended by the arcs CD and AD are $\dfrac{2\pi\alpha}{\alpha+\beta} - \alpha$ and $\dfrac{2\pi\beta}{\alpha+\beta} - \beta$ respectively, given the condition $\dfrac{\overline{AB}}{\overline{AC}} = \dfrac{\overline{CD}}{\overline{CA}} = \dfrac{\alpha}{\alpha+\beta}$. From what is proved in the lemma, the size of angle AEB is $\dfrac{\pi\alpha}{\alpha+\beta}$, which is simply the mean of α and $\dfrac{2\pi\alpha}{\alpha+\beta} - \alpha$. In other words, $A\hat{E}B = \dfrac{1}{2}\left(A\hat{O}B + C\hat{O}D\right)$, where O is the centre of the circle.

Lemma 5. — As in Lemma 4 let there be two chords AC and BD that cut one another, and are such that $\dfrac{\overline{AB}}{\overline{ABC}} = \dfrac{\overline{CD}}{\overline{CDA}}$; and a chord GH parallel to AC. Let there be the points I on the arc GDH and K on the arc GBH such that $\dfrac{\overline{IG}}{\overline{GIH}} = \dfrac{\overline{DA}}{\overline{ADC}}$ and $\dfrac{\overline{KH}}{\overline{HKG}} = \dfrac{\overline{BC}}{\overline{CBA}}$, then the straight line KI is parallel to BD.

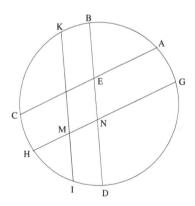

Fig. 1.5

The hypothesis $\dfrac{\overset{\frown}{AB}}{\overline{ABC}} = \dfrac{\overset{\frown}{CD}}{\overline{CDA}}$ is equivalent to $\dfrac{\overset{\frown}{BC}}{\overline{ABC}} = \dfrac{\overset{\frown}{AD}}{\overline{ADC}}$; so we have $\dfrac{\overset{\frown}{IG}}{\overline{GIH}} = \dfrac{\overset{\frown}{KH}}{\overline{GKH}}$; it follows that, if M is the point of intersection of GH and KI, we have from Lemma 4

$$\frac{K\hat{M}H}{\pi} = \frac{\overset{\frown}{KH}}{\overline{GKH}},$$

hence

$$\frac{K\hat{M}H}{\pi} = \frac{\overset{\frown}{BC}}{\overline{CBA}} = \frac{B\hat{E}C}{\pi}$$

and it follows that

$$K\hat{M}H = B\hat{E}C.$$

The straight line BD cuts HG in N (because $AC \parallel GH$), so $B\hat{E}C = E\hat{N}H$, hence $E\hat{N}H = K\hat{M}H$; so the straight lines BD and KI are parallel.

Lemma 6. — In a circle with centre G let there be two chords AC and BD that cut one another in E and are such that $\dfrac{\overset{\frown}{AB}}{\overline{ABC}} = \dfrac{\overset{\frown}{CD}}{\overline{CDA}}$; let IK be the diameter perpendicular to AC at its midpoint H. If $\overset{\frown}{AB} < \overset{\frown}{AK}$,[13] then E lies between H and C and BD cuts GH between G and H.

Let PQ be the diameter parallel to AC. GL, the straight line parallel to BD, cuts AC in M. We have $P\hat{G}L = A\hat{M}L = A\hat{E}B$.

From Lemma 4, we know that

$$\frac{\overset{\frown}{AB}}{\overline{ABC}} = \frac{A\hat{E}B}{\pi};$$

but

$$\frac{\overset{\frown}{LP}}{\overline{PLQ}} = \frac{P\hat{G}L}{\pi},$$

hence

---

[13] A specification added so that E lies between H and C, as required by the statement of the lemma. If $\overset{\frown}{AB} > \overset{\frown}{AK}$, the point E lies H and A, but the point of intersection of BD and GH still lies between G and H.

$$\frac{\overset{\frown}{BA}}{\overset{\frown}{ABC}} = \frac{\overset{\frown}{LP}}{\overset{\frown}{PLQ}}$$

and it follows that

$$\frac{\overset{\frown}{LP}}{\overset{\frown}{PLK}} = \frac{\overset{\frown}{AB}}{\overset{\frown}{ABK}}.$$

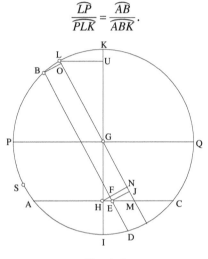

Fig. 1.6

Let S be a point on the arc AP such that $\dfrac{\overset{\frown}{SP}}{\overset{\frown}{PA}} = \dfrac{\overset{\frown}{LP}}{\overset{\frown}{PLK}} = \dfrac{\overset{\frown}{AB}}{\overset{\frown}{ABK}}$, we then

have

$$\frac{\overset{\frown}{AB}}{\overset{\frown}{ABK}} = \frac{\overset{\frown}{SP} + \overset{\frown}{LP}}{\overset{\frown}{AP} + \overset{\frown}{PLK}} = \frac{\overset{\frown}{SL}}{\overset{\frown}{APK}},$$

hence $\overset{\frown}{SL} = \overset{\frown}{AB}$ and $\overset{\frown}{AS} = \overset{\frown}{BL}$. But $\dfrac{\overset{\frown}{SP}}{\overset{\frown}{PA}} = \dfrac{\overset{\frown}{LP}}{\overset{\frown}{PK}}$ gives

$$\frac{\overset{\frown}{PA}}{\overset{\frown}{AS}} = \frac{\overset{\frown}{PK}}{\overset{\frown}{KL}},$$

hence

$$\frac{\overset{\frown}{PK}}{\overset{\frown}{KL}} = \frac{\overset{\frown}{PA}}{\overset{\frown}{BL}};$$

and since $\overset{\frown}{PA} < \overset{\frown}{PK}$, we have

$$\frac{\sin \overset{\frown}{PA}}{\sin \overset{\frown}{BL}} > \frac{\sin \overset{\frown}{PK}}{\sin \overset{\frown}{KL}}, \quad \frac{\sin \overset{\frown}{PA}}{\sin \overset{\frown}{BL}} = \frac{HG}{BO}, \quad \frac{\sin \overset{\frown}{PK}}{\sin \overset{\frown}{KL}} = \frac{GK}{LU} = \frac{GL}{LU},$$

hence

$$\frac{HG}{BO} > \frac{GL}{LU}.$$

But *GLU* and *GHN* are similar, $\frac{GL}{LU} = \frac{GH}{HN}$, hence

$$\frac{HG}{BO} > \frac{GH}{HN}$$

and it follows that $BO < HN$. So the straight line BD cuts GH between G and H and cuts AC in E between H and C.

We may note that the triangles *GHM* and *EJM* are similar, hence

$$\frac{GM}{ME} = \frac{GH}{EJ} = \frac{GH}{BO}$$

and

$$\frac{GM}{ME} = \frac{\sin \widehat{AP}}{\sin \widehat{BL}} = \frac{\sin \widehat{AP}}{\sin \widehat{AS}}.$$

Proposition 7. — Let *ABCD* be the horizon plane at the place E, E being a point in the northern hemisphere. The horizon is cut by the equator in the line BD, by the circle of Cancer in the line IL and by the circle of Capricorn in the line UF. These three straight lines are parallel. The meridian plane for E cuts BD, IL and UF in their respective midpoints E, N and S, which lie on the meridian line AC. The respective centres of these three circles, E, R and R', lie on the axis of the world.

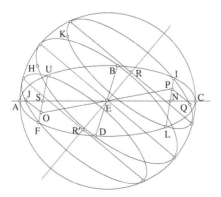

Fig. 1.7.1

Let K be a point of the diurnal arc that is the major arc IL of the circle of Cancer and let Q be the point of the minor arc such that

$$\frac{\overline{QI}}{\overline{IQL}} = \frac{\overline{KL}}{\overline{LKI}} = k \ ;$$

then, from Lemma 6, if $k < \dfrac{1}{2}$, the straight line KQ cuts the straight line IL in P which lies between N and I and cuts the straight line RN between R and N.

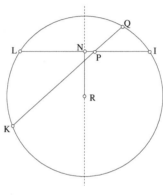

Fig. 1.7.2

If $1 > k > \dfrac{1}{2}$, Ibn al-Haytham proves that the point P lies between N and L.

Ibn al-Haytham's approach is, effectively, to consider the point E as a centre of symmetry in the figure formed by the horizon plane $ABCD$ and the two parallel circles. Thus we have: $EN = ES$, $IL \parallel UF$ and $IL = UF$. The straight line PE cuts UF in O and we have $EP = EO$, $NP = SO$, $IP = OF$. The plane KPO contains the centre of symmetry E; so it cuts the circle FJU in the straight line OJ, which is parallel to QP. In the symmetry with centre E the points Q, P, I, L have as their respective homologues the points J, O, F, U, hence the equalities $J\hat{O}F = Q\hat{P}I$, $\widehat{JF} = \widehat{QI}$ and $\widehat{FJU} = \widehat{IQL}$. So we have

$$\frac{\widehat{JF}}{\widehat{FJU}} = \frac{\overline{QI}}{\overline{IQL}},$$

hence

$$\frac{\widehat{JF}}{\widehat{FJU}} = \frac{\widehat{KL}}{\widehat{LKI}}.$$

The two points J and K are associated with homologous seasonal hours. The plane KPO which contains E is a diametral plane of the sphere of the Universe. It cuts the plane of the equator in the straight line EH and we have $EH \parallel KP \parallel JO$. Moreover, we know that $ED \parallel NL \parallel SF$, hence $\widehat{HED} = \widehat{KPL} = \widehat{JOF}$.

From Lemma 4, we have

$$\frac{\widehat{KL}}{\widehat{LKI}} = \frac{\widehat{KPL}}{\pi};$$

moreover, $\dfrac{\widehat{HED}}{\pi} = \dfrac{\widehat{HD}}{\widehat{DHB}}$, because $\widehat{DHB}$ is a semicircle, so

$$\frac{\widehat{HD}}{\widehat{DHB}} = \frac{\widehat{KL}}{\widehat{LKI}}.$$

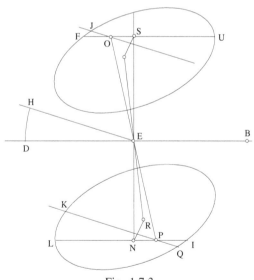

Fig. 1.7.3

So the three points K, J and H correspond to homologous seasonal hours. They lie on a great circle with centre E whose plane cuts the horizon plane in the straight line *OEP* and cuts any plane parallel to the horizon plane in a straight line parallel to *EP*.

If the plane of a sundial is parallel to the horizon at *E*, it is cut by the plane *KHJ* in a straight line *Δ* parallel to *EP*.

When the sun is at one of the points *K*, *H* or *J*, the shadow cast by the tip of the gnomon on the plane of the sundial lies on the straight line *Δ*.

Note: The proof Ibn al-Haytham gives for this Proposition 7 may be compared with the proof by Ibn Sinān.[14] We have just seen that, in order to show that the three points that correspond to the same seasonal hour on the circles of Cancer, the equator and Capricorn, points that lie on the same great circle, Ibn al-Haytham takes a general point *K* on the diurnal arc of Cancer and by using Lemma 4 – a lemma in plane trigonometry – specifies a great circle on the celestial sphere passing through *K*. This great circle cuts the diurnal arc of the equator in *H* and that of Capricorn in *J*. He shows that the three points *J*, *K* and *H* satisfy

$$\frac{\widehat{LK}}{\widehat{LKI}} = \frac{\widehat{FJ}}{\widehat{FJU}} = \frac{\widehat{DH}}{\widehat{DHB}};$$

so they correspond to the same seasonal hour. The fact that the two parallel circles concerned are those of Cancer and Capricorn plays no part in the proof, which is valid for any two parallel circles symmetrical with respect to the equator.

This is the hypothesis Ibn Sinān makes for the first proposition of his second book. He first proved a lemma, a case of an equality for two spherical triangles drawn on the same sphere or on two equal spheres, and uses this lemma to prove that if three points *E*, *I*, *K* correspond to the same seasonal hour on the equator and on the two parallel circles concerned, then *E*, *I* and *K* lie on the same great circle.

In short, Ibn al-Haytham works this way:

point *K* and Lemma 4 → great circle → *J* and *H* giving the same seasonal hour as *K*

whereas Ibn Sinān works this way:

[14] *Fī ālāt al-aẓlāl*, in R. Rashed and H. Bellosta, *Ibrāhīm ibn Sinān. Logique et géométrie au Xᵉ siècle*, pp. 330–1 and pp. 416 ff.

the points E, I, K correspond to the same seasonal hour + lemma (in spherical trigo-nometry) → there exists a great circle passing through E, I, K.

Proposition 8. — Let MVF be an hour circle situated between the circle of Cancer and the equator and let it cut the plane of the horizon in MF. The centre of this circle O lies on the straight line ER, the axis of the Universe; U, the midpoint of MF, lies on EN and we have $OU \perp MF$; the straight line EP cuts MF in W. The point V is defined by $\dfrac{\overparen{FV}}{\overparen{FVM}} = \dfrac{\overparen{LK}}{\overparen{LKI}}$. If T is the point on the arc that completes the arc FVM and satisfies the equation

$$\frac{\overparen{MT}}{\overparen{MTF}} = \frac{\overparen{FV}}{\overparen{FVM}},$$

then the straight line VT cuts FM at the point Q between U and M and we have $V\hat{Q}F = K\hat{P}L$, so $VQ \parallel KP$.

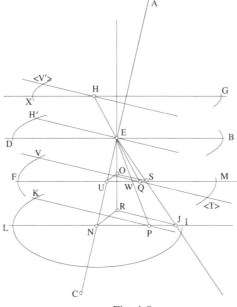

Fig. 1.8

Let the point J on NI be such that $RJ \parallel KP$,[15] then $VQ \parallel RJ$, the straight line EJ cuts MF in S and we have $OS \parallel RJ \parallel VQ$. We assumed that the arc $\overarc{FVM}$ is smaller than the arc similar to $\overarc{LKI}$, meas. $\overarc{FVM}$ < meas. $\overarc{LKI}$.

Let α be a part of $\overarc{VF}$ defined by the relation $\dfrac{\alpha}{\overarc{VF}} = \dfrac{\overarc{LK}}{\frac{1}{2}\overarc{LKI}}$ and β the

part of $\overarc{KL}$ such that $\dfrac{\alpha}{\overarc{VF}} = \dfrac{\beta}{\overarc{LK}}$, then

$$\text{meas.}\left(\overarc{VF} - \alpha\right) < \text{meas.}\left(\overarc{KL} - \beta\right).$$

But

$$\frac{\sin \overarc{VF}}{\sin\left(\overarc{VF} - \alpha\right)} = \frac{OS}{SQ} \quad \text{and} \quad \frac{\sin \overarc{KL}}{\sin\left(\overarc{KL} - \beta\right)} = \frac{RJ}{JP},$$

(from the end of Lemma 6), hence

$$\frac{OS}{SQ} > \frac{RJ}{JP}.$$

Moreover, the right-angled triangles OSU and RJN are similar, hence

$$\frac{OS}{SU} = \frac{RJ}{JN}.$$

So we have

$$\frac{US}{SQ} > \frac{JN}{JP},$$

hence

$$\frac{JN}{NP} > \frac{SU}{UQ},$$

or by permutation

$$\frac{JN}{SU} > \frac{NP}{UQ}.$$

[15] We shall see later (pp. 498–9) that if the diurnal arc LKI measures 210°, the point J coincides with the point I. Here no assumptions are made about the arc LKI.

But

$$\frac{JN}{SU} = \frac{RN}{OU} = \frac{EN}{EU} = \frac{NP}{UW},$$

so

$$\frac{NP}{UW} > \frac{NP}{UQ}$$

and it follows that $UW < UQ$; the point Q does not lie on PE. In the horizon plane, the straight line EP, which is the straight line associated with the seasonal hour that corresponds to the point K, is thus distinct from the straight line EQ associated with the point V.

The plane of the hour circle symmetrical with the circle MVF with respect to E is cut by the horizon plane in the straight line GX, and by the plane VQE in the straight line HV', which is parallel to the straight line VQ and passes through the point H, the point of intersection of QE and GX. The plane VQE cuts the sphere of the Universe in a great circle whose circumference passes through the points V, H' and V' that lie, respectively, on the circles MVF, $DH'B$ and $XV'G$; these points correspond to the same seasonal hour for the days identified by these three circles.

The two planes VQE and KPE that pass, respectively, through the parallel straight lines VQ and KP, cut one another in a straight line parallel to these two straight lines, the straight line EH' that lies in the plane of the equator.

To summarize: if to the hour h there correspond, on the one hand, the points K, H, J of the first figure and, on the other hand, the points V, $H' = H$, V' of the second figure, then with the points K, H, J there is associated the straight line EP, with the points V, H', V' there is associated the straight line EQ, which is distinct from EP.

Proposition 9. — We consider four arcs of the ecliptic cut off between the points γ, γ' (the equinoxes), and σ and σ' (the solstices). If with each degree i, from 0 to 90, of the arc $\gamma\sigma$ we associate a point α_i where $\alpha_0 = \gamma$ and $\alpha_{90} = \sigma$, then there is an hour circle that corresponds to each point α_i.

To a given hour h taken on each of the 91 hour circles there correspond 91 points that are the shadows of the tip of the gnomon on the plane of the sundial, let $w_{0 \cdot h}$ be the associated point when $i = 0$.

For two points α_i and α'_i symmetrical with respect to $\sigma\sigma'$, the hour circle is the same; so the sun describes it twice in the year.

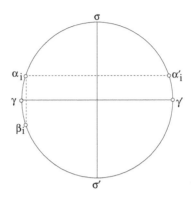

Fig. 1.9.1

For two points α_i and β_i symmetrical with respect to $\gamma\gamma'$, the hour circles C_{α_i} and C_{β_i} are symmetrical with respect to the equator.

For $i \neq 0$, with the 4 points α_i, α'_i, β_i, β'_i of the ecliptic there are associated two circles C_{α_i} and C_{β_i}, and to a given hour h on these circles there corresponds a single straight line $\Delta_{i \cdot h}$ in the plane of the sundial.

For i between 0 and 90, with an hour h there are associated the point $w_{0 \cdot h}$ and 90 straight lines $(\Delta_{i \cdot h})_{i=1}^{90}$ which all pass through the point $w_{0 \cdot h}$.

For a given hour h, $w_{0 \cdot h}$ corresponds to two days, the equinoxes, $\Delta_{90 \cdot h}$ also corresponds to two days, the solstices, and any other straight line $\Delta_{i \cdot h}$ corresponds to four days, one day in each season.

It remains for us to show that, for an hour h, the angle $\Delta_{90 \cdot h}$ makes with any straight line $\Delta_{i \cdot h}$ is negligible.

Ibn al-Haytham returns to the figure for his investigation of the first hour, $h = 1$, when the sun is on the circle of Cancer, to fix the position of the point P.

He supposes that the arc LKJ measures 210°, which he calls fourteen hours; which is to take one hour as corresponding to 15° as is the case for an equinoctial hour

$$210° = \frac{180°}{12} \times 14.$$

But he then takes $\widehat{KL} = \dfrac{210°}{12} = 17.5°$ as the arc for one hour.

The value given for the arc $\overset{\frown}{LKJ} = 210°$ allows us to calculate the latitude of the place concerned.[16] We shall take $\alpha = 23°27'$ for the angle between the equator and the plane of the ecliptic. Later on (p. 507), Ibn al-Haytham says that the latitude of the place concerned is $30°$.

In the calculations that follow, Ibn al-Haytham takes $\alpha = 24°$ as the inclination of the ecliptic.

Let us put $EH = r$ (Fig. 1.9.3). Taking $r = 60$, we have

$$ER = r \sin \alpha = 60 \times 0.4069366 = 24.4022 = 24.24.15$$

$$RH = RI = r \cos \alpha = 60 \times 0.91354545 = 54.812727 = 54.48.46$$

$$RN = RI \sin 15° = 14.186577 = 14.11.11^*.$$

---

[16] Calculation of λ the latitude of the place of observation: the arc of the circle of Cancer above the horizon is given: $210°$, hence $\widehat{HRL} = 105°$, $\widehat{LRN} = 75°$,

(1) $RN = RH \cos 75°$.

Let α the inclination of the ecliptic to the equator.

• If $EH = r$ is the radius of the universe,

$$ER = r \sin \alpha.$$
$$RN = ER \tan \lambda = r \sin \alpha \tan \lambda.$$

(2) $RH = r \cos \alpha.$

(1) and (2) $\Rightarrow \tan \lambda = \dfrac{\cos 75°}{\tan \alpha}.$

• If $\alpha = 23°27'$, we have

$$\tan \lambda = \frac{\cos 75°}{\tan 23°27'} = \frac{0.258819}{0.433775} = 0.59666$$

$\lambda = 30°49'43''.$

• If $\alpha = 24°$, $\tan 24° = 0.445228$

$\tan \lambda = 0.58131788$

$\lambda = 33°18'25''.$

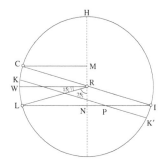

Fig. 1.9.2

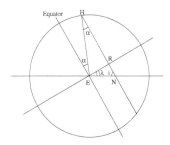

Fig. 1.9.3

Baghdad 33°14′

---

* The text gives 14.10.10.

We have

$$\widehat{LC} = 30°$$
$$\widehat{LW} = \frac{1}{2}(210° - 180°) = 15°$$
$$\widehat{LK} = 17.5°$$
$$\widehat{KC} = \widehat{LW} - \frac{1}{6}\widehat{LW} = 15 - 2.5 = 12.5°,$$

so RW bisects $\widehat{CRL}$, so $RW \parallel IL$ and the straight line CR passes through I.

Position of P: from the preceding, if K' is defined by the equation $\frac{\widehat{LK}}{210} = \frac{\widehat{IK'}}{150}$, the straight line KK' cuts LI in P.

$$\widehat{IK'} = \frac{17.5 \times 5}{7} = 12.5,$$

so $\widehat{IK'} = \widehat{KC}$, hence $KK' \parallel CI$ and we have

$$C\hat{R}W = K\hat{P}L \text{ and } \frac{\widehat{KC}}{\widehat{LK}} = \frac{5}{7}.$$

If $CM \perp RH$, then $RN = RM$ and $CM = IN$.

Calculation of CM: $CM = IN$

$$CM = RH \sin 75° = RI \sin 75° = RI \times 0.96592$$
$$= 54.812727 \times 0.96592 = 52.94471 = 52.56.40.$$

Calculation of IP: from Lemma 6, we have

$$\frac{IP}{IR} = \frac{\sin \widehat{KC}}{\sin \widehat{LW}} = \frac{\sin 12.5°}{\sin 15°} = \frac{0.216439}{0.258819} = 0.836259$$

$$IP = RI \times 0.836259 = 54.812727 \times 0.836259 = 45.8371 = 45.50.13$$

$$PN = NI - IP = 52.56.40 - 45.50.13 = 7.6.27.$$

Note: $\dfrac{IP}{IR} = 0.836259,$ $\dfrac{5}{6} = 0.8333,$ hence $\dfrac{IP}{IR} > \dfrac{5}{6},$ a result Ibn al-Haytham proves later.

Calculation of EN:

$$EN^2 = ER^2 + RN^2 = (24.4042)^2 + (14.186577)^2$$
$$= 595.5650 + 201.2632 = 796.8282$$

$$EN = 28.228.$$

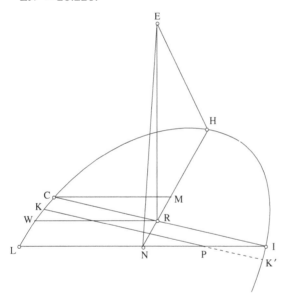

Fig. 1.9.4

Let there be a general hour circle with centre R', lying between the circle of Cancer, centre R, and the circle of the equator, centre E; the points E, R and R' are collinear. The plane ELI cuts the plane of the hour circle with centre R' in the straight line $L'I'$, the straight lines EI, EP and EN cut that straight line in I', P' and N' respectively. The triangles ENI and $EN'I'$ are similar, as are the triangles EPI and $EP'I'$. At the same time, the triangles RNI and $R'N'I'$ are similar. We have

$$\frac{EN'}{EN} = \frac{ER'}{ER} = \frac{EP'}{EP} = \frac{EI'}{EI}$$

and

$$\frac{RI}{IN} = \frac{R'I'}{I'N'} \quad \text{and} \quad \frac{RI}{IP} = \frac{R'I'}{I'P'}.$$

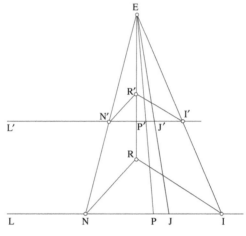

Fig. 1.9.5

But

$$\frac{RI}{IP} = \frac{\sin \widehat{LW}}{\sin \widehat{KC}};$$

we have

$$\widehat{LW} = 15° \quad \text{and} \quad \widehat{KC} = 12.5°, \quad \widehat{LW} = \left(1 + \frac{1}{5}\right)\widehat{KC},$$

hence

$$\frac{\widehat{LW}}{\widehat{KC}} = 1 + \frac{1}{5} \quad \text{and} \quad \frac{RI}{IP} = \frac{\sin \widehat{LW}}{\sin \widehat{KC}} < 1 + \frac{1}{5},$$

hence

$$RI < \left(1 + \frac{1}{5}\right)IP$$

or

$$IP > \left(\frac{1}{2} + \frac{1}{3}\right)RI.$$

This result holds true whatever hour circle we choose.

We had $IP > \left(\frac{1}{2} + \frac{1}{3}\right)RI$. Let J be a point on IN such that $IJ = \left(\frac{1}{2} + \frac{1}{3}\right)RI$; we have $IJ < IP$. The straight line EJ cuts any straight line $I'P'$ homologous to IP, in a point J' and we have

$$I'J' = \left(\frac{1}{2} + \frac{1}{5}\right)R'I' \text{ (Fig. 1.9.5).}$$

Each of the great circles associated with the first seasonal hour cuts the plane of the horizon in a straight line enclosed between EP and EJ, for hour circles between the circle of Cancer and the equator, and in a straight line enclosed between EH and EF for hour circles between the equator and the circle of Capricorn.

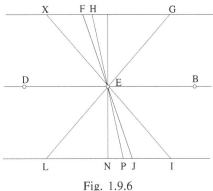

Fig. 1.9.6

Calculation of the ratio $\dfrac{PJ}{PE}$:

In decimal form, we have $RI = 54.8127$, $IJ = 45.6773$, $IP = 45.8371$, hence

$$JP = 0.1598, \quad JP < \frac{1}{6};$$
$$PN = NI - IP = 7.1076,[17]$$

hence
$PN^2 = 50.5180$

[17] Ibn al-Haytham gives $PN = 7.6.54$, that is, 7.11944, hence $PN^2 = 50.686 \cong 50$ and $\frac{2}{3}$.

$EN^2 = 796.8282 \cong 797$

$EP^2 = EN^2 + PN^2 = 847.3462$

$EP = 29.10938 \qquad (EP > 29)$,

hence

$$\frac{JP}{EP} < \frac{1}{29 \times 6} \quad \text{and} \quad \frac{JP}{EP} < \frac{1}{174}.$$

Note: The point P that corresponds to the end of the first hour is thus such that $PN \cong 7$ and $EN \cong 30$

$$\frac{PN}{EN} \cong \frac{7}{30};$$

EN is the meridian line, and we have $(\tan \ N\hat{E}P)_1 \cong \dfrac{7}{30}$.

Ibn al-Haytham next considers the point K on the circle of Cancer such that the arc LK is the arc for the beginning of the fifth hour.

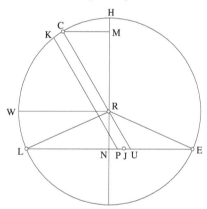

Fig. 1.9.7

$\overset{\frown}{LH} = 105° \qquad \overset{\frown}{LW} = 15° \qquad \overset{\frown}{WC} = 75° \qquad \overset{\frown}{CH} = 15°$

$\overset{\frown}{LK} = 5 \text{ hours} = 17.5° \times 5 = 87.5°, \ \overset{\frown}{WK} = 72.5°, \text{ so } \overset{\frown}{KC} = 2.5°$.

We have

$RH = CR = 54.812727$,

$CM = RN = CR \sin 15° = 14.11.10$ and $MR = CR \sin 75° \ (= 57.57.20)$,

$$\frac{CR}{RU} = \frac{MR}{RN} = \frac{\sin 75°}{\sin 15°} = \frac{0.965926}{0.258819} = 3.732052$$

$CR = 54.812727$,

hence

$$RU = \frac{54.812727}{3.732052} = 14.687029 = 14.41.13$$

$$CM = CR \times 0.258819$$

and

$$\frac{CM}{NU} = \frac{CR}{RU} \Rightarrow RU \times 0.258819 = NU$$

$$NU = 14.687029 \times 0.258819 = 3.801282 = 3.48.6.$$

$$\frac{\sin \widehat{LW}}{\sin \widehat{KC}} = \frac{RU}{UP} = \frac{\sin 15°}{\sin 2.5°} = \frac{0.258819}{0.043619} = 5.933629$$

$$UP = \frac{RU}{5.933629} = \frac{14.687029}{5.933629} = 2.475218 = 2.28.30.$$

Therefore

$$PN = 3.48.6 - 2.38.30 = 1.19.34 = 1.326$$

$$UJ = \frac{1}{6}RU = \frac{14.41.13}{6} = 2.27$$

$$JP = UP - UJ = 2.28.30 - 2.27 = 0.1.30 < 2 \text{ minutes.}$$

We know that $EN^2 \cong 798$ and $PN^2 < 2$, so $EP^2 \cong 800$, $EP = 28,284\ldots$

Ibn al-Haytham gives $EP = 28 + \frac{1}{4} + \frac{1}{7}$, or 28.264, which he converts into minutes $EP \cong 1700$, hence

$$\frac{JP}{EP} < \frac{2}{1700} \quad \text{or} \quad \frac{JP}{EP} < \frac{1}{850}.$$

The ratio $\frac{JP}{EP}$, calculated for the fifth hour, is thus less than $\frac{1}{174}$, which was the ratio associated with the first hour.

The same method can be used for each of the other hours, so, at the latitude concerned, for any hour of the day we have

$$\frac{JP}{EP} < \frac{1}{174}.$$

In applying the same method for any other horizon, that is, for any latitude, we find that the ratio $\frac{JP}{EP}$ has a very small value, and JP is thus negligible in relation to EP.

Notes:

1) We have $PN^2 < 2$, $EN^2 \cong 800$, $\dfrac{PN^2}{EN^2} < \dfrac{1}{400}$, $\dfrac{PN}{EN} < \dfrac{1}{20}$, a result that corresponds to the point P, which is associated with the end of the fifth hour; we then have

$$(\tan N\hat{E}P)_5 < \frac{1}{20}.$$

2) At the end of the sixth hour, the point C on the circle of Capricorn is at H, so it is in the meridian plane. The same holds for the point corresponding to C on any of the hour circles. So all these points lead us to the same straight line EP identified with EN, so $(\tan N\hat{E}P)_6 = 0$.

A different straight line is associated with each hour h.

Proposition 10. — Calculation of the length of the longest shadow in the plane of the sundial.

For the horizon $ABCD$ that we considered before, the diurnal arc of Capricorn is 150°, so one seasonal hour is 12.5°.

So we have $\overset{\frown}{DGB} = 150°$, and if $\overset{\frown}{DG}$ is the first seasonal hour and IK the diameter parallel to BD, we have $\overset{\frown}{DG} = 12.5°$, $\overset{\frown}{BI} = \overset{\frown}{DK} = 15°$; hence $\overset{\frown}{GK} = 27.5°$.

Let $GL \perp IK$; GL cuts BD in M, we have, where r is the radius of the circle of Capricorn:

• $GL = r \sin 27.5° = r \times 0.46174861$; putting $r = 60$, we have $GL = 27.42.18$;

• $ML = r \sin 15° = r \times 0.258819$; putting $r = 60$, we have $ML = 15.31.45$.

From this we have $GM = 12.10.33$ (or $GM = r \times 0.20293$).

Let GN be perpendicular to $(ABCD)$; GN is parallel to the vertical from E, so the plane GEN passes through the zenith of E.

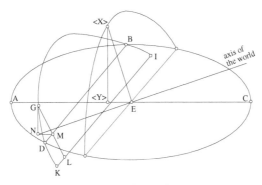

Fig. 1.10

We have $GN \perp (ABCD)$ and $GM \perp BD$, so $MN \perp BD$. So the plane GMN is parallel to the meridian plane whose position is determined by the straight line AC and the point X, the culmination of the sun when it is on the equator, that is, the plane EXY $[XY \perp (ABCD)]$. We have $M\hat{G}N = E\hat{X}Y$; this angle is the latitude of the place E. If the latitude is $30°$, we have $MN = \frac{1}{2}MG$.

If $r = 60$, $GM = 12.1758$, $GM^2 = 148.2501$, $GN^2 = \frac{3}{4}GM^2 = 111.1875$ and $GN = 10.5445 \cong 10 + \frac{1}{3} + \frac{1}{4}$.

But $r \neq 60$, $r = 54.48.16 = 54.812727$,[18] hence

$$\frac{GN}{54.812727} = \frac{10.5445}{60} \Rightarrow GN = 9.6328 \cong 9 + \frac{2}{3}.$$

EG is a radius of the sphere of the Universe, $EG = 60$, so $EG^2 = 3600$; but $GN^2 = \left(9 + \frac{2}{3}\right)^2 = 93.32$, hence $EN^2 = 3506.68$ and $EN = 59.22 \cong 59 + \frac{1}{4}$.

[18] Ibn al-Haytham proved this result in Proposition 9 where he took $24°$ for the inclination of the ecliptic on the equator and $60°$ for the radius of the celestial sphere.

$$\frac{GN}{NE} = \frac{9 + \frac{2}{3}}{59 + \frac{1}{4}} = \frac{116}{711} = \frac{1}{6.13},$$

so

$$GN < \frac{1}{6} NE.$$

The ratio $\frac{GN}{NE}$ is equal to the ratio of the height h of the gnomon of a horizontal sundial to the length l of the shadow of the gnomon, when the sun is at the position G, the first hour of the circle of Capricorn.

$$\frac{h}{l} = \frac{1}{6,13} < \frac{1}{6}, \qquad\qquad l > 6h.$$

We may note that the ratio $\frac{h}{l}$ is the tangent of the angle the ray from the sun, GE, makes with the plane of the horizon.

Proposition 11. — Conclusion of Propositions 9 and 10.

Let there be a horizontal sundial, let CD be its gnomon, AB the line of intersection of the plane of the sundial with the great circle that marks the first hour on the equator and on the two circles of Cancer and Capricorn; and let GEH be the straight line described by the shadow of D, the tip of the stylus of the gnomon, on the days of the equinoxes, that is, the days when the sun describes the equator. The point I, the intersection of AB and EH, corresponds to the end of the first hour for the days of the equinoxes. Let L and M be the shadows of the point D at the end of the first hour on the circles of Capricorn and Cancer respectively. So the shadows of the stylus CD are CL, CI, CM for the first hour of the days in question. We have seen that CL is the longest and that $CL \cong 6\ CD$.

If we turn back to the results proved in Proposition 9 (Fig. 1.9.6), we draw through I the straight line KN such that $L\hat{I}K = J\hat{E}P$ and $I\hat{B}K = I\hat{A}N = E\hat{P}J$.[19]

We have

$$\frac{BK}{BI} = \frac{PJ}{PE} \cong \frac{1}{174}.$$

[19] That is, the angles JEP and EPJ of Proposition 9.

The straight line *AB* corresponds to the straight line *PEH* of Fig. 1.9.6 and the straight line *NIK* corresponds to the straight line *JEF*.

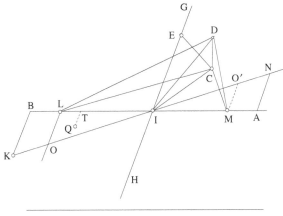

Fig. 1.11

From the result established in Proposition 9 it follows that the great circles that mark the first seasonal hour of the other days of the year cut the plane of the sundial in straight lines that pass through *I* and lie between the straight lines *AB* and *NK*.

The straight line *HG* – associated with the equinoxes – corresponds to the straight line *BD* of Proposition 9, so $\hat{BIE} > 90°$; *a fortiori*, $\hat{BIC} > 90°$, hence $CL > LI$.

If *LO* ∥ *BK*, we have

$$\frac{OL}{LI} = \frac{BK}{BI} \cong \frac{1}{174}.$$

Moreover, $CD \cong \frac{1}{6}CL$, so we have

$$CD > \frac{1}{6}LI, \quad \frac{OL}{CD} < \frac{6 \cdot OL}{LI} < \frac{6}{174} \quad \text{and} \quad \frac{6}{174} < \frac{1}{30},$$

hence

$$OL < \frac{1}{30}CD.$$

If Q is the shadow of D, the tip of the gnomon, for the first hour of any day of the year, the point Q will lie inside one of the triangles LIO or IMO'; so if from Q we draw a line QT parallel to BK, then $QT < LO$, so

$$\frac{QT}{CD} < \frac{1}{30}.$$

If $CD = 3$ fingers, $QT < \dfrac{1}{10}$ of a finger.

If $CD = 18$ sha'ïra, $QT < \dfrac{3}{5}$ sha'ïra.

The point Q the shadow of D, the tip of the gnomon, is thus never at a perceptible distance from the straight line LM, that is from the straight line AB that corresponds to the first hour on the circles of the equator, Cancer and Capricorn.

The proof carried out for the straight line AB that corresponds to the first hour, could also be carried out in the same way for any hour, hence the conclusion: for an 'inclined horizon', that is a place with non-zero latitude, the hour lines on the sundial are straight lines as far as our senses can perceive, which is to say that, if we take as hour line h on all the hour circles C_{α_i} the straight line $\Delta_{90 \cdot h}$ marked by the hour h on the circles of the equator, Cancer and Capricorn, the error we make is negligible; so any straight line $\Delta_{i \cdot h}$ can be regarded as $\Delta_{90 \cdot h}$.

For a place with latitude zero, that is, for any point on the terrestrial equator, the axis of the world is in the plane of the horizon; so the diurnal arcs are all semicircles and a great circle that marks an hour h on the celestial equator marks the same hour h on any hour circle. This great circle cuts the horizon in the meridian line and on the plane of the sundial there corresponds to it a straight line parallel to the meridian line. The same holds true for every hour of the day. In these places the hour lines are parallel straight lines.

Thus Ibn al-Haytham develops a general theory of the sundial and the hour lines drawn on it; he proves that the same sundial can be used in any place if one allows a negligible error.

The theory mainly relies upon theorems in plane trigonometry concerning the variation of functions such as $\dfrac{\sin \varphi}{\varphi}$ and theorems in spherical geometry that lead to certain inequalities of ratios that can be used to check the size of errors.

This concern to keep track of errors seems to be entirely new.

1.3. HISTORY OF THE TEXT

This treatise was cited under the same title – *Fī khuṭūṭ al-sāʿāt* – by the old biobibliographers, al-Qifṭī, Ibn Abī Uṣaybiʿa,[20] in the lists they drew up of the works of Ibn al-Haytham before 1038. We also know that in his book *The Burning Sphere*, Ibn al-Haytham returns to an important proposition from this book, which he cites. Moreover, in *The Configuration of the Motions of Each of the Seven Wandering Stars*, he again returns to a proposition in this book, as we have shown.

This treatise exists in two manuscripts:

1. Collection Askari, no. 3025, fols 1^v–19^v, in the Library of the Military Museum (Askari Müze), Istanbul.

2. Collection ʿĀṭif, no. 1714, fols 57^r–76^v, in the Süleymaniye Library, Istanbul.

We have given a detailed account of the history of these two collections in Volume 3,[21] and we have shown there that the first collection is part of a larger collection, which was separated into two parts, the first of which is to be found in the Staatsbibliothek (Oct. 2970) in Berlin. The original collection, before it was divided into two parts, was copied by the mathematician Qāḍī Zādeh at some time around the 1430s. The collection largely consists of treatises by Ibn al-Haytham. We have established that the ʿĀṭif collection is simply a copy of this original collection, and of it alone; so that it is of no value at all in a family tree independently of its predecessor. So in the ʿĀṭif manuscript, the text of *On the Hour Lines* has been copied from that in the Military Museum and from it alone. We note, moreover, that there are 31 omissions of a word and 5 of a phrase, where nothing is missing in the copy in the Military Museum when it is compared with the ʿĀṭif manuscript.

---

[20] *Ibn al-Haytham and Analytical Mathematics. A history of Arabic sciences and mathematics*, vol. 2, Culture and Civilization in the Middle East, London, 2013, pp. 406–7. The Arabic text is edited in *Les Mathématiques infinitésimales*, vol. V.

[21] *Ibn al-Haytham's Theory of Conics, Geometrical Constructions and Practical Geometry. A history of Arabic sciences and mathematics*, vol. 3, London, 2013, pp. 269–71.

TRANSLATED TEXT

Al-Ḥasan ibn al-Haytham

On the Hour Lines

TREATISE BY AL-ḤASAN IBN AL-ḤASAN IBN AL-HAYTHAM

On the Hour Lines

When we examined the book by the geometer Ibrāhīm ibn Sinān *On Instruments for Shadows*, we noticed that he criticizes the opinion of earlier writers who suppose that the lines that define the edges of seasonal hours on the planes of sundials are straight lines, and who believes that on each day of the year the tip of the shadow of the gnomon, at the end of the same seasonal hour and at the beginning of the hour that follows it, lies close to a straight line. He stated that one straight line in the plane of a horizontal sundial does not define the edge of the same seasonal hour except for three of the seasonal circles – one of which is the equator, while the two others lie on either side of the equator and at equal distances from it; and that the straight line that lies in the plane of a horizontal sundial and defines the edge of the same seasonal hour in the three circles we have just mentioned is the intersection of the plane of the dial and the plane of a great circle that passes through the tip of the gnomon and through the points that indicate the edges of the same seasonal hour on the three circles. This statement is true and cannot be doubted. He went on to state that this great circle does not cut any of the remaining hour circles in a point that marks the edge of the seasonal hour associated with the circle in question. This statement is also a true one; however he was not able to prove it, for when he came to give a proof of his statement, he showed correctly that one great circle cuts the circumferences of the three circles in three points that mark the edges of the same seasonal hour. He next wanted to prove that the great circle that cuts off a seasonal hour on the three circles, does not cut off this same seasonal hour on any other remaining hour circle. He then presents a proof that does not show this idea is true. he has in fact imagined two great circles that cut off two seasonal hours from the three circles; he went on to draw a fourth hour circle and he showed that these two great circles cut off two different arcs on the fourth circle, but he did not show that, of these two different arcs, neither is a seasonal hour; thus the result <established

by> his proof is different from what is set out clearly in his statement; moreover, the result established by the proof does not make it impossible for one of the two different arcs to be a seasonal hour one of the two different arcs from being a seasonal hour. It is as if he had stated that none of the hour lines is straight, and proved that not all the hour lines are straight. So what he said about this idea falls short of what he intended, and furthermore does not show the idea is a true one.

Similarly, he has not shown what is the magnitude of the distance by which the tips of the shadows at the seasonal hours depart from the <straight> line given for that hour. It is possible that the tips of the shadows depart very little from the straight line given for that hour, so that this deviation is insensibly small. And the proof indeed depends on the fact that a mathematical straight line is a length that has no width, whereas the line drawn on the plane surface of the sundial is one that has a noticeable width, which could take in the deviation of the shadows, if this deviation is insensibly small, or is less than it <*sc.* the width of the line> by some negligibly small amount.

In the same way, all instruments constructed for <observing> the sun and the planets are constructed in a manner that is approximate and not absolutely exact. The astrolabe divides its circles into three hundred and sixty parts. If we take a height with this instrument, we obtain it only in whole degrees; now a height is never a whole number of degrees, instead, on most occasions one can have minutes along with the whole degrees; now these minutes do not appear on the astrolabe; it is even possible that the minutes are numerous, but, despite their number, they do not appear. In the same way the lines that serve to divide the circles of the astrolabe each have a perceptible width; this width is a part of the degree cut off by each line, and it is a part that has a magnitude, for the parts of a circle on an astrolabe are small, and especially so if the astrolabe is small. However we do not take into account the width of the lines that mark the divisions on an astrolabe.

These notions apply equally to an armillary sphere, a quadrant used to observe the sun and all the instruments used to observe the sun and the planets. It is possible that our predecessors supposed that the hour lines <on a sundial> are straight lines, while at the same time knowing how far they deviated <from straight lines>, given that what they are aiming to achieve by their assumption is an approximation, and not the ultimate exactitude, that they aimed for in the construction of the astrolabe and of observing instruments. since we found this idea unclear, because Ibrāhīm ibn Sinān had not succeeded in showing it was true; and since it can be accepted by way of approximation, we decided to go deeper into investigating the truth of this idea, and to allow ourselves to discuss it, as well as to find out about the boundaries between seasonal hours on the surfaces of horizontal sundials. So we reflected on these matters and pursued our researches until the truth was clear. It thus became apparent that our predecessors had been right to suppose that the hour lines are straight lines, that this is by way of an approximation, and the best approximation, and that there is no other way of drawing the boundaries between hours on the surfaces of sundials.

From what we have proved it is clear that Ibrāhīm ibn Sinān had been right in one respect, and mistaken in another respect, and this in fact happened because he employed mathematical procedures without thinking about physical ones; so he was right from the point of view of imagination, but wrong from the point of view of sensory perception, because he chose to prove the result he had stated as if the lines drawn on sundials were imagined lines, that is to say, having length without breadth; but the lines drawn on sundials have breadth; thus he did not distinguish an imagined line from one perceived by the senses: so he was completely mistaken.

Once we had come to this idea that we have described, we composed this treatise to provide a justification for our predecessors' opinions on the subject, to give an argument in support of what they had supposed to be true, and to indicate where Ibrāhīm ibn Sinān went astray.

Before the treatise we have given lemmas that are themselves new results, results that none of those who preceded us has mentioned – as it seems to us – and thanks to these lemmas we can go on to derive all the ideas that we have expressed in this treatise. So let us now begin to speak of them, with God's help in everything.

Lemmas

<Lemma 1>: Let there be a circle in which we draw two parallel chords that cut off from the circle two arcs such that the greater is not greater than a semicircle. We take an arbitrary point on the smaller of the two arcs and from this point we draw a line perpendicular to the two chords; thus the ratio of the complete perpendicular to the part of it cut off by the small arc is greater than the ratio of the part cut off from the large arc to the part cut off from the small arc; and the ratio of the part cut off from the large arc to the part cut off between the two chords is greater than the ratio of the perpendicular to the part cut off from it between the two chords.

Example: Let there be the circle *ABCD* in which are drawn the two parallel chords *BD* and *EG*, such that the arc *BAD* is not greater than half the circle *ABCD*. On the arc *EAG* we take an arbitrary point *A* and we draw the perpendicular *AHI*.

I say that the ratio of *IA* to *AH* is greater than the ratio of the arc *DA* to the arc *AG* and that the ratio of the arc *AD* to the arc *DG* is greater than the ratio of the perpendicular *AI* to *IH*.

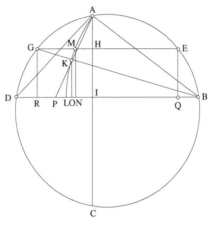

Fig. I.1

Proof: We put in the two straight lines *BA* and *BG*; so *BA* will be smaller than *BG* and *BG* will be smaller than *BD*. We take *B* as centre and with the distance *BA* we draw an arc of a circle; let the arc be *AKL*. This arc cuts the straight line *EG* before reaching the straight line *BG*, because the straight line *GB* is below the straight line *EG*; let it (*sc.* the arc) cut it at the point *M*. We draw the two perpendiculars *KO* and *MN*. Since the point *B* is the centre of the arc *AKL* and it lies on the circumference of the circle *ABCD*, accordingly the arc *AL* is similar to half of the arc *AD* and the arc *KL* is similar to half of the arc *GD*; so the ratio of the arc *AL* to the arc *LK* is equal to the ratio of the arc *AD* to the arc *DG*. We join *AK* and extend it to *P*. The ratio of the sector *BAK* to the sector *BKL* is greater than the ratio of the triangle *BAK* to the triangle *BKP*; so the ratio of the arc *AK* to the arc *KL* is greater than the ratio of the straight line *AK* to the straight line *KP*. By composition, the ratio will be as follows: the ratio of the arc *AL* to the arc *LK* is greater than the ratio of the straight line *AP* to the straight line *KP*, so the ratio of the arc *AL* to the arc *LK* is greater than the ratio of the straight line *AI* to the straight line *KO*. But the ratio of *AI* to *KO* is greater than the ratio of *AI* to *MN*, so the ratio of the arc *AL* to the arc *LK* is greater than the ratio of *AI* to *IH*. So the ratio of the arc *AD* to the arc *DG* is greater than the ratio of *AI* to *IH*. So the ratio of <the arc> *IA* to <the arc> *AH* is greater than the ratio of the arc *DA* to the arc *AG*.

Similarly, we prove that the ratio of *IA* to *AH* is greater than the ratio of the arc *BA* to the arc *AE* and that the ratio of the arc *AB* to the arc *BE* is greater than the ratio of *AI* to *IH*.

If from the point *E* we draw a perpendicular to the straight line *BD*, the ratio of *IB* to the part that the perpendicular cuts off from the straight line *BD*, on the side of the point *B*, is greater than the ratio of the arc *AB* to the arc *BE*, if the perpendicular *AI*, when extended until it cuts the circle, cuts off an arc of it, on the side of the point *B*, which is not greater than a semicircle. If from the point *G* we draw a perpendicular to the straight line *BD*, the ratio of *ID* to the part the perpendicular cuts off from the straight line *BD*, on the side of the point *D*, is greater than the ratio of the arc *AD* to the arc *DG*, if the perpendicular *AI*, when extended until it cuts the circle, cuts off an arc from it, on the side of the point *D*, which is not greater than a semicircle. That is what we wanted to prove.

<Lemma 2>: If we cut off from a circle two different arcs, one being a half of the other and the greater one not being greater than a quarter of the circle, if we then divide the two arcs in the same ratio, then the ratio of the sine of the small arc to the sine of its part is greater than the ratio of the sine of the large arc to the sine of the part of it that corresponds to the part of the small arc.

Example: We cut off from the circle *ABCD* the two arcs *AB* and *AD*, such that the arc *AD* is half the arc *AB* and the arc *AB* is not greater than a quarter of the circle. We put the ratio of <the arc> *BA* to <the arc> *AE* equal to the ratio of <the arc> *DA* to <the arc> *AG*.

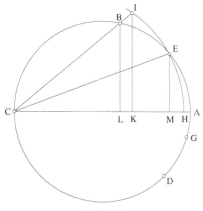

Fig. I.2

I say that the ratio of the sine of the arc *DA* to the sine of the arc *AG* is greater than the ratio of the sine of the arc *BA* to the sine of the arc *AE*.

Proof: From the point *A* we draw a diameter of the circle, let it be *AC*, and we draw the two straight lines *BC*, *CE*. So the straight line *BC* is shorter than the straight line *EC* and the straight line *EC* is shorter than the straight line *AC*. We take the point *C* as centre and with distance *CE* we draw an arc of a circle. That arc cuts the straight line *AC* inside the circle and cuts the straight line *BC* outside the circle; let it cut the straight line *AC* at the point *H* and let it cut the straight line *BC* at the point *I*. We draw the perpendiculars *IK*, *BL*, *EM*. So *IK* is greater than *BL* and *BL* greater than *EM*; so the ratio of *IK* to *EM* is greater than the ratio of *BL* to *EM*. But *IK* is the sine of the arc *IEH*, *EM* is the sine of the arc *EH* and the sine of the arc *EA*, and *BL* is the sine of the arc *BA*, because the straight line *AC* is a diameter common to the two arcs. The ratio of the sine of the arc *IH* to the sine of the arc *EH* is greater than the ratio of the sine of the arc *BA* to the sine of the arc *AE*. But the arc *IEH* is similar to half of the arc *BA*, because the angle *ACI* is at the centre of the circle *IEH* and it is inscribed in the circle *ABC*. In the same way, the arc *EH* is similar to half the arc *EA*; so the arc *IEH* is similar to the arc *AD* and the arc *EH* is similar to the arc *AG*; so the ratio of the sine of the arc *IEH* to the sine of the arc *EH* is equal to the ratio of the sine of the arc *DA* to the sine of the arc *AG*, so the sine of the arc *DA* to the sine of the arc *AG* is greater than the ratio of the sine of the arc *BA* to the sine of the arc *AE*. That is what we wanted to prove.

<Lemma 3>: If we cut off from a circle two different arcs and if we divide the two arcs in the same ratio in such a way that the greater part of the greater arc is not greater than a quarter of the circle, then the ratio of the sine of the greater part of the small arc to the sine of the small part of this latter is greater than the ratio of the sine of the large part of the large arc to the sine of the small part of this latter.

Example: We cut off from the circle *ABC* the arc *ABC* which we divide at the point *B* in such a way that *AB* is greater than *BC* and the arc *AB* is not greater than a quarter of the circle. We put the ratio of <the arc> *DB* to <the arc> *BE* equal to the ratio of <the arc> *AB* to <the arc> *BC*.

I say that the ratio of the sine of the arc *DB* to the sine of the arc *BE* is greater than the ratio of the sine of the arc *AB* to the sine of the arc *BC*.

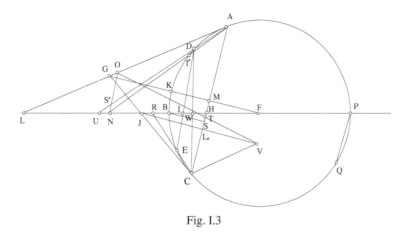

Fig. I.3

Proof: We mark the centre of the circle, let it be F; we join FB and draw AC and DE; let them be cut by the straight line FB at the points H and I. We draw from the two points A and C two tangents to the circle. They meet because the two angles formed at the two points A and C are acute, since each of them stands on the arc ABC, which is smaller than a semicircle; let them meet one another at the point G.

Let the arc AB first be smaller than a quarter of the circle. We join FG; it divides AC into two equal parts; it is perpendicular to it and cuts the arc ABC into two equal parts; let it cut that arc at the point K and let it cut the straight line AC at the point M. We extend BF in the direction of F; let it meet the circle at the point P. Since the arc AB is greater than the arc BC, the arc CP is greater than the arc AP. We cut off the arc CQ equal to the arc AP and we join PQ. PQ will then be parallel to the straight line AC; so the angle BHC is equal to the angle BPQ. But since the arc CQ is equal to the arc AP and the arc AP is greater than a quarter of the circle, because the arc AB is smaller than a quarter of the circle, the arc CQ is greater than the arc AB; thus the arc QCB is greater than the arc ABC. But the arc QCB is intercepted by the angle BPQ and the arc ABC is intercepted by the angle GAC; so the angle BHC is greater than the angle GAC. If we extend them, the straight line AG meets the straight line HB; let them be extended and let them meet one another at the point L. The straight line HL cuts the straight line CG; let it cut it at the point J. Since the straight line PB is a diameter of the circle, the ratio of AH to HC is equal to the ratio of the sine of the arc AB to the sine of the arc BC, because the two sines are the two perpendiculars dropped from the points A and C onto the diameter PB. So they will be parallel and the ratio of the one to the other is equal to the ratio of AH to HC.

In the same way, we prove that the ratio of *DI* to *IE* is equal to the ratio of the sine of the arc *DB* to the sine of the arc *BE*. But the ratio of the arc to the arc is greater than the ratio of the sine to the sine, so the ratio of the arc *AB* to the arc *BC* is greater than the ratio of *AH* to *HC*.

We draw *BS* perpendicular <to *AC*>; then the ratio of *MC* to *CS* is greater than the ratio of the arc *KC* to the arc *CB*, from what has been proved in the first proposition of this treatise. So the ratio of *AC* to *CS* is greater than the ratio of the arc *ABC* to the arc *CB*. So the ratio of *AS* to *SC* is greater than the ratio of the arc *AB* to the arc *BC* and the ratio of *AH* to *HC* is smaller than the ratio of the arc *AB* to the arc *BC*. The point that divides[1] the straight line *AC* in the ratio of the arc *AB* to the arc *BC* will thus lie between the points *H* and *S*; let the point of division be the point *T*. We draw from the point *J* the perpendicular JL_a; so the point L_a will lie between the two points *S* and *C*. The ratio of AL_a to L_aC is much greater than the ratio of *AT* to *TC*. We draw from the point *C* a straight line parallel to the tangent *AG*; let it be *CV*. The angle *ACV* is equal to the angle *ACG*. Let us extend the perpendicular JL_a. It meets the straight line *CV*; let it meet it at the point *V*. So L_aV is equal to L_aJ and *VC* is equal to *CJ*. We join *VT* and we extend it. It meets the straight line *AL*; let it meet it at the point *O*. The ratio of *AO* to *CV* is equal to the ratio of *AT* to *TC*, which is the ratio of the arc *AB* to the arc *BC*; so the ratio of *AO* to *CJ* is equal to the ratio of the arc *AB* to the arc *BC*. But the ratio of the arc *AB* to the arc *BC* is equal to the ratio of the arc *AD* to the arc *CE*, because it is equal to the ratio of the arc *DB* to the arc *BE*. So the ratio of *AO* to *CJ* is equal to the ratio of the arc *AD* to the arc *CE*. We draw the straight line *ON* parallel to the straight line *AC*. It cuts the straight line *HL*; let it cut it at the point *N*. Since *AT* is greater than *TC*, *AO* is greater than *CV*; so it is greater than *CJ* and it cuts the straight line *CJ* above the point *J*; it cuts the straight line *HL* beyond the point *J*; now the angle *ANO* is acute, because it is equal to the acute angle *NAC*, so the angle *AON* is obtuse. We draw the two chords *AD* and *CE* and we join *AN*. It cuts the arc *AB*; let it cut it at the point *I′*. If the point *D* lies between the two points *A* and *I′*, then, if we extend *AD*, it cuts *ON* between the two points *O* and *N*; let it cut it at the point *S′*. So the point *S′* is between the two points *O* and *N* and the straight line *AS′* is greater than the straight line *AO*. We extend *AS′*; it meets the straight line *HL* between the two points *L* and *N*; let it meet it at the point *U*. So *AU* is much greater than the straight line *AO*. The ratio of *AU* to *CJ* is much greater than the ratio of the arc *AD* to the arc *CE*. We extend *CE*; it meets the straight line *BJ*; let it meet it at the point *R*. We join *CB*, then the angle *CBH* is acute; so the angle *CBR* is obtuse, the angle *CRJ* is obtuse, the

---

[1] Lit.: the separation that divides up.

straight line *CJ* is greater than the straight line *CR* and the ratio of *AU* to *CR* is much greater than the ratio of the arc *AD* to the arc *CE*. But the ratio of the arc *AD* to the arc *CE* is greater than the ratio of the chord *AD* to the chord *CE*. If we permute, we have that the ratio of *UA* to *AD* is greater than the ratio of *RC* to *CE*; so the ratio of *UD* to *DA* is greater than the ratio of *RE* to *CE* and the ratio of *DU* to *UA* is greater than the ratio of *ER* to *RC*. We join *DC*; let it cut the straight line *BH* at the point *W*; thus the two straight lines *CD* and *UH* have cut one another, between the two straight lines *UA* and *AC*, at the point *W*. So the ratio of *CH* to *HA* is compounded of the ratio of *CW* to *WD* and the ratio of *DU* to *UA*; so the ratio of *CW*, the third, to *WD*, the fourth, is compounded of the ratio of *CH*, the first, to *HA*, the second, and the ratio of *AU*, the sixth, to *UD*, the fifth. Indeed, if between the first and the second we put in an intermediate magnitude and if we put the ratio of the first magnitude to the intermediate one equal to the ratio of the third to the fourth, then the ratio of the intermediate to the second is equal to the ratio of the fifth to the sixth. So the ratio of the first magnitude to the intermediate one, which is equal to the ratio of the third to the fourth, is compounded of the ratio of the first to the second and the ratio of the second to the intermediate one, which is the ratio of the sixth to the fifth. So the ratio of the third to the fourth is compounded of the ratio of the first to the second and the ratio of the sixth to the fifth. So the ratio of *CW* to *WD* is compounded of the ratio of *CH* to *HA* and the ratio of *AU* to *UD*. So if we invert, the ratio of *DW* to *WC* is compounded of the ratio of *AH* to *HC* and the ratio of *DU* to *UA*.

In the same way, since the two straight lines *DE* and *RW* cut one another between the two straight lines *DC* and *CR*, at the point *I*, then the ratio of *DW* to *WC* is compounded of the ratio of *DI* to *IE* and the ratio of *ER* to *RC*. But the ratio of *DW* to *WC* is compounded of the ratio of *AH* to *HC* and the ratio of *DU* to *UA*; so the ratio compounded of the ratio of *DI* to *IE* and the ratio of *ER* to *RC* is the ratio compounded of the ratio of *AH* to *HC* and the ratio of *DU* to *UA*. But the ratio of *DU* to *UA* is greater than the ratio of *ER* to *RC*, so the ratio of *DI* to *IE* is greater than the ratio of *AH* to *HC*. But the ratio of *DI* to *IE* is the ratio of the sine of the arc *DB* to the sine of the arc *BE* and the ratio of *AH* to *HC* is the ratio of the sine of the arc *AB* to the sine of the arc *BC*; so the ratio of the sine of the arc *DB* to the sine of the arc *DE* is greater than the ratio of the sine of the arc *AB* to the sine of the arc *BC*.

If the point *D* is the point *I′*, then the point *S′* is the point *N* and the straight line *AN* is greater than the straight line *AO*; the proof is completed as before.

If the point D lies between the two points I' and B, which happens if the arc DB is extremely small, then we double the arc DB, we repeatedly double the double until the endpoint of this double lies beyond the point I',[2] and we double the arc BE the same number of times. We reduce this case for the figure to the case for which the proof has been given. So the ratio of the sines of doublings of the arc DB to the sines of the doublings of the arc BE is greater than the ratio of the sine of the arc AB to the sine of the arc BC. But in Proposition 2 of this treatise we have proved that the ratio of the sine of any arc to the sine of a part of it is greater than the ratio of the sine of double this arc to the sine of double the part. So the ratio of the sine of the arc DB to the sine of the arc BE is greater than the ratio of the sine of the doublings of the arc DB to the sine of the doublings of the arc BE. But the ratio of the sine of the doublings of the arc DB to the sine of the doublings of the arc BE is greater than the ratio of the sine of the arc AB to the sine of the arc BC, because the figure for the doublings is like the figure for Proposition 2. So the ratio of the sine of the arc DB to the sine of the arc BE is greater than the ratio of the sine of the arc AB to the sine of the arc BC, whatever the magnitude of the arc DE.

If the arc AB is a quarter of a circle, then the arc AP is also a quarter of a circle. Then the arc QC is equal to a quarter of a circle and the arc QCB is equal to the arc ABC; so angle CAG is equal to the angle BPQ, the angle CAG is equal to the angle CHB and the straight line AG is parallel to the straight line HB. If we extend AD, it meets the straight line HB in all the preceding divisions[3] and the proof is like the previous one. So if the ratio of the arc AB to the arc BC is equal to the ratio of the arc DB to the arc BE and if the arc AB is not greater than a quarter of a circle, then the ratio of the sine of the arc DB to the sine of the arc BE is greater than the ratio of the sine of the arc AB to the sine of the arc BC, in all cases of the figure and for all the forms of division.

It follows necessarily from this ratio between the arcs in different circles that, if two different arcs of the same circle are similar to two arcs of another circle, then the ratio of the sine of one of the two arcs to the sine of the other arc of the same circle is equal to the ratio of the sine of the arc similar to the other one in the previous circle to the sine of the similar arc to the second one.

Consequently, for two different arcs of a circle such that the greater is smaller than a quarter of a circle, the ratio of the sine of the greater of the

---

[2] This condition can be satisfied, but it is not sufficient; in fact we require that the endpoint in question shall lie between I' and A (see Mathematical commentary).

[3] That is to say, for any position of the point D that divides the arc AB and of the point E that divides the arc BC.

two to the sine of the smaller of the two is greater than the ratio of the sine of any arc greater than the one similar to the greater of the two arcs – if it is not greater than a quarter of a circle – to the sine of the arc corresponding to the smaller of the arcs, if they are in the same circle and if they are proportional to the two small arcs.[4] That is what we wanted to prove.

<**Lemma 4**>: If in a circle we draw a chord that divides it into any two parts, if we then divide the two arcs in the same ratio, the two homologous parts being opposite one another, and then if we join the endpoints of the two opposite arcs with a straight line, it then meets the chord at an angle such that its ratio to two right angles is equal to the ratio of each of the two opposite arcs to the arc in which each of them is placed.

Example: In the circle *ABCD* we draw the chord *AC*, which divides it into two parts, we put the ratio of the arc *BA* to the arc *ABC* equal to the ratio of the arc *DC* to the arc *CDA* and we join *BD*.

I say that the ratio of the angle *AEB* to two right angles is equal to the ratio of the arc *BA* to the arc *ABC*, which is the ratio of the arc *DC* to the arc *CDA*.

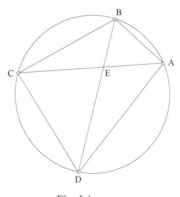

Fig. I.4

Proof: The straight line *BED* cuts the straight line *AC* between the two points *A* and *C*, because the two points *B* and *D* are on either side of the straight line *AC*; let it cut it at the point *E*. We draw the straight lines *AB*, *BC*, *AD*, *DC*. The ratio of the angle *ACB* to the angle *CAB* is equal to the ratio of the arc *AB* to the arc *BC*. In the same way, the ratio of the angle *CAD* to the angle *ACD* is equal to the ratio of the arc *CD* to the arc *DA*. But the ratio of the arc *CD* to the arc *DA* is equal to the ratio of the arc *AB* to the arc *BC*, the angle *CAD* is equal to the angle *DBC* and the angle *ACD* is

---

[4] That is to say, the two arcs of the first circle.

equal to the angle *ABD*; thus the ratio of the angle *ACB* to the angle *CAB* is equal to the ratio of the angle *CBE* to the angle *EBA* and is equal to the ratio of the whole, which is the angle *AEB*, to the whole, which is the angle *BEC*. The ratio of the angle *ACB* to the angle *CAB* is thus equal to the ratio of the angle *AEB* to the angle *BEC*. So the ratio of the angle *AEB* to the angle *BEC* is equal to the ratio of the arc *AB* to the arc *BC*. The ratio of the angle *AEB* to the sum of the two angles *AEB* and *BEC*, which is equal to two right angles, is equal to the ratio of the arc *BA* to the arc *ABC* and the ratio of the angle *BEC* to two right angles is equal to the ratio of the arc *BC* to the arc *ABC*. That is what we wanted to prove.

<Lemma 5>: In the same way, let us again draw the circle and the two arcs. We draw *GH* parallel to the chord *AC*, we put the ratio of the arc[5] *IG* to the arc *GIH* equal to the ratio of the arc *DA* to the arc *ADC*, and we put the ratio of the arc *KH* to the arc *HKG* equal to the ratio of the arc *BC* to the arc *CBA*; the ratio of the arc *IG* to the arc *GIH* is thus equal to the ratio of the arc *KH* to the arc *HKG*. Let us join *KI*.

I say that the straight line *IK* is parallel to the straight line *BD*.

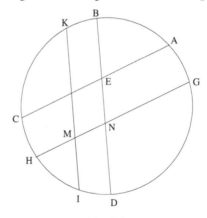

Fig. I.5

Proof: The straight line *IK* cuts the straight line *GH*; let it cut it at the point *M*. The ratio of the angle *KMH* to two right angles is equal to the ratio of the arc *KH* to the arc *HKG*, as has been proved in the preceding proposition. So the ratio of the angle *KMH* to two right angles is equal to the ratio of the arc *CB* to the arc *CBA*. But the ratio of the arc *BC* to the arc *CBA* is equal to the ratio of the angle *BEC* to two right angles. So the ratio of the angle *KMH* to two right angles is equal to the ratio of the angle *BEC*

---

[5] We have added the word 'arc' throughout this paragraph.

to two right angles; so the angle *KMH* is equal to the angle *BEC* and the straight line *BD* cuts the straight line *GH*, because it cuts the straight line *AC*, which is parallel to it. Let the straight line *BD* cut the straight line *GH* at the point *N*. So the angle *BEC* is equal to the angle *ENH*, where the point *N* lies inside or outside the circle. So the angle *ENH* is equal to the angle *GMI*, <since they are> alternate internal angles, and the two straight lines *IK* and *BD* are parallel. That is what we wanted to prove.

<**Lemma 6**>: In the same way, let us redraw the circle and the two arcs; let the point *G* be the centre of the circle. We draw from the point *G* a perpendicular to the straight line *AC*; let it be *GH*, which we extend in both directions to *I* and *K*.

I say that the point *E* lies between the two points *H* and *C*, and the straight line *BD* cuts the straight line *GH* between the points *G* and *H*.

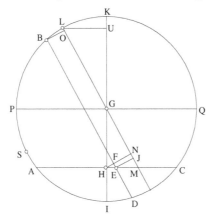

Fig. I.6

Proof: We draw from the point *G* a diameter parallel to the straight line *AC*; let it be *PGQ*. We draw from the point *G* a straight line parallel to the straight line *BD*; let it meet the arc *PKQ* at the point *L* and let it meet the straight line *AC* at the point *M*. The angle *PGL* is equal to the angle *AML* and the angle *AML* is equal to the angle *AEB*, so the angle *PGL* is equal to the angle *AEB*. But the ratio of the angle *AEB* to two right angles is equal to the ratio of the arc *BA* to the arc *ABC*;[6] so the ratio of the angle *PGL* to two right angles is equal to the ratio of the arc *BA* to the arc *ABC*. But the ratio of the angle *PGL* to two right angles is equal to the ratio of the arc *LP* to the arc *PLQ*, because the point *G* is the centre of the circle and

---

[6] From Lemma 4.

the arc PLQ is a semicircle. So the ratio of the arc LP to the arc PLQ is equal to the ratio of the arc BA to the arc ABC. So the ratio of the arc LP to the arc PLK is equal to the ratio of the arc BA to the arc ABK. We put the ratio of the arc SP to the arc PA equal to the ratio of the arc LP to the arc PLK; so the ratio of the arc SL to the arc ABK is equal to the ratio of the arc LP to the arc PK, which is equal to the ratio of the arc BA to the arc ABK. So the ratio of the arc SL to the arc ABK is the ratio of the arc BA to the arc ABK, and the arc SL is equal to the arc BA. So the arc AS is equal to the arc BL, the ratio of the arc PA to the arc AS is the ratio of the arc PA to the arc BL and the ratio of the arc SP to the arc PA is equal to the ratio of the arc LP to the arc PK; so the ratio of the arc PK to the arc KL is equal to the ratio of the arc PA to the arc AS. So the ratio of the arc AP to the arc BL is equal to the ratio of the arc PK to the arc KL. The ratio of the sine of the arc AP to the sine of the arc BL is greater than the ratio of the sine of the arc PK to the sine of the arc KL, since the arc AP is smaller than the arc PK.

We draw the line LU perpendicular <to KH>, HN perpendicular <to LM>, EJ perpendicular <to MN> and BO perpendicular <to LM>; we have that LU is the sine of the arc LK and GK the sine of the arc PK. But GL is equal to GK, HG is the sine of the arc AP and BO is the sine of the arc LB; so the ratio of GH to BO is greater than the ratio of GL to LU. But the ratio of GL to LU is equal to the ratio of GM to MH, and the ratio of GM to MH is equal to the ratio of GH to HN, so the ratio of GH to BO is greater than the ratio of GH to HN. So the straight line BO is smaller than the straight line HN. So the straight line BED cuts the perpendicular HN between the two points H and N; let it cut it at the point F. But if it cuts the perpendicular HN between the two points H and N, it cuts the straight line GH between the two points G and H. Now the straight line BED cuts the straight line AC; if it cuts the straight line GH between the two points G and H, and if it cuts the straight line AC, it accordingly cuts the straight line AC between the two points H and C. That is what we wanted to prove.

From this proof, we go on to prove that the ratio of GM to ME is equal to the ratio of the sine of the arc AP to the sine of the arc BL, because the ratio of GM to ME is equal to the ratio of GH to EJ, because of the similarity of the two triangles GHM and EJM. Now GH is the sine of the arc AP and EJ is the sine of the arc BL; so the ratio of GM to ME is equal to the ratio of the sine of the arc AP to the sine of the arc BL.[7]

---

[7] This paragraph seems to be merely a general comment, not directly related to the result that has been stated.

<Proposition 7>: Having proved these lemmas, let us now begin to find the hour lines.

Let *ABCD* be one of the inclined horizon circles, let *AC* be the meridian line and let *BED* be the <line of> intersection of the horizon and the circle of the equator. The point *E* will then be the centre of the world. Let the arc *BHD* be a semicircle of the equator and let the arc *IKL* be the diurnal arc for the first point of Cancer. So it is greater than a semicircle and its centre is above the horizon; let the point *R* be its centre. Let the intersection with the horizon be the straight line *IL* and let the straight line *IL* cut the straight line *AC* at the point *N*. We join *RN*; it is perpendicular to the straight line *IL*, because the two points *R* and *N* are in the plane of the meridian circle and they are in the plane of the circle *IKL*; thus the straight line *RN* is the intersection of the meridian circle and the circle *IKL*. But each of the circles *ABCD*, *IKL* is perpendicular to the meridian circle; so the straight line *IN* is perpendicular to the plane of the meridian circle and the straight line *RN* is perpendicular to the straight line *INL*. We complete the circle *IKL*; let the complement be the arc *IQL*. Let the point *K* be an endpoint of one of the seasonal hours. We put the ratio of the arc *QI* to the arc *IQL* equal to the ratio of the arc *KL* to the arc *LKI* and we join *QK*. It then cuts the straight line *IL*; let it cut it at the point *P*. Thus the point *P* lies between the points *N* and *I* and the straight line *KP* cuts the straight line *RN* between the two points *R* and *N*, as has been proved in the lemmas.

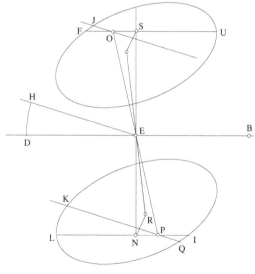

Fig. I.7

Let the <line of> intersection between the tropic of Capricorn and the horizon be the straight line *UF*. *UF* is thus equal to the straight line *IL*; let the straight line *UF* cut the straight line *AC* at the point *S*. Let the diurnal arc for the first point of Capricorn be the arc *UJF*. We join *PE* and we extend it to meet the straight line *UF*; let it meet it at the point *O*. So the straight line *OS* is equal to the straight line *PN*, because *EN* is equal to *ES*. We imagine that the plane containing the straight lines *KP* and *PO* cuts the plane of the circle *UJF*; let the <line of> intersection of the two <planes> be the straight line *OJ*; then the angle *JOF* is equal to the angle *KPL*, because the two straight lines *JO* and *OF* are parallel to the two straight lines *KP* and *PL*. Indeed, the two circles *UJF* and *QKI* are parallel and are cut by the plane of the horizon and by the plane of the two straight lines *KP* and *PO*. But since the straight line *UF* is equal to the straight line *IL*, *SF* is equal to *NI* and *SO* is equal to *NP*, because *NE* is equal to *ES*; so the straight line *OF* is equal to the straight line *PI* and the angle *JOF* is equal to the angle *KPL*. But the angle *KPL* is equal to the angle *IPQ*, so the angle *JOF* is equal to the angle *IPQ*. But the straight line *OF* is equal to the straight line *PI*, the arc *FJU* is equal to the arc *IQL* and they belong to two equal circles, and the arc *JF* is equal to the arc *QI*; the ratio of the arc *JF* to the arc *FJU* is thus equal to the ratio of the arc *QI* to the arc *IQL*. But the ratio of the arc *QI* to the arc *IQL* is equal to the ratio of the arc *KL* to the arc *LKI*; the ratio of the arc *JF* to the arc *FJU* is thus equal to the ratio of the arc *KL* to the arc *LKI*. So the point *J* is the endpoint of the seasonal hour corresponding to the hour whose endpoint is the point *K*, and the straight line *JO* is parallel to the straight line *KQ*; so they lie in the same plane. But the straight line *OEP* lies in their plane; so the three straight lines are in the same plane and the point *E*, which is the centre of the world, is in the plane of these straight lines. The plane of these straight lines cuts <the sphere of> the world and in it forms a great circle that passes through the points *J* and *K*. This circle cuts the circle of the equator; let it cut it along the straight line *EH*. So the straight line *EH* is parallel to each of the straight lines *KP* and *JO*. But the straight line *ED* is parallel to each of the straight lines *NL* and *SF*, so the angle *HED* is equal to each of the angles *KPL* and *JOF*. But the ratio of the angle *KPL* to two right angles is equal to the ratio of the arc *KL* to the arc *LKI*, from what was proved in in Lemma 4 of this treatise. So the ratio of the angle *HED* to two right angles is equal to the ratio of the arc *KL* to the arc *LKI*. But the ratio of the angle *HED* to two right angles is equal to the ratio of the arc *HD* to the arc *DHB*, and since the arc *DHB* is a semicircle whose centre is the point *E*, accordingly the ratio of the arc *HD* to the arc *DHB* is equal to the ratio of the arc *KL* to the arc *LKI*. So the point *H* is the endpoint of the seasonal hour corresponding to the hour whose endpoint is the point K. But the point H lies on the circumference of the great circle that passes through the two points K and J, and thus through the points K, H and J, there passes a single great circle whose centre is the point E; let it be the circle JHK. So the <lines of> intersection of this circle with the circles UJF, BHD and IKL are the parallel straight lines JO, HE and KP, and the intersection of the circle JHK with the horizon is the straight line OEP; so the circle JHK cuts any plane parallel to the horizon. So if in the place whose horizon is the circle $ABCD$ we have a plane sundial parallel to the horizon, then the circle JHK cuts this sundial along a straight line parallel to the straight line PO and the ends of the shadows of the gnomon of the sundial, whose tip is the point E, fall on this straight line if the sun is in the plane of the circle JHK. So if the sun is at the point K, then the ray coming from the point K is directed towards the point E along a straight line in the plane of the circle JHK; and if it reaches the plane of the sundial that is parallel to the horizon, then the end of the ray, that is the tip of the shadow of the gnomon, lies on the <line of> intersection, which is the straight line formed by the circle JHK on the plane of the sundial. In the same way, if the sun is at the point H, its ray travels along the straight line HE and it ends on the line of intersection of the circle JHK with the plane of the sundial. In the same way, if the sun is at the point J, its ray travels to the point E and then travels from the point E to the plane of the sundial while always remaining in the plane of the circle JHK; so it ends on the straight line that is the intersection. The straight line that is the intersection of the plane of the sundial with the great circle that passes through the points J, H, K marks a single seasonal hour on the three days during which the sun moves along the three circles which are the paths of Cancer, of Aries and of Capricorn, if the sun comes to the points on these circles that mark the same hour in regard to each of the three circles. That is what we wanted to prove.

This is a proposition Ibrāhīm ibn Sinān established, but he established it in a different way.

<**Proposition 8**>: Having proved this, we say that the straight line that lies in the plane of the sundial and marks the same seasonal hour on the three days on which the sun moves along the paths of Cancer, of Aries and of Capricorn, does not mark this hour for any other day except these three days, that is to say that, at the end of the hour corresponding to the hour whose end was marked by the straight line we described before, the tip of the shadow of the gnomon is not on that straight line on any day other than the three days we mentioned before; and that the point which is the endpoint of the hour corresponding to the hour marked by that straight line

in regard to any hour circle other than the three circles we mentioned before, will not be on the circumference of the circle that in the example is the circle *JHK*, but will lie outside it: for the hour circles that lie between the tropic of Cancer and the circle of the equator, the points that are the ends of this hour are closer to the meridian circle than the circle corresponding to the circle *JHK*; for the hour circles that lie between the circle of the equator and the tropic of Capricorn, the points that are the ends of this hour are closer to the horizon circle than the circle corresponding to the circle *JHK*.

This is a proposition Ibrāhīm ibn Sinān tried to establish, but he was not able to establish it for every hour line, as we establish it here now.

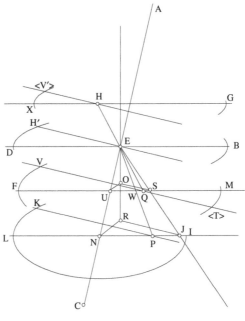

Fig. I.8

Let us return to the figure, without the circle of Capricorn. Let us draw a diurnal arc of one of the hour circles that lie between the circle of Cancer and the circle of the equator, let it be the arc *MVF*; the arc *MVF* is thus smaller than the arc similar[8] to the arc *IKL*. We join *ER*; thus the straight line *ER* is the axis of the Universe and the centre of the circle *MVF* lies on the straight line *ER*; let that be the point *O*. Let us draw the <line of>

---

[8] Lit: 'the similar', which we translate throughout as 'the arc similar'.

intersection of the circle of the horizon and the circle *MVF*; let it be *MF*. Then the straight line *EC* cuts it into two equal parts; let it cut it at the point *U*. We join *OU*; it is perpendicular to the straight line *MF*. We join *EP*; it then cuts the straight line *MF*; let it cut it at the point *W*. From the point *R* let us draw a straight line parallel to the straight line *KP* – let it be <the straight line> *RJ* [9] – and let us put the ratio of the arc *VF* to the arc *FVM* equal to the ratio of the arc *KL* to the arc *LKI*. If we complete the circle *MVF*, if we cut off from the <added> arc, which is the complement, an arc <lying> beyond the point *M*, if we put its ratio to the complement of the circle equal to the ratio of the arc *VF* to the arc *FVM* and if we join the end of the arc to the point *V* with a straight line, this cuts the straight line *MU* in a point between the two points *M* and *U*; let this point be the point *Q*. Let us join *VQ*; the angle *VQF* is then equal to the angle *KPL*, because the ratio of each of the two angles to two right angles is the same ratio as the ratio of the arc *KL* to the arc *LKI*; so *VQ* is parallel to the straight line *KP* and it is parallel to the straight line *RJ*. We join *EJ*; let it cut the straight line *MU* at the point *S*. We join *OS*; it is parallel to the straight line *RJ*, because *OS* is the <line of> intersection of the circle *MVF* and the plane of the triangle *EJR* which cuts the circle *MVF* and the tropic of Cancer; so *OS* is parallel to the straight line *VQ*. Since the arc *MVF* is smaller than the arc similar to the arc *IKL*, the arc *VF* is smaller than the arc similar to the arc *KL*, the arc cut off from it and whose ratio to it is equal to the ratio of *KL* to half of *LKI* is smaller than the arc similar to the corresponding arc cut off from the arc *KL*, and the remaining arc is smaller than the arc remaining from the arc *KL*. But the ratio of the sine of the arc *VF* to the sine of the arc that remains from it is equal to the ratio of *OS* to *SQ*; so the ratio of *OS* to *SQ* is greater than the ratio of *RJ* to *JP*, as was proved at the end of Lemma 6. But the ratio of *OS* to *SU* is equal to the ratio of *RJ* to *JN* because the two triangles *OSU* and *RJN* are similar. In fact, the angles at the points *S* and *J* are equal, because they are equal to the angles that are at the two points *Q* and *P*, which are equal; now the angles that are at the points *U* and *N* are right angles; so the ratio of *OS* to *SU* is equal to the ratio of *RJ* to *JN*, so the ratio of *US* to *SQ* is greater than the ratio of *NJ* to *JP* and the ratio of *JN* to *NP* is greater than the ratio of *SU* to *UQ*. If we permute, we have that the ratio of *JN* to *SU* is greater than the ratio of *NP* to *UQ*. But the ratio of *NJ* to *SU* is equal to the ratio of *RN* to *OU*, because of the similarity of the triangles *RJN* and *OSU*; so the ratio of *RN* to *OU* is greater than the ratio of *NP* to *UQ*. But the ratio of *RN* to *OU* is equal to the ratio of *NE* to *EU* and the ratio of *NE* to *EU* is equal to the ratio of *NP* to *UW*; so the ratio of *NP* to *UW* is greater than the ratio of *NP* to *UQ*, the straight line *UW* is smaller

---

[9] *J* is on *LI*.

than the straight line *UQ*, the point *Q* is at the end of the straight line *PE* and the straight line *PE* is the diameter of the circle that marks the seasonal hour in the course of the days when the sun is moving on the paths of Cancer, of Aries and of Capricorn. We join *QE* and in the plane of the horizon we draw a straight line equal and parallel to the straight line *MF*, let it be *GX*; we extend *QE* until it meets it; let it meet it at the point *H*.[10] If in its plane we draw the hour circle as far as the straight line *GX* and if we construct the plane that contains the two straight lines *VQ* and *QE*, in the hour circle a straight line is formed that is parallel to the straight line *VQ* and on the surface of the <sphere of> the world we form a great circle. This great circle cuts the circumference of the circle of the equator, so this circle cuts off from the two circles separated[11] by *MF* and *GX* and the circle of the equator three arcs which are the same single seasonal hour corresponding to the seasonal hour for the three days we have already mentioned, as has been proved in the preceding proposition.

It is clear that the circle whose diameter is *QEH* is distinct from the circle whose diameter is *PE*, that the circle whose diameter is *QEH* cuts the equator, and that if the hour whose endpoint is *H'*[12] is the hour whose endpoint is the point *K* on the tropic of Cancer, the circle whose diameter is *QEH* cuts the equator along the same straight line <as that> along which it is cut by the first circle corresponding to the circle *JHK* and whose diameter is *PE*, because the angle formed at the point *E* is equal to the angle *VQF* which is equal to the angle *KPL*. The point at which the circle with diameter *QEH* cuts the circumference of the tropic of Cancer is closer to the meridian circle than the point *K*. If the point *K* lies between the horizon and the meridian circle, then the point *H'* is closer to the meridian circle than the first circle whose diameter is *PE* and the point of the hour circle that is equal to the circle *MVF*, <the point> in which it is cut by the circle whose diameter is *QEH*, is closer to the horizon than the circle whose diameter is *PE*. The circle whose diameter is *QEH* cuts the plane of the sundial in a straight line parallel to the straight line *QEH*. This straight line cuts the first straight line parallel to the straight line *PE* in a point corresponding to the point *E* which is on the first straight line. This second straight line marks the same hour, corresponding to the hour marked by the first straight line and in the course of the three days during which the sun moves on the circle *MVF*, and on the circle equal to it, as well as on the circle of the equator, if the sun comes to the three points that are the endpoints of that hour, then the tips of the shadows of the gnomon lie on a

---

[10] The point *H* is not the same one as in Proposition 7.

[11] The straight lines *MF* and *GX* are their diameters.

[12] The point *H'* is the point *H* in Proposition 7. Ibn al-Haytham says so later.

straight line parallel to the straight line *QEH*. That is what we wanted to prove.

If we draw *RI* and *EI*,[13] we prove, as we proved for the straight line *EJ*, that the straight lines drawn from the centres of the hour circles parallel to the straight line *RI* all reach the straight line *EI*, because they all lie in the plane of the triangle *ERI*.

It thus becomes clear, from what we have proved in the last two propositions, that the same seasonal hour is not marked for every day of the year by a <particular> single straight line lying in the plane of the horizontal sundial, but by many <different> straight lines; and that for two circles on either side of the equator a seasonal hour marks out, from them and from the circle of the equator, a great circle that cuts off from the two circles two arcs, each of which is a seasonal hour. On the equator let us cut off an hour like that same hour, then throughout the whole year the same seasonal hour marks ninety-one <great> circles; if we put in a circle for each part of the circle of the ecliptic, all these <great> circles cut one another in a single point on the circle of the equator. These circles generate, in the plane of the horizontal sundial, ninety-one straight lines that all cut one another in a point on the <line of> intersection of the plane of the sundial and the plane of the equator; it is this point that marks the seasonal hour for the two days of the equinoxes. And the straight line generated by the circle which intersects the tropic of Cancer and the tropic of Capricorn marks the seasonal hour for the two days of the solstices. Each of the remaining straight lines marks a seasonal hour for four days of the year: two days from the motion of the sun in the northern half of the circle of the ecliptic and two days from its motion in the southern half, because each of these circles, apart from the circles of Cancer and Capricorn, cuts the circle of the ecliptic in two points; the sun then moves on each of these circles in the course of one day in the year. For each of the seasonal hours, we have straight lines as has been described, <straight lines> whose number is that number and which cut one another in a point on the <line of> intersection of the plane of the sundial and the plane of the circle of the equator. All these straight lines are imaginary straight lines, so each is a length without breadth. This is the way the hour lines that mark out the seasonal hours in the planes of horizontal sundials are generated.

<**Proposition 9**>: It remains for us to show the distance between the tips of the shadows at a <specific seasonal> hour for the different days and the straight line *EP* that marks the same hour on the tropics of Cancer and Capricorn and on the circle of the equator.

---

[13] Here *I* designates the point that earlier was *J*.

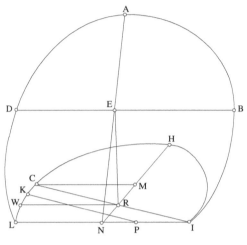

Fig. I.9.1

Let us return to the last figure for the horizon and the tropic of Cancer. Let the diurnal arc of Cancer be fourteen hours, it is two hundred and ten parts; let the point *H* lie on the meridian circle; we join *HL*; then the arc *HL* is one hundred and five parts and the arc *LW*, which is half the amount by which the diurnal arc exceeds half the circle, is fifteen parts. Let the arc *KL* be the first seasonal hour, then it will be seventeen parts and a half; the arc *KC* is twelve parts and a half because if we cut off a sixth from the arc *LW*, the remainder is equal to the arc *KC*, since that has been proved in the sixth lemma. But the straight line *RE* is the sine of the maximum inclination, so it is 24.24.15[14] in terms of the magnitude by which the straight line *EH*, which is the semidiameter of the universe, is sixty parts. But the angle *EHN* is equal to the angle subtended by the maximum inclination at the centre of the world, so it is approximately twenty-four parts in terms of the magnitude by which four right angles are three hundred and sixty parts; so it is forty-eight parts in terms of the magnitude by which two right angles are three hundred and sixty parts. The arc cut off by the straight line *ER* from the circle that circumscribes the triangle *EHR* is forty-eight parts,[15] so the arc cut off by the straight line *RH* is one hundred and thirty-two parts. So the straight line *RH* is one hundred and nine parts, thirty-seven minutes and thirty-two seconds in terms of the magnitude by which the straight line *EH* is one hundred and twenty parts. In terms of the magnitude by which the straight line *EH* is sixty parts, the straight line *RH* is 54.48.46. But the

---

[14] This notation designates 24°24′15″ in all the text.
[15] If we take angle *EHR* = 24°, we have arc *ER* = 48°.

straight line *RN* is equal to the sine of the arc *L W*. We draw the perpendicular *CM*; then *MR* is the sine of the arc *WC* and the arc *WC* is 15 parts, because its ratio to the semicircle is equal to the ratio of the arc *KL* to the arc *LKI*, since the angle *CRW* is equal to the angle *KPL*, so the ratio of each of them to two right angles is the same ratio. The arc *WC* is fifteen parts, its sine, which is the straight line *RM*, is equal to the sine of the arc *LW*, which is *RN*, so, if we extend the straight line *CR* until it reaches the straight line *IN*, it will be equal to the semidiameter of the circle. It accordingly meets the straight line *IN* at the point *I*; let it be like the straight line *CRI*. Then the point *I* in that figure will be the point *J* of the preceding figure and the straight line *IN* is equal to the straight line *CM*, because *IR* is equal to *RC*. But the straight line *CM* is the sine of the arc *CH* which is seventy-five parts and its sine is 57.57.20 in terms of the magnitude by which the straight line *RI* is sixty parts. In terms of the magnitude by which the straight line *RI* is 54.48.46, the straight line *CM* is thus 52.56.50.[16] But the ratio of the sine of the arc *LW* to the sine of the arc *KC* is equal to the ratio of *RI* to *IP*, the arc *LW* is fifteen parts and its sine is 15.31.45 in terms of the magnitude by which the straight line *RI* is sixty parts. So in terms of the magnitude by which the straight line *RI* is 54.48.46, the sine of the arc *LW*, which is the straight line *RN* is 14.11.10. But the arc *KC* is twelve parts and a half and its sine is 12.59.21[17] in terms of the magnitude by which the straight line *RI* is sixty; consequently, in terms of the magnitude by which the straight line *RI* is 54.48.46, the sine of the arc *KC* is 11.51.45; so the ratio of *RI* to *IP* is equal to the ratio of 14.11.10 to 11.51.45. But the straight line *RI* is 54.48.46, so the straight line *IP* is 45.50.6. But the straight line *IN* is 52.57, so the straight line *PN* is seven parts, six minutes and fifty-four seconds. But since *ER* is 24.24.15 and its square is five hundred and ninety-six, approximately, and *RN* is 14.11.10 and its square is two hundred and two, approximately, and their sum is 798, whose root is 28 parts and a quarter, accordingly the straight line *EN* is 28 parts and a quarter in terms of the magnitude by which the semidiameter of the universe is sixty parts. Now it is clear that the straight lines drawn from the centres of the hour circles and which are parallel to the straight line *RI* reach the straight line *EI*, because we have proved this in the preceding proposition. These straight lines cut off from the planes of the hour circles triangles similar <to one another> and similar to the triangle *RIN*, the ratios of the straight lines that are the bases of the

---

[16] Calculation gives us sin 75 = 0.96592; with $r = 60$, we obtain *CM* = 57°57′18″; with $r = 54°48′46″ = 54.812727$, we obtain *CM* = 52.94471 = 52°56′40″. Ibn al-Haytham later takes *CM* = 52°57′.

[17] Calculation gives 12°59′24″, 11°51′48″, 45°50′13″.

triangles are equal to the ratio of *RI* to *IN* and their ratio to what is cut off from their bases by the straight line *EP* is equal to the ratio of *RI* to *IP*. But the ratio of *RI* to *IP* is equal to the ratio of the sine of the arc *LW* to the sine of the arc *KC*. But each of the arcs of the hour circles corresponding to the arc *LW* that belongs to one of the hour circles that lie between the tropic of Cancer and the equator is smaller than the arc *LW*. The same holds for every arc that is homologous with the arc *KC*. The ratio of the sine of each of the arcs of the hour circles corresponding to the arc *LW* to the sine of each of the arcs corresponding to the arc *KC* is thus greater than the ratio of *RI* to *IP* and the tips of the shadows for the same seasonal hour corresponding to the arc *LK* will not lie on the straight line *EP*, that is to say they will lie between the two straight lines *EI* and *EP*. But since the arc *LW* is equal to the arc *KC* plus a fifth of it – the arc *KC* is in fact a half plus a third of the arc *LW* – the straight line *RI* is smaller than one and a fifth times the straight line *IP*. All the straight lines homologous to the straight line *RI*, in all the seasonal circles, are smaller than one and a fifth times the straight lines homologous to the straight line *IP*. But since the straight line *RI* is smaller than one and a fifth times the straight line *IP*, the straight line *IP* is greater than a half plus a third of the straight line *RI*; in the same way, each of the straight lines homologous to the straight line *IP* in all the hour circles are greater than a half plus a third of the straight line homologous to the straight line *RI*.

We put <the straight line> *IJ* equal to a half plus a third of the straight line *RI* and we join *EJ*; it cuts all the straight lines homologous to the straight line *IP*, the segment that the straight line *EJ* cuts off from each of the straight lines homologous to the straight line *IP* is a half plus a third of the straight line homologous to the straight line *RI*. The straight line cut off by the great circles that cut off the same seasonal hour, <which is one> of the straight lines homologous to the straight line *IP* on the side nearer the straight line *EI*, is always greater than the straight lines cut off by the straight line *EJ*, <one> of the straight lines homologous to the straight line *IP* on the same side as the straight line *EI*. The points in which the great circles – that mark out the first hour on the hour circles lying between the tropic of Cancer and the circle of the equator – cut off the straight lines homologous to the straight line *IP*, always lie between the straight lines *EP* and *EJ*.

Let the <line of> intersection of the horizon and the tropic of Capricorn be the straight line *GX*. We extend the straight lines *IE, JE, PE* in the direction towards it; let the straight line *IE* end at the point *X*, let the straight line *JE* end at the point *F* and let the straight line *PE* end at the point *H* on the straight line *GX*; then all the intersections of the great circles that mark off the first seasonal hour with the horizon lie between the two straight lines *PH* and *JF* and they all cut one another at the point *E*. But since the straight line *RI* is 54.48.46, the straight line *IJ* is 45.40.28 and the straight line *IP* 45.50.6, we have that the straight line *JP* is 0.9.28, the square of the straight line *EN* is seven hundred and ninety-eight, *PN* is seven parts and six minutes and fifty-four seconds, its square is fifty parts and two thirds approximately, and their sum is eight hundred and forty-eight parts and two thirds, whose root, which is the straight line *PE*, is twenty-nine parts plus an eighth approximately; so the straight line *PE* is twenty-nine parts plus an eighth approximately, and the straight line *JP* is nine minutes and twenty-eight seconds; so it is less than a sixth of a part and the ratio of the straight line *JP* to the straight line *PE* is smaller than the ratio of a sixth to twenty-nine parts plus an eighth. If we convert the twenty-nine parts plus an eighth into sixths, the number of the parts is greater than one hundred and seventy-four; so the ratio of *JP* to *PE* is less than the ratio of unity to one hundred and seventy-four.

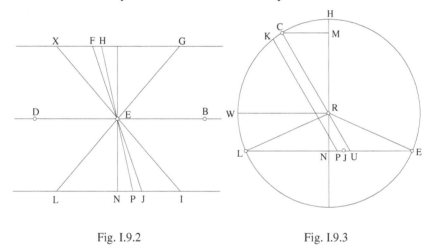

Fig. I.9.2   Fig. I.9.3

In the same way, let us return to the figure. Let the arc *LK* be equal to five hours, so that the point *K* is the start of the sixth hour. We take *WC* as seventy-five parts, we draw the perpendicular *CM* and we extend *CR*; let it meet the straight line *IN* at the point *U*; the straight line *MR* is thus the sine of seventy-five, the straight line *CM* is the sine of fifteen, the straight line *RN* is the sine of the arc *LW* and *LW* is fifteen, as it was <before>. The ratio of *CR* to *RU* is equal to the ratio of *MR* to *RN*. But *MR* is 57.57.20 in terms of the magnitude by which the semidiameter of the universe is sixty parts, and in terms of this magnitude the straight line *RN* is 14.11.10; so the ratio of *CR* to *RU* is equal to the ratio of 57.57.20 to 14.11.10. But the straight line *CR* is 54.48.46, so the straight line *RU* is 14.41.15. But the ratio of *CM* to *UN* is equal to the ratio of 57.57.20 to 14.11.10 and the straight line *CM* is 14.11.10, so the straight line *UN* is 3.48.14. But the ratio of *RU* to *UP* is equal to the ratio of the sine of the arc *LW* to the sine of the arc *KC*, the arc *KC* is a sixth of the arc *LW* since *CH* is a sixth of *WH* and the arc *LW* is fifteen parts; so the arc *KC* is 2.30 and its sine is 2.37.17 in terms of the magnitude by which the straight line *RC* is sixty parts, and thus in terms of the magnitude by which the straight line *CR* is 54.48.46; the sine of the arc *KC* is thus 2.23.27 and the ratio of *RU* to *UP* is equal to the ratio of 14.11.10 to 2.23.27. But the straight line *RU* is 14.41.15, so the straight line *UP* is 2.28.9 and the straight line *PN* is 1.19.14. Let us take *UJ* as a sixth of *RU*; then *UJ* is equal to 2.27. So the straight line *JP* is less than two minutes, the straight line *EN* is twenty-eight and a quarter, its square is seven hundred and ninety-eight, the straight line *PN* is 1.19.14, its square is less than two parts and their sum is less than eight hundred, whose root, which is the straight line *PE*, is less than twenty-eight plus a quarter and a seventh; if we convert it to minutes, we have approximately one thousand seven hundred; so we have that the ratio of *JP* to *PE* is equal to the ratio of two to one thousand seven hundred, so it is one part of eight hundred and fifty parts.

So if, for the <hour> lines for each of the remaining hours, we follow the method we adopted for the two hour lines for these hours, that is to say, the first and the fifth, we prove that the ratio of the straight line corresponding to the straight line *JP* to the straight line corresponding to the straight line *PE* is equal to a small ratio that is less than the ratio of unity to one hundred and seventy-four, which was <that for> the first hour.

In the same way, if for any of the inclined horizons, we follow the method we adopted for the <previous> horizon, we show that the ratio of the straight line corresponding to the straight line *JP* to the straight line corresponding to the straight line *PE* is a small ratio, such that, because it is small, the magnitude of the straight line corresponding to the straight line *JP* vanishes in comparison with <the magnitude of> the straight line corresponding to the straight line *PE*.

<Proposition 10>: In the same way, let us return to the <previous> horizon and the tropic of Capricorn; let the horizon be *ABCD* and its centre *E*; let the diurnal arc of Capricorn, be <the arc> *BGD*; let the <line of> intersection between that circle and the horizon be the straight line *BD*; let the arc *GD* be the first seasonal hour; so the arc *GD* is twelve parts and a half, because the arc *BGD* is one hundred and fifty parts since it is equal to the remainder of the tropic of Cancer.

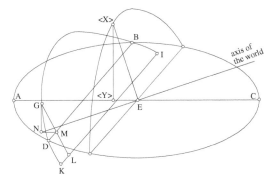

Fig. I.10

We imagine a diameter of this circle *BGD* drawn through its centre and parallel to the straight line *BD*; let it be *IK*. We complete the semicircle, let it be *IGK*; we have that the sum of the two arcs *BI* and *DK* is equal to thirty parts and they [the arcs] are equal; so the arc *DK* is fifteen parts and the arc *GK* is twenty-seven parts and a half, so its sine is 27.42.18 in terms of the magnitude by which the semidiameter of the tropic of Capricorn is sixty parts. From the point *G* we draw the perpendicular *GML*, then *GL* is the sine of the arc *GK* and *ML* is equal to the sine of the arc *DK*; so the straight line *GL* is 27.42.18 and the straight line *ML* is 15.31.45; so the straight line *GM* is 12.10.33 in terms of the magnitude by which the semidiameter of the tropic of Capricorn is sixty parts. Since the arc *BGD* is inclined to the horizon, the straight line *GM* is inclined to the horizon. From the point *G* we draw a perpendicular to the plane of the horizon, let it be *GN*, and we draw *EG* and *MN*. The plane that contains the two straight lines *EG* and *GN* is the plane of the zenith circle that passes through the point *G*; the perpendicular *GN* is the sine of the height of the point *G*. But since the straight line *GM* is perpendicular to the straight line *BD* and the straight line *GN* is perpendicular to the horizon, the straight line *NM* is perpendicular to straight line *BD*; so the angle *GMN* is the angle of inclination and the triangle *GMN* is similar to the triangle formed in the plane of the meridian circle, bounded by the semidiameter of the equator and the sine of the height on the meridian of the first point of Aries; the part of the straight meridian line cut off between them, that is to say, cut off from the straight meridian line between the two straight lines we mentioned, is equal to the sine of the latitude of the place whose longest day is fourteen hours; but the place whose longest day is fourteen hours has a latitude of thirty parts in terms of the magnitude by which the semidiameter of the equator is twice the straight line cut off from the straight meridian line; so the straight line *GM* is twice the straight line *MN* and the square of *MN* is a quarter of the square of *GM*; but the straight line *GM* is 12.10.33 and its square is one hundred and forty-eight parts plus a half plus a quarter.[18] If we take away a quarter of it, what remains is approximately one hundred and eleven and a quarter, whose root is approximately ten plus a third plus a quarter. So the perpendicular *GN* is ten parts plus a third plus a quarter in terms of the magnitude by which the semidiameter of the circle *BGD* is sixty parts. In terms of the magnitude by which the semidiameter of the circle *BGD* is 54.48.46, and by which half of the diameter of the world is sixty parts, the perpendicular *GN* is thus nine parts and two thirds approximately. We join *EG*; *EG* is then the semidiameter of the world because *E* is the centre of the world and the point *G* lies on the surface of the sphere of the world. So the perpendicular *GN* is nine parts and two thirds in terms of the magnitude by which the straight line *EG* is sixty parts and the square of *GN* is ninety-three parts and four ninths of a part and the square of *EG* is three thousand six hundred; there remains the square of *EN*, <which is> three thousand five hundred and six and five ninths; so its root, which is the straight line *EN*, is approximately fifty-nine parts and a quarter. Thus in terms of the magnitude by which the straight line *GN* is nine parts and two thirds, the straight line *EN* is fifty-nine parts and a quarter. So the ratio of *GN* to *NE* is the ratio of nine plus two thirds to fifty-nine plus a quarter; multiply everything by twelve, thus *GN* becomes one hundred and sixteen and *NE* seven hundred and eleven. So the ratio of *GN* to *NE* is the ratio of one hundred and sixteen to seven hundred and eleven. Let us divide everything by one hundred and sixteen, we have that *GN* is equal to one and *NE* is equal to six parts and an eighth approximately. So we have that *GN* is smaller than approximately a sixth of *NE*. But the ratio of *GN* to *NE* is the ratio of the gnomon <that is> perpendicular to the plane of the sundial <that is> parallel to the horizon to the shadow of the gnomon at the end of the first hour while the motion of the sun is on the tropic of Capricorn, which is the longest shadow of the gnomon for any time in the year.

<Proposition 11>: Now that all the above has been proved, let the plane of the horizontal sundial be the plane of <the triangle> *ABC*, and let the gnomon perpendicular to this <plane> be the straight line *CD*. let the <line of> intersection of the great circle which marks the first hour on the paths of Cancer, Aries and Capricorn, with the plane of the sundial be the straight line *AB*. Let the base of the gnomon be the point *C* and the tip of the gnomon the point *D*, the meridian line the straight line *CE*, and let the straight line along which the tip of the shadow moves in the course of the two days of the equinoxes be the straight line *GEH*; then the straight line *AB* cuts this

---

[18] Calculation gives 148.2501.

straight line, because the great circle which marks the first hour cuts the equator and cuts the plane of the sundial, accordingly cuts their <line of> intersection; let these two straight lines cut one another at the point *I*; the tips of the shadows of the gnomon at the end of the first hour thus lie on the straight line *AB* and the tip of the shadow of the gnomon at the end of the first hour on the two days of the equinoxes will be on the point *I*. Let the tip of the shadow at the end of the first hour <when the sun is> on the tropic of Capricorn be the point *L* and the tip of the shadow at the end of that hour <when the sun is> on the tropic of Cancer be the point *M*. We draw the straight lines *DL, DI, DM, CL, CI, CM*; the straight lines *DL, DI, DM* thus lie in the plane of the great circle that marks the first hour, and these straight lines are the straight lines of the rays that travel from the tip of the gnomon to the plane of the sundial at the end of the first hour, on the days when the sun moves along the paths of Capricorn, Aries and Cancer. But the straight lines *CL, CI, CM* are the lines of the shadow for the three days, they are called the lines of azimuths.

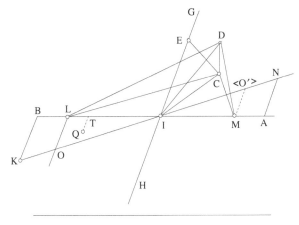

Fig. I.11

It has been shown in the previous proposition that the ratio of *DC* to *CL* is the ratio of one to six. From the point *I* let us draw the straight line *NIK* in the plane of the sundial and such that the angle *BIK* is equal to the angle *JEP* of Proposition 9, which is in the plane of the horizon to which the plane of the sundial is parallel. We make the angle *IBK* equal to the angle *EPJ* of that proposition; and similarly for the angle *IAN*; so the triangle *IBK* is similar to the triangle *EPJ* of the proposition we mentioned. So we have that the ratio of *KB* to *BI* is equal to the ratio of one to a hundred and seventy-four. But since the plane of the sundial is parallel to the plane of the horizon, the straight line *AB* is parallel to the straight line formed by the great circle that marks the first hour in the plane of the horizon and which is the straight line *PEH* in the figure we mentioned earlier. So the straight line *NIK* is parallel to the straight line *JEF* of the proposition we mentioned. But it has been proved in Proposition 9 that all the great circles that cut off the first hour cut the horizon along straight lines that all lie between the two straight lines *PEH* and *JEF*; so all the great circles that mark the first hour intersect the plane of the sundial along straight lines that all lie between the two straight lines *AB* and *NK*. But the straight line *GH* is parallel to the straight line *BD* of the proposition we mentioned earlier. So the angle *BIE* is obtuse and the angle *BIC* is even more obtuse; so the straight line *CL* is greater than the straight line *LI*. We draw *LO* parallel to the straight line *BK*, so we have that the ratio of *OL* to *LI* is equal to the ratio of *KB* to *BI*, which is equal to the ratio of one to one hundred and seventy-four. But the straight line *LC* is greater than the straight line *LI* and the straight line *CD* is a sixth of the straight line *CL*; so the straight line *CD* is much greater than a sixth of *LI*. But the ratio of *LO* to *LI* is the ratio of one to one hundred and seventy-four; so the ratio of *LO* to *CD* is smaller than the ratio of one to a sixth of one hundred and seventy-four, so the straight line *LO* is smaller than a third of a tenth of *CD*. But since *CL* is the shadow for the first hour on the tropic of Capricorn, the tip of the shadow is at the point *L* itself; consequently, the tips of the shadows for the first hour on the remaining days, and which lie outside the straight line *BI*, are always closer to the point *I*; so any one of the straight lines drawn from their tips parallel to the straight line *BK* is smaller than <the straight line> *LO*. So the ratio of each of these straight lines to the straight line *CD* is smaller than a third of a tenth. So if the length of the gnomon is the width of three fingers of a hand, the straight line *LO* is smaller than a third of a tenth of three fingers. But a single finger of the hand does not reach a width of six *sha'īra*; so the length of the gnomon does not reach the width <of a third of a tenth of> eighteen *sha'īra*, the straight line *LO* does not reach three fifths of the width of a *sha'īra* and the straight line *LO* is of a magnitude that is not perceptible in relation to the length of the straight line *ML* which is the line of the first hour.

Now the tip of the shadow for the first hour <when the sun is on> the tropic of Capricorn is also at the point *L* itself, and does not lie between the two straight lines *LI* and *OI*; every shadow of the gnomon *CD* whose tip is between the two straight lines *LI* and *OI* is thus closer to the point *I* than to the straight line *LO* and the straight line drawn from the tip of the shadow, lying between the two straight lines *LI* and *OI* and parallel to the straight line *LO* is smaller than <the straight line> *LO*; so this straight line is a part of *LO* and it is a small part of the length of the gnomon. Each of the straight lines drawn from the tips of the shadows to the straight line *LI* parallel to the straight line *LO* thus has no <perceptible> magnitude in relation to the length of the gnomon. Thus none of these straight lines has a magnitude that is perceptible in relation to the straight line *LM*.

If the separation between the tips of the shadows and the straight line *LM* has no <perceptible> magnitude, then the tips of the shadows do not separate themselves from the width of the perceptible line that is drawn in the plane of the sundial, and if some of them do separate themselves from it, then it is by a magnitude that the sense does not perceive and its magnitude does not affect the time of the hour if the length of the gnomon is three fingers. But for the majority of sundials, the length of the gnomon is less than three fingers; thus the lengths of the shadows are shorter and the distance between the tips of the shadows and the straight line *LM* is thus smaller because the ratio of these widths to the length of the gnomon is the same ratio.

So we have to establish, starting from this proof, that the distance between the tips of the shadows and the straight line *LM* that is the line of the first hour – if we imagine the straight line *LM* to be a length without width – is a distance whose magnitude is not such as the senses can perceive and which does not exceed the width of the perceptible straight line in a way that can affect the time of the hour.

Using a method like the previous one, we can prove that, for each of the hour lines, the distance between the tips of the shadows and the line is an imperceptible distance, because the distance between the tips of the shadows and the line for each of the remaining hours is smaller than the distance between the tips of the shadows and the line for the first hour, since the straight line corresponding to the straight line *LO* has a smaller ratio to the length of the gnomon, from what was proved in Proposition 10 of this treatise.

Once this is proved, we have then proved that the hour lines are straight lines as far as the senses can perceive and that earlier scholars were right when they supposed that these lines are straight lines; and we have proved that Ibrāhīm ibn Sinān made an error when he claimed that earlier scholars were mistaken in regard to the hour lines; we have proved that his error occurred because he carried out an imaginary, mathematical investigation and did not carry out a physical investigation of what is perceived.

All that we have proved holds only for inclined horizons. As for horizons on the equator, the hour lines for them are straight lines, and for each seasonal hour <there is> a single straight line both in imagination and in perception. In fact, for horizons on the equator seasonal hours are equinoctial hours, because their horizons pass through the two poles and their diurnal arcs are semicircles of the seasonal arcs; the great circle that passes through the two poles, and cuts off a seasonal hour from the semicircle of the equator, [this great circle] thus cuts off from halves of the hour circles, which are parts of the day,[19] arcs that are similar to the arc it cuts off from the circle of the equator. So the same circle that passes through the two poles marks the same hour for all the days of the year; but this single circle [that bisects the diurnal arcs] cuts the horizon along the meridian line for that horizon, because the two poles of the world lie on the circumference of the horizon; so this circle always cuts the plane of a sundial parallel to the horizon along the same imaginary straight line parallel to the meridian line in the plane of the horizon. In the same way, each of the seasonal hours is cut off by a single great circle that passes through the two poles and cuts off similar arcs from all the seasonal circles; each of them is the same seasonal hour that is the same equinoctial hour. The <line of> intersection of each of these circles with the horizon is the meridian line. All the great circles that cut off the seasonal hours intersect one another on the meridian line of the horizon and these circles cut the plane of the sundial along straight lines, each of which marks a single one of the seasonal hours for all the days of the year. Each of these straight lines is parallel to the meridian line that lies in the plane of the horizon. The hour lines in horizontal sundials positioned in horizons on the equator are all straight lines, both in perception and in imagination, and they are all parallel. These are the results we wished to prove in this treatise.

This treatise is completed.
Thanks be given to God, Lord of the worlds.

---

[19] Lit.: partition.

CHAPTER II

HORIZONTAL SUNDIALS

2.1. INTRODUCTION

Horizontal sundials were among the most widely known instruments, the easiest to make, and also the most effective, since they told the time as long as the sun was shining. This may explain why Ibn al-Haytham began his investigations into sundials by considering this type of instrument. His treatise on horizontal sundials was written before the much more learned one about hour lines. In any case, there seems to be every indication that Ibn al-Haytham wanted to straighten out matters concerning this kind of dial before, in his capacity as a mathematician, embarking upon much more advanced researches on a general theory of sundials.

The two sundial texts by Ibn al-Haytham are not the same either in purpose or style. *Horizontal Sundials (Fī al-rukhāmāt al-ufuqiyya)* is a workman's handbook composed by a mathematician who provides no more explanation than the artisan will actually need in order to construct a sundial. Composing such handbooks was nothing new. Ibn al-Haytham's predecessor, Ibn Sinān,[1] had also written a manual on sundials for artisans. Ibn al-Haytham himself was not above writing handbooks for craftsmen, such as the work on geometry that he addressed to people who made measurements[2] (today we would call them surveyors). Here, as in the latter book, Ibn al-Haytham obviously wanted to provide a sound scientific basis for the artisan's practice, so that he would have a proper grasp of how the

---

[1] Ibn Sinān, *Fī ālāt al-aẓlāl*, edited with translation and commentary in R. Rashed and H. Bellosta, *Ibrāhīm ibn Sinān. Logique et géométrie au Xe siècle*, Leiden, 2000, Chap. IV.

[2] Ibn al-Haytham, *Fī uṣūl al-misāha*, edited with French translation and commentary in R. Rashed, *Les Mathématiques infinitésimales du IXe au XIe siècle*, vol. III: *Ibn al-Haytham. Théorie des coniques, constructions géométriques et géométrie pratique*, London, 2000, Chap. IV. English translation: *Ibn al-Haytham's Theory of Conics, Geometrical Constructions and Practical Geometry. A History of Arabic Sciences and Mathematics*, vol. 3, London, 2013.

instrument was constructed. In the following few pages we make some comments on Ibn al-Haytham's text.

2.2. MATHEMATICAL COMMENTARY

1. In this treatise, Ibn al-Haytham first turns to the procedures used to construct horizontal sundials, notably those for drawing the lines on the dial. Then, from his investigation of the craft practices, he proposes a short simplified procedure, a method the artisan can adopt, perfectly securely, to construct a sundial for a place of given latitude. Throughout his discussion, Ibn al-Haytham tacitly assumes that he has

$$\delta_m < \lambda < 90° - \delta_m,$$

where δ_m is the maximum declination of the sun and λ the latitude of the place considered.

First let us explain this condition, before we examine the procedures for making sundials.

Let us draw a diagram to represent the celestial sphere and let us show the circle centred on the place O, taken as the centre of the world, the zenith Z and the axis of the world OP.

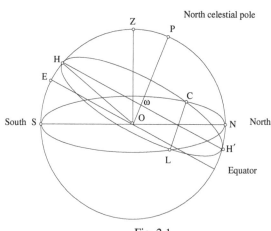

Fig. 2.1

The circle described by the sun in the course of its diurnal motion – uniform circular motion – lies in a plane parallel to the equator. Let L, C, H be, respectively, the points of sunrise, sunset and meridian passage corresponding to a declination δ; we have

$$E\hat{O}H = \delta, \ N\hat{O}P = Z\hat{O}E = \lambda.$$

The term *day* designates the interval of time that elapses between sunrise L and sunset C; its length varies over the course of the year. The day is determined by the length J of the arc LHC described by the sun above the horizon. It is divided into twelve seasonal hours, which are of equal length for a given day. But, from one day to another, the length of the seasonal hour varies. Each day we have hours with the same *number*. There are twelve numbers in order: thus sunrise is numbered 0, sunset is numbered 12, meridian passage, numbered 6, is called *midday*.

Ibn al-Haytham states – see below – that the arc LHC is known when we know δ and λ. He does not prove this result, he merely notes that one may find 'proofs in books on astronomy'.[3]

Thus the artisan needs to know the result but not necessarily how to prove it. In fact we need to prove that

$$\left|\cos\frac{J}{2}\right| = \tan\delta \cdot \tan\lambda.$$

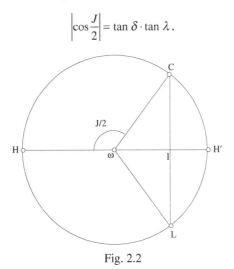

Fig. 2.2

[3] See below, p. 571; Arabic text in *Les Mathématiques infinitésimales*, vol. V, p. 827, 26.

Let ω be the centre of the circle described by the apparent motion of the sun. HH', the diameter of this circle in the plane of the meridian, cuts NS, the diameter of the horizon, in I. Let R be the radius of the celestial sphere, then we have

$$\omega H = R \cos \delta, \; O\omega = R \sin \delta, \; \omega I = O\omega \tan \lambda = R \sin \delta \tan \lambda.$$

The arc HC has the same size $\dfrac{J}{2}$ as the angle at the centre $H\omega C$. We have

$$\left| \cos \frac{J}{2} \right| = \frac{\omega I}{\omega C} = \frac{\omega I}{H\omega} = \frac{R \sin \delta \tan \lambda}{R \cos \delta} = \tan \delta \cdot \tan \lambda.$$

If $\delta > 0$, $J > \pi$, $\cos \dfrac{J}{2} < 0$; and if $\delta < 0$, $J < \pi$, $\cos \dfrac{J}{2} > 0$.

When the condition imposed on λ is satisfied, we have $\tan \lambda > 0$, hence

$$\cos \frac{J}{2} = -\tan \delta \tan \lambda.$$

For example, in a place with latitude $\lambda = 45°$, we have $\tan \lambda = 1$ and $\cos \dfrac{J}{2} = -\tan \delta$; if $\delta = -\delta_m = -23°27'$, we have $\cos \dfrac{J}{2} = 0.43378$ and $\dfrac{J}{2} = 64°18'$. The shortest day is measured by $J_{\min} = 128°36'$ and the longest day by $J_{\max} = 231°24'$.

Having explained this implicit condition and having given a proof for it, let us turn to the text by Ibn al-Haytham. From the first paragraphs, he is concerned with the construction of horizontal sundials. This type of sundial must be drawn on a surface that is perfectly plane and parallel to the horizon of the chosen place. The gnomon must be set up to be perpendicular to this plane. We then find the line for midday, that is the straight line on which the shadow of the tip of the gnomon falls, on every day of the year, when the sun passes through the meridian. Finally, in the course of a day we observe the sun at each seasonal hour of that day and make a mark to show the position of the shadow of the tip of the gnomon.

In the course of two *similar days* – days for which the declination of the sun is the same – the length of the day is the same and the positions of the points L and C on the horizon, as well as the position of the point H on the meridian, are also the same. Observations show that the point we obtain on the sundial is the same for any hour with the same number n; for $1 \leq n \leq 11$. But for days that correspond to different declinations, the points we obtain for the hour numbered n are different. This time, observations show that if we join up these points we obtain a curve that is only very little different from a straight line. Makers of sundials accordingly considered they could replace this curve with a straight line, which could be determined by two points. So that the distance of the other points from the straight line we draw shall be as little as possible, we determine the straight line not by means of any two points but by the two end points, which are the points obtained on the days of the summer solstice and the winter solstice, which correspond to $\delta = \delta_m$ and $\delta = -\delta_m$.

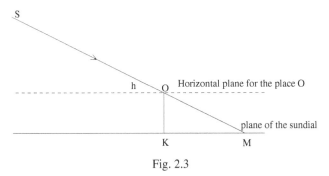

Fig. 2.3

It remains to draw the straight lines for the hours. For this we need to determine the lengths of the shadows that correspond to the extreme points. If we consider SO, the ray from the sun that, at any moment, passes through the tip of the gnomon OK – the point O being taken to be the centre of the world – this ray cuts the plane of the sundial at the point M, the tip of the shadow. If at that same moment, the height of the sun above the horizon is h, we have $h = O\hat{M}K$ in the right-angled triangle OKM. If $OK = d$, the height of the gnomon, we have

$$KM = l = \frac{d}{\tan h},$$

where l is the length of the shadow, or as Ibn al-Haytham expresses it 'the ratio of any gnomon to its shadow is equal to the ratio of the sine of the height of the sun, at that instant, to its cosine';[4] a formula that, as Ibn al-Haytham says, is known to those who have written about sundials. We can indeed find it in Thābit ibn Qurra, for instance in his book on sundials,[5] as well as in the works of many others.

Let H be the point of meridian passage of the sun, for a declination δ that can be positive or negative. Let us suppose the circle is in the direction $SEZPN$; we have $\widehat{PN} = \widehat{EZ} = \lambda$, $\widehat{EH} = \delta$, $\widehat{HZ} = \lambda - \delta$, and it follows that the required height $\widehat{SH} = \frac{\pi}{2} - (\lambda - \delta)$. So at the summer solstice, that is on the day when the sun is at the first point of Cancer, we have $\delta = \delta_m$ and $\widehat{SH} = \frac{\pi}{2} - \lambda + \delta_m$. At the winter solstice, on the day when the sun is at the first point of Capricorn, we have $\delta = -\delta_m$ and $\widehat{SH'} = \frac{\pi}{2} - \lambda - \delta_m$.

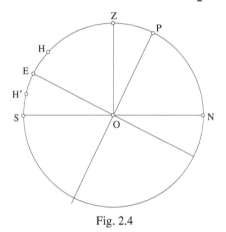

Fig. 2.4

Ibn al-Haytham next explains that, for each seasonal hour that corresponds to a declination δ, we need to know the right ascension and the azimuth of the sun. We then take as origins, for measuring right ascension

---

[4] See below, p. 569; Arabic text in *Les Mathématiques infinitésimales*, vol. V, p. 825, 1–2.

[5] *Fī ālāt al-sā'āt allatī tusammā rukhāmāt*, in *Thābit ibn Qurra, Œuvres d'astronomie*, text edited and translated by R. Morelon, Paris, 1987, pp. 133–4.

and azimuth, the points of intersection of the circle of the equator and the circle of the horizon with the local meridian; let these be the points E and S.

We suppose the sun is at the first point of Cancer and that H is the point at which it crosses the meridian ZSZ'.

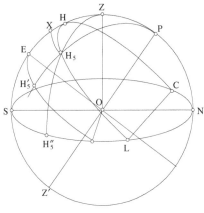

Fig. 2.5

We have seen that we know h, the height of this point; its right ascension and its azimuth are zero. Ibn al-Haytham then considers the situation when the position of the sun is H_5, that is its position at the fifth seasonal hour. The size of the arc $\widehat{HH_5}$ on the circle parallel to the equator is known. In fact, we have $\widehat{HH_5} = \frac{1}{6}\widehat{LH}$ where $\widehat{LH} = \frac{J}{2}$ is defined by

$$\cos\frac{J}{2} = -\tan\delta \cdot \tan\lambda.$$

The great circle PH_5 cuts the equator in H_5', so the arc $\widehat{EH_5'}$ is the right ascension of H_5; and the great circle ZH_5 cuts the horizon circle in H_5'', so the arc $\widehat{SH_5''}$ is the azimuth of the point H_5.

The parallel arcs $\widehat{HH_5}$ and $\widehat{EH_5'}$ have the same measure in degrees: $\widehat{EH_5'} = \widehat{HH_5} = \alpha_5$. So we know the right ascension of H_5.

When the first point of Cancer is at the point H_5, another point of the ecliptic is on the meridian circle, say at X. So there is an arc of the ecliptic, XH_5, that has a right ascension $\widehat{EH_5'}$, which is known. To a known right ascension there corresponds a known arc on the ecliptic. So the arc H_5X is a known arc and the point H_5 is known; so SX, its height above the horizon, is known. We can then determine the azimuth and the height of the point H_5 by applying Menelaus' theorem. Here again, Ibn al-Haytham does not consider it necessary to show the use of the theorem or to provide the proof, neither of which is required by the artisan.

We shall restore the missing proof. Let O be the place concerned; the circle of the ecliptic and the circle of the horizon for that place cut one another in a diameter MM'. The arc MXM' of the ecliptic is a semicircle that is cut in a point X by the meridian of O, the arc ZS. The point X is determined by H_5, the position of the first point of Cancer at the fifth hour, and we know the arcs SX and XH_5. We also know the arcs MS and MX.

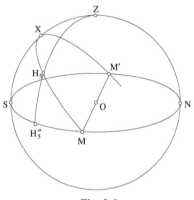

Fig. 2.6

We can apply Menelaus' theorem in two different ways to determine the arcs $\widehat{SH_5''}$ and $\widehat{H_5H_5''}$.

Let us put $\widehat{SH_5''} = x$ the azimuth of H_5, $\widehat{H_5H_5''} = y$ the height of H_5, $\widehat{MS} = a$, $\tan\widehat{XS} = c$; we have $\widehat{MH_5''} = a - x$, $\widehat{H_5Z} = \dfrac{\pi}{2} - y$ and $\widehat{XZ} = \dfrac{\pi}{2} - \widehat{XS}$. In the figure MH_5ZS Menelaus' theorem gives

$$\frac{\sin\widehat{MS}}{\sin\widehat{MH_5''}} \cdot \frac{\sin\widehat{H_5H_5''}}{\sin\widehat{H_5Z}} \cdot \frac{\sin\widehat{XZ}}{\sin\widehat{XS}} = 1,$$

that is,

$$(1) \qquad \frac{\sin a}{\sin(a-x)} \cdot \tan y = c;$$

and by applying the same theorem to the circle ZH_5H_5'' we have

$$\frac{\sin \widehat{ZS}}{\sin \widehat{ZX}} \cdot \frac{\sin \widehat{H_5X}}{\sin \widehat{H_5M}} \cdot \frac{\sin \widehat{MH_5''}}{\sin \widehat{H_5''S}} = 1;$$

now $\widehat{ZS} = \dfrac{\pi}{2}$ and $\widehat{SX}$ is known; so $\widehat{ZX}$ is also known. Moreover $\widehat{H_5X}$ and $\widehat{MX}$ are known; so $\widehat{H_5M}$ is also known. So the first two ratios are known; let $\dfrac{1}{k}$ be their product; the previous equality can be rewritten:

$$\sin (a-x) = k \cdot \sin x,$$

which can be rewritten

$$(2) \qquad \tan x = \frac{\sin a}{k + \cos a}$$

and from that we have the azimuth $x = \alpha_5$.

Thus we know $(a-x)$, and from (1) we have

$$\tan y = \frac{c\sin(a-x)}{\sin a} = c \cdot k \frac{\sin x}{\sin a};$$

from this we can find h_5 the height of the position $H_5 = y$.

So we may draw the conclusion that, on the day when the sun is at the first point of Cancer (the summer solstice), we know the right ascension and we know how to draw the azimuth and the height of H_5, the position of the sun at the fifth hour, that is one hour before it crosses the meridian of a given place. For the point H_5, we thus know: the right ascension a_5, the azimuth α_5 and the height h_5. Using the same method, we can find these three coordinates a_i, α_i, h_i for all the points H_i, where $1 \leq i \leq 5$. The positions of the points H_i, for $7 \leq i \leq 11$, which correspond to the positions of the first point of Cancer at the hours after midday, are symmetrical with the points H_5, H_4, ... , H_1 about the meridian circle ZS. Their right ascensions are arcs from the origin S and their heights are ZH_i; so two symmetrical points such as H_5 and H_7 have equal coordinates.

The same method can be used to investigate the points of the circle described by each of the points of the ecliptic that corresponds to the beginning of one of the signs of the Zodiac. So we can construct a sundial for a given horizon, since we can determine the azimuth and the height for the position of each of the relevant points of the ecliptic at each of the twelve seasonal hours.

After this investigation of right ascension and height for Cancer (or Capricorn) at a given hour, Ibn al-Haytham considers the ortive amplitude.

The figure shows the celestial sphere. The circle of the equator and the circle of the horizon are perpendicular to the meridian plane PZS; so their line of intersection $O\Sigma$ is perpendicular to this plane and to any straight line in it. So we have that ΣO is perpendicular to NS. The point Σ marks due East for the horizon of the place O; and we have $\widehat{\Sigma N} = \widehat{\Sigma S} = \dfrac{\pi}{2}$.

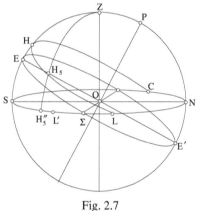

Fig. 2.7

The first point of Cancer, with positive declination, rises at the point L to the North of Σ. The arc ΣL is the ortive amplitude, and we have $\widehat{LN} = \dfrac{\pi}{2} - \widehat{\Sigma L}$ and $\widehat{SL} = \dfrac{\pi}{2} + \widehat{\Sigma L}$. This arc SL is the azimuth of the point L where Cancer rises. For the first point of Capricorn, which has negative declination, the point of rising L' lies South of Σ and we have

$$\widehat{L'N} = \widehat{L'\Sigma} + \frac{\pi}{2} \text{ and } \widehat{SL'} = \frac{\pi}{2} - \widehat{L'\Sigma};$$

this arc SL' is the azimuth of the point L' where Capricorn rises.

2. Once he has explained how to find the azimuth and the height of the position of the sun at the seasonal hours of each of the days that correspond to the first point of each of the signs of the Zodiac, Ibn al-Haytham is then ready to set out the rules of the method the artisan should employ to construct a horizontal sundial for a given place. He twice refers back to how to use the results already established to find the position of the tip of the shadow of the gnomon.

Constructing a sundial in fact reduces to finding the straight lines for the hours, the *hour lines*. So it is enough to know two points to fix each of them. It is better if these two points are extreme ones, that is if, for each hour, one of the points corresponds to the positions of the first point of Cancer, and the other to positions of the first point of Capricorn. Ibn al-Haytham gives a detailed description of the successive stages of making a dial. First of all, we must draw up a table which, for each of the first points of signs of the Zodiac, shows its azimuth and its height at each seasonal hour of the day that corresponds to it.

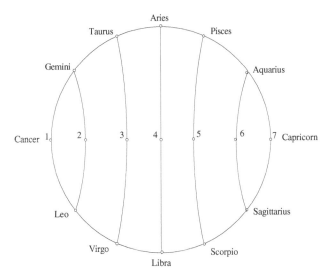

Fig. 2.8.1

6th hour	5th hour	4th hour	3rd hour	2nd hour	1st hour
$h_{1,6}$	$\alpha_{1,5}$	$h_{1,5}$			Cancer
					...
	$\alpha_{3,5}$	$h_{3,5}$			Taurus Virgo
					...

Fig. 2.8.2

The table has six columns and seven rows. On the rows we mark the signs of the Zodiac in order of their declinations, starting with Cancer, which corresponds to the declination δ_m, and ending with Capricorn, which has declination $-\delta_m$. The columns correspond to the seasonal hours. Each column – apart from the sixth – has two parts; in one is written the azimuth and in the other the height. At the sixth hour, the azimuth is zero because this is meridian passage (see also Fig. II.1, p. 575).

We then make a 'master circle': on a copper plate we draw a circle with centre E and two perpendicular diameters AC and BD.

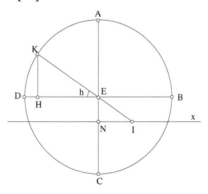

Fig. 2.9

The circle is graduated in degrees. Let d be the height of the gnomon to be used in the sundial. On AC we mark the point N such that $EN = d$, and through N we draw a line parallel to DB; let it be Nx.

This 'master circle' can be used to find the length of the shadow of the gnomon at a given moment, as follows:

Let h be the height of the sun at the moment in question and K a point on the circumference such that $\overset{\frown}{DK} = h$. We draw $KH \perp DB$; when extended, the straight line KE cuts Nx at the point I. The right-angled triangles KHE and ENI are similar; we have $\dfrac{KH}{HE} = \dfrac{EN}{NI} = \tan h$. So if KE represents the ray of light from the sun, at height h, that reaches E, the tip of the gnomon, and meets the straight line Nx that represents the plane of the dial, then NI is the length of the shadow

$$NI = l = d \cot h.$$

To construct the dial, we take a perfectly plane plate that we fix in position so that its surface is parallel to the horizontal plane of the place. On the plate we find the straight line that will be the meridian line. To do this, we choose a point on the plate, say J; at that point we set up a gnomon and in the course of the day we find two shadows that are of the same length, let them be JM and JM'. The line JU that bisects the angle MJM' is the meridian line for the point J.

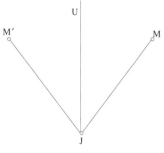

Fig. 2.10

We then take the point J and on the plate we draw a circle with centre J equal to the master circle, let UJO be its diameter (the meridian line). It is clear that, if a gnomon GJ is perpendicular to the plate at the point J, at the sixth hour the shadow of its tip G will fall on JU.

We first take the sun to be at the first point of Cancer and we investigate the shadow associated with the fifth hour for a gnomon of height d equal to EN the length used for the master circle. We then take from the table the azimuth $\alpha_{1,5}$ and the height $h_{1,5}$ that correspond to this case (Cancer, fifth hour) and we place on the circle (J, JU) the point P such that $\widehat{UP} = \alpha_{1,5}$. The straight line JP is the straight line of the azimuth.

On the master circle we then carry out the required construction taking $h = h_{1,5}$ and we obtain l the length of the shadow of the gnomon of height d. We place that length l on the azimuth line JP, let it be $JN_5 = l$. The point N_5 we obtain in this way is the tip of the shortest shadow for a fifth hour.

We repeat the same process for the fifth hour for Capricorn. The table gives us the azimuth $\alpha_{7,5}$ for this case. On the circle on the dial we mark the point S such that $\widehat{US} = \alpha_{7,5}$ and we draw JS on which we mark out the length $JM_5 = l'$, a length we obtain on the master circle by taking $h = h_{7,5}$. The point M_5 is the tip of the longest shadow for a fifth hour. We then draw the straight line M_5N_5 on the dial; this is the straight line for the hours numbered five. At the fifth hour of every day, the tip of the shadow of a gnomon of height d falls at a point very close to that straight line.

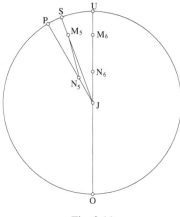

Fig. 2.11

We repeat the same process for every hour of the day for Cancer and for Capricorn. One by one we mark the points M we obtain. We find that the points which correspond to two successive hours are very close to one another. Thus we may regard the points as forming a locus that is a curve.

Ibn al-Haytham points out that this curve is a conic, but does not pause to justify the remark.

Let us show that the curve is a hyperbola.

Each day the sun, S, describes a circle parallel to the equator. The point G, the tip of the gnomon, is taken as the centre of the universe; so, as it moves, the ray from the sun, SG, sweeps out the surface of a cone whose axis is the diameter that passes through the poles of the celestial sphere. For two opposite declinations, for example for the first points of Cancer and Capricorn, we obtain the two sheets of the surface of the same cone. For the chosen place, we have, on the intersection of the plane of the sundial with one of the sheets, the set of points N associated with the first point of Cancer; and on the intersection with the other sheet we have the points M.

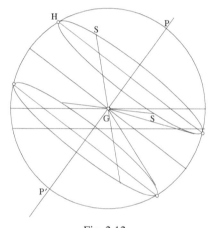

Fig. 2.12

If the place considered is at a Northern latitude greater than δ_m, the maximum declination of the sun, and less than its complement, $\delta_m < \lambda < \dfrac{\pi}{2} - \delta_m$, that is, if we are considering a place that lies between the tropic of Cancer and the Arctic circle, the conic sections we obtain, for the days when the sun is at the first point of Cancer and that of Capricorn are the two branches of a hyperbola. We can repeat this procedure for the first point of each of the signs of the Zodiac. The first point of Leo and that of

Gemini lie on either side of Cancer and have the same declination δ. The line of the points N that correspond to these two first points is the same.

The first point of Aquarius and that of Sagittarius lie on either side of Aries and have the same declination $-\delta$. The line of the points M that correspond to them is the same.

The lines we obtain in this way for Leo and Gemini, on the one hand, and for Aquarius and Sagittarius on the other, are the two branches of the same hyperbola. The same holds for Taurus and Virgo on one side and Pisces and Scorpio on the other.

The first point of Aries and that of Libra have declination zero. When the sun is at one of these positions, its diurnal motion lies in the plane of the equator and in the course of each of these two days the tip of the shadow of the gnomon describes on the horizontal plane of the sundial a straight line perpendicular to the straight line UJ.

So we see that each of the hyperbolas we have considered corresponds to a Solar declination δ, which is the declination of the first point of one of the signs of the Zodiac. The point N corresponds to a positive declination δ and the point M to the opposite declination $-\delta$. It remains to find the lengths JN and JM associated with δ and $-\delta$ for latitude λ.

Let h and h' be the heights of H and H', the points of meridian passage, where $\widehat{EH} = \delta$, $\widehat{EH'} = -\delta$ and $\widehat{EZ} = \lambda$.

We have

$$h = \widehat{SH} = \frac{\pi}{2} - \widehat{HZ} = \frac{\pi}{2} - (\lambda - \delta),$$

so $JN = d \cot h = d \tan (\lambda - \delta)$, where d is the height of the gnomon; moreover

$$h' = \widehat{SH'} = \frac{\pi}{2} - \widehat{H'Z} = \frac{\pi}{2} - (\lambda + \delta),$$

so

$$JM = d \cot h' = d \tan (\lambda + \delta).$$

For Aries and Libra we have $\delta = 0$; so M and N coincide at the point M_0 such that $JM_0 = d \tan \lambda$.

For the values of $|\delta| \neq 0$ in question, the points M and N lie on opposite sides of M_0.

Finally, in the plane perpendicular to the dial, on the straight line JU, which contains the gnomon JG with tip G, we have $J\hat{G}M_0 = \lambda$, $N\hat{G}M_0 = M_0\hat{G}M = \delta$ and the ray GM_0 bisects the angle between the two rays GN and GM.

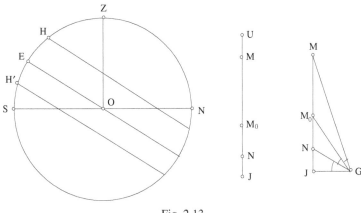

Fig. 2.13

So Ibn al-Haytham has just set out, as he himself expresses it:

> the subject as a whole and the principles on which the construction of sundials is based, and to draw attention to the manner of the construction and the parts where we need these ideas to which reference is not often made except in the books of specialists on shadows.[6]

In fact it was a matter of setting out the geometrical principles that underlie, and validate, the instrument makers' practice. This mathematico-astronomical understanding is required for making such instruments. Thus Ibn al-Haytham sets out what is strictly necessary, and does not concern himself with the mathematical theory of sundials, on which he intended to write a separate book. Ibn al-Haytham kept this promise and wrote *On the Hour Lines*, in which – as we shall see – he provides a magisterial development of the theory of the subject.

---

[6] See below, p. 581; Arabic text in *Les Mathématiques infinitésimales*, vol. V, p. 849, 1–4.

2.3. HISTORY OF THE TEXT

Fī al-rukhāmāt al-ufuqiyya (*On Horizontal Sundials*) is the title used by the old biobibliographers – al-Qifṭī, Ibn Abī Uṣaybiʿa and the anonymous scholar of Lahore – to designate this treatise in the list of works by Ibn al-Haytham before 1038.[7] Further, Ibn al-Haytham tells us that this treatise was composed before *On the Hour Lines*, by referring forward to the second at the end of the first. So both of them were composed before *The Burning Sphere* (*Fī al-kura al-muḥriqa*).

The treatise has come down to us in two manuscripts:

1. Collection 2970/9, fols 153ᵛ–161ʳ, in the Staatsbibliothek in Berlin,[8] copied by Qāḍī Zādeh during the 1430s. This manuscript is identified as B.

2. Collection Tugābunī 110, fols 1–19, in the Majlis Shūrā in Teheran. This is a 581-page collection of scientific writings. We know very little about the history of this collection. This manuscript is identified as I.

A comparison of B and I shows that B is missing a sentence and four words, whereas in I there are six omissions, each of one word. So we know that B cannot be the immediate ancestor of I.

We have established the Arabic text from these two manuscripts in *Les Mathématiques infinitésimales*, vol. V, pp. 821–49.

---

[7] *Ibn al-Haytham and Analytical Mathematics.* A History of Arabic Sciences and Mathematics, vol. 2, Culture and Civilization in the Middle East, London, 2013, pp. 416–17.

[8] We have traced the history of this collection in *Ibn al-Haytham's Theory of Conics, Geometrical Constructions and Practical Geometry*, vol. 3, pp. 423–5.

TRANSLATED TEXT

Al-Ḥasan ibn al-Haytham

On Horizontal Sundials

In the name of God, the Compassionate the Merciful

TREATISE BY ABŪ ʿALĪ AL-ḤASAN IBN AL-ḤASAN IBN AL-HAYTHAM

On Horizontal Sundials

A sundial is a plane of known position[1] that has a raised gnomon and lines such that, if the tips of the shadows of the gnomon fall on these lines, they indicate the seasonal hours that have elapsed so far during that day. The purpose of having a sundial is to know the hour <of day>. One can put it to other uses if one adds markings other than hour lines, except that it is agreed that, among its ordinary uses, one is to let us know, at any moment, how much of the day has elapsed and how much remains, and the moments of midday. A seasonal hour is one of the twelve <equal> parts of the duration of the length of the day to which this hour belongs. Seasonal hours have a different length each day because the duration of the day is not the same for every day. The way in which, to begin with, experts on shadows constructed sundials was as follows: they set up a plane parallel to the horizon, on which they found the meridian line,[2] erected a gnomon perpendicular [to the plane] and positioned on the meridian line; if its shadow fell on the meridian line, they deduced that the sun had reached the meridian circle and that the part of the day that had elapsed was equal to what remained; they then observed the sun every day with an astrolabe, or something that could take the place of one, and watched the sun attentively until one seasonal hour of the day had elapsed, a seasonal hour being one part of the twelve <equal> parts of the diurnal arc for that day; they examined the shadow of the gnomon erected on the sundial and made a mark at the position of the tip of the shadow; then they watched the sun and the shadow attentively until two hours of the day had passed; then they similarly made another mark at the tip of the shadow; they proceeded in the same way for the other hours until they obtained marks for the remaining

---

[1] Later it is specified that the plane is parallel to the horizon.

[2] See Mathematical commentary.

hours on the surface of the sundial. They then did this for every day until
the sun reached its extreme position closest to the zenith and its extreme
position furthest from it [at noon]. In this way they obtained many points,
each of which indicates an hour of one of the days. Then they considered
the shadow and they found that for each day, as an hour passed, the tip of
the shadow fell on the mark they had made on the surface in the course of
the day that was similar [in length] to this day of the year. So these marks
provided a rule that allowed them to know which <seasonal> hour of the
day has <just> elapsed. They next examined these points and they then
found that the points that defined corresponding hours[3] for all the days lie
on a line which does not depart much from a straight line; moreover, all the
points defining corresponding hours lay on a line that, as far as could be
perceived, was very close to a straight line; accordingly they later drew
corresponding hours as straight lines. So if the tips of the shadows fell on
them [the lines], that indicated the hours. Since they used that a great deal,
it led them, when they wanted to construct a sundial, to carry out
observations like this for one day of <each of> the parts of the year and for
all the corresponding hours to determine two points that they joined with a
straight line, which they took as the line for corresponding hours; indeed
for a straight line, if we have two points on it, we have the whole line. So
they relied upon this method.

But since they imagined that the points for corresponding hours do not
lie exactly on a straight line, and that if we join two neighbouring
corresponding points by a straight line, and if we extend that straight line,
there is no guarantee that, as it is extended further, the discrepancy does not
increase, thus it seems they looked for the two points that mark the extreme
positions of the corresponding points for the tips <of two shadows> and
they joined them with a straight line which they took as the line for
corresponding hours. Once this was established, then, for each of the
straight lines for hours, the person making the sundial[4] looked for the two
points that are the two endpoints of the line. But since it was difficult for
the maker of a dial to wait until the sun reached the extreme point of its
declination and returned to the other extreme point – because it is at these
two extreme points of the sun's declination that we find the extreme points
on which the tips of the shadows fall – makers of dials turned to
geometrical investigation to find the points that are the ends of the hour

---

[3] Corresponding hours for two different days correspond to the same number of
twelfths of the length of the day for each of the days concerned. So we are considering
hours with the same number.

[4] Lit: dial of hours.

lines. They recognized, by means of a general proof, that the ratio of any gnomon to its shadow is equal to the ratio of the sine of the height of the sun, at that instant, to its cosine. So the straight line on which the shadow lies is either the meridian line or a straight line that makes with the meridian line an angle that intercepts the arc of the horizon lying between the meridian line and the arc of the height <of the sun>, which is called the arc of the azimuth.

Proof: The sun always lies on one of the circles of azimuth[5] and the gnomon is always in <the plane of> each of the azimuth circles, because its tip serves as the centre of the world;[6] it lies on the extension of the straight line of the mid heaven[7] and the ray from the sun to the tip of the gnomon is always a diameter of the azimuth circle that passes through the sun at that moment. But since the gnomon and the ray at that moment both lie in the plane of the azimuth circle, the shadow will lie along a straight line in the plane of the azimuth circle, because it lies in the same plane with these two straight lines, that is the gnomon and the ray; so the shadow always lies along the straight line that is in the plane of the azimuth circle and in the plane of the horizon; so it[8] ends at the endpoint of the arc of the altitude, because it is the line of intersection of the horizon and the azimuth circle, which is either the meridian line, or a line such that the arc lying between its endpoint and the meridian line is the arc that lies between the meridian line and the arc of the altitude; that is the arc called the arc of the azimuth.

Moreover, since the sine of the height is the perpendicular dropped from the sun onto the diameter of the azimuth circle that passes through the tip of the gnomon, and it [this perpendicular] is parallel to the straight line from the mid heaven drawn as the extension of the gnomon, and since the diameter of the azimuth circle that passes through the tip of the gnomon is parallel to the shadow – because they lie in two parallel planes, namely the

---

[5] The shadow lies on the line of intersection of the horizontal plane of the sundial and the plane of the circle through the zenith defined by *S* the position of the sun. This circle through the zenith is sometimes called the altitude circle, because the arc that measures altitude is part of it; and sometimes the azimuth circle because it cuts the circle of the horizon in a point that defines the azimuth. Thus, if *S* is in the plane of the meridian, the shadow is on the meridian line, if not, it is on a straight line that defines the azimuth. The azimuth circle is a horizontal circle along which the azimuths of points on the sphere are measured. We have translated *al-dā'ira al-samtiyya* as circle of azimuth or azimuth circle.

[6] That is, it is the centre of our imagined celestial sphere – making the usual approximations, *e.g.* taking the radius of the Earth as negligibly small.

[7] The vertical at the place concerned.

[8] That is the straight line along which the shadow lies.

horizon, which passes through the tip of the gnomon, and the plane of the
dial – and they lie in the plane of the azimuth circle, accordingly the
straight line of each ray forms with these straight lines two triangles that
are similar, because their sides are parallel; the ratio of the sine of the
height to the cosine, that is the straight line between the foot of the
perpendicular and the tip of the gnomon, is thus equal to the ratio of the
gnomon to the shadow.

Once this had been proved, they wrote it into their books, together with
the proofs, and to find the <positions of the> tips of the shadows they made
use of these two lemmas,[9] because these were sufficient for their purpose.
So, to construct the dial, the artisan who made sundials needed to know the
height of the sun at the hours for which he wanted to find the <positions of
the> tips of the shadows, and to know the arc of the circumference of the
horizon that is cut off between the line of the shadow and the meridian line,
[the arc] that is called the arc of the azimuth. This is why someone who
wanted to construct a sundial began by knowing the height and the arcs of
azimuth for each of the moments for which he wanted to make marks on
the dial. The method of finding this is to suppose, to employ the method of
analysis, that the sun is at the extreme point of its declination, that is to say,
at the first point of Cancer or of Capricorn, and that we are at midday. So
the shadow falls on the meridian line. We begin with midday because that
is easier. It is then that the first point of Cancer or of Capricorn lies on the
meridian circle and the height of the sun at this moment is the height of the
first point of Cancer or of Capricorn; but the height of the first point of
Cancer or of Capricorn in the assumed position on the Earth at midday is
known, because it is the arc of the meridian circle that lies between the
point through which the first point of Cancer or Capricorn passes and the
horizon. This arc is known, because the inclination of the first point of
Cancer or Capricorn with respect to the circle of the equator is known, the
distance from the zenith of the known horizon to the equator is known, so
their sum – or the amount by which one exceeds the other – is known,
which is the distance of the first point of Cancer or Capricorn at this
moment in relation to the zenith. If a quarter of a circle is taken away from
it, the remainder is the height of the first point of Cancer or Capricorn at
that moment. If the sun is on the meridian circle[10] and it is at the first point
of Cancer or Capricorn, the height of the sun is known. And this is one of

---

[9] First lemma: the points marking the tips of the shadows for corresponding hours
lie on a straight line. Second lemma: the ratio of the height of the gnomon to the length
of its shadow is equal to the ratio of sin h to cos h, where h is the height of the sun at the
moment concerned.

[10] Lit.: on the line of midday.

the moments for which we try to find the shadows. We then suppose that the sun is at the first point of Cancer and we imagine there is an interval of a single seasonal hour between it and the meridian circle; now the seasonal hour for this moment is one of the known parts, because it is one part of the twelve parts of the diurnal arc at the first point of Cancer for this horizon; but the diurnal arc for a known degree <of the Zodiac> for a known horizon is known, because it has been established through proofs in books on astronomy. The distance between the sun and the meridian circle <measured> along the circle parallel to the equator is known and is the seasonal hour; and at this moment, between the sun and the meridian circle there is an arc of the circle of the ecliptic; so the arc that represents the seasonal hour of known magnitude is the ascension of this arc[11] of the circle of the ecliptic, which lies between the sun and the meridian circle, at a place that has the meridian circle as a horizon, <a place> that lies on the equator.

But the ascensions of the parts of the circle of the ecliptic on the equator are known; so the known ascensions are ascensions of known parts of the circle of the ecliptic, on it. Consequently, the arc between the sun and the meridian circle along the circle of the ecliptic is known, and the sun is taken to be at the first point of Cancer; so the point of the circle of the ecliptic that lies on the meridian circle is known from its height which is the arc of the meridian circle that lies between this point and the horizon, and which is known, because the inclination of this point is known and the distance of the zenith from the equator is known. But this point of the circle of the ecliptic is on the line of the mid heaven[12] at this moment. If, on the line of the mid heaven for a known horizon, a part of the circle of the ecliptic is known, the ascendant at this moment is known. The arc of the circle of the ecliptic that lies between the horizon and the meridian circle for this moment is known and this arc is divided by the position of the sun in a known ratio; so the arc of the height at this moment is known – that has been proved from the 'sector figure'.[13] The height of the sun at the moment when there is a single hour between it and the meridian circle, when the sun is at the first point of Cancer, is thus known.

---

[11] Ascension of an arc: the difference between the right ascensions of its two endpoints measured along the circle of the equator from a fixed origin. In this text, one of the endpoints of the arc concerned is taken as the origin (the point of intersection of the equator and the meridian south of the place in question): so the ascension of the arc is the right ascension of its second endpoint.

[12] It lies on the meridian circle through the zenith and the pole of the equator; this circle is called the mid heaven.

[13] A proof by Menelaus.

In the same way, it can be shown that the height is known if there are two hours, or more, between the sun and the meridian circle, because <for each> of the hours between the sun and the meridian circle, if the hour is known, the arc of the circle of the ecliptic that lies between the sun and the meridian circle is known, since these hours are its right ascensions. The mid heaven for the circle of the ecliptic is thus a known point, the ascendant is also known and, as has been proved, the height is known.

In the same way, if we suppose the sun is at the first point of Capricorn or at an arbitrary given point of the circle of the ecliptic, the heights of the hours will be known because the inclinations of known points of the circle of the ecliptic are known and their ascensions are known, and from these one can find the magnitude of the height. It is by this method that all the heights have been determined for the moments for which we want to know the <positions of the> tips of the shadows. As for the azimuths, since it has been proved that the ascendant, <found> from the circle of the ecliptic, for the given hour corresponds to a given point, then the ortive amplitude is known and it is the arc of the horizon between this point and the meridian circle;[14] thus if we take away this arc from a quarter of a circle, what remains is the arc of the horizon between this point and the meridian circle, towards the north or the south. And since the circular arc of azimuth,[15] which runs between the zenith and the horizon, is a quarter of a circle and since it is divided by the position of the sun in a known ratio; since the arc of the circle of the ecliptic that we have introduced is also known and is divided by the position of the sun in a known ratio, as we have proved, accordingly the arc of the horizon that lies between the circle of the ecliptic and the meridian circle – we have proved that it is known – is divided by the azimuth circle in a known ratio, because this can also be proved by the 'sector figure'.[16] Thus the arc between the azimuth circle and the meridian circle is known; it is the arc called the azimuth and the straight line drawn from the centre of the horizon to the endpoint of that arc is what is called the straight line of the azimuth; this method was also used to find all the azimuths for the given hours.

Now, when they knew the azimuths and the arcs for heights, they drew a circle with its centre at the base of the gnomon, which they divided into three hundred and sixty parts, and <working> from the meridian line, in the direction towards the position of the azimuth circle at the moment for

[14] The meridian circle towards the east.

[15] Here, the azimuth circle is the circle through the zenith that passes through the position of the sun, meeting the horizon at the point that defines the arc that is called the azimuth.

[16] That is to say by Menelaus' theorem.

which they took the sun to be at the first point of Cancer, and whose distance from the meridian circle is a single hour, they marked off the magnitude of the arc of azimuth for this moment, an arc they had found by proof[17] and calculations. They then joined the centre of the base of the gnomon to the endpoint of this arc with a straight line; so this straight line is the line of the azimuth in the plane of the dial for this moment, and it is on this line that the shadow of the gnomon falls, at this moment, because it [the shadow] is in the plane of the azimuth circle and the sun and the gnomon are also in the plane of the azimuth circle. Then, starting from the centre of the base of the gnomon, they cut off from the azimuth line a straight line whose ratio to the gnomon was equal to the ratio of the sine of the height to its cosine – we shall show later how to do this – and thus, by this procedure, they obtained a point in the plane of the dial, a point that is the tip of the shadow that marks the fifth hour, because they had assumed there was a single hour between the sun and the meridian circle; then, also starting from the meridian line, they took the azimuth arc for the fifth hour when the sun is at the first point of Capricorn, they drew the straight line for the azimuth and found the straight line whose ratio to the gnomon is equal to the ratio of the sine of the height, at this moment, to its cosine; they then obtained another point on the plane of the dial which also marks the fifth hour.

We have already said that they believed that the points that mark corresponding hours lie on a single straight line, as far as the senses can perceive, and that they set out to find the endpoints of this straight line, which are the two points that are the endpoints of the straight line that indicates the fifth hours, and that they joined these points with a straight line which they put in place as a marker for the fifth hours. They did the same for each of the remaining hours and in this way they obtained straight lines indicating the twelve hours.[18]

This is an account of the method that was used to find the hour lines in the planes of sundials; now it has been shown that to construct the dial, we need to know the inclinations, the ortive amplitude,[19] the arc of the height, the arc of the azimuth and the straight azimuth line. These things can be found by calculation and they can also be found by means of an instrument.

---

[17] The sense may seem to require a term such as 'mathematical argument', but the Arabic term used here is *burhān*, which means 'proof'.

[18] The twelfth hour corresponds to sunset; the ray joining the sun to the top of the gnomon is horizontal so that point has no shadow on the plane of the dial.

[19] The ortive amplitude, that is, the arc from the point of rising, has not been used in the calculations (see Mathematical commentary).

This has been mentioned in their books by many specialists in the science of shadows and they have referred to it.

But since what is said about this is repeated in books, we shall dispense with presenting it again here.

Let us now summarize the method of constructing sundials and let us set it out so as to make it easily accessible to anyone who wants to apply it. The first thing, with which the maker of the sundial must start, is to find the arcs for the height [*sc.* of the sun] for each of the hours of the day, when the sun is at the first point of Cancer, for the horizon for which we wish to construct the sundial, this by means of calculation, as has been shown in the astronomical tables; he finds that, similarly, when the sun is at the first point of Capricorn, he also finds, by means of calculation, the azimuth arcs for these hours and he finds the height and the azimuth for each hour <of the days> when the sun is taken to be at the first points of all the signs of the Zodiac. And to make the construction easier, he draws up a table, whose length he divides into seven parts, and he places in the first part the first point of Cancer, in the second the first point of Gemini and that of Leo, in the third the first point of Taurus and that of Virgo, in the fourth the first point of Aries and that of Libra, in the fifth the first point of Pisces and that of Scorpio, in the sixth the first point of Aquarius and that of Sagittarius and in the seventh the first point of Capricorn. The width of the table is divided into six parts, in which the hours are written, in order, from the first hour to the sixth hour, which is midday, then each of the parts of the width, except the sixth part, is divided into two parts; in one of them we write the height and in the other the azimuth and in the part reserved for the sixth hour only the height, because for the height for the sixth hour there is no azimuth arc, given that it is midday. Then this table is filled in with all the heights and azimuths that have been found by calculation, each in its place; so we write opposite the first point of Cancer, under the first hour and in the part reserved for the height, the parts of the arc of the height,[20] which has been found for the first hour, when the sun was at the first point of Cancer. But the first hour is the hour whose distance from the meridian circle is five hours. Under the first hour we also write, in the part reserved for the azimuth, the parts of the arc of the azimuth,[21] which has been found for the first hour. We write under the second hour the height and the azimuth that have been found for it; similarly under the third, fourth and fifth hours. We write under the sixth hour the height on the meridian; we

---

[20] The measure of the height, in degrees.
[21] The measure of the azimuth, in degrees.

do the same thing for each of the signs of the Zodiac. This is the form of table we need to use in constructing sundials, which is for example:

sixth hour	fifth hour		fourth hour		third hour		second hour		first hour		Signs of the Zodiac
height	azimuth	height	azimuth	height	azimuth	height	azimuth	height	azimuth	height	
											first point of Cancer
											Gemini and Leo
											Taurus and Virgo
											Aries and Libra
											Pisces and Scorpio
											Aquarius and Sagittarius
											first point of Capricorn

Fig. II.1

We draw a circle on a flat piece of copper or on a solid body, in the circle we draw two diameters that cut one another at right angles, we divide the circle into three hundred and sixty parts and starting from the centre of the circle we cut off on one of the diameters that cut one another [at right angles] a straight line equal to the magnitude of the length of the gnomon that we wish to erect on the plane of the sundial; from the point where the cut was made we draw a straight line parallel to the other diameter and we call this circle the master circle.[22] Once all this is done, we then begin by setting up a smooth surface parallel to the horizon[23] or to a known horizon, and we adjust it as exactly as possible; in this surface we find the meridian line in the usual way that is by taking <note of> two equal shadows, on the same day and for a single gnomon, one to the east and the other to the west; we divide the angle they enclose into two equal parts with a straight line. That straight line is thus the meridian line, as has been proved, in its place, in the books of mathematicians. We then choose a point on the meridian line, and in this plane we draw a circle equal to the master circle, then we return to the table to find the number of parts in the arc of the azimuth at the fifth hour when the sun is at the first point of Cancer; then from the master circle we take off parts equal to these parts, which we measure with the opening of compasses, then, using these compasses with this opening,

[22] For the master circle, our figure will be drawn to show the case in the example considered by Ibn al-Haytham.

[23] Lit.: his horizon. The maker's horizon.

we cut off an arc on the circle drawn on the sundial, starting from the meridian line, using this opening; this arc will be equal to the arc of the azimuth we found in the table. We join the centre of the circle to the endpoint of the arc with a straight line, that straight line will then be the azimuth line for the fifth hour when the sun is at the first point of Cancer, and it is on this line that the shadow of the gnomon falls at that moment. We then return to the table, as before, to find the arc of the height for the fifth hour when the sun is at the first point of Cancer, starting from the diameter we have not divided, on the circumference of the master circle we cut off an arc whose magnitude is equal to the number of parts that are in the height[24] at that moment. We then join the endpoint of that arc and the centre of the circle with a straight line that we extend until it meets the straight line drawn from the point of division parallel to the diameter. The ratio of the length of the gnomon to the straight line cut off from the straight line parallel to the diameter is thus equal to the ratio of the sine of the height to its cosine.

Proof: On the master circle let us use the letters *ABCD*, and at the centre *E*, and at the point that marks the length of the gnomon, *N*; at the point in which the straight line parallel to the diameter was divided, *I*; at the endpoint of the arc that is equal to the arc of the height on the azimuth circle, the point *K*, and starting from *K* the perpendicular *KH*; then *KH* is the sine of the height and *HE* its cosine; but the ratio of *KH* to *HE* is equal to the ratio of *EN* to *NI*, because the two triangles are similar, *EN* is the length of the gnomon, and *NI* is the length of the shadow because the point *K* stands for the position of the sun, the straight line *KEI* indicates the ray and the straight line *EN* indicates the gnomon; so the straight line *NI* is the shadow.

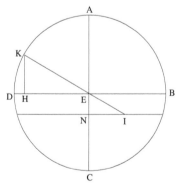

Fig. II.2

---

[24] That is the measure of the height in degrees.

So if we find that straight line, we take its measure with compasses and from the azimuth line in the <plane of> the sundial, starting from the centre of the circle drawn on the surface of the sundial, using the two legs of the compasses, we cut off a straight line equal to the straight line that has been found; that straight line is accordingly the length of the shadow at that moment. Thus we obtain on the surface of the dial a point that is one end of the straight line that marks the fifth hours, because this shadow is the shadow associated with one of the two extreme points for the declination of the sun, and the sun does not exceed this declination. This straight line is the shortest shadow for the fifth hours because at this moment the sun is closest to the zenith and the shadow shortens only if the sun moves closer to the zenith. Then, again on the circumference of the master circle, we take an arc equal to the arc of the azimuth, at the fifth hour, when the sun is at the first point of Capricorn; we measure it with compasses and, starting from the meridian line, we cut off an equal arc from the circle on the dial. Starting from the centre we draw a straight line to the endpoint of the arc. This straight line will be the line of the azimuth at that moment; we then return to the master circle, from which we cut off an arc equal, <measured> in parts, to the arc of the height at this moment, parts which are to be found in the table. We join the endpoint of the arc to the centre, thus we have the straight line that is the length of the shadow. We cut off a piece equal to it from the line of the azimuth, we thus obtain the final point of the straight line on which the shadow falls at the fifth hours. Now we join the two points with a straight line; accordingly this straight line will be the straight line that marks the fifth hours. We do the same thing for the remaining hours, so as to find the straight line that marks each of the fourth, third, second and first hours. In this way we obtain, on one of the two sides of the sundial, five straight lines that indicate five hours and the meridian line that indicates the sixth hours.

In the same way, on the master circle, we again take a <number of> parts equal to the <number of> parts of the height on the meridian, starting from the first point of Cancer, an arc that is in the table, and we join its endpoint to the centre of the master circle. We find the shadow for noon of the day starting from the first point of Cancer; we cut off this same length on the meridian line, which on the sundial runs from the centre of the <master> circle.

Similarly, we find the shadow for noon of the day starting from the first point of Capricorn. We cut off this same length on the meridian line, by this means we obtain the straight line that defines the two extreme positions of <the tips of> the midday shadows on the sundial.

Then, working in the other direction on the dial, we cut off, from the circle in the plane of the dial, arcs equal to the azimuth arcs that lie in the first direction and we join their endpoints to the centre of the circle with straight lines; these give us the remaining hours because the distance of each of the hours to the west from the meridian circle is equal to the distance <from the meridian> of the corresponding hour among those to the east. So the azimuth, for the hour to the east, is equal to the azimuth for the hour to the west that corresponds to it. Similarly, it necessarily follows that the two heights are equal and the two shadows are of equal length. So, on the azimuth lines in the second direction, we cut off lengths equal to the lengths of the shadows in the first direction. Again the extreme positions of each pair of <tips of> shadows, the pairs which mark the endpoints of corresponding hours, are joined with a straight line. So these straight lines are those that mark corresponding hours. By this procedure the artisan obtains the eleven straight lines that mark all the hours in the plane of the sundial.

For example, for this let there be a circle UPO; let this be the circle drawn in the plane of the sundial whose centre is J; let the straight lines drawn in it be the straight lines whose endpoints are M, their other endpoints N and let the meridian line be UO. Let the arc of the azimuth for the fifth hour when the sun is at the first point of Cancer be UP; let the straight line of the azimuth, on which the shadow falls, be JP and let the length of the shadow determined from the master circle be JN; let the arc of the azimuth for the fifth hour, again when the sun is at the first point of Capricorn, let the arc of the azimuth for the fifth hour be US, the straight line of the azimuth, on which the shadow falls, is JS and the length of the shadow obtained from the master circle is, for the second time, JN. The two points that mark the two extreme positions <of the tip of the shadow> for the fifth hours are N and M; so the straight line MN is the line on which the <tip of the> shadow falls, approximately, for the fifth hour for every day. In the same way, each of the remaining straight lines corresponds, successively, to one of the <following> hours. Let us then join the endpoints of the hour lines at which we have M, <and then> the endpoints of the lines at which we have N, using straight lines.[25] Thus in a single day, the tip of the shadow passes through all the points we have marked M, and this <happens> if the sun is at the first point of Capricorn; in the course of

---

[25] We join successive points M with line segments and similarly for points N. These are very short segments so the line we obtain from the points M or the points N may be considered as a curve. This curve, the locus of the points we obtained, is a hyperbola for places in northern latitudes (if $23°27' < \lambda < 66°33'$) (see Mathematical commentary).

that day, and in the course of every day, the tip of the shadow moves along the outline of a conic section, except on the day of the equinox. So all the points *M* lie on the outline of that conic section. But since there was no easy way of drawing the outline of a conic section on the surface of the sundial, they merely drew straight lines to join the points *M*, lines which, together, are taken as a substitute for the conic section; thus the whole line on which the points *M* appear is called the tropic of Capricorn, because if the sun is at the first point of Capricorn, the tip of the shadow moves approximately on that line. In the same way, the whole line on which the points *N* appear is, in addition, called the tropic of Cancer, because the tip of the shadow moves approximately along it if the sun is at the first point of Cancer.

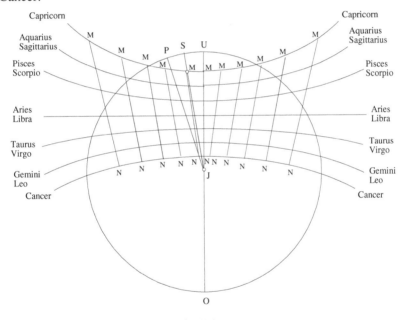

Fig. II.3

We then also find the points on the hour lines through which the tip of the shadow passes when the sun is at the first point of <each of> the remaining signs of the Zodiac, and this <is done> by returning to the tables;[26] in this way we shall know the azimuth for the fifth hour when the sun is at the first point of Leo. So, starting from the meridian line, we take an arc on the circumference of the circle equal to that arc, we place a ruler

[26] The reference is to tables, in the plural, because the table that has been drawn up consists of several tables.

on the centre of the circle that is in the sundial; and at the endpoint of that
arc, where the ruler cuts the line for the fifth hour, we mark the point; then,
again from the tables, we shall know the azimuth for the fourth hour when
the sun is at the first point of Leo; on the circle we also take an arc equal to
this azimuth and we place a ruler on its endpoint and on the centre; where
the ruler cuts the line for the fourth hour we mark a point. We do the same
for the third hour and for all the remaining hours. By this procedure we
obtain points on the hour lines through which the tip of the shadow passes
on the day when the sun is at the first point of Leo or at the first point of
Gemini. We join them up with straight lines. That is called the path of
Gemini and Leo. We do the same thing for each of the signs of the Zodiac;
we thus obtain in the surface of the sundial, and on the hour lines, seven
paths that are the paths of the first points of the signs of the Zodiac, as in
the figure.

As for the path of Aries and Libra, it is a straight line; indeed, the sun is
on the equator that day,[27] the tip of the gnomon is at the centre of the circle
of the equator and on that day all the rays drawn to the tip of the gnomon
are diameters of the circle of the equator; so they all lie in the plane of the
circle of the equator. But they all fall on the surface of the dial and end at
the tips of the shadows; so on that day all the tips of the shadows lie in the
plane of the circle of the equator and they are in the plane of the dial; so
they lie on the intersection of the circle of the equator and the plane of the
dial, and thus lie on a straight line. This straight line cuts the meridian line
at right angles, because it is perpendicular to the plane of the meridian
circle. In fact, each of the planes of the circles of the equator and of the
horizon is perpendicular to the plane of the meridian circle. Their
intersection, which is the straight line on which the tips of the shadows fall,
is perpendicular to any straight line lying in the meridian circle; so it makes
right angles with the meridian line; that is why, for finding that line, the
line that is the path of Aries and of Libra, it is enough to know the height
on the meridian of the first point of Aries, from tables. A height equal to
that one is taken on the master circle and the length of the shadow is found
in the manner already described. We take a length equal to that on the
meridian line that is on the surface of the dial, starting from the centre of
the circle, then from the endpoint of that straight line we draw a
perpendicular straight line to meet the lines for all the hours. This straight
line will be the path of Aries and Libra.

Once all this has been completed, we try to have a conical gnomon
made of material that is solid and not susceptible to decay and we make its

---

[27] That is the day when the sun is at the first point of Aries or of Libra.

height the same size as the length of the straight line that has been cut off from the diameter of the master circle; we put a small additional piece on the side of its circular end [*i.e.* its base], then we fix it at the centre of the circle on the sundial, as the gnomon *PJ*, and we set it up so that the base of the gnomon rests on the plane of the circle and the centre of its base rests on the centre of the circle and we drive the additional piece into the body of the sundial so that it [the gnomon] stands perpendicularly and perfectly securely.

What we have explained is what specialists on shadows use to construct sundials, and they find it sufficient. Using the same construction, we can draw lines that indicate parts of hours, and also mark on all the lines points that indicate the paths of all the parts of the circle of the ecliptic. We can also add to these sundials lines that indicate equinoctial hours, right ascension, the mid heaven [meridian], as well as other kinds of constructions considered by specialists on shadows. It is also possible to give details for the construction of sundials, and explanations of how they work, for every plane, for every horizon [*i.e.* geographical latitude] and for every position of the plane <of the dial> for each of the horizons. But our purpose, in this treatise, is to explain the subject as a whole and the principles on which the construction of sundials is based, and to draw attention to the manner of the construction and the parts where we need these ideas to which reference is not often made except in the books of specialists on shadows. After that, we shall begin with a book concerning shadow instruments in which we shall give an exhaustive treatment of all the ideas, as well as their purposes and the constructions required in this art. It is from God that there comes help and success in this <endeavour>. God is all we need, the excellent protector!

The treatise of Abū 'Alī al-Ḥasan ibn al-Ḥasan ibn al-Haytham on horizontal sundials is completed.

Praise be to God, Lord of the worlds. May the blessing of God be upon his messenger Muḥammad and all that are his.

CHAPTER III

COMPASSES FOR LARGE CIRCLES

3.1. INTRODUCTION

The compasses for large circles are a mathematical instrument invented by Ibn al-Haytham for drawing circles with various radii that may be rather large. The instrument was, according to its inventor, intended to respond to the needs of astronomers and engineers. So it was appropriate, in addressing such readers, to set out the geometrical basis for the invention and to explain how the instrument was made. This is exactly what the author has in mind in the organisation of his short treatise, in which, as he himself puts it, Ibn al-Haytham consciously combines 'theory' (*'ilm*) and 'practice' (*'amal*).

That is not the only interesting thing about the treatise. Like other treatises on mathematical instruments composed by eminent mathematicians, such as al-Qūhī, Ibn Sahl, al-Sijzī and Ibn al-Haytham himself, this one sheds light on the geometrical ideas of its time. We have shown that, from the mid ninth century onwards, geometry was not limited to the study of figures, but was more and more concerned with transformations of figures. Not only was Ibn al-Haytham actively engaged in developing this new outlook, but, more importantly, he also conceived of a new discipline to provide a solid basis for the use of transformations. This discipline was given the name 'The Knowns'.[1] The short treatise on his new instrument confirms the direction in which geometrical research is going.

3.2. MATHEMATICAL COMMENTARY

Ibn al-Haytham begins by proving three geometrical propositions before turning to the instrument itself. These propositions bear witness to the part Ibn al-Haytham allows motion to play in geometry. He makes use of a rotation about a point in the plane as well as a rotation about an axis in

---

[1] R. Rashed, *Les Mathématiques infinitésimales du IXᵉ au XIᵉ siècle*, vol. IV: *Méthodes géométriques, transformations ponctuelles et philosophie des mathématiques*, London, 2002, Chap. II.

space, and he determines the quantities that are invariant under these rotations as well as the paths of points.

In the first proposition, Ibn al-Haytham proves that the displacement
that takes *EB* to *GC* (as defined in Fig. 3.1) is a rotation about a certain
centre. As the endpoints *E*, *G* on the one hand and *B*, *C* on the other lie,
respectively, on two circles with common centre *I*, the centre of the rotation
is *I*.

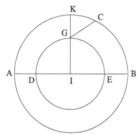

Fig. 3.1

Ibn al-Haytham in fact uses a rotation about the centre *I* that transforms
E to *G*; it thus takes *B* to *K* and *EB* to *GK*. Moreover, by a displacement
that is not defined, *EB* is taken to *GC*; so *GK* = *GC*, which makes *C* = *K*.

In Proposition 2, Ibn al-Haytham proves that, in a continuous rotation
with centre *I*, each point of the plane remains at a constant distance from
the fixed point *I*. So it traces out a circle with centre *I*.

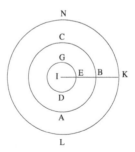

Fig. 3.2

In Proposition 3, he considers a continuous rotation of space, with axis
IO. Any point in space *M* describes a circle whose centre is a point *O* on
the axis. In fact the plane perpendicular to the axis and passing through *M*
is invariant under the rotation (a property left implicit in the text); so the

distance from *M* to the point *O*, where this plane meets the axis, remains constant and *M* describes a circle with centre *O* in the plane concerned.

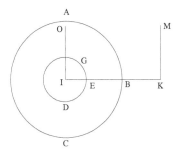

Fig. 3.3

These three propositions are evidence of the part Ibn al-Haytham assigns to motion in geometry. Here he uses a rotation about a point in the plane as well as one about an axis in space and he determines quantities that are invariant under these rotations and the paths of points.

In the remaining two propositions, Ibn al-Haytham explains the process of constructing these compasses, as well as the manner of using the instrument.

In Proposition 4, he explains that the instrument is made up of four pieces:
• a circular ring with a small diameter;
• a cylindrical rod whose thickness is equal to that of the ring at whose end there are fixed two conical pins whose separation is the diameter of the ring;
• two identical plates in the shape of parts of rings with width equal to the diameter of the first ring. To the two ends of one of them there are fixed two cylindrical pins whose length is equal to that of the conical pins and whose tips are filed into the shape of nails; at the tips of the other we bore two holes corresponding to the two nails just mentioned.

The cylindrical rod is soldered to the ring so as to be aligned along a diameter. We set up the instrument by attaching the ring onto the plate with holes bored through it, so that the pins are underneath; so the plate is on top. We then put the second plate (the lower plate) onto the plane of the drawing and, using the nails, we fix the upper plate over it.

When the ring slides along the upper plate, its diameter, in line with the direction of the rod, always points towards the centre of the circles of which the edges of the upper plate are arcs, from Proposition 1. From Proposition 2, any point of the rod describes a circle with the same centre in the plane of the upper plate. From proposition 3, the tips of the conical pins trace out circles with the same centre in the plane of the lower plate, which is the plane of the drawing.

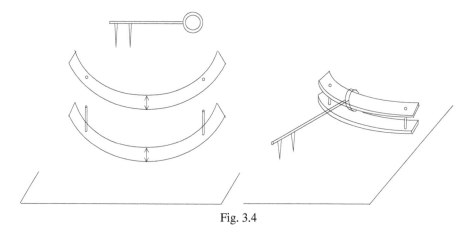

Fig. 3.4

In Proposition 5, Ibn al-Haytham proposes to construct a set of annular plates that will allow us to draw circles with radii that are any multiple of the length of the rod. We cut out the first plate by pivoting the rod about the point O of the ring, the point opposite the point O' where the rod is attached; the two radii of the ring (inner and outer) are the distances from the two pins to this point of the ring; let them be r_1 and r_2. Using this first ring, now in duplicate, we can then draw two new circular arcs with radii $2r_1$ and $r_1 + r_2$, hence we have a new ring that will allow us to draw new circles, with radii $3r_1$ and $2r_1 + r_2$. After n iterations, we have a ring of radii nr_1 and $(n-1)r_1 + r_2$. We can see that is a recursive procedure based on using Archimedes' axiom;[2] in fact the values of r_1 and $r_2 = r_1 + a$ (where a is the diameter of the small ring) are fixed in advance and we know that nr_1 and $(n-1)r_1 + r_2 = nr_1 + a$, can exceed any length given in advance.

---

[2] Ibn al-Haytham is referring to the fifth postulate in Book I of Archimedes' *On the Sphere and Cylinder*, which is of unequal lines and unequal surfaces and unequal solids, the greater exceeds the less by such a magnitude as, when added to itself, can be made to exceed any assigned magnitude among those comparable with one another. (*Greek Mathematical Works*, vol. II, *Aristarchus to Pappus*, trans. Ivor Thomas, Loeb Classical Library, LCL 362, Harvard University Press, 1941, reprinted, pp. 46, 47.) In *On the Sphere and Cylinder* Archimedes distinguishes axioms from postulates and lists them separately, but this postulate has nevertheless come to be known as Archimedes' axiom.

$$OO' = a = SS'$$
$$O'S' = r_1 = OS$$
$$OS' = r_2 = r_1 + a.$$

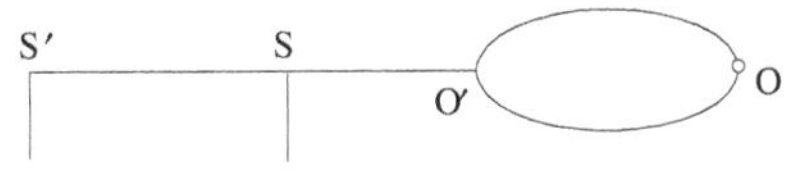

Fig. 3.5

To draw a whole circle, we mount the instrument on an annular upper plate identical with the lower annular plate. After drawing an arc with the instrument in a first position, we mark on the plane of the drawing three points G, M, C, all on BC, the inner edge of the lower plate. We then reposition the instrument so that B is transferred to G and the points M and C of the plane still lie on the inner edge of the lower plate. Accordingly, this inner edge, GH, remains on the same circle in the new position and the new arc of a circle it allows us to draw is an extension of the previous one, since the two annular plates are fixed together. We repeat this procedure until the circle is completed.

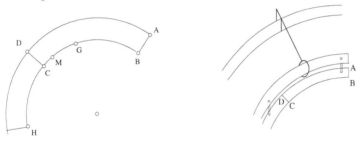

Fig. 3.6

This construction derives from the fact that three points serve to define a circle.

3.3. HISTORY OF THE TEXT

Ibn al-Haytham's treatise on compasses for drawing large circles has come down to us in five manuscripts, three of which were transcribed at the end of the thirteenth century and in the course of the first half of the fourteenth century. Three of these manuscripts belong to collections that include other works by Ibn al-Haytham. We have described these collections in the second volume of *A History of Arabic Sciences and Mathematics* (*Ibn al-Haytham and Analytical Mathematics*). Here we simply refer the reader to that volume. These manuscripts are:

[A] Aligarh 678, copied on 5 Jumādā I, in the year 721 of the Hegira, that is, 2 June 1321.[3] The folios of the text of the treatise are out of order: 29^r, 8^r–8^v–9^r. Examining the texts shows that thirteen words, as well as seven sentences, are missing specifically in this text. Further, lines 9–13 of page 9 are merely a summary.

[B] India Office 734, fols 116^v–118^r.[4] We do not know the date of transcription, which may have taken place in the tenth century of the Hegira. This manuscript presents no omissions that are peculiar to it, but (on the other hand) includes a very large number of transcription errors.

[R] Rampur 3666, fols 436–442. A page is missing from this manuscript, between the page numbered 436 and the one numbered 437. It was probably lost in the course of binding. This manuscript is late; it was copied on 9 Rabīʿ I, in the year 1281 of the Hegira, that is, 12 August 1864, in India.

Again there are no omissions that are peculiar to it, but it includes a very large number of transcription errors.

[L] St. Petersburg B1030. We have made the point[5] that this collection of writings by Ibn al-Haytham is particularly valuable, not only for the treatises it contains, but also because it was checked against the original in 750 H/1349. The text of the treatise appears on fols 125^v–131^r. This text is individual in lacking two phrases, one of two words and the other of seven words.

[N] Leiden 133/6, fols 106–111, copied in 692 of the Hegira, that is in 1293. Apart from the one he was transcribing, the copyist of this manuscript had at his disposal another manuscript that he had consulted and from which he cites four readings. This manuscript is individual in lacking seven words and for expressions of two or more words.

An investigation of the omissions, additions and other accidents in copying in these manuscripts, taken two by two, allows us to divide them into three groups. The first consists only of a single manuscript, [L]; the second also contains no more than a single manuscript, [N]; and the third includes manuscripts [A], [B] and [R].

Moreover we can see that [L] and [N] belong to a manuscript tradition in which the first paragraph, the one in which the author addresses a particular Prince, is absent. Now this paragraph is of some importance, at least sociologically, of a kind moreover that copyists enjoyed. It is present in the

---

[3] R. Rashed, *Ibn al-Haytham and Analytical Mathematics. A History of Arabic Sciences and Mathematics*, vol. 2, Culture and Civilization in the Middle East, London, 2013, pp. 30-1.

[4] *Ibid.*, p. 33.

[5] *Ibid.*, pp. 32-3.

three other manuscripts, including [A], which was transcribed in the same period as [L] and [N]. Should we see this as the production of two editions of the text by Ibn al-Haytham himself, or the effect of an accident to the text that resulted in this paragraph being cut out, or being one more addition by a copyist in the course of the history of the text? In the current state of our knowledge, we have no means of constructing a serious argument in support of one or another of these suggestions. We only know that, if Ibn al-Haytham frequently wrote for his colleagues, he was not in the habit of dedicating his writings to princes. Finally, the fact that this prince is anonymous makes it impossible for us to assess whether the dedication is plausible. So we shall go no further than to suggest that we have two textual traditions for this treatise.

Comparing these manuscripts permits us to propose the following *stemma*, based on all the variants (which have also been noted in the *apparatus criticus*).[6]

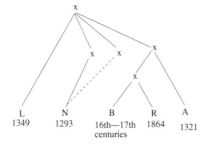

[6] See the Arabic edition of this treatise in *Les Mathématiques infinitésimales*, vol. V, pp. 861–79.

TRANSLATED TEXT

Al-Ḥasan ibn al-Haytham

On Compasses for Large Circles

In the name of God, the Compassionate the Merciful

TREATISE BY AL-ḤASAN IBN AL-ḤASAN IBN AL-HAYTHAM

On Compasses for Large Circles

One of the ingenious geometrical procedures that occurred to the servant of our Lord the Minister, the illustrious Prince – may God cause his power to endure – is the invention of an instrument, small in size, that serves as a pair of compasses, and which, despite its small size, draws extremely large circles, circles such that their diameters are many times the size of its opening. I shall begin by describing how it may be used, and then [explain] how to make it.

Among the variety of ingenious procedures, one example, even if distinguished by its scientific rank and in this respect equal to another example, may nevertheless not be its equal in the importance of the benefits it confers. It may indeed happen that some examples bring greater benefits than others. The science of astronomy, to understand what is the reality of the motions of the wandering stars and of the form of their orbs and what are the shapes of celestial bodies, that science is of the highest rank in honour; and the benefit derived from the instruments designed for understanding these matters is of no mean importance. Now, in instruments for observation, there is a great need to set out large circles, or arcs of large circles that are determinate, so as to end up by dividing them into the smallest parts. But it will be difficult, if not impossible, to set out the circles if they become too large, because the distance between the centre and the circumference must be kept constant; and achieving that is the aim of the instrument with which circles are drawn, if the distance between its two points does not change. Thus, if the size of the circle we need reaches the limit of a very great distance [*i.e.* radius], it will perhaps be impossible to find an instrument whose size is that large. Certainly, instruments can be designed only to work up to a certain limit, and not for any distance or for any extreme case. Even if it were possible, we should not be able to take advantage of the fact, because the instruments[1] would need an interval the same size as the distance [radius] to be free from any obstruction, so that they could turn – even if what we need is an extremely small arc. If we obtain an interval corresponding to a very great distance [radius] having the required property, and an instrument that is extremely long, even if the one and the other are impossibly perfect, the movement will not be perfectly precise; indeed, if the instrument becomes too big, one cannot avoid its being subject to some disturbance when it is set in motion. We may perhaps need to draw an arc of a large circle on a surface that is not very wide, so we should not be able to draw it with that instrument.

The need for such circles becomes apparent not only for observation instruments but also for other purposes, indispensable ones such as the geometry of architectural constructions and similar practical arts, and such as [making] spherical mirrors and other ingenious instruments of that kind associated with ingenious procedures, when the given conditions involve [the use of] large circles. So it is necessary for us to attempt to use some ingenious procedure to devise an instrument for drawing circles, or arcs of circles, whose diameters are of any size, with extreme accuracy and precision, and that by the simplest and easiest method. This instrument, and the ease of handling it, are of no mean interest in the procedures we mentioned earlier. Observers, and those who pay close attention to the science of ingenious procedures and the geometry of constructions, assuredly know this.

Let us start by proving the truth concerning what we are looking for, then we shall come to the manner of constructing an instrument with which we may draw any circle whatsoever, employing the property we have introduced.

Proposition 1: For two parallel circles, if from their centre we draw a straight line to the circumference, then we cut off from it the part that lies between the two parallel circles, and if this part moves in such a way that the two points at its ends move on the circumferences of the two circles, then, if we extend it [the part cut off between the circles], it will always go to the centre.

Example: Let the two circles be ABC and DEG and let their common centre be the point I, from which we draw the straight line IEB. From it we cut off EB. We move EB until it becomes <a line> such as CG. I say that, if we extend CG, it necessarily goes to the point I.

---

[1] Lit.: that instrument.

Proof: If it does not go to the point *I*, then we join *IG* and we extend it to *K*. So the straight line *IK* will be a semidiameter on which there lies the point *G*; so the straight line *CG* is greater than the straight line *GK*; but *CG* is equal to *EB*, so the straight line *EB* would be greater than *GK*. But *BI* is equal to *KI*. So we should have that *EI* is smaller than *GI*, which is impossible. So, if we extend *CG*, it goes to the point *I*. That is what we wanted to prove.

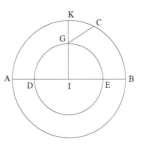

Fig. III.1

Proposition 2: If in two parallel circles we draw a straight line from the centre to the circumference, if we extend it outside the circle and if we cut off from it a piece, a part of which lies outside the circle while its other part is the straight line between the two circles; if we move this segment of the straight line in such a way that the two points move on the two circumferences of the circles, then any point of that straight line traces out a circle.

Example: Let *ABC*, *DEG* be two circles, with centre the point *I*, in which we draw the straight line *IEB* that we extend to *K*. We cut off *KE* and move it in such a way that the points *B* and *E* move on the circumferences of the two circles. I say that any point on the straight line *KE* traces out a circle. In the course of the movement of the straight line, let the point *K* trace out the circumference *KLN*. I say that the line *KLN* is a circle.

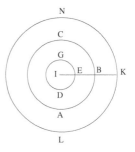

Fig. III.2

Proof: During all its movement, the straight line *KE* is directed towards the point *I*; so if we extend it, in all of its positions, it goes to the point *I*.

So the distance between the point *K* and the point *I* is always the same, and the line *KLN* is a circle. In the same way, any point on the straight line *KE* traces out a circle. That is what we wanted to prove.

Proposition 3: If the set-up we have described remains as before, if at the point *K* we erect a perpendicular to the plane of the circle and if we move the straight line *KE* as before, then any point on the perpendicular traces out a circle.

Let the perpendicular be *KM* and let us move the straight line *KE* in such a way that its ends are on the circumferences of the two circles. I say that any point on *KM* traces out a circle.

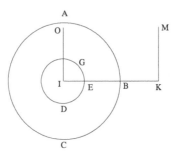

Fig. III.3

Proof: At the centre of the circle, which is *I*, we erect the perpendicular *IO*, as a vertical line. It will thus be parallel to *KM*. We cut off from it *IO* equal to *KM*. We have shown in the previous proposition that if we move *EK*, then the distance from the point *K* to the point *I* is a constant distance. But the straight line *KM* is equal to the straight line *IO* and is parallel to it. So if the straight line *EK* moves, and with it the perpendicular *KM*, and if the perpendicular *IO* is fixed, then the distance between the point *M* and the point *O* will always be the same. So the point *M* traces out a circle and the same holds for any point on the perpendicular *KM*. That is what we wanted to prove.

This part of the proof is enough for our [present] purpose.

<Proposition 4>: Let us now show how to construct an instrument with which we can draw circles whose diameters are whatever size we please.

We take an iron ring, circular and of perfect circularity, and of a moderate size, such that the surface that surrounds its body is a smooth circular surface and such that its diameter is small in size. We take an iron rod, cylindrical in shape and uniformly straight and such that its thickness is the thickness of the body of the ring; we join it to the ring by soldering; let it lie in the direction of one of the diameters of the ring. Next we take two points, at the end of it, and at these two points we set up two conical steel pins that we fix onto the rod with solder; then we file their tips; we reinforce it [each tip] and we polish it so that it will cut everything over which it passes, in the same way as we fashion the compasses that cut the plates of an astrolabe, [and all this is done] in such a way that the distance between the points of the two tips is equal in size to the diameter of the ring. Next we make a piece of yellow copper[2] or of red copper, or of some similar material, and rather thin, then we draw on it [the plate] two arcs of two parallel circles such that the distance between them, that is to say, the amount by which the semidiameter of one exceeds the semidiameter of the other, is equal to the diameter of the ring.[3] We cut the plate along these two arcs using sharp-ended compasses, so that we obtain a piece of the plate similar to a piece of the ring, such that its width is equal to the diameter of the original [iron] ring. We take a plate of hard material, a rectangular piece, such that its length is equal to the length of the piece of ring. On its two ends, we set up <perpendicularly> two small cylindrical pins that we solder firmly in place, we take away from the tops of them two small parts so as to obtain two cylindrical nails such that what remains of the height of these two small cylinders is the same size as the height of the two pins on the tip of the rod. Next, at the two ends of the original plate, we cut two holes whose size is the thickness of the two nails and [placed] so that the distance between them is the size of the distance between the two cylinders.

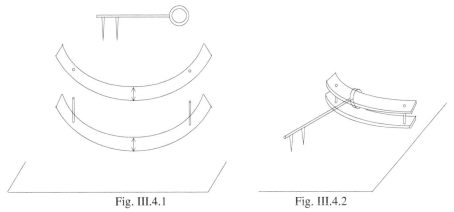

Fig. III.4.1   Fig. III.4.2

If we wish to draw a circle with this instrument, then we thread the plate that has the shape of a piece of the ring through the circular iron ring, then we mount the plate on the two cylinders so that the ends of the two cylinders thread through the two holes and they fit symmetrically. We put the lower plate on the surface on which we wish to draw the circle and we move the rod

---

[2] *I.e.* brass.

[3] He means the diameter of the inside circle of the ring.

that has the two pins at the end of it. The tips of the two pins then draw two circles on this surface because the diameter of the ring, which is the greatest dimension it has, is equal to the width of the plate, which is the smallest dimension it has. So the position of the ring around the plate is always the same, never changing. Thus the diameter of the ring is always identical with the width of the plate, which lies along the line of the diameter of its circle. So the rod, which lies along the line of the diameter of the circular ring, always lies along the line of the diameter of the circle of the plate, along which the rod moves. So any point on the rod, and any point on the pin that is perpendicular to the rod, traces out a circle.[4]

<Proposition 5>: If we want to draw a circle such that its diameter is equal to a given distance, we shall need to make a plate of copper similar to a piece of a ring whose diameter is of known magnitude.

For that we shall need to make numerous rings, in the manner we have described above, until we arrive at the ring we want.

By means of this instrument we can thus make these rings in the most economical and convenient way.

We take a plate of copper, then we set up the instrument and we fix the plate on a surface so that the two tips of the two pins are in contact with the surface; then we hold the circular ring with one hand and the end of the rod with the other hand,[5] we move the rod, we press the two tips of the two pins onto the plate and we continue with this until the plate is cut along the two circles. While it may become difficult to cut the plate [right through] by this movement, it will not be difficult to engrave on it with the two points of the two pins, and if it is engraved, we then use compasses to cut away the part beyond the two arcs scribed on the plate in an incomplete manner; then we use a fine and flexible file [to remove] what remains of the excess and we file gently and firmly until the file has removed all that is beyond the two arcs, so as to reach the circumferences of the two arcs. if we work free the piece enclosed between the two arcs, we are left with a piece of a ring, at whose two ends have been cut, [and] on which we then mount the instrument and it is pressed onto another plate, then that produces another piece of a ring such that the semidiameter of this second ring exceeds the semidiameter of the first ring

---

[4] This follows from Proposition 3.

[5] The ring is held in position so as to keep fixed the endpoint of the diameter that is an extension of the rod. The instrument is rotated about the tangent at that endpoint. See figure.

by the length of the rod. If we do this repeatedly, then the diameter of the ring that is formed is multiplied up, until it reaches the sizes we want. Using this method, we can make numerous rings in the most economical way, and this so as to complete the ring we wanted and to achieve our aim.

If we want to draw a complete circle with this instrument, instead of some piece of a ring we construct two pieces or one substantial piece that we divide into two halves. We mount one of the two pieces on the two tops of the two cylinders and we mount the other under the bases of the two cylinders with the help of two little pegs that are in the base and the two holes that are in the piece, and we mount them so that their surfaces are parallel and their arcs are superimposed.[6] Each arc of one must be superimposed upon the corresponding arc of the other; we then apply the surface of the lower piece of the ring to the given surface, on which we wish to draw the circle.

For example, let us have the piece $ABCD$. We set up the instrument and start to make it move. It accordingly draws an arc of a circle. On the given surface we mark, on the perimeter of the arc BC, on the side towards the point C, three points close to one another, as the points G, M, C, then we move the lower ring away from its position and move it along until a part of the arc BC is applied to the points G, M, C in such a way that the remainder of the arc lies outside it, as the arc $GMCH$. Since the arc BC is applied to the points G, M, H and it will be equal to the arc $GMCH$, the points G, M, C, H will lie on the circumference of a circle that is equal to the circle BC. But since the arc GH met the circumference of the circle that is equal to its circle in three points, thus the whole arc GH is applied to the circumference of the circle and the arc GH will have the same curvature as the circumference of the same circle to which the arc BC is applied. So if we move the rod with one pin in a second trial too, then the tip of the pin completes the circle that it had [partly] drawn.

Then we also mark on the surface, and on the arc CH, three points on the side towards the point H. We move the lower ring until a part of the arc is applied to these points, as we did in the first trial. We continue like this and we move the instrument until it returns to its initial position. Using this procedure we draw a complete circle.

---

[6] That is, the arcs are related to one another by an orthogonal projection.

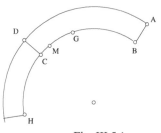

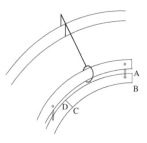

Fig. III.5.1   Fig. III.5.2

If we want to draw an arc of a circle of known diameter and such that the ratio of the arc to the <whole> circle is known, then we make the arc of the first ring similar to the arc we require and we complete the procedure, that is to say, the construction of multiple rings.

If the required arc is large in proportion to the <complete> circle and belongs to a circle that is large in size, we arrange matters so that the first ring is a very small part of the arc we require, to which it bears a known ratio, and that it is a measure of it; and we fashion the rings using the method we described for constructing rings. If we reach the circle we require, we shall then have a ring that is a known part of the required arc. We mount it in the instrument and by means of it we fashion the required arc, using the method that we have explained to make the complete circle.

For all these rings the piece must be of greater size than the length of the two arcs on one side and on the other so that the field of the circular ring grows larger and so that there is a given point in the circular ring on which to apply the endpoint of the arc at the start of the movement and on the other endpoint of the arc when it finishes, in order that the arc produced by the movement of the tip of the pin shall be similar to the arc of the ring.

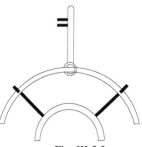

Fig. III.5.3

We have thus completed the explanation of the way to construct the instrument with which we draw large circles, <that is explained in both> theoretical and practical terms. Which is what we sought to show. And here is the form of the instrument.

APPENDIX

IBN AL-HAYTHAM'S INSTRUMENT

We have noted that in *The Configuration of the Motions* Ibn al-Haytham returns to certain results and problems that he had already examined in his other writings. Thus we find a theorem that he had already proved in *On the Hour Lines*, as well as a question considered earlier in *The Variety of Hours that Appears for the Heights of the Wandering Stars*. There is every indication that, there, Ibn al-Haytham is also returning to the construction of a useful instrument for determining the heights of wandering stars, considered in his book *On the Correction of Astrological Operations* (*Fī taṣḥīḥ al-a'māl al-nujūmiyya*, ms. Oxford, Bodleian Library, Seld. A32).

In the introductory section of *The Configuration of the Motions*, Ibn al-Haytham indicates that the third part of his book deals with an instrument that allows one to make a precise calculation, in minutes and parts of minutes, of the heights of the wandering stars. He writes:

> Then we accomplish what this art requires and we spare the specialists in it the trouble of writing essays on the observation of minutes [of arc] and small parts for the altitude of the sun and of all the wandering stars, by presenting an instrument that is easy to handle and can be understood by everyone, by means of which we find the height of the sun and of each of the [wandering] stars using the minutes and small parts [of the height]. Thanks to this instrument and to the procedures we explain, all the procedures followed by astronomers are shown to be correct and an end is put to all the disputes that arise over principles, because of the fractions that are missed by observers and that they find it hard to see, on account of the design of the instruments.[1]

In his book *On the Correction of Astrological Operations*, he refers to this same instrument, and does so in the same terms. Unfortunately, in this book, as well as in *The Configuration of the Motions*, the description of the instrument is missing – but, on the other hand, probably not for the same reasons. In *The Configuration of the Motions*, the book concerned with it has been lost, whereas the copyist of *On the Correction of Astrological Operations* says that in his autograph Ibn al-Haytham omitted to describe

---

[1] See above, p. 260.

the instrument. Perhaps that was because he had not yet perfected it. Perhaps he had decided to add a description to *The Configuration of the Motions*, which he was already planning to write. Or perhaps, more simply, the passage had disappeared from the autograph manuscript. We have no means of deciding such matters. For lack of better evidence, let us look at what Ibn al-Haytham says about the instrument in the introductory section of *On the Correction of Astrological Operations*. There Ibn al-Haytham says:

Second Book of the Correction of Astrological Operations

We have proved in the first book that many of the operations to which astronomers have recourse in astronomy depart from [perfect] correctness, and that many of the results they give as approximate are far from exact; and we have moreover proved things of which none of our predecessors, or contemporaries, had any idea. We prove in this book how to determine precisely the results on which astronomers rely in an approximate way; how to grasp what has escaped them and how to grasp in complete detail the small fractions that they neglect. The results in astronomy that we wish to determine precisely are the positions of the stars on the orbs that are proper to them, their distance from the equator, the magnitude of their heights with respect to the given horizons and at the given instants also the ascendant from the circle of the ecliptic on the eastern horizon; how to determine the ascendant from the height of each of the seven wandering stars, and how to determine from heights of these stars the circular <line> of the orb and the hour <lines>. In such matters for which it was possible to establish [results] with extreme precision, we have set out, without using approximation; and for matters in which one cannot avoid approximation, we have improved the approximate value until we arrived at the limit where between it [the approximate value] and perfect precision there is no difference that might affect their truth.

We pursue this [purpose] by constructing an instrument of small dimensions, convenient to handle and easy to use, by means of which we determine the heights and the positions of the wandering stars to minutes and seconds – namely the instrument to which we referred in the introduction to the first book; and we give details of its construction, how it functions and how it is used.

This instrument is the one that allows the user to find most of what astronomers have not succeeded in obtaining and were incapable of determining, in regard to which they resorted to approximation, because they could not <obtain> the minutes and small fractions, either for heights or for positions of the wandering stars, when making observations, for lack of small fractions on their instruments.

None of the ancients or the moderns had thought of this instrument, nor had imagined it, and none of them made it a target for investigation. This instrument is very useful for all the operations in astronomy that are deduced from the height and from observational instruments, thanks to which we obtain the positions of the stars at the moments when we observe them.

Towards the end of the book, where Ibn al-Haytham should describe the instrument and explain how to construct it, he writes:

It remains for us to explain this instrument by means of which we obtain the height to minutes and seconds, and to explain both how to construct it and how to use it. We say.

The text stops there and the copyist writes:

The book is completed. This is the end of what I have found that is from his hand. May God be merciful to him. He did not complete the [text concerning] the construction of the instrument. Thanks be rendered to God alone. <The copy> has been checked against the original.[2]

We note that the manuscript is old. It was copied at the latest before the beginning of the thirteenth century (before 1235), from an autograph manuscript by Ibn al-Haytham.

Reading this introduction together with that of *The Configuration of the Motions*, points up a commonality of problems and a quasi-identity in the expressions relating to the instrument, to its construction and to its purpose. It was to be used to register the minutes and seconds or small fractions when observing and calculating heights of the wandering stars on their orbits, for a given horizon. This would allow the user to carry out an exact calculation, when such a thing was possible, or at least to make a notable improvement in the approximation. The loss of the second, and above all the loss of the third book of *The Configuration of the Motions*, taken together with the rather abrupt ending of the manuscript of *On the Correction of Astrological Operations* and the brevity of Ibn al-Haytham's remarks about the instrument in the introductory paragraphs of the two treatises, leave us with no possibility of working out what the instrument was like, even very roughly.

---

[2] Ms. Oxford, Bodleian Library, Seld. A32, fol. 132ᵛ.

SUPPLEMENTARY NOTES

[1] If 2α, $2\alpha_1$, 2β, $2\beta_1$ are respectively the measures of the angles at the centre corresponding to the arcs AB, BC, DE, EG, we have by hypothesis:

$$\frac{\pi}{2} > \alpha + \alpha_1 > \beta + \beta_1 \ \text{ and } \ \frac{\alpha}{\alpha_1} = \frac{\beta}{\beta_1};$$

so we have

$$\frac{\alpha}{\alpha + \alpha_1} = \frac{\beta}{\beta + \beta_1}$$

and consequently $\alpha > \beta$.

But the angle AB makes with the tangent to the arc AB at the point A is α and the angle DE makes with the tangent to the arc DE at the point D is β, the angle AB makes with the tangent to the arc AIB (which is similar to DE) at the point A will also be β; we have $\alpha > \beta$, hence the position of the arc AIB.

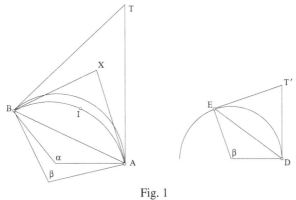

Fig. 1

If T is the point of intersection of the tangents to the arc AB at A and B and T' the point of intersection of the tangents to the arc DE at D and E, we have $A\hat{T}B = \pi - 2\alpha$ and $D\hat{T}'E = \pi - 2\beta$, so $D\hat{T}'E > A\hat{T}B$.

The arc AIB lies in the angle AXB, the angle between the tangents at A and B and we have $A\hat{X}B = D\hat{T}'E > A\hat{T}B$, so 'the angle within which it falls is thus greater than the angle within which the arc AB falls' (above, pp. 263–4).

[2] Several times, Ibn al-Haytham makes use of two arcs, commensurable or incommensurable with a quarter circle, or two arcs commensurable with one another, or two arcs that are not commensurable.
    • If each of the two arcs is commensurable with the quarter circle, they are commensurable with one another.
    • If one of the two arcs is commensurable with the quarter circle and the other is not commensurable with it, then they are incommensurable with one another.
    • But if each of the two arcs and the quarter circle are incommensurable, then the two arcs may be commensurable or incommensurable.

[3] By hypothesis, if $\overset{\frown}{BE} < \overset{\frown}{BD} < \frac{1}{2}\overset{\frown}{BDA}$, and if we suppose that the diameters d_1 and d_2 of the circles (EI) and (DC) satisfy the condition $d_1 > d_2$ and that the circle (DC) cuts the plane through B parallel to the horizon.
    This requires that the sphere be inclined to the south, that is, in the direction of B, which is our hypothesis. If the sphere were right or inclined in the direction of A, we could not have $d_1 > d_2$ with $\overset{\frown}{BE} < \overset{\frown}{BD}$ except in the case where $\overset{\frown}{BD} > \frac{1}{2}\overset{\frown}{ADB}$.

[4] The proof will distinguish two cases for the arc LDC:
    1) $\overset{\frown}{LDC} \leq \frac{1}{2}$ circle.
    2) $\overset{\frown}{LDC} > \frac{1}{2}$ circle.
    In the first case, the centre of the circle, ω, is below the horizontal plane AB, which is not possible unless this plane is not the horizon itself.
    In the second case, the centre ω is above the horizontal plane AB, which then can be the horizon or be above the horizon.
    If $\overset{\frown}{LDC} > \frac{1}{2}$ circle, the proof requires a supplementary hypothesis:

$$\frac{HE}{HJ} \geq \frac{d_1}{d_2}.$$

Investigation of the inequality $\dfrac{HE}{HJ} \geq \dfrac{d_1}{d_2}$, in the case where ω, the centre

of the circle DC, is above the horizontal plane AB. Let O be the place concerned, $O\Pi$ the axis of the world, $\Sigma\Sigma'$ the diameter of the equator.

Let us put $\omega'E = r_1$, $\omega D = r_2$ ($d_1 = 2r_1$, $d_2 = 2r_2$).

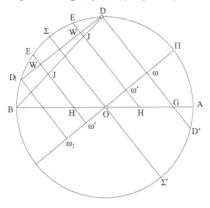

Fig. 2

1) Case where AB is the horizon. By hypothesis ω is above the horizon.

Let $DD_1 \perp O\Sigma$, E must lie on the arc $D_1\Sigma D$. The chord HE cuts DD_1 in the point W, between J and E, because the angle GDB is acute and the angle GDD_1 is a right angle. We have $\omega'E = r_1$ and $\omega'W = \omega D = r_2 \Rightarrow$
$\dfrac{\omega'E}{\omega'W} = \dfrac{r_1}{r_2}$.

In all cases, we have

$$\frac{\omega'E}{\omega'J} > \frac{\omega'E}{\omega'W},$$

so

$$\frac{\omega'E}{\omega'J} > \frac{r_1}{r_2}.$$

We have

$$\omega'E > \omega'J > \omega'H.$$

In this case, we apply the lemma

$$a > b > c \Rightarrow \frac{a-c}{b-c} > \frac{a}{b} > \frac{a+c}{b+c}.$$

- If E lies on the arc ΣD_1, O is between ω' and ω

$$HE = \omega' E - \omega' H$$
$$HJ = \omega' J - \omega' H,$$

so

$$\frac{HE}{HJ} > \frac{\omega' E}{\omega' J},$$

so

$$\frac{HE}{HJ} > \frac{r_1}{r_2}.$$

- If E lies at the point Σ, ω' is in O, $H = \omega' = O$, so $\dfrac{HE}{HJ} > \dfrac{r_1}{r_2}$.

- If E lies on the arc ΣD, ω' is between O and ω, we have

$$HE = \omega' E + \omega' H$$
$$HJ = \omega' J + \omega' H,$$

so

$$\frac{\omega' E}{\omega' J} > \frac{HE}{HJ} \quad \text{and} \quad \frac{\omega' E}{\omega' J} > \frac{r_1}{r_2}.$$

We can draw no conclusions regarding the ratio $\dfrac{HE}{HJ}$.

Limiting case: $\omega' = \omega$. $H = G$, $E = J = D$, $r_1 = r_2$, $\dfrac{HE}{HJ} = \dfrac{r_1}{r_2} = 1$.

2) In the case where AB is above the horizon, the angle GDB may be acute, a right angle or obtuse.

a) If $G\hat{D}B \leq 1$ right angle, we have $\omega' W \geq \omega' J$ and $\dfrac{\omega' E}{\omega' J} \geq \dfrac{\omega' E}{\omega' W}$, so $\dfrac{\omega' E}{\omega' J} \geq \dfrac{r_1}{r_2}$ and we continue as in 1). So

- if ω' is below the plane AB, we have $\dfrac{HE}{HJ} > \dfrac{r_1}{r_2}$;

- but if ω' is above the plane AB, we cannot draw any conclusion.

b) If $G\hat{D}B > 1$ right angle, J is between E and W, so $\dfrac{\omega' E}{\omega' J} < \dfrac{r_1}{r_2}$; we cannot draw any conclusion.

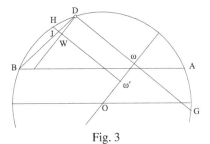

Fig. 3

[5] The proof requires the additional condition $\widehat{EI} \leq \widehat{CD'L}$, which does not always hold.

Investigation of this condition in the case where ABC is the horizon and the circle EI the equator. We then have $\widehat{EI} = \dfrac{\pi}{2}$.

$$\widehat{EI} \leq \widehat{CD'L} \Leftrightarrow \frac{\pi}{2} \leq C\hat{\omega}L \Leftrightarrow \frac{\pi}{4} \leq C\hat{\omega}G \Leftrightarrow \frac{\sqrt{2}}{2} \geq \cos C\hat{\omega}G$$

$$\cos C\hat{\omega}G = \frac{\omega G}{\omega C} = \frac{\omega G}{\omega D}.$$

If we put $\widehat{ED} = \beta$ and make the latitude λ, we have

$$\omega C = \omega D = R\cos \beta, \quad O\omega = R\sin \beta, \quad \omega G = O\omega \tan \lambda = R\sin \beta \tan \lambda,$$

so

$$\cos C\hat{\omega}G = \tan \beta \cdot \tan \lambda.$$

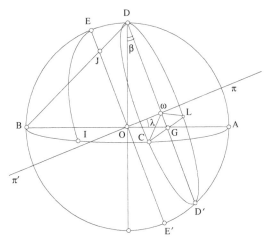

Fig. 4

In this special case, we have

$$\widehat{El} \leq \widehat{CD'L} \iff \frac{\sqrt{2}}{2} > \tan\beta \cdot \tan\lambda. \ (*),$$

a condition which is not always satisfied.

Example: If $\lambda = 60°$, the circle CD cuts the horizon when $\beta < 30°$; but the condition (*) is satisfied only if

$$\tan\beta \leq \frac{\sqrt{2}}{2\sqrt{3}}, \ \tan\beta \leq \frac{\sqrt{6}}{6},$$

that is $\beta < 22°12'$.

[6] *Position referred to the ecliptic*

If P and P' are the poles of the ecliptic, to any point M of the sphere there corresponds a half of a great circle PMP' which cuts the ecliptic in M'. Thus we define a transformation f in which $M' = f(M)$; M' is the 'position of M referred to the ecliptic'. If l and λ are respectively the ecliptic longitude and latitude of M, M' has longitude l and latitude 0. So the transformation f preserves the longitude and makes the latitude zero $M'(l, 0) = f[M(l, \lambda)]$.

[7] The points lying between the points A and B, which play a part in this paragraph concerned with the motion of the moon, correspond to the points lying between the point where it rises, at B, and the point it crosses the meridian, at N, which play a part in the investigation of the motion of the moon. The points identified here as A, B, D, G, K correspond to the points B, N, I, M, S in the figures on p. 180. In the first investigation, the point M is, in general, in a position below or above the hour circle that passes through the position called B and the author has noted that it can, exceptionally, lie on this circle; in this case $M = L$. The same observation is made here for the point G: 'So the point G is to the south of the circle AD or to the north of it' (above, p. 340), so the great circle CG in general cuts the circle AD in a point R which is different from G. Exceptionally, R and G may be identical. The point R corresponds to the point L. Here the argument is given for $R \neq G$, because the point K lies on the hour circle AD

and Ibn al-Haytham writes: 'the arc *GK* is parallel to the circle of the ecliptic' and 'it has the magnitude of the motion of the node in the known time' (above, p. 341). So the point *G* is not taken to be on the hour circle of the point *A*.

[8] *Inclination of the orb of Mercury or Venus with respect to the plane of the ecliptic*

The inclination is at its maximum when the centre of the epicycle is at its apogee *A* or at its perigee *P* on the eccentric. This inclination is known (Ptolemy).

The inclination becomes zero when the centre of the epicycle reaches *n* on the line of intersection of the orb with the plane of the ecliptic; the orb then coincides with the plane of the ecliptic.

Let *m* be an intermediate position of the centre of the epicycle on the eccentric. To the points *A*, *m*, *n* there correspond the points *A*, *M*, *N* on the orb with centre ω, the centre of the Universe.

When the orb moves from the position of maximum inclination to the position of zero inclination, the point *m* describes the arc *An* of the eccentric, with centre ω', and the point *M* describes a quarter circle *AN* on the circle with centre ω.

The arc *NA* is a quarter circle, and the arc *nA* that corresponds to it on the eccentric is known. $\widehat{nA} = \alpha$.

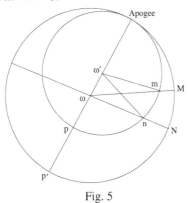

Fig. 5

At any instant the arcs *NM* and *nm* are known.

Let i_m be the maximum inclination, which corresponds to the apogee and let i be the inclination of the orb when the centre of the epicycle is at m; we have $\dfrac{i}{i_m} = \dfrac{\widehat{nm}}{\widehat{nA}}$ (see p. 201).

Here the text gives $\dfrac{i}{i_m} = \dfrac{\widehat{NM}}{\widehat{NA}}$; which would be correct if the deferent $AmnP$ were concentric, with ω and ω' identical; in this case we should have

$$\frac{\widehat{nm}}{\widehat{nA}} = \frac{\widehat{NM}}{\widehat{NA}}.$$

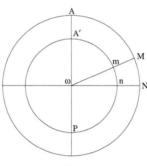

Fig. 6

Ibn al-Haytham then gives the ratio of the arcs on the eccentric.

So the ratio $\dfrac{i}{i_m}$ is a known ratio and the inclination i of the planet (Mercury or Venus) to the plane of the ecliptic is known at any known instant.

BIBLIOGRAPHY

I. MANUSCRIPTS

Al-Ḥasan ibn al-Haytham
Fī birkār al-dawā'ir al-'iẓām
> Aligarh, 678, fols 29^r, 8^{r-v}, 9^{r-v}, 10^r.
> Leiden, Or. 133/6, fols 106–111.
> London, India Office, Loth 734, fols 116^v–118^r.
> Rampur 3666, fols 436–442.
> St. Petersburg, B 1030, fols 125^v–131^r.

Fī hay'at ḥarakāt kull wāḥid min al-kawākib al-sab'a
> St. Petersburg, 600 (formerly Kuibychev, Library V.I. Lenin),
> fols 368^v, 397^v, 397^r–401^v, 402^v, 402^r, 403^r–408^v, 369^r–396^v, 409^r–420^v.

Fī al-ikhtilāf fī irtifā'āt al-kawākib
> Istanbul, Süleymaniye, Fātiḥ 3439, fols 151^r–155^r.

Fī khuṭūṭ al-sa'āt
> Istanbul, Askari Müze, 3025, fols 1^v–19^v.
> Istanbul, Süleymaniye, 'Āṭif 1714, fols 57^r–76^v.

Fī al-rukhāmāt al-ufuqiyya
> Berlin, Staatsbibliothek, Oct. 2970.
> Teheran, Majlis Shūrā, Tugābūnī 110.

II. OTHER MANUSCRIPTS CONSULTED FOR THE ANALYSIS

Abū al-Wafā' al-Būzjānī
Fīmā yaḥtāju ilayhi al-ṣāni' min a'māl al-handasa
> Istanbul, Aya Sofia 2753.

Ibn al-Haytham
Fī ḥall shukūk ḥarakat al-iltifāf
> St. Petersburg, B1030/1.

Fī taṣḥīḥ al-a'māl al-nujūmiyya
> Oxford, Bodleian Library, Arch. Seld A30.

Fī ḥall shukūk al-Majisṭī
> Aligarh, 'Abd al-Ḥayy 21.
> Istanbul, Beyazit, 2304.

[*Fī hay'at al-'ālam*]
> Kastamonu, 2298.
> London, India Office, Loth 734.
> Rabat, Royal Library, 8691.

Ibn al-Shāṭir
al-Zīj al-jadīd
    Oxford, Bodleian Library, Arch. Seld. A30.

al-Kharaqī
Muntahā al-idrāk fī taqāsīm al-aflāk
    Paris, BNF, Arabe 2499.

III. BOOKS AND ARTICLES

F. Bancel, 'Les centres de gravité d'Abū Sahl al-Qūhī', *Arabic Sciences and Philosophy*, 11.1, 2001, pp. 45–78.

C. Brockelmann, *Geschichte der arabischen Literatur*, 2nd ed., Leiden, 1937–1942.

A. Dallal, 'Ibn al-Haytham's universal solution for finding the direction of the *Qibla* by calculation', *Arabic Sciences and Philosophy*, 5.2, 1995, pp. 145–93.

M.-Th. Debarnot, *Al-Bīrūnī: Kitāb Maqālīd 'ilm al-hay'a. La Trigonométrie sphérique chez les Arabes de l'Est à la fin du Xe siècle*, Institut Français de Damas, Damascus, 1985.

Al-Dhahabī, *Siyar a'lām al-nubalā'*, ed. Sh. al-Arna'ūṭ *et al.*, Beirut, Mu'assasat al-Risāla, 1984, vol. XVIII.

P. Duhem, *Le Système du monde*, t. II: *Histoire des doctrines cosmologiques de Platon à Copernic*, Paris, Hermann, 1965.

A. Heinen, 'Ibn al-Haiṭams Autobiographie in einer Handschrift aus dem Jahr 556 H./1161 A.D.', in Ulrich Haarmann and Peter Bachmann (eds), *Die islamische Welt zwischen Mittelalter und Neuzeit, Fetschrift für Hans Robert Roemer zum 65. Geburtstag*, Beiruter Texte und Studien, Band 22, Beirut, Franz Steiner Verlag, 1979, pp. 254–77.

Ibn Abī Uṣaybi'a, *'Uyūn al-anbā' fī ṭabaqāt al-aṭibbā'*, ed. N. Riḍā, Beirut, Dār Maktabat al-Ḥayāt, 1965.

Ibn 'Asākir, *Tārīkh Madīnat Dimashq,* vol. 43, ed. Sakīna al-Shīrabī, Damascus, Academy of the Arabic Language of Damascus, 1993.

Ibn al-Haytham
Majmū' al-rasā'il, Osmānia Oriental Publications Bureau, Hyderabad, 1938–1939.

Al-Shukūk 'alā Baṭlamiyūs, ed. A. I. Sabra and N. Shehaby, Cairo, Dār al-Kutub, 1971.

Ibn al-'Imād, *Shadharāt al-dhahab fī akhbār man dhahab*, Beirut, n.d.

Ibn Taghrībardī, *al-Nujūm al-zāhira fī muluk Miṣr wa-al-Qāhira*, 12 vols, Beirut, 1992, vol. V.

S. A. Jazbi, *Applied Geometry*, Teheran, Soroush Press, 1991.

Al-Khāzinī, *Kitāb Mīzān al-ḥikma*, Hyderabad, 1941.

Y. T. Langermann, *Ibn al-Haytham's On the Configuration of the World*, New York/London, Gerland Publishing, 1990.

R. Morelon, 'Eastern Arabic astronomy between the eighth and the eleventh century', in R. Rashed (ed.), *Encyclopedia of the History of Arabic Science*, 3 vols, London/New York, Routeledge, 1996, vol. I, pp. 20–57.

Al-Nadīm, *Kitāb al-Fihrist*, ed. R. Tajaddud, Teheran, 1971.

M. Naẓīf, *Al-Ḥasan ibn al-Haytham, Buḥūthuhu wa-kushūfuhu al-baṣariyya*, 2 vols, Cairo, University of Fou'ad 1, 1942–1943.

Al-Nu'aymī, *al-Dāris fī ta'rīkh al-madāris*, ed. Ja'far al-Ḥasanī, Damascus, Academy of the Arabic Language of Damascus, 1951, vol. II.

S. Pines
'Ibn al-Haytham's critique of Ptolemy', in *Actes du dixième Congrès international d'histoire des sciences*, 1, no. 10, Paris, 1964, pp. 547–50

'What was original in Arabic science', in A. C. Crombie (ed.), *Scientific Change*, Leiden, Heinemann, 1963, pp. 181–205.

Ptolemy, *Composition mathématique de Claude Ptolémée*, French transl. N. Halma, vol. I, Paris, Imprimerie de J.-M. Eberhart, 1813; vol. II, 1816.

Al-Qifṭī, *Ta'rīkh al-ḥukamā'*, ed. J. Lippert, Leipzig, Dieterich'sche Verlagsbuchhandlung, 1903.

F. J. Ragep, *Naṣīr al-Dīn al-Ṭūsī: Memoir on Astronomy (al-Tadhkira fī 'ilm al-hay'a)*, 2 vols, New York, Springer Verlag, 1993.

R. Rashed
Sharaf al-Dīn al-Ṭūsī, Œuvres mathématiques. Algèbre et géométrie au XIIᵉ siècle, Collection 'Sciences et philosophie arabes – textes et études', 2 vols, Paris, Les Belles Lettres, 1986.

'Optique géométrique et doctrine optique chez Ibn al-Haytham', *Archive for History of Exact Sciences*, 6.4, 1970, pp. 271–98; repr. *Optique et Mathématiques: Recherches sur l'histoire de la pensée scientifique en arabe*, Variorum reprints, Aldershot, 1992, II.

Géométrie et Dioptrique au Xe siècle: Ibn Sahl – al-Qūhī et Ibn al-Haytham, Paris, Les Belles Lettres, 1993.

Les Mathématiques infinitésimales du IXe au XIe siècle, vol. I: *Fondateurs et commentateurs: Banū Mūsā, Thābit ibn Qurra, Ibn Sinān, al-Khāzin, al-Qūhī, Ibn al-Samḥ, Ibn Hūd*, London, al-Furqān, 1996; English translation: *Founding Figures and Commentators in Arabic Mathematics*. A History of Arabic Sciences and Mathematics, vol. 1, Culture and Civilization in the Middle East, London, Centre for Arab Unity Studies/Routledge, 2012.

Les Mathématiques infinitésimales du IXe au XIe siècle, vol. II: *Ibn al-Haytham*, London, al-Furqān, 1993. English translation: *Ibn al-Haytham and Analytical Mathematics*. A History of Arabic Sciences and Mathematics, vol. 2, Culture and Civilization in the Middle East, London, Centre for Arab Unity Studies/Routledge, 2012.

Les Mathématiques infinitésimales du IXe au XIe siècle, vol. III: *Ibn al-Haytham. Théorie des coniques, constructions géométriques*, London, al-Furqān, 2000; English translation: *Ibn al-Haytham's Theory of Conics, Geometrical Constructions and Practical Geometry*. A History of Arabic Sciences and Mathematics, vol. 3, Culture and Civilization in the Middle East, London, Centre for Arab Unity Studies/Routledge, 2013.

Les Mathématiques infinitésimales du IXe au XIe siècle, vol. IV: *Méthodes géométriques, transformations ponctuelles et philosophie des mathématiques*, London, al-Furqān, 2002.

Œuvre mathématique d'al-Sijzī. Vol. I: *Géométrie des coniques et théorie des nombres au Xe siècle*, Les Cahiers du Mideo, 3, Louvain-Paris, Éditions Peeters, 2004.

Geometry and Dioptrics in Classical Islam, London, al-Furqān, 2005.

Les Mathématiques infinitésimales du IXe au XIe siècle, vol. V: *Ibn al-Haytham*: *Astronomie, géométrie sphérique et trigonométrie*, London, al-Furqān, 2006.

'The Configuration of the Universe: a Book by al-Ḥasan ibn al-Haytham?', *Revue d'histoire des sciences*, t. 60, no. 1, 2007, pp. 47–63.

R. Rashed and H. Bellosta, *Ibrāhīm ibn Sinān. Logique et géométrie au Xe siècle*, Leiden, E. J. Brill, 2000.

R. Rashed and B. Vahabzadeh, *Al-Khayyām mathématicien*, Paris, Librairie Blanchard, 1999; English translation: *Omar Khayyam. The Mathematician*, Persian Heritage Series no. 40, New York, Bibliotheca Persica Press, 2000.

A. I. Sabra

'Ibn al-Haytham', *Dictionary of Scientific Biography*, vol. VI, New York, Charles Scribner's sons, 1972, pp. 189–210.

'Maqālat al-Ḥasan ibn al-Haytham fī ḥall shukūk ḥarakat al-iltifāf', *Journal for the History of Arabic Science*, 3.2, 1979, pp. 183–212, 388–92.

'One Ibn al-Haytham or Two? An Exercise in Reading the Bio-Bibliographical Sources', *Zeitschrift für Geschichte der arabischen-islamischen Wissenschaften*, Band 12, 1998, pp. 1–50.

A. S. Saidan, *Arabic Arithmetic ('Ilm al-ḥisāb al-'arabī)*, Amman, University of Jordania, 1971.

M. Schramm, *Ibn al-Haythams Weg zur Physik*, Wiesbaden, Franz Steiner Verlag, 1963.

F. Sezgin, *Geschichte des arabischen Schrifttums*, vol. V, Leiden, E. J. Brill, 1976; vol. VI, 1978.

Thābit ibn Qurra, Œuvres d'astronomie, texte établi et traduit par R. Morelon, Paris, Les Belles Lettres, 1987.

I. Thomas (transl.), *Greek Mathematical Works*, vol. II: *Aristarchus to Pappus*, Loeb Classical Library, LCL 362, Harvard University Press, 1941.

Al-'Urḍī, *The Astronomical Work of Mu'ayyad al-Dīn al-'Urḍī: Kitāb al-Hay'ah* edition with English and Arabic introductions by G. Saliba, Tārīkh al-'ulūm 'inda al-'Arab 2, Beirut, Markaz Dirasāt al-Waḥda al-'Arabiyya, 1990.

Yāqūt, *Mu'jam al-buldān*, Beirut, n.d., vol. III.

INDEX OF PROPER NAMES

SUBJECT INDEX

INDEX OF WORKS

INDEX OF MANUSCRIPTS

Milton Keynes UK
Ingram Content Group UK Ltd.
UKHW030901141024
449569UK00025B/1287